PROBLEM SUPPLEMENT #1

Jearl Walker

Cleveland State University

to accompany   **SIXTH EDITION**

Fundamentals of Physics

David Halliday

Robert Resnick

Jearl Walker

John Wiley & Sons, Inc.

New York / Chichester / Weinheim / Brisbane
Singapore / Toronto

ACQUISITIONS EDITOR   Stuart Johnson
DEVELOPMENTAL EDITOR   Ellen Ford
MARKETING MANAGER   Sue Lyons and Bob Smith
ASSOCIATE PRODUCTION DIRECTOR   Lucille Buonocore
SENIOR PRODUCTION EDITOR   Monique Calello
TEXT/COVER DESIGNER   Madelyn Lesure
COVER PHOTO   Tsuyoshi Nishiinoue/Orion Press
PHOTO MANAGER   Hilary Newman
ILLUSTRATION EDITORS   Edward Starr and Anna Melhorn
ILLUSTRATION   Radiant/Precision Graphics

This book was set in 10/12 Times Roman by Progressive Information
Technologies and printed and bound by Port City Press.
The cover was printed by Brady Palmer Printing Company.

This book is printed on acid-free paper. ∞

The paper in this book was manufactured by a mill whose forest
management programs include sustained yield harvesting of its
timberlands. Sustained yield harvesting principles ensure that the
numbers of trees cut each year does not exceed the amount of new growth.

To order books or for customer service call 1-800-CALL-WILEY (225-5945).

ISBN 0-471-36031-7 (paperback)

Printed in the United States of America

10   9   8   7   6

PREFACE

This collection, a supplement to Halliday, Resnick, and Walker, *Fundamentals of Physics,* Sixth Edition, contains the following items, with the numbering systems continuing from those in the main textbook:

ADDITIONAL SAMPLE PROBLEMS. These include ones that were shifted from the main book (such as Sample Problem 2-10 on page 5) plus many new ones (such as Sample Problem 2-9 on page 4). All begin with one or more basic Key Ideas and then build step by step to a solution.

QUESTIONS

1. Checkpoint-style questions, as in the main book, require reasoning, decisions, and mental calculations of the reader rather than use of a calculator. They are written with the implicit question, "If you really understand the physics, can you do this?" Some have the quality of a game in which the solution is plain to view but at first hard to see (such as Question 23 on page 29). Some can require much thought (such as Question 15 on page 37).

2. Organizing questions focus on either a general organization of the physics in a chapter or a type of problem common to the chapter. (See Question 11 on page 36 and Question 11 on page 51.) Many of these questions help bridge the gap between the conceptual Questions and the data-filled Exercises & Problems.

3. Discussion questions from the fourth and earlier editions of the book (back by request). See Questions 20 through 33 on page 67.

EXERCISES & PROBLEMS. More homework problems, including many shifted from the main book. These are *not* ordered according to difficulty, section titles, or appearance of the associated physics in the chapter, and worked-out solutions are provided only to instructors. Some of the new homework problems involve applied physics (see Problem 52 on page 39 and Problem 59 on page 68).

In some chapters the homework problems end with **Clustered Problems** (written by Fred F. Tomblin of the New Jersey Institute of Technology), in which similar problems are grouped together. In other chapters, the homework problems end with **Tutorial Problems** (written by Laurent Hodges of Iowa State University), in which solutions are worked out.

I am responsible for all the material in this supplement and the main HRW textbook. You can send suggestions, corrections, and positive or negative criticisms directly to me (mail address: Jearl Walker, Physics Department, Cleveland State University, Cleveland OH 44115 USA; fax number: (USA) (216) 687-2424; email address: physics@wiley.com) or to John Wiley & Sons (Web site: www.wiley.com/college/hrw). I may not be able to respond to all suggestions, but I keep and study each of them.

Jearl Walker

CONTENTS

Measurement

Sample Problem 1-5

The research submersible vessel ALVIN is diving at a speed of 36.5 fathoms per minute.

(a) Express this speed in meters per second. A *fathom* (fath) is precisely 6 ft.

SOLUTION: The **Key Idea** in chain-link conversions is to write the conversion factors as ratios that will eliminate unwanted units. Here we write

$$36.5 \text{ fath/min} = \left(36.5 \frac{\text{fath}}{\text{min}}\right)\left(\frac{1 \text{ min}}{60 \text{ s}}\right)\left(\frac{6 \text{ ft}}{1 \text{ fath}}\right)\left(\frac{1 \text{ m}}{3.28 \text{ ft}}\right)$$
$$= 1.11 \text{ m/s}. \qquad \text{(Answer)}$$

(b) What is this speed in miles per hour?

SOLUTION: Using the same **Key Idea**, now we have

$$36.5 \text{ fath/min} = \left(36.5 \frac{\text{fath}}{\text{min}}\right)\left(\frac{60 \text{ min}}{1 \text{ h}}\right)\left(\frac{6 \text{ ft}}{1 \text{ fath}}\right)\left(\frac{1 \text{ mi}}{5280 \text{ ft}}\right)$$
$$= 2.49 \text{ mi/h}. \qquad \text{(Answer)}$$

(c) What is this speed in light-years per year?

SOLUTION: A light-year (ly) is the distance that light travels in 1 year, 9.46×10^{12} km. Using the **Key Idea** and the result of (a), we find

$$1.11 \text{ m/s} = \left(1.11 \frac{\text{m}}{\text{s}}\right)\left(\frac{1 \text{ ly}}{9.46 \times 10^{12} \text{ km}}\right)$$
$$\times \left(\frac{1 \text{ km}}{1000 \text{ m}}\right)\left(\frac{3.16 \times 10^7 \text{ s}}{1 \text{ y}}\right)$$
$$= 3.71 \times 10^{-9} \text{ ly/y}. \qquad \text{(Answer)}$$

We can write this in the even more unlikely form of 3.71 nly/y, where "nly" is an abbreviation for nanolight-year.

Sample Problem 1-6

How many square meters are in an area of 6.0 km²?

SOLUTION: The **Key Idea** here is to convert *square* kilometers to *square* meters. One sure way of doing so is to convert each kilometer in the original measure:

$$6.0 \text{ km}^2 = 6.0 \text{ (km)(km)}$$
$$= 6.0 \text{ (km)(km)} \left(\frac{1000 \text{ m}}{1 \text{ km}}\right)\left(\frac{1000 \text{ m}}{1 \text{ km}}\right)$$
$$= 6.0 \times 10^6 \text{ m}^2.$$

1. How would you criticize this statement: "Once you have picked a standard, by the very meaning of 'standard' it is invariable"?

2. List characteristics other than accessibility and invariability that you would consider desirable for a physical standard.

3. Can you devise a system of base units (Table 1-1) in which time is not included? Explain.

4. Of the three base units listed in Table 1-1, only one—the kilogram—has a prefix (see Table 1-2). Would it be wise to redefine the mass of the standard platinum–iridium cylinder at the International Bureau of Weights and Measures (Fig. 1-4) as 1 g rather than 1 kg?

5. Why are there no SI base units for area and volume?

6. The meter was originally defined to be one ten-millionth of a meridian line that extends from the north pole to the equator and that passes through Paris. This definition disagrees with the standard meter bar by 0.023%. Does this mean that the standard meter bar is inaccurate to this extent? Explain.

7. In defining the meter bar as the standard of length, the temperature of the bar is specified. Can length be called a base quantity if another physical quantity, such as temperature, must be specified in choosing a standard?

8. In redefining the meter in terms of the speed of light, why did not the delegates to the 1983 General Conference on Weights and Measures simplify matters by defining 3×10^8 m/s to be the exact speed of light? For that matter, why did they not define it to be 1 m/s exactly? Were both of these possibilities open to them? If so, why did they reject them?

9. What does the prefix "micro-" signify in the words "microwave oven"? It has been proposed that food that has been irradiated by

gamma rays to lengthen its shelf life be marked "picowaved." What do you suppose that means?

10. Suggest a way to measure (a) the radius of Earth, (b) the distance between the Sun and Earth, and (c) the radius of the Sun.

11. Suggest a way to measure (a) the thickness of a sheet of paper, (b) the thickness of a soap bubble film, and (c) the diameter of an atom.

12. Name several repetitive phenomena occurring in nature that could serve as reasonable time standards.

13. You could define "1 s" to be one pulse beat of the current president of the American Physical Society. Galileo used his pulse as a timing device in some of his work. Why is a definition based on the atomic clock better?

14. What criteria should be satisfied by a good clock?

15. Cite the drawbacks to using the period of a pendulum, such as that for a grandfather clock, as a time standard.

16. On June 30, 1981, the "minute" extending from 10:59 to 11:00 A.M. was arbitrarily lengthened to contain 61 s. The extra second—the *leap second*—was introduced to compensate for the fact that, as measured by our atomic time standard, Earth's rotation rate is slowly decreasing. Why is it desirable to readjust our clocks in this way?

17. Why do we find it useful to have two standards of mass, the kilogram and the carbon-12 atom?

18. Is the current standard kilogram of mass accessible and invariable? Does it have simplicity for comparison purposes? Would an atomic standard be better in any respect?

19. Suggest objects whose masses would fall in the wide range in Table 1-5 between those of an ocean linear and a small mountain, and estimate their masses.

20. Opponents of a switch to the metric system often cloud the issue by saying things such as "Instead of buying one pound of butter you will have to ask for 0.454 kg of butter." The implication is that life would be more complicated. How might you refute this?

EXERCISES & PROBLEMS

30. To the nearest order of magnitude, how many times per year do you take a breath?

31. The fastest growing plant on record is a *Hesperoyucca whipplei* that grew 3.7 m in 14 days. What was its growth rate in micrometers per second?

32. On a spending binge in Malaysia, you buy an ox with a weight of 28.9 piculs in the local unit of weights: 1 picul = 100 gin, 1 gin = 16 tahils, 1 tahil = 10 chee, and 1 chee = 10 hoon. The weight of 1 hoon corresponds to a mass of 0.3779 g. When you arrange to ship the ox home to your astonished family, how much mass in kilograms must you declare on the shipping manifest?

33. One molecule of water (H_2O) contains two atoms of hydrogen and one atom of oxygen. A hydrogen atom has a mass of 1.0 u and an atom of oxygen has a mass of 16 u, approximately. (a) What is the mass in kilograms of one molecule of water? (b) How many molecules of water are in the world's oceans, which have an estimated total mass of 1.4×10^{21} kg?

34. The record for the largest glass bottle was set in 1992 by a team in Millville, New Jersey—they blew a bottle with a volume of 193 U.S. fluid gallons. (a) How much short of 1.0 million cubic centimeters is that? (b) If the bottle were filled with water of a density (mass per volume) of 1000 kg/m^3 at the leisurely rate of 1.8 g/min, how long would the filling process take?

35. During a total solar eclipse, your view of the Sun is almost exactly replaced by your view of the Moon. Assuming that the distance from you to the Sun is about 400 times the distance from you to the Moon, (a) find the ratio of the Sun's diameter to the Moon's diameter. (b) What is the ratio of the Sun's volume to the Moon's volume? (c) Position a small coin in your view so that it just eclipses the full Moon, and measure the angle it subtends at the eye. From this experimental result and the given Earth–Moon distance (= 3.8×10^5 km), estimate the Moon's diameter.

36. The cubit is an ancient unit of length based on the distance between the elbow and the tip of the middle finger of the measurer,

usually 43 to 53 cm. If ancient drawings indicate that a cylindrical pillar in a tomb was to have the length 9 cubits, what would have been the length in (a) meters and (b) millimeters? (c) If the diameter of the pillar was 2 cubits, what was the volume of the pillar in cubic meters?

37. A traditional unit of length in Japan is the ken (1 ken = 1.97 m). What are the ratios of (a) square kens to square meters and (b) cubic kens to cubic meters? What is the volume of a cylindrical water tank of height 5.50 ken and radius 3.00 ken in (c) cubic kens and (d) cubic meters?

38. To the nearest order of magnitude, how many standard toilet paper sheets would be needed to "TP" the shortest path between Cleveland and Los Angeles?

39. A *cord* is a volume of cut wood equal to a stack 8 ft long, 4 ft wide, and 4 ft high. How many cords of wood are in 1.0 m^3?

40. An *astronomical unit* (AU) is equal to the average distance from Earth to the Sun, about 92.9×10^6 mi. A *parsec* (pc) is the distance at which 1 AU would subtend an angle of exactly 1 second of arc (Fig. 1-9). A *light-year* (ly) is the distance that light, traveling through a vacuum with a speed of 186,000 mi/s, would cover in 1.0 year. Express the distance from Earth to the Sun in (a) parsecs and (b) light-years. Express (c) ly and (d) pc in miles. Although "light-year" appears frequently in science fiction, the parsec is preferred by astronomers.

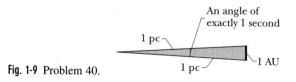

Fig. 1-9 Problem 40.

41. An old English cookbook carries this recipe for cream of nettle soup: "Boil stock of the following amount: 1 breakfastcup plus 1 teacup plus 6 tablespoons plus 1 dessertspoon. Using gloves, separate nettle tops until you have 0.5 quart; add the tops to the boiling stock. Add 1 tablespoon of cooked rice and 1 saltspoon of salt. Simmer for 15 min." The following table gives some of the conversions among old (premetric) British measures and among common (still premetric) U.S. measures. (These measures scream for metrication.) For liquid measures, 1 British teaspoon = 1 U.S. teaspoon. For dry measures, 1 British teaspoon = 2 U.S. teaspoons and 1 British quart = 1 U.S. quart. Translate the old recipe into U.S. measures.

Old British Measures	U.S. Measures
teaspoon = 2 saltspoons	tablespoon = 3 teaspoons
dessertspoon = 2 teaspoons	half cup = 8 tablespoons
tablespoon = 2 dessertspoons	cup = 2 half cups
teacup = 8 tablespoons	
breakfastcup = 2 teacups	

42. Use the prefixes in Table 1-2 to express (a) 10^6 phones; (b) 10^{-6} phone; (c) 10^1 cards; (d) 10^9 lows; (e) 10^{12} bulls; (f) 10^{-1} mate; (g) 10^{-2} pede; (h) 10^{-9} Nannette; (i) 10^{-12} boo; (j) 10^{-18} boy; (k) 2×10^2 withits; (l) 2×10^3 mockingbirds. Now that you have the idea, invent a few similar expressions. (See, in this connection, p. 61 of *A Random Walk in Science,* compiled by R. L. Weber; Crane, Russak & Co., New York, 1974.)

43. A unit of area, often used in measuring land areas, is the *hectare,* defined as 10^4 m^2. An open-pit coal mine consumes 75 hectares of land, down to a depth of 26 m, each year. What volume of earth, in cubic kilometers, is removed in this time?

44. The description for a certain brand of house paint claims a coverage of 460 ft^2/gal. (a) Express this quantity in square meters per liter. (b) Express this quantity in SI base units (see Appendices A and D). (c) What is the inverse of the original quantity, and what is its physical significance?

2
Motion Along a Straight Line

ADDITIONAL SAMPLE PROBLEMS

Sample Problem 2-8

In 1977, Kitty O'Neal set a dragster record by reaching 663.85 km/h from rest in a sizzling time of 3.725 s. In 1958, Eli Beeding Jr. rode a rocket sled along a track, from rest to a speed of 117 km/h in 0.040 s (less than an eye blink). Which ride was more exciting (i.e., more frightening)?

SOLUTION: Because the human body senses accelerations and not speeds, we should compare the average accelerations a_{avg} of O'Neal and Beeding. A **Key Idea** here is to relate the average accelerations to the given final speeds v and elapsed times Δt via the definition of a_{avg} (Eq. 2-7). For O'Neal, assumed to be moving in the positive direction of an axis, we find

$$a_{avg} = \frac{\Delta v}{\Delta t} = \frac{663.85 \text{ km/h} - 0}{3.725 \text{ s}}$$

$$= 178.21 \text{ km/h} \cdot \text{s} \approx 49.5 \text{ m/s}^2 = 5.1g. \quad \text{(Answer)}$$

Similarly, for Beeding

$$a_{avg} = \frac{\Delta v}{\Delta t} = \frac{117 \text{ km/h} - 0}{0.040 \text{ s}}$$

$$= 2925 \text{ km/h} \cdot \text{s} \approx 810 \text{ m/s}^2 = 83g. \quad \text{(Answer)}$$

Beeding clearly had the more exciting ride even though his final speed was considerably slower than O'Neal's. In fact, Beeding's acceleration would have been lethal had it continued for much longer.

Sample Problem 2-9

On a movie set, a red car and a green car are to move toward each other on opposite sides of a highway lane divider line (Fig. 2-25). When the drivers pass each other, the driver of the red car is to toss a package of contraband to the other driver. To catch the toss on film, one of the cameras must be set up near the point where the toss will be made. When "action" is yelled out on the set, the initial separation between the drivers will be 200 m, the red car will accelerate from rest at a constant 6.12 m/s², and the green car will already be in motion with a constant speed of 60 km/h. How far from the initial position of the red car will the toss occur?

SOLUTION: One **Key Idea** here is that both cars will move at constant acceleration (6.12 m/s² for the red car and 0 for the green car); so Table 2-1 applies. As shown, we superimpose an x axis along the divider line, with the origin at the initial location of the red car's driver. Another **Key Idea** is that when the drivers pass each other, they have the *same* coordinate x_P at the same time t_P. We seek x_P.

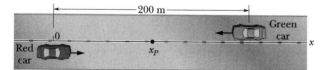

Fig. 2-25 Sample Problem 2-9. Initial separation of two cars approaching each other. The drivers will pass each other at position x_P.

From Table 2-1, we choose Eq. 2-15 ($x - x_0 = v_0 t + \frac{1}{2}at^2$) because it includes position x, initial velocity v_0, and acceleration a, all of which are involved here. Then we substitute known data for each driver separately. For the driver of the red car, with initial position $x_0 = 0$, initial velocity $v_0 = 0$, and acceleration $a = +6.12$ m/s², we find that x_P and t_P are related by

$$x_P - 0 = 0t_P + \tfrac{1}{2}(6.12 \text{ m/s}^2)t_P^2. \quad (2\text{-}22)$$

For the driver of the green car, with initial position $x_0 = 200$ m, constant velocity $v_0 = -60$ km/h ($= -16.67$ m/s), and acceleration $a = 0$, we find that x_P and t_P are related by

$$x_P - 200 \text{ m} = -(16.67 \text{ m/s})t_P + \tfrac{1}{2}(0)t_P^2. \quad (2\text{-}23)$$

Because these equations contain the same two unknowns, we can solve them simultaneously. Perhaps the easiest way is to substitute the right side of Eq. 2-22 for x_P in Eq. 2-23. We find that

$$3.06t_P^2 + 16.67t_P - 200 = 0,$$

which gives us

$$t_P = 5.807 \text{ s}.$$

Inserting this result into Eq. 2-22 then yields

$$x_P = \tfrac{1}{2}(6.12 \text{ m/s}^2)(5.807 \text{ s})^2 = 103 \text{ m}. \quad \text{(Answer)}$$

Thus, the toss will occur at a point 103 m from the initial position of the red car.

Sample Problem 2-10

In 1939, Joe Sprinz of the San Francisco Baseball Club attempted to break the record for catching a baseball dropped from the greatest height. Members of the Cleveland Indians had set the record the preceding year when they caught baseballs dropped about 210 m from atop a building. Sprinz used a blimp at 240 m. Ignore the effects of air on the ball and assume that the ball free-falls that distance.

(a) Find its time of fall.

SOLUTION: Mentally erect a vertical y axis along the ball's path. Then the ball's displacement along the axis is $y - y_0 = -240$ m (negative because the ball falls *down* the axis, not up it), and the ball's initial velocity is $v_0 = 0$. A **Key Idea** here is that the ball falls with the free-fall acceleration $a = -g$. Because this is constant, the equations of Table 2-1 apply. We choose Eq. 2-15 because the only unknown quantity in it is the desired time of fall t. Writing the equation in y notation, we get

$$y - y_0 = v_0 t + \tfrac{1}{2}at^2.$$

Thus, $t = \sqrt{\dfrac{2(y - y_0) - 2v_0}{a}} = \sqrt{\dfrac{2(-240 \text{ m}) - 0}{-9.8 \text{ m/s}^2}}$

$$= 7.0 \text{ s.} \qquad \text{(Answer)}$$

(We choose the positive, not the negative, square root because the ball reaches the ground *after* it is released.)

(b) What is the ball's velocity v just before it is caught?

SOLUTION: Our **Key Idea** remains the same, but we now know enough about the ball's motion to use any equation in Table 2-1 that involves v. Let's choose Eq. 2-11. Then we have

$$v = v_0 + at = 0 + (-9.8 \text{ m/s}^2)(7.0 \text{ s})$$
$$= -68.6 \text{ m/s} \approx -69 \text{ m/s,} \qquad \text{(Answer)}$$

which is about 250 km/h.

Neglecting the effects of air is actually unwarranted in such a fall. If you included them, you would find that the fall time was longer and the final speed was smaller than the values calculated above. Still, the speed must have been considerable, because when Sprinz finally managed to get a ball in his glove (on his fifth attempt) the impact slammed the glove and hand into his face, fracturing the upper jaw in 12 places, breaking five teeth, and knocking him unconscious. And he dropped the ball.

PROBLEM-SOLVING TACTICS

Tactic 9: *Derivatives and Slopes*
Every derivative can be interpreted as the slope of a curve at a point. In Sample Problem 2-2, for example, the velocity of the cab at any instant (a derivative; see Eq. 2-4) is the slope of the $x(t)$ curve of Fig. 2-6a at that instant. Here's how you can find a slope at a point (and thus a derivative) graphically.

Figure 2-26 shows an $x(t)$ plot for a moving particle. To find the velocity of the particle at $t = 1$ s, put a dot on the curve at the point that represents $t = 1$ s. Then draw a line tangent to the curve through the dot (*tangent* means *touching;* the tangent line touches the curve at a single point, the dot), judging carefully by eye. Then construct a right triangle ABC with sides parallel to the axes. (Although the slope is the same no matter what the size of this triangle, the larger the triangle, the more precise will be your measurement of the slope.) Find Δx and Δt, using the vertical and horizontal scales. The slope (derivative) is the quotient $\Delta x/\Delta t$. In Fig. 2-26,

$$\text{slope} = \frac{\Delta x}{\Delta t} = \frac{5.5 \text{ m} - 2.3 \text{ m}}{1.8 \text{ s} - 0.3 \text{ s}} = \frac{3.2 \text{ m}}{1.5 \text{ s}} = +2.1 \text{ m/s.}$$

As Eq. 2-4 tells you, this slope is the velocity of the particle at $t = 1$ s. If you change the scale on either axis of Fig. 2-26, the appearance of the curve and the angle θ will change, but the value you find for the velocity at $t = 1$ s will not.

If you have a mathematical expression for the function $x(t)$, as in Sample Problem 2-3, you can find the derivative dx/dt by the methods of calculus and avoid this graphical method.

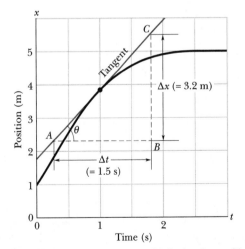

Fig. 2-26 The slope of a curve at any point is the slope of its tangent line at that point. At $t = 1.0$ s, the slope of the tangent line (and thus of the curve) is $\Delta x/\Delta t = +2.1$ m/s. That is also the instantaneous velocity dx/dt at that time.

QUESTIONS

10. Figure 2-27 is a graph of a particle's position along an x axis versus time. (a) At time $t = 0$, what is the sign of the particle's position? Is the particle's velocity positive, negative, or zero at (b) $t = 1$ s, (c) $t = 2$ s, and (d) $t = 3$ s? (e) How many times does the particle go through the point $x = 0$?

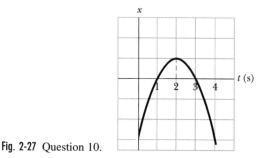

Fig. 2-27 Question 10.

11. Figure 2-28 shows that a particle moving along an x axis undergoes three periods of acceleration. Without written computation, rank the acceleration periods according to the increases they produce in the particle's velocity, greatest first.

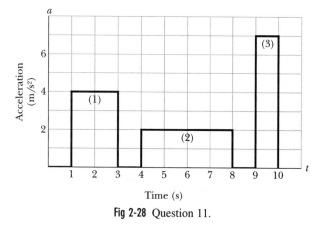

Fig 2-28 Question 11.

12. In Fig. 2-29, assume that the vertical axis pertains to the velocity v of an object moving along an x axis. Then determine which of the 10 plots of v versus time t best describes the motion for the four situations listed in the table. (Plots 2, 3, 7, and 9 are straight; the others are curved.)

Situation	a	b	c	d
Initial x (m)	+10	−10	+10	−10
Initial v (m/s)	+5	−5	−5	+5
Constant a (m/s^2)	+2	−2	+2	−2

13. Question 12 continued: If, instead, the vertical axis pertains to the position x of the object, then which of the 10 plots best describes the motion for the four situations given in the table?

14. Suppose that the position of a particle is given by the expression $x = (1.0)t^n$, where n is an integer, x is in meters, and t is in seconds. That expression is graphed in Fig. 2-30a for $n = 1, 2, 3,$ and 4.

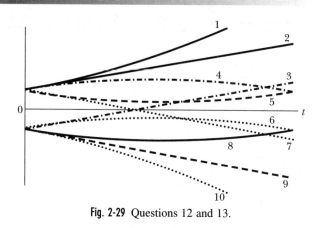

Fig. 2-29 Questions 12 and 13.

(a) Which graphed curve corresponds to which value of n, and what are the corresponding units of the coefficient 1.0 in the expression for $x(t)$? Rank the values of n according to the particle's displacement during (b) the first 0.5 s of motion ($t = 0$ to $t = 0.5$ s) and (c) the first 1.5 s of motion, greatest first.

The acceleration $a(t)$ of the particle is graphed in Fig. 2-30b for the four values of n. (Curve H lies along the horizontal axis.) (d) Which $a(t)$ curve corresponds to which $x(t)$ curve? (e) Rank the $a(t)$ curves according to the rate at which the velocity is changing at time $t = 0.20$ s, greatest first.

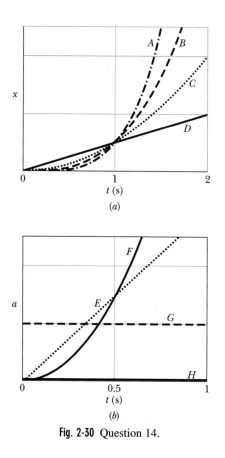

Fig. 2-30 Question 14.

15. Suppose that a hot-air balloonist drops an apple over the side while the balloon is accelerating upward at 4.0 m/s² during liftoff. (a) What is the apple's acceleration once it is released? (b) If the velocity of the balloon is 2 m/s upward at the instant of release, what is the apple's velocity just then?

16. (a) Graph y, v, and a versus t for a cream tangerine that is thrown straight up from the edge of a cliff and that, upon falling back down, barely passes by the cliff edge. (b) On the same figures, graph the same quantities for a cream tangerine that is dropped (released with no initial speed) from the cliff edge.

17. *Organizing question:* At the National Enchilada Proving Grounds, enchiladas are launched either directly upward or directly downward from a high overhang with an initial speed of 10 m/s. Set up equations, complete with known data, to find the time t an upward-launched enchilada takes to reach a point (a) 2 m above the launch point and (b) 2 m below the launch point. (c) Rewrite the equation for (b) for a downward launch. Next, set up equations, complete with known data, to find the enchilada's displacement $y - y_0$ from the launch point when it is traveling at a speed of 12 m/s after being launched (d) directly upward and (e) directly downward.

18. *Math tool time.* What are the results of differentiating with respect to time t the position functions (a) $x = 3t^2 + 4t + 5$ and (b) $x = 3t^{-2}$? What are the results of the integrations (c) $\int 3t^2 \, dt$ and (d) $\int (2t + 5) \, dt$?

DISCUSSION QUESTIONS

19. What are some physical phenomena involving Earth in which Earth cannot be treated as a particle?

20. Can the speed of a particle ever be negative? If so, give an example; if not, explain why.

21. Each second a rabbit moves half the remaining distance from its nose to a head of lettuce. Does the rabbit ever get to the lettuce? What is the limiting value of the rabbit's velocity? Draw graphs showing the rabbit's position and average velocity versus time.

22. In this book, "average speed" means the ratio of the total distance covered to the elapsed time. However, sometimes the phrase is used to mean the magnitude of the average velocity. How, and in what cases, do the meanings differ?

23. In a qualifying two-lap heat, a racing car covers the first lap with an average speed of 140 km/h. Can the driver take the second lap at a speed such that the average speed for the two laps is 280 km/h? Explain.

24. Bob beats Judy by 10 m in a 100 m dash. Bob, claiming to want to give Judy an equal chance, agrees to race her again but to begin 10 m behind the starting line. Does this really give Judy an equal chance?

25. When the velocity is constant, can the average velocity over any time interval differ from the instantaneous velocity at any instant? If so, give an example; if not, explain why.

26. Can the average velocity of a particle moving along an x axis ever be $\frac{1}{2}(v_0 + v)$ if the acceleration is not uniform? Prove your answer with graphs.

27. (a) Can an object have zero velocity and still be accelerating? (b) Can an object have a constant velocity and still have a varying speed? In each case, give an example if your answer is yes; explain why if your answer is no.

28. Can the velocity of an object reverse direction when its acceleration is constant? If so, give an example; if not, explain why.

29. Can an object be increasing in speed as its acceleration decreases? If so, give an example; if not, explain why.

30. If a particle is accelerated from rest ($v_0 = 0$) at $x_0 = 0$, beginning at the time $t = 0$, Eq. 2-15 says that it is at a certain position x at two different times—namely, $+\sqrt{2x/a}$ and $-\sqrt{2x/a}$. Does the negative root have meaning? If, instead, the particle had been moving before $t = 0$, does the negative root have more meaning?

31. What are some examples of falling objects for which it would be unreasonable to neglect the effects of air?

32. On a planet where the value of g is one-half the value on Earth, an object is dropped from rest and falls to the ground. How is the time needed for it to reach the ground from rest related to the time required to fall the same distance on Earth?

EXERCISES & PROBLEMS

Additional Problems

66. From January 26, 1977 to September 18, 1983, George Meegan of Great Britain walked from Ushuaia, at the southern tip of South America, to Prudhoe Bay in north Alaska, covering 30,600 km. In meters per second, what was his average speed during that time period?

67. The sport with the fastest moving ball is jai alai, where measured speeds have reached 303 km/h. If a professional jai alai player faces a ball at that speed and involuntarily blinks, he blacks out the scene for 100 ms. How far does the ball move during the blackout?

68. At a military drill team contest at Fort Meade, Maryland, the home team moves forward with a sequence of four rates, given here with their U.S. definitions: (1) quick time (120 steps per minute, 30 in. per step), (2) half-step in quick time (120 steps per minute, 15 in. per step), (3) double time (180 steps per minute, 36 in. per step), (4) half-step in double time (180 steps per minute, 18 in. per step). If the team moves at each rate for 5.0 s, (a) how far in meters does the team move during the sequence and (b) what is the magnitude of their average velocity in meters per second? If, instead, the team moves at each rate for a distance of 8.00 m, (c) how long does the sequence take and (d) what is the magnitude of their average velocity in meters per second?

69. The wings on a stonefly do not flap, and thus the insect cannot fly. However, when the insect is on a water surface, it can sail across the surface by lifting its wings into a breeze (Fig. 2-31). Suppose that you time stoneflies as they move at constant speed along a straight path of a certain length. On average, the trips each take 7.1 s with the wings set as sails and 25.0 s with the wings tucked in. (a) What is the ratio of the sailing speed v_s to the non-sailing speed v_{ns}? (b) In terms of v_s, what is the difference in the times the insects take to travel the first 2.0 m along the path with and without sailing?

Fig. 2-31 Problem 69. Stonefly on water.

70. Most important in an investigation of an airplane crash by the U.S. National Transportation Safety Board is the data stored on the airplane's flight-data recorder, commonly called the "black box" in spite of its orange coloring and reflective tape (Fig. 2-32). The recorder is engineered to withstand a crash with an average deceleration of magnitude $3400g$ during a time interval of 6.50 ms. In such a crash, if the recorder and airplane have zero speed at the

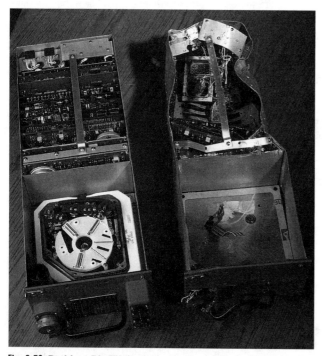

Fig. 2-32 Problem 70. Flight-data recorders: one ready to be installed and the other after a crash, with the data tape removed.

end of that time interval, what is their speed at the beginning of the interval?

71. A motorcyclist starts from rest and accelerates along a horizontal straight track. Photogates are attached to six posts evenly spaced every 10.0 m along the track; the first post is at the starting point. The photogates on each post measure the time required by the motorcyclist to reach that post. The following table gives the results for one test. (a) Find an expression for the distance d to each post in terms of the time t to reach that post and the acceleration a of the motorcyclist (assumed constant). (b) Using the data of the table, graph d versus t^2. (c) Using a linear regression fit of the data, find the acceleration of the motorcyclist.

Post number	1	2	3	4	5	6
Distance traveled (m)	0	10.0	20.0	30.0	40.0	50.0
Time required (s)	0	1.63	2.33	2.83	3.31	3.79

72. A graph of x versus t for a particle in straight-line motion is shown in Fig. 2-33. (a) What is the average velocity of the particle between $t = 0.50$ s and $t = 4.5$ s? (b) What is the instantaneous velocity of the particle at $t = 4.5$ s? (c) What is the average acceleration of the particle between $t = 0.50$ s and $t = 4.5$ s? (d) What is the instantaneous acceleration of the particle at $t = 4.5$ s?

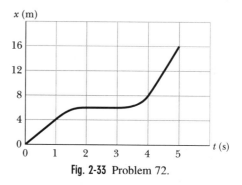

Fig. 2-33 Problem 72.

73. If the position of an object is given by $x = 2.0t^3$, where x is measured in meters and t in seconds, find (a) the average velocity and (b) the average acceleration between $t = 1.0$ s and $t = 2.0$ s. Then find (c) the instantaneous velocity v and (d) the instantaneous acceleration a at $t = 1.0$ s. Next find (e) v and (f) a at $t = 2.0$ s. (g) Compare the average and instantaneous quantities and in each case explain why the larger one is larger. (h) Graph x versus t and v versus t, and indicate on the graphs your answers to (a) through (f).

74. An object falls a distance h from rest. If it travels $0.50h$ in the last 1.00 s, find (a) the time and (b) the height of its fall. (c) Explain the physically unacceptable solution of the quadratic equation in t that you obtain when you solve this problem.

75. A lead ball is dropped in a lake from a diving board 5.20 m above the water. It hits the water with a certain velocity and then sinks to the bottom with this same constant velocity. It reaches the bottom 4.80 s after it is dropped. (a) How deep is the lake? (b) What is the average velocity of the ball for the entire fall? (c) Suppose that all the water is drained from the lake. The ball is now thrown from the diving board so that it again reaches the bottom in 4.80 s. What is the initial velocity of the ball?

76. In 1889, at Jubbulpore, India, a tug-of-war was finally won after 2 h 41 min, with the winning team displacing the center of the rope 3.7 m. In centimeters per minute, what was the magnitude of the average velocity of that center point during the contest?

77. (a) If the position of a particle is given by $x = 20t - 5t^3$, where x is in meters and t is in seconds, when, if ever, is the particle's velocity zero? (b) When is its acceleration a zero? (c) When is a negative? Positive? (d) Graph $x(t)$, $v(t)$, and $a(t)$.

78. A graph of acceleration versus time for a particle as it moves along an x axis is shown in Fig. 2-34. At $t = 0$ the coordinate of the particle is 4.0 m and the velocity is 2.0 m/s. (a) What is the velocity of the particle at $t = 2.0$ s? (b) Write an expression for the velocity as a function of the time that is valid for the interval $2.0 \text{ s} \le t \le 4.0 \text{ s}$.

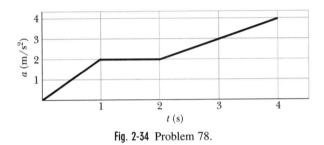

Fig. 2-34 Problem 78.

79. (a) If the maximum acceleration that is tolerable for passengers in a subway train is 1.34 m/s^2, and subway stations are located 806 m apart, what is the maximum speed a subway train can attain between stations? (b) What is the travel time between stations? (c) If a subway train stops for 20 s at each station, what is the maximum average speed of the train, from one start-up to the next? (d) Graph x, v, and a versus t for the interval from one start-up to the next.

80. A car travels up a hill at a constant speed of 40 km/h and returns down the hill at a constant speed of 60 km/h. Calculate the average speed for the round trip.

81. A car can be braked to a stop from the autobahn-like speed of 200 km/h in 170 m. Assuming the acceleration is constant, find its magnitude in (a) SI units and (b) g units. (c) How much time T_b is required for the braking? Your *reaction time* T_r is the time you require to perceive an emergency, move your foot to the brake, and begin the braking. (d) If $T_r = 400$ ms, then what is T_b in terms of T_r, and is most of the full time required to stop spent in reacting or braking? Dark sunglasses delay the visual signals sent from the eyes to the visual cortex in the brain, increasing T_r. (e) In the extreme case in which T_r is increased by 100 ms, how much farther along the road does the car travel during your reaction time?

82. A train started from rest and moved with constant acceleration. At one time it was traveling 30 m/s, and 160 m farther on it was traveling 50 m/s. Calculate (a) the acceleration, (b) the time required to travel the 160 m mentioned, (c) the time required to attain the speed of 30 m/s, and (d) the distance moved from rest to the time the train had a speed of 30 m/s. (e) Graph x versus t and v versus t for the train, from rest.

83. Two subway stops are separated by 1100 m. If a subway train accelerates at $+1.2$ m/s^2 from rest through the first half of the distance and decelerates at -1.2 m/s^2 through the second half, what are (a) its travel time and (b) its maximum speed? (c) Graph x, v, and a versus t for the trip.

84. Carl Lewis ran the 100 m dash in about 10 s, and Bill Rodgers ran the marathon (42 km) in about 2 h 10 min. (a) What are their average speeds? (b) If Lewis could have maintained his sprint speed during a marathon, how long would he have taken to finish?

85. The single cable supporting an unoccupied construction elevator breaks when the elevator is at rest at the top of a 120-m-high building. (a) With what speed does the elevator strike the ground? (b) How long is it falling? (c) What is its speed when it passes the halfway point on the way down? (d) How long has it been falling when it passes the halfway point?

86. A rock is thrown vertically upward from ground level at time $t = 0$. At $t = 1.5$ s it passes the top of a tall tower, and 1.0 s later it reaches its maximum height. What is the height of the tower?

87. You are driving toward a traffic signal when it turns yellow. Your speed is the legal speed limit of $v_0 = 55$ km/h; your best deceleration rate has the magnitude $a = 5.18$ m/s^2. Your best reaction time to begin braking is $T = 0.75$ s. To avoid having the front of your car enter the intersection after the light turns red, should you brake to a stop or continue to move at 55 km/h if the distance to the intersection and the duration of the yellow light are (a) 40 m and 2.8 s, and (b) 32 m and 1.8 s?

88. Sketch a $v(t)$ graph that would be associated with the $a(t)$ graph shown in Fig. 2-35.

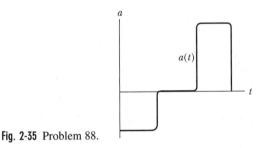

Fig. 2-35 Problem 88.

89. A rocket-driven sled running on a straight, level track is used to investigate the physiological effects of large accelerations on humans. One such sled can attain a speed of 1600 km/h in 1.8 s, starting from rest. Find (a) the acceleration (assumed constant) in g units and (b) the distance traveled.

90. Sketch a graph of x versus t for a mouse that is constrained in a narrow corridor (the x axis) and that scurries in the following sequence: (1) runs leftward (the negative direction of x) with a constant speed of 1.2 m/s, (2) gradually slows to 0.6 m/s toward the left, (3) gradually speeds up to 2.0 m/s toward the left, (4) gradually slows to a stop and then speeds up to 1.2 m/s toward the right. Where is the graph steepest? Least steep?

91. The position function $x(t)$ of a particle moving along an x axis is $x = 4.0 - 6.0t^2$, with x in meters and t in seconds. (a) At what time and (b) where does the particle (momentarily) stop? (c) At what two times does the particle pass through the origin? (d) Graph x versus t. (e) To shift the curve rightward on the graph, should we include the term $+20t$ or the term $-20t$ in $x(t)$? (f) Does that inclusion increase or decrease the value of x at which the particle momentarily stops?

92. The graph of x versus t in Fig. 2-36a is for a particle in straight-line motion. (a) State, for each of the intervals AB, BC, CD, and DE, whether the velocity v is positive, negative, or 0 and whether the acceleration a is positive, negative, or 0. (Ignore the end points of the intervals.) (b) From the curve, is there any interval over which the acceleration is obviously not constant? (c) If the axes are shifted upward together such that the time axis ends up running along the dashed line, do any of your answers change?

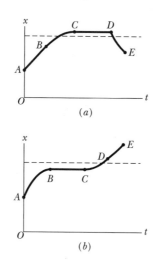

(a)

(b)

Fig. 2-36 Problems 92 and 93.

93. Repeat Problem 92 for the motion described by the graph of Fig. 2-36b.

94. A certain sprinter has a top speed of 11.0 m/s. If the sprinter starts from rest and accelerates at a constant rate, he is able to reach his top speed in a distance of 12.0 m. He is then able to maintain this top speed for the remainder of a 100 m race. (a) What is his time for the 100 m race? (b) In order to improve his time, the sprinter tries to decrease the distance required for him to reach his top speed. What must this distance be if he is to achieve a time of 10.0 s for the race?

95. On a dry road a car with good tires may be able to brake with a constant deceleration of 4.92 m/s². (a) How long does such a car, initially traveling at 24.6 m/s, take to stop? (b) How far does it travel in this time? (c) Graph x versus t and v versus t for the deceleration.

96. A hot rod can accelerate from 0 to 60 km/h in 5.4 s. (a) What is its average acceleration, in m/s², during this time? (b) How far will it travel during the 5.4 s, assuming its acceleration is constant? (c) How much time would it require to go a distance of 0.25 km if its acceleration could be maintained at the value in (a)?

97. On average, an eye blink lasts about 100 ms. How far does a MiG-25 "Foxbat" fighter travel during a pilot's blink if the plane's average velocity is 3400 km/h?

98. When the legal speed limit for the New York Thruway was increased from 55 mi/h to 65 mi/h, how much time was saved by a motorist who drove the 700 km between the Buffalo entrance and the New York City exit at the legal speed limit?

99. Suppose the coordinate of a particle moving along the x axis is given as a function of time t by

$$x(t) = -32.0 + 24.0t^2 e^{-0.0300t},$$

where x is in meters and t is in seconds. (a) Write expressions for the velocity v and acceleration a as functions of the time. (b) Plot graphs of the coordinate, velocity, and acceleration as functions of time from $t = 0$ to $t = 100$ s. (c) Find the time at which the coordinate of the particle is zero; then find the velocity and acceleration at that time. (d) Find the time at which the velocity of the particle is zero; then find the coordinate and acceleration at that time.

100. A motorcycle is moving at 30 m/s when the rider applies the brakes, giving the motorcycle a constant deceleration. During the 3.0 s interval immediately after braking begins, the speed decreases to 15 m/s. What distance does the motorcycle travel from the instant braking begins until the motorcycle stops?

101. In Sample Problem 2-5a and with a list-capable calculator, produce a list of the accelerations required (in kilometers per hour-squared) if the initial and final speeds are, respectively, (a) 85 km/h and 65 km/h, (b) 80 km/h and 60 km/h, and (c) 50 km/h and 40 km/h. The displacement is still 88 m.

102. As Fig. 2-37 shows, Clara jumped from a bridge, followed closely by Jim. How long did Jim wait after Clara jumped? Assume that Jim is 170 cm tall and that the jumping-off level is at the top of the figure. Make scale measurements directly on the figure.

Fig. 2-37 Problem 102.

103. A steel ball is dropped from a building's roof and passes a window, taking 0.125 s to fall from the top to the bottom of the window, a distance of 1.20 m. It then falls to a sidewalk and bounces back past the window, moving from bottom to top in 0.125 s. Assume that the upward flight is an exact reverse of the fall. The time the ball spends below the bottom of the window is 2.00 s. How tall is the building?

104. A particle starts from the origin at $t = 0$ and moves along the positive x axis. A graph of the velocity of the particle as a function of the time is shown in Fig. 2-38. (a) What is the coordinate of the particle at $t = 5.0$ s? (b) What is the velocity of the particle at $t = 5.0$ s? (c) What is the acceleration of the particle at $t = 5.0$ s? (d) What is the average velocity of the particle between $t = 1.0$ s and $t = 5.0$ s? (e) What is the average acceleration of the particle between $t = 1.0$ s and $t = 5.0$ s?

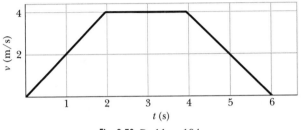

Fig. 2-38 Problem 104.

CLUSTERED PROBLEMS

In these problems, points A, B, and C lie, in that order, along a straight line.

105. A car speeds up at a constant rate from rest beginning at point A. It achieves a speed of 10.0 m/s at point B, which is 40.0 m from point A. From point B the car travels at constant speed, arriving at point C 10.0 s after leaving point B. (a) Find the acceleration of the car from point A to point B. (b) Find the time required to travel from point A to point B. (c) Find the distance from point B to point C. (d) Find the average velocity of the car as it moves from point A to point C.

106. A car passes point A moving at 20.0 m/s. Accelerating at a constant rate for 10.0 s, it passes point B moving at 30.0 m/s. After passing point B it slows at a constant rate to a speed of 15.0 m/s at point C, which is 150 m from point B. (a) Find the distance from point A to point B. (b) Find the time required to travel from point B to point C. (c) Find the average velocity of the car as it moves from point A to point C. (d) Find the average acceleration of the car between points A and C.

107. A car passes point A moving at 20 m/s. It travels for 5.0 s at constant velocity to point B and then slows at a constant rate, stopping at point C in 10 s. (a) How far does the car travel from point A to point C? (b) What is the deceleration of the car as it moves from point B to point C?

108. A car travels from point A to point B in 5.00 s at constant velocity and then decelerates at a constant rate, reaching point C in 20.0 s, where it is traveling 10.0 m/s. Point C is 300 m from point A. (a) What is the velocity of the car at point A? (b) What is the rate of change of the velocity from point B to point C?

109. A car travels from point A to point B in 5.00 s at constant velocity and then slows at a constant rate of -0.500 m/s^2, stopping at point C. Point C is 250 m from point A. (a) What is the velocity of the car at point A? (b) How long does the car take to go from point A to point C?

110. A car starts from rest at point A, accelerates at a constant rate to reach point B in 20 s, then accelerates at a different constant rate to reach point C in 40 s. The car is going 50 m/s at point C, and point A is 1300 m from point C. (a) Find the velocity of the car at point B. (b) Find the distance from point A to point B. Find the acceleration (c) from point A to point B and (d) from point B to point C.

Vectors

3

11. In a game within a three-dimensional maze, you must move your game piece from *start,* at *xyz* coordinates (0, 0, 0), to *finish,* at coordinates (−2 cm, 4 cm, −4 cm). The game piece can undergo only the displacements (in centimeters) given below. If, along the way, the game piece lands at coordinates (−5 cm, −1 cm, −1 cm) or (5 cm, 2 cm, −1 cm), you lose the game. Which displacements and in what sequence will get your game piece to *finish*?

$$\vec{p} = -7\hat{i} + 2\hat{j} - 3\hat{k} \qquad \vec{r} = 2\hat{i} - 3\hat{j} + 2\hat{k}$$
$$\vec{q} = 2\hat{i} - \hat{j} + 4\hat{k} \qquad \vec{s} = 3\hat{i} + 5\hat{j} - 3\hat{k}.$$

12. Can the sum of the magnitudes of two vectors ever be equal to the magnitude of the sum of the same two vectors? If no, why not? If yes, when?

13. The *x* and *y* components of four vectors $\vec{a}$, $\vec{b}$, $\vec{c}$, and $\vec{d}$ are given below. For which vectors will your calculator give you the correct angle θ when you use it to find θ with Eq. 3-6? Answer first by examining Fig. 3-13, and then check your answers with your calculator.

$$a_x = 3 \qquad a_y = 3 \qquad c_x = -3 \qquad c_y = -3$$
$$b_x = -3 \qquad b_y = 3 \qquad d_x = 3 \qquad d_y = -3.$$

14. Figure 3-32 shows a vector $\vec{R}$ and two coordinate systems (the primed system $x'y'$ and the double-primed system $x''y''$, with a common origin). (a) Is component $R_{x'}$ greater than, less than, or equal to component $R_{x''}$? (b) Is component $R_{y'}$ greater than, less than, or equal to component $R_{y''}$? (c) Is the angle that $\vec{R}$ makes with the x' axis greater than, less than, or equal to the angle it makes with the x'' axis? (d) Is the value of $\sqrt{(R_{x'})^2 + (R_{y'})^2}$ greater than, less than, or equal to the value of $\sqrt{(R_{x''})^2 + (R_{y''})^2}$?

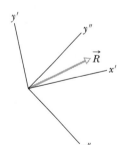

Fig. 3-32 Question 14.

15. If $\vec{A} = 2\hat{i} + 4\hat{j}$ and $\vec{B} = 7\hat{k}$, what is $\vec{A} \cdot \vec{B}$?

16. An *area vector* $\vec{A}$ is a vector assigned to a surface; its magnitude is equal to the area of the surface, and its direction is perpendicular to the surface. When the surface is part of a closed structure, $\vec{A}$ is directed away from the interior. Figure 3-33 shows a cube and a

vector $\vec{B}$ that extends in the positive *x* direction; the area of each face is *A*. What is $\vec{B} \cdot \vec{A}$ for each face?

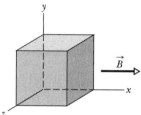

Fig. 3-33 Question 16.

17. *Organizing question:* Which of the following are correct (meaningful) vector expressions? What is wrong with any incorrect expression?

(a) $\vec{A} \cdot (\vec{B} \cdot \vec{C})$;
(f) $\vec{A} + (\vec{B} \times \vec{C})$;
(b) $\vec{A} \times (\vec{B} \cdot \vec{C})$;
(g) $5 + \vec{A}$;
(c) $\vec{A} \cdot (\vec{B} \times \vec{C})$;
(h) $5 + (\vec{B} \cdot \vec{C})$;
(d) $\vec{A} \times (\vec{B} \times \vec{C})$;
(i) $5 + (\vec{B} \times \vec{C})$;
(e) $\vec{A} + (\vec{B} \cdot \vec{C})$;
(j) $(\vec{A} \cdot \vec{B}) + (\vec{B} \times \vec{C})$

18. In Fig. 3-34, what is the direction of (a) $\vec{E} \times \vec{F}$, (b) $\vec{F} \times \vec{E}$, and (c) $\vec{G} \times \vec{E}$, where $\vec{G}$ is in the *xy* plane? (d) Does the answer to (c) change if $\vec{G}$ is shifted parallel to the *z* axis without any change in its direction?

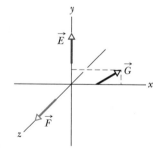

Fig. 3-34 Question 18.

19. Figure 3-35 shows vectors $\vec{r}$ and $\vec{F}$, both in the *xy* plane. What is the direction of (a) $\vec{r} \times \vec{F}$ and (b) $\vec{r} \times (-\vec{F})$?

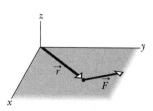

Fig. 3-35 Question 19.

20. Suppose that a vector $\vec{F}$ is given by

$$\vec{F} = q(\vec{v} \times \vec{B}),$$

where q is a constant and $\vec{v}$ and $\vec{B}$ are vectors. (a) What are the directions of the cross product $\vec{v} \times \vec{B}$ for the three situations in Fig. 3-36? What are the directions of $\vec{F}$ for those situations if q is (b) a positive quantity and (c) a negative quantity?

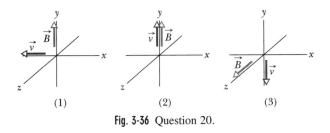

Fig. 3-36 Question 20.

21. Using the expression for $\vec{F}$ given in Question 20, determine the direction of $\vec{B}$ for the three situations of Fig. 3-37 if q is (a) a positive quantity and (b) a negative quantity. If the situation is impossible, state that fact.

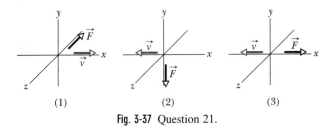

Fig. 3-37 Question 21.

22. Figure 3-38 shows vector $\vec{A}$ and four choices for vector $\vec{B}$ that differ only in orientation. Rank the choices according to the absolute value of $\vec{A} \cdot \vec{B}$ that they give, greatest first.

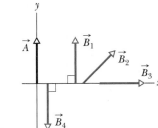

Fig. 3-38 Question 22.

23. The two vectors shown in Fig. 3-39 lie in an xy plane. What are the signs of the x and y components, respectively, of (a) $\vec{d}_1 + \vec{d}_2$, (b) $\vec{d}_1 - \vec{d}_2$, and (c) $\vec{d}_2 - \vec{d}_1$?

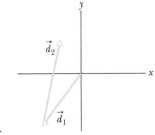

Fig. 3-39 Question 23.

DISCUSSION QUESTIONS

24. In 1969, three Apollo astronauts left Cape Canaveral, went to the Moon and back, and splashed down in the Pacific Ocean. An admiral bid them good-bye at the Cape and then sailed to the Pacific Ocean in an aircraft carrier where he picked them up. Compare the displacements of the astronauts and the admiral.

25. Can two vectors having different magnitudes be combined to give a zero resultant? Can three vectors?

26. If three vectors add up to zero, must they all be in the same plane?

27. Explain in what sense a vector equation contains more information than a scalar equation.

28. Why do the unit vectors $\hat{i}$, $\hat{j}$, and $\hat{k}$ have no units?

29. Name several scalar quantities. Does the value of a scalar quantity depend on the coordinate system you choose?

30. You can order events in time. For example, event b may precede event c but follow event a, giving us a time order of events a, b, c. Hence there is a sense of time, distinguishing past, present, and future. Is time a vector therefore? If not, why not?

31. Do the commutative and associative laws apply to vector subtraction?

32. Can a scalar product be a negative quantity?

33. If $\vec{a} \cdot \vec{b} = 0$, does it follow that $\vec{a}$ and $\vec{b}$ are perpendicular to one another?

34. If $\vec{a} \times \vec{b} = 0$, must $\vec{a}$ and $\vec{b}$ be parallel to each other? Is the converse true?

35. Must you specify a coordinate system when you (a) add two vectors, (b) form their scalar product, (c) form their vector product, or (d) find their components?

EXERCISES & PROBLEMS

39. Find the angle between each pair of body diagonals in a cube with edge length a. See Problem 27.

40. (a) In unit-vector notation, what is $\vec{r} = \vec{a} - \vec{b} + \vec{c}$ if $\vec{a} = 5.0\hat{i} + 4.0\hat{j} - 6.0\hat{k}$, $\vec{b} = -2.0\hat{i} + 2.0\hat{j} + 3.0\hat{k}$, and $\vec{c} = 4.0\hat{i} + 3.0\hat{j} + 2.0\hat{k}$? (b) Calculate the angle between $\vec{r}$ and the positive z axis. (c) What is the component of $\vec{a}$ along the direction of $\vec{b}$?

(d) What is the component of $\vec{a}$ perpendicular to the direction of $\vec{b}$ but in the plane of $\vec{a}$ and $\vec{b}$? (*Hint:* For (c), consider Eq. 3-20 and Fig. 3-19; for (d), consider Eq. 3-27.)

41. Show that $\vec{a} \cdot (\vec{b} \times \vec{c})$ is equal in magnitude to the volume of the parallelepiped formed on the three vectors $\vec{a}, \vec{b}$, and $\vec{c}$ as shown in Fig. 3-40.

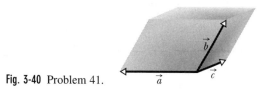

Fig. 3-40 Problem 41.

42. Two vectors are given by $\vec{a} = 3.0\hat{i} + 5.0\hat{j}$ and $\vec{b} = 2.0\hat{i} + 4.0\hat{j}$. Find (a) $\vec{a} \times \vec{b}$, (b) $\vec{a} \cdot \vec{b}$, (c) $(\vec{a} + \vec{b}) \cdot \vec{b}$, and (d) the component of $\vec{a}$ along the direction of $\vec{b}$. (*Hint:* For (d), consider Eq. 3-20 and Fig. 3-19.)

43. Three vectors are given by $\vec{a} = 3.0\hat{i} + 3.0\hat{j} - 2.0\hat{k}$, $\vec{b} = -1.0\hat{i} - 4.0\hat{j} + 2.0\hat{k}$, and $\vec{c} = 2.0\hat{i} + 2.0\hat{j} + 1.0\hat{k}$. Find (a) $\vec{a} \cdot (\vec{b} \times \vec{c})$, (b) $\vec{a} \cdot (\vec{b} + \vec{c})$, and (c) $\vec{a} \times (\vec{b} + \vec{c})$.

44. Vector $\vec{a}$ lies in the yz plane $63°$ from the positive direction of the y axis, has a positive z component, and has magnitude 3.20 units. Vector $\vec{b}$ lies in the xz plane $48°$ from the positive direction of the x axis, has a positive z component, and has magnitude 1.40 units. Find (a) $\vec{a} \cdot \vec{b}$, (b) $\vec{a} \times \vec{b}$, and (c) the angle between $\vec{a}$ and $\vec{b}$.

45. In a right-handed coordinate system show that

$$\hat{i} \times \hat{i} = \hat{j} \times \hat{j} = \hat{k} \times \hat{k} = 0$$

and $\hat{i} \times \hat{j} = \hat{k};$ $\hat{k} \times \hat{i} = \hat{j};$ $\hat{j} \times \hat{k} = \hat{i}.$

If the coordinate system is rectangular but not right-handed, do the results change?

46. Find (a) "north cross west," (b) "down dot south," (c) "east cross up," (d) "west dot west," and (e) "south cross south." Let each "vector" have unit magnitude.

47. A vector $\vec{B}$, with a magnitude of 8.0 m, is added to a vector $\vec{A}$, which lies along an x axis. The sum of these two vectors is a third vector that lies along the y axis and has a magnitude that is twice the magnitude of $\vec{A}$. What is the magnitude of $\vec{A}$?

48. A particle undergoes three successive displacements in a plane, as follows: $\vec{d}_1$, 4.00 m southwest; then $\vec{d}_2$, 5.00 m east; and finally $\vec{d}_3$, 6.00 m in a direction $60.0°$ north of east. Choose a coordinate system with the y axis pointing north and the x axis pointing east. What are (a) the x component and (b) the y component of $\vec{d}_1$? What are (c) the x component and (d) the y component of $\vec{d}_2$? What are (e) the x component and (f) the y component of $\vec{d}_3$?

What are (g) the x component, (h) the y component, (i) the magnitude, and (j) the direction of the particle's net displacement? If the particle is to return directly to the starting point, (k) how far and (l) in what direction should it move?

49. An ant, crazed by the Sun on a hot Texas afternoon, darts over an xy plane scratched in the dirt. The x and y components of four consecutive darts are the following, all in centimeters: (30.0, 40.0), (b_x, −70.0), (−20.0, c_y), (−80.0, −70.0). The overall displacement of the four darts has the xy components (−140, −20.0). What are (a) component b_x and (b) component c_y? What are (c) the magnitude and (d) the angle of the overall displacement?

50. A person desires to reach a point that is 3.40 km from her present location and in a direction that is $35.0°$ north of east. However, she must travel along streets that are oriented either north–south or east–west. What is the minimum distance she could travel to reach her destination?

51. A golfer takes three putts to get the ball into the hole. The first putt displaces the ball 3.66 m north, the second 1.83 m southeast, and the third 0.91 m southwest. What are (a) the magnitude and (b) the direction of the displacement needed to get the ball into the hole on the first putt?

52. Consider $\vec{a}$ in the positive direction of x, $\vec{b}$ in the positive direction of y, and a scalar d. What is the direction of $\vec{b}/d$ if d is (a) positive and (b) negative? What is the magnitude of (c) $\vec{a} \cdot \vec{b}$ and (d) $\vec{a} \cdot \vec{b}/d$? What is the direction of the vector resulting from (e) $\vec{a} \times \vec{b}$ and (f) $\vec{b} \times \vec{a}$? (g) What are the magnitudes of the cross products in (e) and (f)? What are (h) the magnitude and (i) the direction of $\vec{a} \times \vec{b}/d$ if d is positive?

53. In Fig. 3-41, a vector $\vec{a}$ with a magnitude of 17.0 m is directed $56.0°$ counterclockwise from the $+x$ axis, as shown. What are the components (a) a_x and (b) a_y of the vector? A second coordinate system is inclined by $18.0°$ with respect to the first. What are the components (c) a_x' and (d) a_y' in this primed coordinate system?

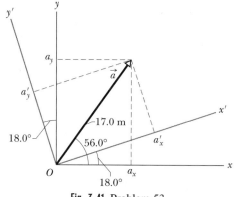

Fig. 3-41 Problem 53.

54. Show for any vector $\vec{a}$ that $\vec{a} \cdot \vec{a} = a^2$ and that $\vec{a} \times \vec{a} = 0$.

55. A vector $\vec{d}$ has a magnitude of 2.5 m and points north. What are (a) the magnitude and (b) the direction of the vector $4.0\vec{d}$? What are (c) the magnitude and (d) the direction of the vector $-3.0\vec{d}$?

56. A ship sets out to sail to a point 120 km due north. An unexpected storm blows the ship to a point 100 km due east of its starting point. (a) How far and (b) in what direction must it now sail to reach its original destination?

57. A heavy piece of machinery is raised by sliding it 12.5 m along a plank oriented at $20.0°$ to the horizontal, as shown in Fig. 3-42. (a) How high above its original position is it raised? (b) How far is it moved horizontally?

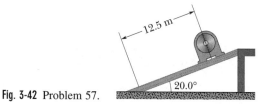

Fig. 3-42 Problem 57.

58. (a) A man leaves his front door, walks 1000 m east, 2000 m north, and then takes a penny from his pocket and drops it from a cliff 500 m high. Set up a coordinate system and write down an expression, using unit vectors, for the displacement of the penny from the front door to its landing point. (b) The man then returns to his front door, following a different path. What is his resultant displacement for the round trip?

59. If $\vec{a} - \vec{b} = 2\vec{c}$, $\vec{a} + \vec{b} = 4\vec{c}$, and $\vec{c} = 3\hat{i} + 4\hat{j}$, then what are (a) $\vec{a}$ and (b) $\vec{b}$?

CLUSTERED PROBLEMS

60. The magnitude and angle of $\vec{A}$, which lies in an xy plane, are 4.00 and 130°, respectively. What are the components (a) A_x and (b) A_y? Vector $\vec{B}$ also lies in the xy plane, and it has components $B_x = -3.86$ and $B_y = -4.60$. What is $\vec{A} + \vec{B}$ in (c) magnitude-angle notation and (d) unit-vector notation? In (e) unit-vector notation and (f) magnitude-angle notation, find $\vec{C}$ such that $\vec{A} - \vec{C} = \vec{B}$. (g) Which of the vector diagrams in Fig. 3-43 correctly show the relationship between those three vectors?

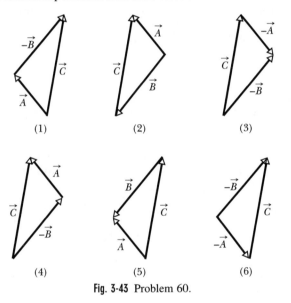

(1)   (2)   (3)

(4)   (5)   (6)

Fig. 3-43 Problem 60.

61. Use $\vec{A}$ and $\vec{B}$ of Problem 60 to answer the following questions. (a) What is $3\vec{A} \cdot \vec{B}$? What is $4\vec{A} \times 3\vec{B}$ in (b) unit-vector notation and (c) magnitude-angle notation with spherical coordinates (see Fig. 3-44)? (d) What is the angle between the directions of $\vec{A}$ and $4\vec{A} \times 3\vec{B}$? (*Hint:* Think a bit before you resort to a calculation.) What is $\vec{A} + 3.00\hat{k}$ in (e) unit-vector notation and (f) magnitude-angle notation with spherical coordinates?

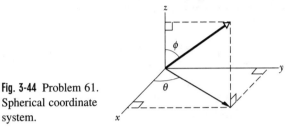

Fig. 3-44 Problem 61. Spherical coordinate system.

62. For $\vec{A}$ and $\vec{B}$ as given in Problem 60, what are the angles between the negative direction of the y axis and (a) the direction of $\vec{A}$, (b) the direction of the product $\vec{A} \times \vec{B}$, and (c) the direction of $\vec{A} \times (\vec{B} + 3\hat{k})$?

4
Motion in Two and Three Dimensions

ADDITIONAL SAMPLE PROBLEMS

Sample Problem 4-12

In Fig. 4-40, a movie stuntman is to run across and directly off a rooftop, to land on the roof of the next building. Can he land there if he runs at 4.5 m/s?

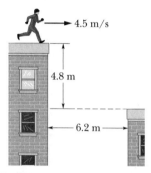

Fig. 4-40 Sample Problem 4-12. Can the stuntman make the jump?

SOLUTION: The **Key Idea** here is that, while in flight, the stuntman is a projectile whose initial velocity $\vec{v}_0$ is horizontal ($\theta_0 = 0°$) and has magnitude $v_0 = 4.5$ m/s. For him to land on the second roof, his horizontal displacement $x - x_0$ during the flight must be at least 6.2 m. We should be able to find the actual value of $x - x_0$ for the given initial speed by using Eq. 4-21,

$$x - x_0 = (v_0 \cos \theta_0)t, \tag{4-45}$$

but for that we need his time of flight t.

To find t, we consider the vertical motion and use Eq. 4-22:

$$y - y_0 = (v_0 \sin \theta_0)t - \tfrac{1}{2}gt^2. \tag{4-46}$$

Here his vertical displacement $y - y_0$ during the flight is -4.8 m. Putting this and other known values into Eq. 4-46 gives us

$$-4.8 \text{ m} = (4.5 \text{ m/s})(\sin 0°)t - \tfrac{1}{2}(9.8 \text{ m/s}^2)t^2,$$

which yields $t = 0.990$ s. Using this in Eq. 4-45, we have

$$x - x_0 = (4.5 \text{ m/s})(\cos 0°)(0.990 \text{ s})$$
$$= 4.5 \text{ m}.$$

Thus, with his given running speed, he cannot land on the roof of the next building.

Sample Problem 4-13

A satellite is in circular Earth orbit, at altitude $h = 200$ km above Earth's surface. There the free-fall acceleration g is 9.20 m/s². What is the orbital speed v of the satellite?

SOLUTION: The **Key Ideas** here are that the satellite is in uniform circular motion about Earth and the centripetal acceleration is the free-fall acceleration $g = 9.20$ m/s². We can find v from Eq. 4-32 ($a = v^2/r$), with $a = g$ and with $r = R_E + h$, where R_E is Earth's radius (see inside front cover or Appendix C). With those substitutions, we have

$$g = \frac{v^2}{R_E + h}.$$

Solving for v gives

$$v = \sqrt{g(R_E + h)}$$
$$= \sqrt{(9.20 \text{ m/s}^2)(6.37 \times 10^6 \text{ m} + 200 \times 10^3 \text{ m})}$$
$$= 7770 \text{ m/s} = 7.77 \text{ km/s}. \qquad \text{(Answer)}$$

You can show that this is equivalent to 28,000 km/h and that the satellite would take 1.47 h to complete one orbital revolution; that is, the period T of the motion is 1.47 h.

Sample Problem 4-14

In Fig. 4-41a, a bat detects an insect (lunch) while the two are flying. The bat has velocity $\vec{v}_{BG}$ relative to the ground, and the insect has velocity $\vec{v}_{IG}$ relative to the ground. What is the velocity $\vec{v}_{IB}$ of the insect relative to the bat, in unit-vector notation and as a magnitude and an angle?

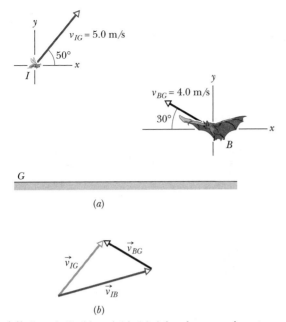

(a)

(b)

Fig. 4-41 Sample Problem 4-14. (*a*) A bat detects an insect. (*b*) Velocity vectors of insect and bat.

SOLUTION: The **Key Idea** here is that we can relate these three relative velocities with a triangle like that in Fig. 4-21 and with an equation like Eq. 4-42. The moving particle *P* is now the insect (*I*), frame

A is now the ground (*G*), and frame *B* is now the bat (*B*). We first construct a sentence that relates the three vectors:

$$\begin{pmatrix} \text{velocity of insect} \\ \text{relative to ground} \\ (IG) \end{pmatrix} = \begin{pmatrix} \text{velocity of insect} \\ \text{relative to bat} \\ (IB) \end{pmatrix} + \begin{pmatrix} \text{velocity of bat} \\ \text{relative to ground} \\ (BG) \end{pmatrix}.$$

Now we can draw this relation as in Fig. 4-41*b* and write it as

$$\vec{v}_{IG} = \vec{v}_{IB} + \vec{v}_{BG}$$

or

$$\vec{v}_{IB} = \vec{v}_{IG} - \vec{v}_{BG}.$$

We can evaluate the right side of this last equation directly on a vector-capable calculator by keying in the given vectors. Alternatively, we can resolve the given vectors into vector components and then add $\vec{v}_{IG}$ and $-\vec{v}_{BG}$. We have

$$\vec{v}_{IG} = (5.0 \text{ m/s})(\cos 50°)\hat{i} + (5.0 \text{ m/s})(\sin 50°)\hat{j}$$
$$= (3.21 \text{ m/s})\hat{i} + (3.83 \text{ m/s})\hat{j}$$

and

$$-\vec{v}_{BG} = (4.0 \text{ m/s})(\cos 30°)\hat{i} - (4.0 \text{ m/s})(\sin 30°)\hat{j}$$
$$= (3.46 \text{ m/s})\hat{i} - (2.00 \text{ m/s})\hat{j}.$$

Thus,

$$\vec{v}_{IB} \approx (6.7 \text{ m/s})\hat{i} + (1.8 \text{ m/s})\hat{j}. \qquad \text{(Answer)}$$

Following Eq. 3-6, we find that $\vec{v}_{IB}$ has

$$\text{magnitude} = 6.9 \text{ m/s}$$

and   angle counterclockwise from $+x \approx 15°$.   (Answer)

QUESTIONS

14. Figure 4-42 shows four paths along which a mechanical rabbit can move from point *i* to point *f*. Paths 1 and 3 are different portions of the same circle; path 2 is half of a different circle; and path 4 is straight. (a) Suppose the rabbit must make the move in a certain given time, no matter which path is used; then rank the paths according to the average velocity of the rabbit, greatest first. (b) Suppose, instead, that the rabbit must travel at a certain constant speed; rank the paths according to the magnitude of the acceleration the rabbit will have, greatest first.

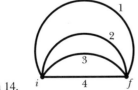

Fig. 4-42 Question 14.

15. *Organizing question:* A marble is launched from the edge of a high cliff with a velocity of 20 m/s at an angle of 30° above the horizontal. (a) Set up equations, complete with known data, to find the horizontal displacement of the marble from the launch point at the instant when the vertical displacement is 2 m below the launch level. (b) Next, set up equations, complete with known data, to find the marble's velocity components v_x and v_y at the same instant.

16. While riding in a moving car, you toss an egg directly upward. Does the egg tend to land behind you, in front of you, or back in your hands if the car is (a) traveling at a constant speed, (b) increasing in speed, and (c) decreasing in speed?

17. A rock can be launched toward a wall from ground level with initial velocity of $\vec{v} = (3 \text{ m/s})\hat{i} + (4 \text{ m/s})\hat{j}$. You can adjust the horizontal distance *d* between the launch site and the wall. With that adjustment, what are (a) the least speed and (b) the greatest speed at which the rock can hit the wall?

Suppose that at a certain value of *d* the rock's velocity is horizontal as it hits the wall. If you move the launch site closer to the wall, do (c) the height at which the rock hits the wall and (d) the speed with which it hits increase, decrease, or remain the same? (e) Does the angle between the velocity vector and the horizontal increase upward or increase downward?

18. When the Germans shelled Paris from 70 mi away with the WWI long-range artillery piece nicknamed "Big Bertha," the shells were fired at an angle greater than 45°; the Germans had discovered that a greater angle gave their gun a greater range, possibly even twice as long as that with a 45° angle. Does that result mean that the density of the air at high altitudes increases with altitude or decreases with altitude?

19. The position vectors of four infant wombats are given in the following table in terms of a coordinate system. Without using a calculator, rank the wombats according to their distances from the origin of the coordinate system, greatest first.

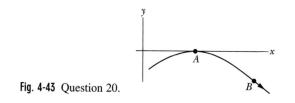

Wombat	Position Vector
a	$(-2 \text{ m})\hat{i} + (3 \text{ m})\hat{j} + (4 \text{ m})\hat{k}$
b	$(2 \text{ m})\hat{i} + (-3 \text{ m})\hat{j} + (4 \text{ m})\hat{k}$
c	$(3 \text{ m})\hat{i} + (2 \text{ m})\hat{j} + (4 \text{ m})\hat{k}$
d	$(4 \text{ m})\hat{i} + (-1 \text{ m})\hat{j} + (2 \text{ m})\hat{k}$

20. Figure 4-43 shows an overhead view of the path taken by a water strider as it darts over the surface of a pond. What is the direction of its velocity vector at (a) point A and (b) point B?

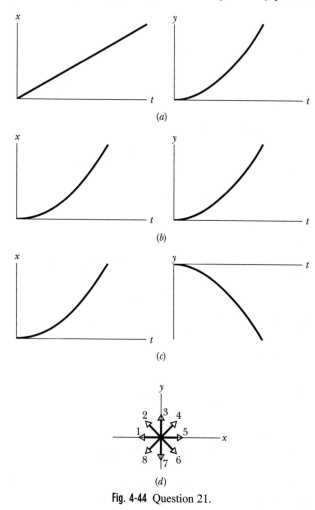

Fig. 4-43 Question 20.

21. Figure 4-44a, b, and c give, for three situations, graphs of the coordinates $x(t)$ and $y(t)$ of a particle moving in the xy plane. In

each situation, the acceleration of the particle is constant and the graphs are drawn to the same scale. For each situation, which of the vectors shown in Fig. 4-44d best represents the acceleration of the particle?

22. A snowball is thrown from ground level (by someone in a hole) with initial speed v_0 at an angle of 45° relative to the (level) ground, on which the snowball later lands. If the launch angle is increased, do (a) the range and (b) the flight time increase, decrease, or stay the same?

23. You are driving directly behind a pickup truck, going at the same speed as the truck. A crate falls from the bed of the truck to the road. (a) Will your car hit the crate before the crate hits the road if you neither brake nor swerve? (b) During the fall, is the horizontal speed of the crate more than, less than, or the same as that of the truck?

24. A Chihuahua, hungry for attention, leaps upward. During its free flight, describe the angle between its velocity vector and its acceleration vector if it jumps (a) directly upward and (b) at an initial angle of 45°.

DISCUSSION QUESTIONS

25. Can the acceleration of a body change direction (a) without the displacement immediately changing direction also and (b) without the velocity immediately changing direction? If so, give an example.

26. In broad jumping, sometimes called long jumping, what factors determine the length of the jump?

27. At what point in the path of a projectile is the speed a minimum? A maximum?

28. Figure 4-45 shows the path followed by a NASA Learjet in a run designed to simulate low-gravity conditions for a short period of time. Make an argument to show that, if the plane follows a particular parabolic path, the passengers will experience weightlessness.

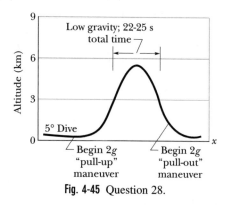

Fig. 4-45 Question 28.

29. A shot is put (thrown) from above the athlete's shoulder level. The launch angle that will produce the longest range is less than 45°; that is, a flatter trajectory has a longer range. Explain why.

30. A rifle is "sighted-in" against a target, with both at the same elevation. Show that, at the same range, the rifle will shoot too high when shooting either uphill or downhill. (See "A Puzzle in Elementary Ballistics," by Ole Anton Haugland, *The Physics Teacher,* April 1983.)

Fig. 4-44 Question 21.

31. In projectile motion when air resistance is negligible, is it ever necessary to consider three-dimensional motion rather than two-dimensional?

32. Show that, taking Earth's rotation and revolution into account, a book resting on your table moves faster at night than it does during the daytime. In what reference frame is this statement true?

33. An aviator, pulling out of a dive, follows the arc of a circle and is said to have "experienced $3g$'s" in pulling out of the dive. Explain what this statement means.

34. If the acceleration of a body is constant as measured from a given reference frame, is it necessarily constant when measured from all other reference frames?

EXERCISES & PROBLEMS

65. The New Hampshire State Police use aircraft to enforce highway speed limits. Suppose that one of the airplanes has a speed of 135 mi/h in still air. It is flying straight north so that it is at all times directly above a north–south highway. A ground observer tells the pilot by radio that a 70.0 mi/h wind is blowing, but neglects to give the wind direction. The pilot observes that in spite of the wind the plane can travel 135 mi along the highway in 1.00 h. In other words, the ground speed is the same as if there were no wind. (a) From what direction is the wind blowing? (b) What is the heading of the plane; that is, in what direction does it point?

66. In a computerized golf simulator, a player hits a golf ball toward the image of a golf course that is projected on a large screen at the end of a playing field. A computer calculates the flight of the ball as if the ball were hit on the imaged field; the computer then brings up a new image for the calculated landing position of the ball. Thus, the golfer can play through a round of golf without leaving the screen.

The flight of the ball is measured by infrared emitters and sensors that lie on two parallel, vertical rings through which the ball is hit. By measuring where in the two rings the ball passes and how long the ball takes to travel between the rings, the computer can determine the flight that the ball would take on an actual golf course. Assume that the rings are separated horizontally by 0.910 m and close enough to the player so that the initial curvature of the ball's path can be neglected. If, for a certain drive, the travel time between the rings is 16.5 ms and the ball passes through the second ring 0.193 m higher than through the first ring, what are the ball's (a) launch angle, (b) launch speed, and (c) theoretical horizontal range? (d) The launch values here are typical values. Why then is the answer to (c) so different from the range attained in an actual game?

67. In a common movie stunt, a performer is thrown through the air because of an explosion, a motorcycle crash, a tornado blast, or such. Actually, the performer is launched by an Air Ramp, which consists of a plate on the end of a launching arm, powered by pneumatic cylinders of pressurized air. As the performer runs onto the plate, electric switches suddenly allow the pressurized air to whip the launching arm, sending the plate upward and throwing the performer across the movie set. Approximate the performer's flight as that of a particle returning to its initial elevation. In g units, find the magnitude of the average acceleration of the performer during the launch for the following situations: (a) The acceleration lasts 0.25 s, and the performer is thrown through a shallow arc (for an indoor set) with a range of 4.0 m and an initial launch angle of 30° with the horizontal. (b) The acceleration lasts 0.29 s, and the performer is thrown through a higher arc (for an outdoor set) with a range of 12 m and an initial launch angle of 45°.

68. A positron undergoes a displacement $\Delta \vec{r} = 2.0\hat{i} - 3.0\hat{j} + 6.0\hat{k}$, ending with the position vector $\vec{r} = 3.0\hat{j} - 4.0\hat{k}$, in meters. What was the positron's former position vector?

69. At one instant a bicyclist is 40.0 m due east of a park's flagpole, going due south with a speed of 10.0 m/s. Then 30.0 s later, the cyclist is 40.0 m due north of the flagpole, going due east with a speed of 10.0 m/s. Find (a) the displacement, (b) the average velocity, and (c) the average acceleration of the cyclist during the 30.0 s interval.

70. The airport terminal in Geneva, Switzerland, has a "moving sidewalk" to speed passengers through a long corridor. Larry does not use the moving sidewalk; he takes 150 s to walk through the corridor. Curly, who simply stands on the moving sidewalk, covers the same distance in 70 s. Moe boards the sidewalk and walks along it. How long does Moe take to move through the corridor? Assume that Larry and Moe walk at the same speed.

71. A projectile is launched with an initial speed of 30 m/s at an angle of 60° above the horizontal. Calculate the magnitude and direction of its velocity (a) 2.0 s and (b) 5.0 s after launch.

72. A helicopter is flying in a straight line over a level field at a constant speed of 6.2 m/s and at a constant altitude of 9.5 m. A package is ejected horizontally from the helicopter with an initial velocity of 12 m/s relative to the helicopter and in a direction opposite the helicopter's motion. (a) Find the initial speed of the package relative to the ground. (b) What is the horizontal distance between the helicopter and the package at the instant the package strikes the ground? (c) What angle does the velocity vector of the package make with the ground at the instant before impact, as seen from the ground?

73. (a) Prove that for a projectile fired from level ground at an angle θ_0 above the horizontal, the ratio of the maximum height H to the range R is given by $H/R = \frac{1}{4} \tan \theta_0$. See Fig. 4-46. (b) For what angle θ_0 does $H = R$?

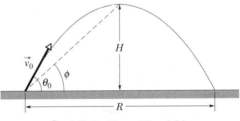

Fig. 4-46 Problems 73 and 74.

74. A projectile is fired from level ground at an angle θ_0 above the horizontal. (a) Show that the elevation angle ϕ of the highest point

of its trajectory, as seen from the launch point, is related to θ_0, the launch angle, by $\tan \phi = \frac{1}{2} \tan \theta_0$. See Fig. 4-46 and Problem 73. (b) Calculate ϕ for $\theta_0 = 45°$.

75. A ball is thrown horizontally from a height of 20 m and hits the ground with a speed that is three times its initial speed. What is the initial speed?

76. In 3.50 h, a balloon drifts 21.5 km north, 9.70 km east, and 2.88 km in upward elevation from its release point on the ground. Find (a) the magnitude of its average velocity and (b) the angle its average velocity makes with the horizontal.

77. A frightened rabbit runs onto a large area of level ice that offers no resistance to sliding, with an initial velocity of 6.0 m/s toward the east. As the rabbit slides across the ice, the force of the wind causes it to have a constant acceleration of 1.4 m/s², directed due north. Choose a coordinate system with the origin at the rabbit's initial position on the ice and the positive x axis directed toward the east. In unit-vector notation, what are the rabbit's (a) velocity and (b) position when it has slid for 3.0 s?

78. The minute hand of a wall clock measures 10 cm from the axis about which it rotates to its tip. What is the displacement vector of its tip (a) from a quarter after the hour to half past, (b) the next half hour, and (c) in the next hour?

79. A track meet is held on a planet in a distant solar system. A shot-putter releases a shot at a point 2.0 m above ground level. A stroboscopic plot of the position of the shot is shown in Fig. 4-47, where the readings are 0.50 s apart and the shot is released at time $t = 0$. (a) What is the initial velocity of the shot in unit-vector notation? (b) What is the magnitude of the free-fall acceleration on the planet? (c) How long after it is released does the shot reach the ground? (d) If an identical throw of the shot is made on the surface of Earth, how long after it is released does it reach the ground?

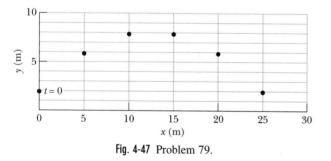

Fig. 4-47 Problem 79.

80. The magnitude of the velocity of a projectile when it is at its maximum height above ground level is 10 m/s. (a) What is the magnitude of the velocity of the projectile 1.0 s before it achieves its maximum height? (b) What is the magnitude of the velocity of the projectile 1.0 s after it achieves its maximum height? If we take $x = 0$ and $y = 0$ to be at the point of maximum height and positive x to be in the direction of the velocity there, where is the projectile (c) 1.0 s before and (d) 1.0 s after it achieves its maximum height?

81. The range of a projectile depends not only on v_0 and θ_0 but also on the value g of the free-fall acceleration, which varies from place to place. In 1936, Jesse Owens established a world's running broad jump record of 8.09 m at the Olympic Games at Berlin (where $g = 9.8128$ m/s²). Assuming the same values of v_0 and θ_0, by

how much would his record have differed if he had competed instead in 1956 at Melbourne (where $g = 9.7999$ m/s²)?

82. During volcanic eruptions, chunks of solid rock can be blasted out of the volcano; these projectiles are called *volcanic bombs*. Figure 4-48 shows a cross section of Mt. Fuji, in Japan. (a) At what initial speed would a bomb have to be ejected, at 35° to the horizontal, from the vent at A in order to fall at the foot of the volcano at B? Ignore, for the moment, the effects of air on the bomb's travel. (b) What would be the time of flight? (c) Would the effect of the air increase or decrease your answer in (a)?

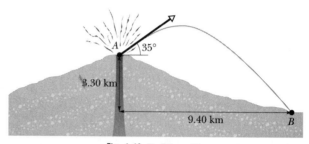

Fig. 4-48 Problem 82.

83. In Fig. 4-49, a stone is projected at a cliff of height h with an initial speed of 42.0 m/s directed 60.0° above the horizontal. The stone strikes at A, 5.50 s after launching. Find (a) the height h of the cliff, (b) the speed of the stone just before impact at A and (c) the maximum height H reached above the ground.

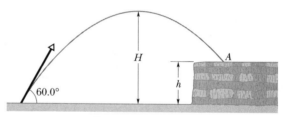

Fig. 4-49 Problem 83.

84. (a) If an electron is projected horizontally with a speed of 3.0×10^6 m/s, how far will it fall in traversing 1.0 m of horizontal distance? (b) Does the answer increase or decrease if the initial speed is increased?

85. A golfer chips balls toward a vertical wall 20.0 m straight ahead, trying to hit a 30.0-cm-diameter red circle painted on the wall. The target is centered about a point 1.20 m above the point where the wall intersects the horizontal ground. On one try, the ball leaves the ground with a speed of 15.0 m/s and at an angle of 35.0° above the horizontal and then hits the wall on a vertical line through the circle's center. (a) How long does the ball take to reach the wall? (b) Does the ball hit the red circle (give the distance to the circle's center)? (c) What is the speed of the ball just before it hits? (d) Has the ball passed the highest point of its trajectory when it hits?

86. A projectile is fired horizontally from a gun that is 45.0 m above flat ground, emerging with a speed of 250 m/s. (a) How long does the projectile remain in the air? (b) At what horizontal distance from the firing point does it strike the ground? (c) What is the magnitude of the vertical component of its velocity as it strikes the ground?

87. In Sample Problem 4-7, suppose that a second identical harbor defense cannon is 30 m above sea level, rather than at sea level. How much longer is the horizontal distance from launch to impact of the second cannon than that of the first, which is found to be 690 m, if the elevation angle is 45°?

88. A magnetic field can force a charged particle to move in a circular path. Suppose that an electron moving in a circle experiences a radial acceleration of magnitude 3.0×10^{14} m/s² in a particular magnetic field. (a) What is the speed of the electron if the radius of its circular path is 15 cm? (b) What is the period of the motion?

89. A car travels around a flat circle on the ground, at a constant speed of 12 m/s. At a certain instant the car has an acceleration of 3 m/s² toward the east. What are its distance and direction from the center of the circle at that instant if it is traveling (a) clockwise around the circle and (b) counterclockwise around the circle?

90. Suppose that a space probe can withstand the stresses of a $20g$ acceleration. (a) What is the minimum turning radius of such a craft moving at a speed of one-tenth the speed of light? (b) How long would it take to complete a 90° turn at this speed?

91. You throw a ball from a cliff with an initial velocity of 15.0 m/s at an angle of 20.0° below the horizontal. Find (a) its horizontal displacement and (b) its vertical displacement 2.30 s later.

92. A 200-m-wide river flows due east at a uniform speed of 2.0 m/s. A boat with a speed of 8.0 m/s relative to the water leaves the south bank pointed in a direction of 30° west of north. (a) What is the velocity of the boat relative to the ground? (b) How long does the boat take to cross the river?

93. The pilot of an aircraft flies due east relative to the ground in a wind blowing 20 km/h toward the south. If the speed of the aircraft in the absence of wind is 70 km/h, what is the speed of the aircraft relative to the ground?

94. A plane flies 483 km east from city A to city B in 45.0 min and then 966 km south from city B to city C in 1.50 h. In magnitude-angle notation, what are (a) the plane's displacement and (b) its average velocity, both for the total trip? (c) What is the plane's average speed for the trip?

95. A baseball is hit at Fenway Park in Boston at a point 0.762 m above home plate with an initial velocity of 33.53 m/s directed 55.0° above the horizontal. The ball is observed to clear the 11.28-m-high wall in left field (known as the "green monster") 5.00 s after it is hit, at a point just inside the left-field foul-line pole. Find (a) the horizontal distance down the left-field foul line from home plate to the wall; (b) the vertical distance by which the ball clears the wall; (c) the horizontal and vertical displacements of the ball with respect to home plate 0.500 s before it clears the wall.

96. A transcontinental flight of 4350 km is scheduled to take 50 min longer westward than eastward. The airspeed of the airplane is 966 km/h, and the jet stream it will fly through is presumed to move due east. What is the assumed speed of the jet stream?

97. At what initial speed must the basketball player in Fig. 4-50 throw the ball, at 55° above the horizontal, to make the foul shot?

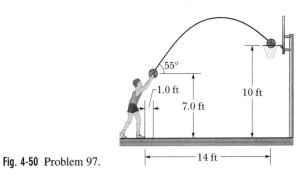

Fig. 4-50 Problem 97.

98. A woman can row a boat at 6.4 km/h in still water. (a) If she is crossing a river where the current is 3.2 km/h, in what direction must her boat be headed if she wants to reach a point directly opposite her starting point? (b) If the river is 6.4 km wide, how long will she take to cross the river? (c) Suppose that instead of crossing the river she rows 3.2 km *down* the river and then back to her starting point. How long will she take? (d) How long will she take to row 3.2 km *up* the river and then back to her starting point? (e) In what direction should she head the boat if she wants to cross in the shortest possible time, and what is that time?

99. A projectile is fired with initial speed $v_0 = 30.0$ m/s from level ground at a target that is on the ground, at distance $R = 20.0$ m, as shown in Fig. 4-51. Find the launch angles that will allow the projectile to hit the target.

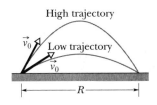

Fig. 4-51 Problem 99.

100. About 4000 car–train collisions occur in the United States each year. Many (perhaps most) occur when a driver spots an oncoming train and then attempts to beat it to a railroad crossing. (This is beyond foolish, and way past stupid as well.) Consider the following situation: One of those 4000 drivers spots a train on a track that is perpendicular to the roadway and judges the train's speed to be equal to his speed of 30.0 m/s. The train and the car are both approaching the rail crossing (Fig. 4-52). The driver figures that by accelerating his car at a constant 1.50 m/s² toward the crossing, he should reach it first. The distance between the car and

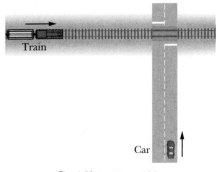

Fig. 4-52 Problem 100.

the center of the crossing and the distance between the train and the center of the crossing are both equal to 40.0 m. The length of the car is 5.0 m, and the width of the train is 3.00 m. (a) Is there a collision? (b) If not, by how much time does the driver miss a collision; if so, how much more time is needed to barely miss the collision? (c) On the same graph, plot the location of the train and the location of the car, both versus time.

101. When balls are juggled through a high arc from one hand to the other, launching all balls with the same speed and angle is crucial; otherwise, catching them becomes improbable. Assume that a ball is launched with a speed of 6.3 m/s and then caught at the launch height. (a) At what optimum angle θ_{opt} relative to horizontal (in degrees) should it be launched if the hands are to have a comfortable horizontal separation of 40 cm? (Is the answer displayed when you use a calculator here the proper angle? See Sample Problem 4-7a.) (b) Using the parametric graphing capability on a calculator, plot the ball's flight.

Using the list capability of the calculator, set up list L_1 to contain the values of five angles: θ_{opt}, $\theta_{opt} \pm 1°$, and $\theta_{opt} \pm 2°$. Then substitute this list for the symbol of the launch angle that you used in step (b), and replot the ball's flight. (c) For each of the angles in the list, estimate from the plots how far horizontally the ball travels to return to the launch height. (d) Also from the plots, estimate the error in catching the ball—that is, the horizontal distance between the returning ball and the catching hand. (e) Using list L_1 in a calculation, generate a list of the errors for the five launch angles. You will find that even a small error in the launch angle can throw the ball off the mark.

CLUSTERED PROBLEMS

Cluster 1

102. In Fig. 4-53a, a ball is fired horizontally through a horizontal distance of 150 m, hitting the ground in 3.00 s. (a) At what height h above the ground is the ball fired? (b) What is the ball's speed just as it hits the ground?

103. In Fig. 4-53b, a ball is fired at an angle of 30.0° from the horizontal, landing 3.00 s later after traveling a horizontal distance of 100 m. (a) At what height h above the firing level does the ball land? (b) At what speed is it fired? (c) At what speed does it land?

104. In Fig. 4-53c, a ball is thrown up onto a building's roof, landing 4.00 s later, 20.0 m above where it was released. Its path just before landing is angled at 60.0° with the roof. (a) What horizontal distance d does it travel? (*Hint:* One way to obtain the answer is to reverse the motion, as if a videotape of the throw is reversed. Then the landing, for which the angle is given, becomes the launch.) (b) What are the magnitude and direction of the velocity at which the ball is thrown?

105. In Fig. 4-53d, a ball that is thrown from the edge of a building hits the ground at an angle of 60.0° with the horizontal, 25.0 m from the building and 1.50 s after it is thrown. (a) From what height h is the ball thrown? (*Hint:* One way to obtain the answer is to reverse the motion, as if a videotape of the throw is reversed. Then the landing, for which the angle is given, becomes the launch.) (b) What are the magnitude and direction of the velocity at which the ball is thrown?

106. In Fig. 4-53e, a baseball is hit at a height of 1.0 m and then caught at the same height. It travels alongside a wall, moving up past the top level 1.00 s after it is hit and then down past the top level 4.00 s later, 50.0 m farther along the wall. (a) What horizontal distance is traveled by the ball from the hit to the catch? (b) What are the magnitude and direction of the ball's velocity just after being hit? (c) How high is the wall?

Cluster 2

107. You are to throw a ball with a speed of 12.0 m/s at a target that is 5.00 m above the level at which you release the ball (Fig. 4-54a). You want the ball's velocity to be horizontal at the instant it reaches the target. (a) At what angle θ from the horizontal

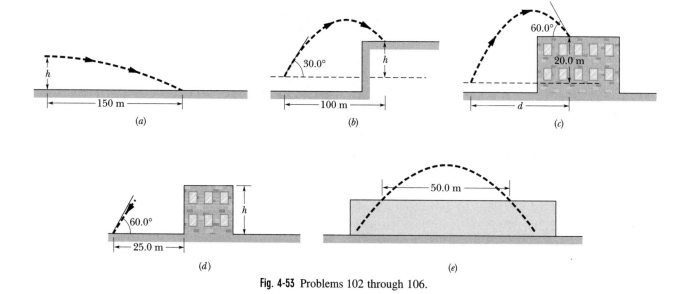

Fig. 4-53 Problems 102 through 106.

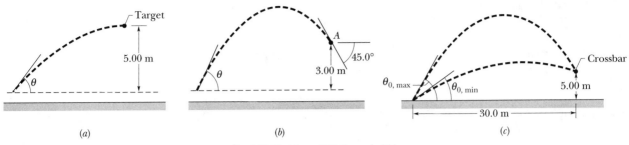

Fig. 4-54 Problems 107 through 111.

must you throw the ball? (b) What is the horizontal distance from the release point to the target? (c) What is the speed of the ball just as it reaches the target?

108. You are to throw a ball with a speed of 12.0 m/s. When the ball passes through point A in Fig. 4-54b, 3.00 m above the point at which you release it, its velocity is to be angled downward by 45.0° from the horizontal. (a) What is the horizontal distance between point A and the release point? (*Hint:* What does the angle of 45.0° imply about the horizontal and vertical components of the ball's velocity at point A?) (b) What is the speed of the ball at point A?

109. In a game of American football, a kicker must kick a field goal by sending the football over a crossbar that is horizontally 30.0 m and vertically 5.00 m from the point of the kick (Fig. 4-54c). Let v_0 represent the speed at which the football is launched. (a) Set up an equation, complete with v_0 and known data, to find the minimum angle $\theta_{0,\min}$ and the maximum angle $\theta_{0,\max}$ at which the football must be launched if it is to just barely clear the crossbar. Use Eq. 4-25 and the trigonometric identity

$$\frac{1}{\cos^2 \theta} = \sec^2 \theta = 1 + \tan^2 \theta.$$

(b) Plot $\theta_{0,\min}$ and $\theta_{0,\max}$ as a function of v_0 for the values of v_0 from 18 m/s to 25 m/s.

110. In the situation of Problem 109, what is the minimum initial velocity (magnitude and direction) at which the football can be kicked and still pass over the crossbar?

(*Hint:* The minimum initial velocity occurs when $\theta_{0,\min} = \theta_{0,\max}$; that is, only one launch angle then allows the football to barely pass over the crossbar. To find the answer graphically, examine the curves in part (b) of Problem 109. To find the answer analytically, use the equation requested in part (a) for the launch angle. First, set the expression whose square root is to be taken equal to zero and solve the result for the launch speed v_0 [an equation-solving program on a calculator helps]. Then substitute that value of v_0 into the equation for the launch angle and solve for the launch angle.)

111. (a) If the kicker of Problem 109 can kick a football with an initial speed of only 15.0 m/s, what is the maximum horizontal distance $x_{\max}$ at which the football can be kicked over the crossbar? (b) At what angle must the ball be kicked? (*Hint:* Use the hint of Problem 110.)

5
Force and Motion—I

Sample Problem 5-10

In Fig. 5-50a, a student (with cleated boots) pushes a loaded sled of mass $m = 240$ kg through a displacement of magnitude $d = 2.3$ m along an x axis, over the frictionless surface of a frozen lake.

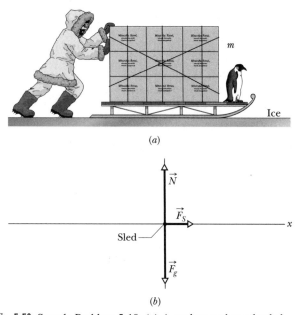

(a)

(b)

Fig. 5-50 Sample Problem 5-10. (a) A student pushes a loaded sled over frictionless ice. (b) A free-body diagram for the sled, showing the gravitational force $\vec{F}_g$ on the sled, the normal force $\vec{N}$ on the sled from the ice, and the student's force $\vec{F}_S$. (This last vector is not drawn to scale; if it were, it would be much shorter than the other two vectors.)

The student's force $\vec{F}_S$ on the sled is horizontal and has a magnitude of 130 N, and the sled starts from rest. What is the sled's velocity at the end of the displacement?

SOLUTION: One **Key Idea** here is that the sled's velocity changes due to the net force on the sled. The individual forces on the sled are the gravitational force $\vec{F}_g$, the normal force $\vec{N}$ from the ice, and the student's force $\vec{F}_S$, as shown in the free-body diagram in Fig 5-50b. The first two balance each other, so the net force on the sled is equal to $\vec{F}_S$.

A second **Key Idea** is that because the student's force is constant, so is the acceleration. Thus, we can use the constant-acceleration equations of Table 2-1. To find the sled's velocity v at the end of the displacement, we try Eq. 2-16,

$$v^2 = v_0^2 + 2a(x - x_0). \tag{5-31}$$

We know that the initial velocity v_0 is 0 and the displacement $x - x_0$ is $d = 2.3$ m. However, we do not know the sled's acceleration a.

To find a, we write Newton's second law for x components $(F_{net,x} = ma_x)$ as

$$F_S = ma.$$

Solving for a and substituting known values give us

$$a = \frac{F_S}{m} = \frac{130 \text{ N}}{240 \text{ kg}} = 0.542 \text{ m/s}^2.$$

Substituting this value and the other known values into Eq. 5-31 gives us

$$v^2 = 0 + 2(0.542 \text{ m/s}^2)(2.3 \text{ m})$$

and

$$v = 1.6 \text{ m/s.} \tag{Answer}$$

Sample Problem 5-11

In Fig. 5-51a, a hand H pulls on a taut horizontal rope R (of mass $m = 0.200$ kg) that is attached to a block B (of mass $M = 5.00$ kg). The resulting acceleration $\vec{a}$ of the rope and block across the frictionless surface has constant magnitude 0.300 m/s^2 and is directed in the positive direction of the x axis indicated in the figure. Note that this rope is not "massless" like those in Fig. 5-10 and elsewhere in this chapter. We return to this feature in part (d).

(a) Identify all the third-law force pairs for the horizontal forces in Fig. 5-51a and show how the vectors in each pair are related.

SOLUTION: The **Key Idea** here is that a third-law force pair arises when two bodies interact; the forces of the pair are equal in magnitude and opposite in direction, and the force on each body is due to the other body. The "exploded view" of Fig. 5-51b shows that here

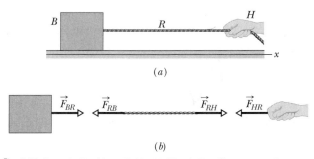

Fig. 5-51 Sample Problem 5-11. (*a*) Hand *H* pulls on rope *R*, which is attached to block *B*. (*b*) An exploded view of block, rope, and hand, with the forces between block and rope and between rope and hand.

there are two such force pairs for the horizontal forces. At the hand–rope boundary, we have the force $\vec{F}_{RH}$ on the rope from the hand and the force $\vec{F}_{HR}$ on the hand from the rope. These forces are equal and opposite in direction, so they are related by

$$\vec{F}_{RH} = -\vec{F}_{HR}. \qquad \text{(Answer)}$$

Similarly, at the rope–block boundary we have

$$\vec{F}_{BR} = -\vec{F}_{RB}. \qquad \text{(Answer)}$$

(b) What is the magnitude of the force $\vec{F}_{BR}$ on the block from the rope?

SOLUTION: We know that the block has an acceleration $\vec{a}$ in the positive direction of the *x* axis. The only force acting on the block along that axis is $\vec{F}_{BR}$. The **Key Idea** here is that we can relate force $\vec{F}_{BR}$ to acceleration $\vec{a}$ by Newton's second law. Because both vectors are along the *x* axis, we use the *x* component version of the law ($F_{\text{net},x} = ma_x$), writing

$$F_{BR} = Ma.$$

Substituting known values, we find that the magnitude of $\vec{F}_{BR}$ is

$$F_{BR} = (5.00 \text{ kg})(0.300 \text{ m/s}^2) = 1.50 \text{ N}. \qquad \text{(Answer)}$$

(c) What is the magnitude of the force $\vec{F}_{RB}$ on the rope from the block?

SOLUTION: From (a), we know that $\vec{F}_{RB} = -\vec{F}_{BR}$, so $\vec{F}_{RB}$ has the magnitude

$$F_{RB} = F_{BR} = 1.50 \text{ N}. \qquad \text{(Answer)}$$

(d) What is the magnitude of the force $\vec{F}_{RH}$ on the rope from the hand?

SOLUTION: A **Key Idea** here is that, with the rope taut, the rope and block form a system on which $\vec{F}_{RH}$ acts. The mass of the system is $m + M$. For this system, Newton's second law for *x* components gives us

$$F_{RH} = (m + M)a \qquad \text{(5-32)}$$
$$= (0.200 \text{ kg} + 5.00 \text{ kg})(0.300 \text{ m/s}^2) = 1.56 \text{ N}. \qquad \text{(Answer)}$$

Now note that the magnitude of the force $\vec{F}_{RH}$ on the rope from the hand (1.56 N) is greater than the magnitude of the force $\vec{F}_{BR}$ on the block from the rope (1.50 N, from part (b) above). The reason is that $\vec{F}_{BR}$ must accelerate only the block but $\vec{F}_{RH}$ must accelerate both the block and the rope, and the rope's mass *m* is *not* negligible. If we let $m \rightarrow 0$ in Eq. 5-32, then we find 1.50 N, the same magnitude as at the other end. We often assume that an interconnecting rope is massless so that we can approximate the forces at its two ends as having the same magnitude.

Sample Problem 5-12

The acceleration $\vec{a}$ of the blocks of Sample Problem 5-5 and Fig. 5-14 can be found in two lines of algebra if we (a) employ an unconventional axis, call it *u*, that runs through *both* blocks and along the cord as shown in Fig. 5-52a, and then (b) mentally straighten out the *u* axis as in Fig. 5-52b and treat the blocks as being portions of a single composite body with mass $M + m$. A free-body diagram for the two-block system is shown in Fig. 5-52c.

SOLUTION: Note that there is only one force acting on the composite body along the *u* axis, and that is the gravitational force $\vec{F}_{gH}$ on block *H*, in the positive direction of the axis. The force $\vec{T}$ shown in Fig. 5-15 is now internal to the composite body and so does not enter into Newton's second law. The force on the cord from the pulley is perpendicular to the *u* axis, so it too does not enter.

Using Eqs. 5-2 as a guide, we can write Newton's law for the *u* axis as

$$F_{\text{net},u} = (M + m)a_u, \qquad \text{(5-33)}$$

where the mass of the composite body is $M + m$. The acceleration of the composite body along the *u* axis has magnitude *a*. The only force on the composite body along the *u* axis is $\vec{F}_{gH}$, with magnitude equal to *mg*. Thus, Eq. 5-33 becomes

$$mg = (M + m)a,$$

or

$$a = \frac{m}{M + m} g, \qquad \text{(5-34)}$$

which matches Eq. 5-21.

To find *T*, we apply Newton's second law to either block, obtaining either Eq. 5-9 or Eq. 5-20. We then substitute for *a* from Eq. 5-34 and solve for *T*, getting Eq. 5-22.

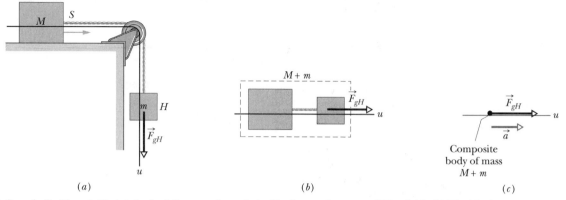

(a)   (b)   (c)

Fig. 5-52 Sample Problem 5-12. (a) An "axis" u runs through the blocks–cord system of Fig. 5-14. (b) The blocks are rearranged to straighten u and then are treated as a single body with mass M + m. (c) The associated free-body diagram, considering only forces along u. There is one such force.

Sample Problem 5-13

Figure 5-53a shows two blocks connected by a cord that passes over a massless, frictionless pulley (the arrangement is known as *Atwood's machine*). The lighter block L has mass m = 1.3 kg and the heavier block H has mass M = 2.8 kg. Find the magnitudes of the accelerations of the two blocks and the magnitude T of the force on each block from the cord.

SOLUTION: We know that the heavier block H will fall and the lighter block L will rise. A **Key Idea** for this problem is that the acceleration of each block is related to the net force on the block via Newton's second law. Moreover, because the blocks are connected by a taut cord, the blocks must have the same acceleration magnitude a.

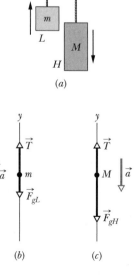

(a)

Fig. 5-53 Sample Problem 5-13. (a) A heavier block H of mass M and a lighter block L of lesser mass m are connected by a cord that passes over a pulley. The directions in which the system will accelerate from rest are shown by the arrows. Free-body diagrams for (b) block L and (c) block H.

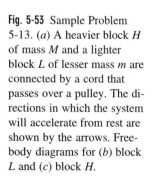

(b)   (c)

The free-body diagram for block L is shown in Fig. 5-53b: The only forces on the block are the upward force $\vec{T}$ from the cord and the downward gravitational force $\vec{F}_{gL}$, with magnitude equal to mg. The acceleration must be upward, in the positive direction of the y axis shown. We can write Newton's second law for y components ($F_{net,y} = ma_y$) as

$$T - F_{gL} = ma$$

or $$T - mg = ma. \qquad (5\text{-}35)$$

Similarly, from Fig. 5-53c for block H, which has downward (negative) acceleration, we can write

$$T - Mg = -Ma. \qquad (5\text{-}36)$$

By solving Eqs. 5-35 and 5-36 simultaneously, we obtain

$$a = \frac{M - m}{M + m}g. \qquad (5\text{-}37)$$

Substituting this result in either Eq. 5-35 or Eq. 5-36 and solving for T yield

$$T = \frac{2mM}{M + m}g. \qquad (5\text{-}38)$$

Inserting the given data in Eqs. 5-37 and 5-38, we obtain

$$a = \frac{M - m}{M + m}g = \frac{2.8 \text{ kg} - 1.3 \text{ kg}}{2.8 \text{ kg} + 1.3 \text{ kg}}(9.8 \text{ m/s}^2)$$

$$= 3.6 \text{ m/s}^2 \qquad \text{(Answer)}$$

and $$T = \frac{2Mm}{M + m}g = \frac{(2)(2.8 \text{ kg})(1.3 \text{ kg})}{2.8 \text{ kg} + 1.3 \text{ kg}}(9.8 \text{ m/s}^2)$$

$$= 17 \text{ N}. \qquad \text{(Answer)}$$

You can show that the magnitudes of the gravitational forces on the two blocks are 13 N (= mg) and 27 N (= Mg), so the magnitude T (= 17 N) of the upward force on the blocks lies between these two values. That makes sense if the lighter block, in Fig. 5-53b, is to rise and the heavier block, in Fig. 5-53c, is to fall.

Sample Problem 5-13—Another Way

Just as we redid Sample Problem 5-5 with an unconventional axis u in Sample Problem 5-12, we can redo Sample Problem 5-13 with u.

SOLUTION: Run the u axis through the system as shown in Fig. 5-54a. Straighten out the axis as in Fig. 5-54b, and treat the blocks as a single body with a mass of $M + m$. Then draw a free-body diagram as in Fig. 5-54c. Note that along the u axis there are two forces on the composite body: $\vec{F}_{gL}$ (with magnitude mg) in the negative direction of the u axis and $\vec{F}_{gH}$ (with magnitude Mg) in the positive direction. (The force on the cord from the pulley is perpendicular to the u axis and so doesn't enter our calculation.) The two forces along the u axis give the composite body (and each block) an acceleration $\vec{a}$. We can write Newton's second law for the u axis as

$$Mg - mg = (M + m)a, \tag{5-39}$$

which yields

$$a = \frac{M - m}{M + m}g,$$

as previously. To get T we apply Newton's second law to either block, using a conventional axis y. For the block with mass m, we obtain Eq. 5-35. With the above result for a substituted into Eq. 5-35, we obtain Eq. 5-38.

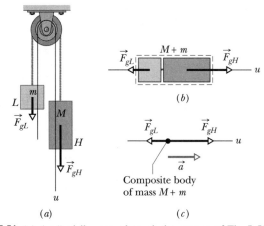

Fig. 5-54 (a) An "axis" u runs through the system of Fig 5-53. (b) The blocks are arranged to straighten u and then treated as a single body with mass $M + m$. (c) The associated free-body diagram, considering only forces with components along u. There are two such forces.

QUESTIONS

13. *Organizing question:* Figure 5-55 shows overhead views of six situations in which a 2 kg box is pulled over a frictionless floor by one or more forces. The magnitudes of the forces are $F_1 = 1$ N, $F_2 = 2$ N, and $F_3 = 3$ N. For each situation, set up equations, complete with known data, to find the acceleration components a_x and a_y; use the first two component equations of Eq. 5-2 ($F_{net,x} = ma_x$ and $F_{net,y} = ma_y$).

14. Two containers of sand S and H are arranged like the blocks in Sample Problem 5-5. The containers alone have negligible mass; the sand in them has a total mass M_{tot}; the sand in the hanging container H has mass m. You are to measure the magnitude a of the acceleration of the system in a series of experiments where m varies from experiment to experiment but M_{tot} does not; that is, you will shift sand between the containers before each trial.

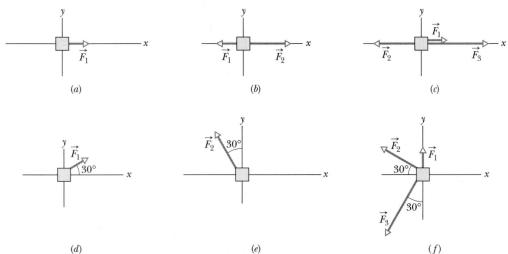

Fig. 5-55 Question 13.

(a) Which of the curves in Fig. 5-56 gives the acceleration magnitude as a function of the ratio m/M_{tot} (the vertical axis is for acceleration)? (b) Which of them gives the tension in the connecting cord (the vertical axis is for tension)?

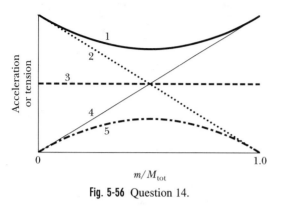

Fig. 5-56 Question 14.

15. *Organizing question:* In each situation of Fig. 5-57, a force $\vec{F}$ of magnitude 5 N is applied to a 0.2 kg tin of sardines that can slide along a frictionless plane tilted at 30°. For each situation, set up equations, complete with known data, to find the acceleration a of the tin and the magnitude N of the normal force on the tin; use the tilted xy coordinate system shown in Fig. 5-18 and the first two component equations of Eq. 5-2 ($F_{net,x} = ma_x$ and $F_{net,y} = ma_y$).

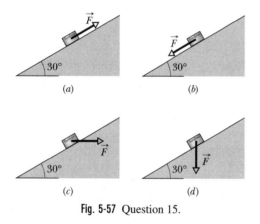

Fig. 5-57 Question 15.

16. (a) In Fig. 5-58, does the vertical component of force $\vec{F}$ help support the box, or does it press the box against the floor? (b) Suppose the mass of the box is m. Then is the magnitude of the normal force on the box equal to, greater than, or less than mg? (c) Is the vertical component of the force $F \sin \theta$ or $F \cos \theta$?

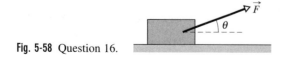

Fig. 5-58 Question 16.

17. *Organizing question:* Figure 5-59 gives overhead views of six situations in which a 2 kg box is accelerated at a magnitude of 1 m/s² along a frictionless floor by one or more forces. The magnitude of force $\vec{F}_4$ differs for each situation, but the magnitudes of the other forces are always $F_1 = 1$ N, $F_2 = 2$ N, and

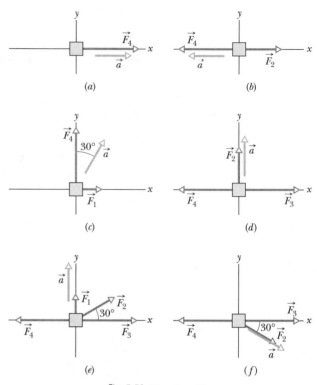

Fig. 5-59 Question 17.

$F_3 = 3$ N. For each situation, set up the first two component equations of Eq. 5-2 ($F_{net,x} = ma_x$ and $F_{net,y} = ma_y$), complete with known data.

18. During a space walk, three astronauts race one another by using their jet backpacks. The race begins at $t = 0$ and with the astronauts at rest relative to the racecourse alongside their ship. During the race, their position functions $x(t)$ along the racecourse are the following, with x in meters and t in seconds: (a) $x = 2t^2$, (b) $x = 8t - 3$, (c) $x = 4t^2 + 2$. Assuming that the astronauts have the same mass, rank them according to the magnitudes of the forces on them from their jet backpacks, greatest first.

19. Figure 5-60 shows four choices for the direction of a force of magnitude F to be applied to a block on an inclined plane. The directions are either horizontal or vertical. (For choices a and b, the force is not enough to lift the block off the plane.) Rank the choices according to the magnitude of the normal force on the block from the plane, greatest first.

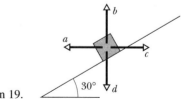

Fig. 5-60 Question 19.

20. Rank the following according to the magnitude of the normal force on you when you stand on them, greatest first: a concrete floor, a floor with a thick carpet, a floor with a thin carpet, and a wood plank suspended between two supports at a construction site.

21. July 17, 1981, Kansas City: the newly opened Hyatt Regency is packed with people listening and dancing to a band playing favorites from the 1940s. Many of the people are crowded onto the walkways that hang like bridges across the wide atrium. Suddenly two of the walkways collapse, falling onto the merrymakers on the main floor.

The walkways were suspended one above another on vertical rods and held in place by nuts threaded onto the rods. In the original design, only two long rods were to be used, each extending through all three walkways (Fig. 5-61a). If each walkway and the merrymakers on it have a combined mass of M, what is the total mass supported by the threads and two nuts on (a) the lowest walkway and (b) the highest walkway?

Threading nuts on a rod is impossible except at the ends, so the design was changed: instead, six rods were used, each connecting two walkways (Fig. 5-61b). What now is the total mass supported by the threads and two nuts on (c) the lowest walkway and (d) the highest walkway? It was this design that failed.

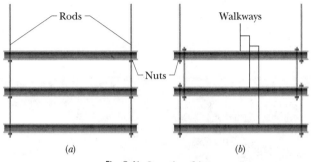

Fig. **5-61** Question 21.

22. An elevator is supported by a single cable and there is no counterweight. The elevator receives passengers at the ground floor and takes them to the top floor, where they disembark. New passengers enter and are taken down to the ground floor. During this round trip, (a) when is the tension in the cable equal to the weight of the elevator plus the passengers, (b) when is it greater, and (c) when is it less?

23. Seven identical dominoes are to be stacked in three columns (Fig. 5-62 gives an example) and pushed across a frictionless ice rink by a horizontal 10 N force. How many dominoes should be in each column, with a minimum of one, (a) to maximize the acceleration of the dominoes, (b) to maximize the force on column C due to column B, (c) to maximize the *net* force on column B due to columns A and C, and (d) to maximize the force on column B due to column A?

Fig. **5-62** Question 23.

24. Figure 5-63 shows overhead views of six situations in which a pirate's chest is pulled over a frictionless surface by pirates who apply horizontal forces $\vec{F}_1$ and $\vec{F}_2$ in the directions indicated. No other horizontal forces act on the chest in these situations. An acceleration vector is also indicated. In which of the situations, if any, is the direction of that vector physically impossible to achieve

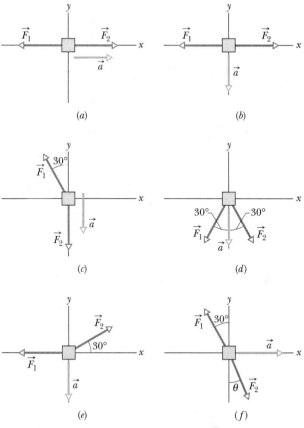

Fig. **5-63** Question 24.

for any possible values of the magnitudes F_1 and F_2 (and, in part (f), the angle θ)?

DISCUSSION QUESTIONS

25. If you stand facing forward during a bus or subway ride, why does a quick deceleration topple you forward and a quick increase in speed throw you backward? Why do you have better balance if you face toward the side of the bus or subway train?

26. Using Newton's first law, explain what happens to a child sitting in the front seat of a car if the seat belt is not used and the driver suddenly slams on the brakes. Suppose, instead, that the child is held by an adult who neglects the seat belt; if the car suddenly stops, is the child safer in the arms of the adult, or is the child actually in more danger? What happens to a person who rides at the back of a pickup truck if the truck suddenly stops?

27. If two forces act on a moving body, is there any way that the body can move at (a) a constant speed or (b) a constant velocity? Is there any way that the velocity can be zero (c) for an instant or (d) continuously?

28. The owner's manual for a certain car suggests that your seat belt should be adjusted "to fit snugly" and that the front seat headrest should be adjusted *not* so it fits comfortably at the back of your neck but so "the top of the headrest is level with the top of your ears." Explain the wisdom of these instructions in terms of Newton's laws.

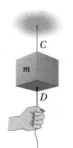

Fig. 5-64 Question 29.

29. A block with mass m is supported by a cord C from the ceiling, and a similar cord D is attached to the bottom of the block (Fig. 5-64). Explain this: If you give a sudden jerk to D, it will break; but if you pull on D steadily, C will break.

30. A Frenchman, filling out a form, writes "78 kg" in the space marked Poids (weight). However, weight has a force unit and kilogram is a mass unit. What do Frenchmen (among others) have in mind when they use a mass unit to report their weight? Why don't they report their weight in newtons? How many newtons does this Frenchman weigh? How many pounds?

31. What is your mass in slugs? Your weight in newtons?

32. Using force, length, and time as fundamental quantities, find the dimensions of mass.

33. A horse is urged to pull a wagon. The horse refuses to try, citing Newton's third law as a defense: The pull of the horse on the wagon is equal to but opposite the pull of the wagon on the horse. "If I can never exert a greater force on the wagon than it exerts on me, how can I ever start the wagon moving?" asks the horse. How would you reply?

34. Comment on whether the following pairs of forces are examples of third-law force pairs: (a) Earth attracts a brick; the brick attracts Earth. (b) An airplane propeller pushes air in toward the tail; the air pushes the plane forward. (c) A horse pulls forward on a cart, accelerating it; the cart pulls backward on the horse. (d) A horse pulls forward on a cart without moving it; the cart pulls back on the horse. (e) A horse pulls forward on a cart without moving it; Earth exerts an equal and opposite force on the cart. (f) Earth pulls down on the cart; the ground pushes up on the cart with an equal and opposite force.

35. Comment on the following statements about mass and weight taken from examination papers. (a) Mass and weight are the same physical quantities expressed in different units. (b) Mass is a property of one object alone, whereas weight results from the interaction of two objects. (c) An object's weight is proportional to its mass. (d) A body's mass varies with changes in its local weight.

36. Describe several ways in which you could, even briefly, experience weightlessness.

37. The mechanical arm on a space shuttle can handle a 2200 kg satellite when extended to 12 m. However, on the ground, this remote manipulator system (RMS) cannot support its own weight. In the weightlessness of an orbiting shuttle, why would the RMS have to be able to exert any force at all?

38. You are on the flight deck of the orbiting space shuttle *Dis-*

covery and someone hands you two wooden balls, outwardly identical. However, one has a lead core and the other does not. Describe several ways of telling them apart.

39. You are an astronaut in the lounge of an orbiting space station and you remove the cover from a long thin jar containing a single olive. Describe several ways—all taking advantage of the mass of either the olive or the jar—to remove the olive from the jar.

40. A horizontal force acts on a body that is free to move. Can it produce an acceleration if it is less than the body's weight?

41. Why does the acceleration of a freely falling object not depend on the weight of the object?

42. What's the relation—if any—between the force acting on an object and the direction in which the object is moving?

43. A bird alights on a stretched telegraph wire. Does this change the tension in the wire? If so, is it by an amount less than, equal to, or greater than the weight of the bird?

44. In November 1984, astronauts Joe Allen and Dale Gardner salvaged a Westar-6 communications satellite in space and placed it into the cargo bay of the space shuttle *Discovery*; see Fig. 5-65. Describing the experience, Joe Allen said of the satellite, "It's not heavy; it's massive." What did he mean?

Fig. 5-65 Question 44.

45. In a tug-of-war, three men pull on a rope to the left at A and three men pull to the right at B with forces of equal magnitude. Then a 3 kg object is hung from the center of the rope. (a) Can the men get the rope AB to be horizontal? (b) If not, explain. If so, determine the required magnitudes of the forces at A and B.

46. The following statement is true; explain it. Two teams are having a tug-of-war; the team that pushes harder (horizontally) against the ground wins.

47. You stand on the large platform of a spring scale and note your weight. You then take a step on this platform and note that the scale reads less than your weight at the beginning of the step, and more than your weight at the end of the step. Explain.

48. Could you weigh yourself on a scale whose maximum reading is less than your weight? If so, how?

49. A woman stands on a spring scale in an elevator. Order the following situations according to the reading they produce on the scale, greatest first: (a) elevator stationary; (b) elevator cable breaks (free fall); (c) elevator accelerates upward; (d) elevator accelerates downward; (e) elevator moves at constant velocity.

50. Under what conditions could objects with unequal masses be strung over a pulley without the pulley having any tendency to turn?

51. In Fig. 5-66, a needle has been placed in each end of a broomstick, with the tips of the needles resting on the edges of filled wine glasses. The experimenter strikes the broomstick a swift and sturdy blow with a stout rod. The broomstick breaks and falls to the floor but the wine glasses remain in place and no wine is spilled. This impressive parlor stunt was popular at the end of the nineteenth century. What is the physics behind it? (If you try it, practice first with empty soft drink cans.)

Fig. 5-66 Question 51.

EXERCISES & PROBLEMS

57. You need to lower a bundle of old roofing material weighing 449 N to the ground with a rope that will snap if the tension in it exceeds 387 N. (a) How can you avoid snapping the rope during the descent? (b) If the descent is 6.1 m and you just barely avoid snapping the rope, with what speed does the bundle hit the ground?

58. In Fig. 5-67, a 3.0 kg box of dirty money on a 30° frictionless incline is connected to a 2.0 kg box of laundered money on a 60° frictionless incline. The pulley is frictionless and massless. What is the tension in the connecting cord?

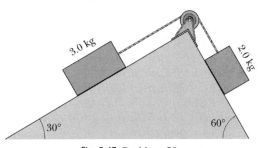

Fig. 5-67 Problem 58.

59. Some insects can walk below a thin rod (such as a twig) by hanging from it. Suppose that such an insect has mass m and hangs from a horizontal rod as shown in Fig. 5-68. All its six legs are under the same tension, and the leg sections nearest the body are horizontal. (a) What is the tension in each tibia (forepart of a leg)? (b) If the insect straightens out the legs somewhat, does the tension in each tibia increase, decrease, or stay the same?

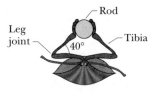

Fig. 5-68 Problem 59.

60. A motorcycle of weight 2.0 kN accelerates from 0 to 88.5 km/h in 6.0s. (a) What is the magnitude of the motorcycle's constant acceleration? (b) What is the magnitude of the net force causing the acceleration?

61. Refer to Fig. 5-14 and suppose that the two blocks have the masses $m = 2.0$ kg and $M = 4.0$ kg. (a) Decide without any calculations which block should be hanging if the magnitude of the acceleration is to be largest. What then are (b) the magnitude of the acceleration and (c) the associated tension in the cord?

62. A 15 000 kg helicopter lifts a 4500 kg truck with an upward acceleration of 1.4 m/s². Calculate (a) the net upward force on the helicopter blades from the air and (b) the tension in the cable between the helicopter and truck.

63. If the 1 kg standard body is accelerated by $\vec{F}_1 = (3.0 \text{ N})\hat{i} + (4.0 \text{ N})\hat{j}$ and $\vec{F}_2 = (-2.0 \text{ N})\hat{i} + (-6.0 \text{ N})\hat{j}$, then (a) what is the net force in unit-vector notation, and what are the magnitude and direction of (b) the net force and (c) the acceleration?

64. There are two horizontal forces on the 2.0 kg box in the overhead view of Fig. 5-69 but only one is shown. The box moves strictly along the x axis. For each of the following values for the acceleration a_x of the box, find the second force: (a) 10 m/s², (b) 20 m/s², (c) 0, (d) −10 m/s², and (e) −20 m/s².

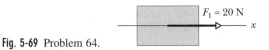

Fig. 5-69 Problem 64.

65. A 52 kg circus performer is to slide down a rope that will break if the tension exceeds 425 N. (a) What happens if the performer hangs stationary on the rope? (b) At what magnitude of acceleration does the performer just avoid breaking the rope?

66. A spaceship lifts off vertically from the Moon, where the free-fall acceleration is 1.6 m/s². If the spaceship has an upward acceleration of 1.0 m/s² as it lifts off, what is the magnitude of the force of the spaceship on its pilot, who weighs 735 N on Earth?

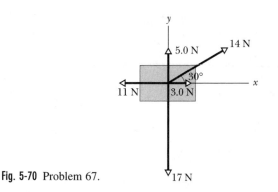

Fig. 5-70 Problem 67.

67. Five forces pull on the 4.0 kg box in the overhead view of Fig. 5-70. Find the box's acceleration (a) in unit-vector notation and (b) as a magnitude and a direction.

68. Only two forces act on a 3.0 kg object that moves with an acceleration of 3.0 m/s² in the positive y direction. If one of the forces acts in the positive x direction and has a magnitude of 8.0 N, what is the magnitude of the other force?

69. A 0.20 kg hockey puck has a velocity of 2.0 m/s toward the east as it slides over the frictionless surface of a frozen lake. What are the magnitude and direction of the average force that must act on the puck during a 0.50 s interval to change its velocity to (a) 5.0 m/s, due west, and (b) 5.0 m/s, due south?

70. A certain force gives an object of mass m_1 an acceleration of 12.0 m/s² and an object of mass m_2 an acceleration of 3.30 m/s². What acceleration would the force give to an object with a mass of (a) $m_2 - m_1$ and (b) $m_2 + m_1$?

71. When the system in Fig. 5-71 is released from rest, the 3.0 kg kimchi container has an acceleration of 1.0 m/s² to the right. The surfaces and the pulley are frictionless. (a) What is the tension in the connecting cord? (b) What is the value of M?

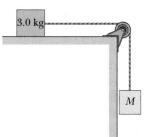

Fig. 5-71 Problem 71.

72. A 1.0 kg FedEx box on a 37° incline is connected to a 3.0 kg UPS box on a horizontal surface (Fig. 5-72). The surfaces are frictionless; the pulley is frictionless and massless. If F = 12 N, what is the tension in the connecting cord?

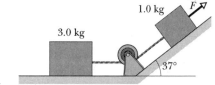

Fig. 5-72 Problem 72.

73. In a laboratory experiment, an initially stationary electron (mass = 9.11 × 10⁻³¹ kg) undergoes a constant acceleration through 1.5 cm, reaching a speed of 6.0 × 10⁶ m/s at the end of

that distance. (a) What is the magnitude of the force accelerating the electron? (b) What is the weight of the electron?

74. Compute the initial upward acceleration of a rocket of mass 1.3 × 10⁴ kg if the initial upward force produced by its engine (the thrust) is 2.6 × 10⁵ N. Do not neglect the gravitational force on the rocket.

75. A 45 kg woman is ice-skating toward the east on a frictionless frozen lake when she collides with a 90 kg man who is ice-skating toward the west. The maximum force exerted on the woman by the man during the collision is 180 N, west. (a) What is the maximum force exerted on the man by the woman during the collision? (b) What is the maximum acceleration experienced by the woman during the collision? (c) What is the maximum acceleration experienced by the man during the collision?

76. A weight-conscious penguin with a mass of 15.0 kg rests on a bathroom scale (Fig. 5-73). What are (a) the penguin's weight W and (b) the normal force $\vec{N}$ on the penguin? (c) What is the reading on the scale, assuming it is calibrated in weight units?

Fig. 5-73 Problem 76.

77. A 100 kg crate sits on the floor of a freight elevator that starts from rest on the ground floor of a building at time $t = 0$ and rises to the top floor during an 8.0 s interval. The velocity of the elevator as a function of time is shown in Fig. 5-74. What are the magnitude and direction of the force on the crate from the elevator floor at (a) $t = 1.8$ s, (b) $t = 4.4$ s, and (c) $t = 6.8$ s?

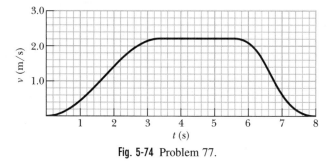

Fig. 5-74 Problem 77.

78. An object is hung from a spring balance attached to the ceiling of an elevator. The balance reads 65 N when the elevator is standing still. What is the reading when the elevator is moving upward (a) with a constant speed of 7.6 m/s and (b) with a speed of 7.6 m/s while decelerating at a rate of 2.4 m/s²?

79. A 12 kg penguin runs onto a large area of level, frictionless ice with an initial velocity of 6.0 m/s toward the east. As the penguin slides across the ice, it is pushed by the wind with a force that is constant in magnitude and direction. Figure 5-75 shows the position of the penguin, at 1.0 s intervals, as it slides on the frictionless ice surface; the positive direction of the x axis is toward the east. The penguin first makes contact with the ice at $t = 0$. What are the

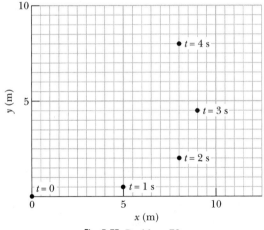

Fig. 5-75 Problem 79.

magnitude and direction of the force of the wind on the penguin as it slides across the ice?

80. In Fig. 5-76, a 1.0 kg tin of anti-oxidants on a frictionless inclined surface is connected to a 2.0 kg tin of corned beef. The pulley is massless and frictionless. An upward force $\vec{F}$ of 6.0 N acts on the corned beef tin, which has a downward acceleration of 5.5 m/s^2. (a) What is the tension in the connecting cord? (b) What is the angle β?

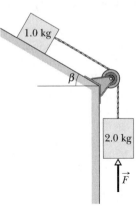

Fig. 5-76 Problem 80.

81. Figure 5-77a shows a crude mobile hanging from a ceiling; it consists of two metal pieces strung together by cords of negligible mass. The masses of the pieces are given. What is the tension in

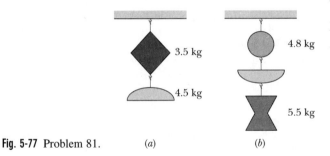

Fig. 5-77 Problem 81.   (a)   (b)

(a) the bottom cord and (b) the top cord? Figure 5-77b shows a similar mobile consisting of three metal pieces. Two masses are given. The tension in the top cord is 199 N. What is the tension in (c) the lowest cord and (d) the middle cord?

82. When an automobile with a weight of 17.0 kN accelerates at 3.66 m/s^2, what is the magnitude of the net force on the car?

CLUSTERED PROBLEMS

In each problem of this cluster, a 10.0 kg block is on a frictionless surface inclined by 30.0° to the horizontal.

83. In Fig. 5-78a, the block slides down the plane. What are (a) the magnitude of the block's acceleration and (b) the magnitude of the normal force on the block from the inclined surface?

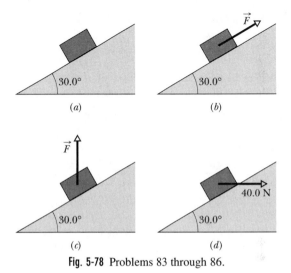

Fig. 5-78 Problems 83 through 86.

84. In Fig. 5-78b, force $\vec{F}$ is applied to the block, directed upward along the inclined surface. What are the magnitude and direction of the block's acceleration if the magnitude of $\vec{F}$ is (a) 40.0 N and (b) 60.0 N?

85. In Fig. 5-78c, a 60.0 N force $\vec{F}$, directed upward, is applied to the block. (a) What are the magnitude and direction of the block's acceleration? (b) What is the magnitude of the normal force on the block from the inclined surface? (c) What must the magnitude of the applied force be if the block is to be stationary?

86. In Fig. 5-78d, a 40.0 N force is applied horizontally to the block. (a) What are the magnitude and direction of the block's acceleration? (b) What is the magnitude of the normal force on the block from the inclined surface? (c) What must the magnitude of the applied force be if the block is to be stationary?

6

Force and Motion—II

Sample Problem 6-10

In Fig 6-40a, a crate of dilled pickles with mass $m_1 = 14$ kg moves along a plane that makes an angle of $\theta = 30°$ with the horizontal. That crate is connected to a crate of pickled dills with mass $m_2 = 14$ kg by a taut, massless cord that runs over a frictionless, massless pulley. The dills descend with constant velocity.

(a) What are the magnitude and direction of the frictional force on the pickles from the plane?

SOLUTION: A **Key Idea** here is that because the dills descend, the pickles must move *up* the plane, and so a *kinetic* frictional force $\vec{f}_k$ must act on the pickles and be directed *down* the plane. However, we cannot find the magnitude of $\vec{f}_k$ with Eq. 6-2 ($f_k = \mu_k N$) because we do not know the coefficient of kinetic friction μ_k between the pickles and the plane.

Another **Key Idea** helps: By Newton's second law, $\vec{f}_k$ and the other forces on the pickles determine the acceleration of the pickles, but we know that the acceleration is zero because the dills and pickles move together at constant velocity. Therefore, let us apply Newton's second law to the pickles.

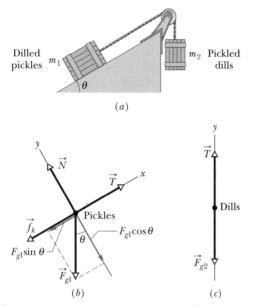

(a)

(b)

(c)

Fig. 6-40 Sample Problem 6-10. (a) A crate of dilled pickles moves up the plane while a crate of pickled dills descends at constant velocity. (b) A free-body diagram for the pickles. (c) A free-body diagram for the dills.

The free-body diagram of Fig. 6-40b shows a reasonable axis system and the forces on the pickles: the desired $\vec{f}_k$, the force $\vec{T}$ from the cord, and the gravitational force $\vec{F}_{g1}$ (of magnitude $m_1 g$). The x and y components of $\vec{F}_{g1}$ are shown; the x component has the magnitude $m_1 g \sin\theta$ and is directed down the plane. We know that the acceleration along the plane is zero, so we write Newton's second law for components along the x axis ($F_{net,x} = ma_x$) as

$$T - f_k - m_1 g \sin\theta = m_1(0),$$

from which

$$f_k = T - m_1 g \sin\theta. \qquad (6\text{-}25)$$

We cannot solve this equation for f_k, because we do not know the value of T.

We next consider the dills, writing Newton's second law for components along the vertical y axis shown in the free-body diagram in Fig. 6-40c. Since $m_2 g$ is the magnitude of $\vec{F}_{g2}$, that law ($F_{net,y} = ma_y$) becomes

$$T - m_2 g = m_2(0),$$

so

$$T = m_2 g. \qquad (6\text{-}26)$$

Equations 6-25 and 6-26 are simultaneous equations in two unknowns (T and our desired f_k). Substituting T from Eq. 6-26 into Eq. 6-25 and using given data, we obtain

$$f_k = m_2 g - m_1 g \sin\theta$$
$$= (14\text{ kg})(9.8\text{ m/s}^2) - (14\text{ kg})(9.8\text{ m/s}^2)(\sin 30°)$$
$$= 68.6\text{ N} \approx 69\text{ N}. \qquad \text{(Answer)}$$

(b) What is μ_k?

SOLUTION: We can use Eq. 6-2 to find μ_k, but first we need the magnitude N of the normal force acting on the pickles. To find N from Fig. 6-40b, we can write Newton's second law for components along the y axis ($F_{net,y} = ma_y$) as

$$N - m_1 g \cos\theta = m_1(0),$$

or

$$N = m_1 g \cos\theta.$$

From Eq. 6-2 we now have

$$\mu_k = \frac{f_k}{N} = \frac{f_k}{m_1 g \cos\theta}$$
$$= \frac{68.6\text{ N}}{(14\text{ kg})(9.8\text{ m/s}^2)(\cos 30°)} = 0.58. \qquad \text{(Answer)}$$

Sample Problem 6-11

You cannot always count on friction to get your car around a curve, especially if the road is icy or wet. That is why highway curves are banked (tilted). As in Sample Problem 6-9, suppose that a car of mass m moves at a constant speed v of 20 m/s around a now banked curved with radius $R = 190$ m (Fig. 6-41a). What bank angle θ makes reliance on friction unnecessary?

SOLUTION: As in Sample Problem 6-9, a centripetal force must act on the car if the car is to move along the circular path. Here, however, the **Key Idea** is that the track is banked so as to tilt the normal force $\vec{N}$ on the car toward the center of the circle (Fig. 6-41b). Thus, $\vec{N}$ now has a centripetal component of magnitude N_r, directed in-

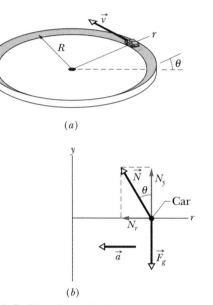

(a)

(b)

Fig. 6-41 Sample Problem 6-11. (a) A car moves around a curved banked road at constant speed. The bank angle is exaggerated for clarity. (b) A free-body diagram for the car, assuming that friction between tires and road is zero. The radially inward component of the normal force (along radial axis r) provides the necessary centripetal force. The resulting acceleration is also radially inward.

ward along a radial axis r. We want to find the value of the bank angle θ such that this centripetal component keeps the car on the circular track without need of friction.

As Fig. 6-41b shows (and as you should verify), the angle that $\vec{N}$ makes with the vertical is equal to the bank angle θ of the track. Thus, the radial component N_r is equal to $N \sin \theta$. We can now write Newton's second law for components along the r axis ($F_{\text{net},r} = ma_r$) as

$$-N \sin \theta = m\left(-\frac{v^2}{R}\right). \tag{6-27}$$

We cannot solve this equation for the value of θ because it also contains the unknowns N and m.

We next consider the forces and acceleration along the y axis in Fig. 6-41b. The vertical component of the normal force is $N_y = N \cos \theta$, the gravitational force $\vec{F}_g$ on the car has the magnitude mg, and the acceleration of the car along the y axis is zero. Thus, we can write Newton's second law for components along the y axis ($F_{\text{net},y} = ma_y$) as

$$N \cos \theta - mg = m(0),$$

from which

$$N \cos \theta = mg. \tag{6-28}$$

This too contains the unknowns N and m, but note that dividing Eq. 6-27 by Eq. 6-28 neatly eliminates both those unknowns. Doing so, and replacing (sin θ)/(cos θ) with tan θ, and solving for θ then yield

$$\theta = \tan^{-1} \frac{v^2}{gR} \tag{6-29}$$

$$= \tan^{-1} \frac{(20 \text{ m/s})^2}{(9.8 \text{ m/s}^2)(190 \text{ m})} = 12°. \quad \text{(Answer)}$$

Equation 6-24 (in Sample Problem 6-9) and Eq. 6-29 (here) tell us that the critical coefficient of friction for an unbanked road is the same as the tangent of the bank angle for a banked road. The road must produce a certain centripetal force one way or the other—either with friction or by being banked.

Sample Problem 6-12

Figure 6-42a shows a *conical pendulum*, in which the bob (the small object at the lower end of the cord) moves in a horizontal circle at constant speed. (The cord sweeps out a cone as the bob rotates.) The distance L between the top of the cord and the center of the bob is 1.7 m. The cord makes an angle θ of 37° with the vertical. Find the period τ of the motion. (We use τ instead of T, because we need T to symbolize a force in the solution.)

SOLUTION: One **Key Idea** here is that the period depends on the speed v of the bob via Eq. 4-33:

$$\tau = \frac{2\pi R}{v}, \tag{6-30}$$

where R is the radius of the circle and v is the constant speed of the bob. To solve this equation for τ, we need first to find expressions for R and v in terms of the given quantities. From Fig. 6-42a, we see that

$$R = L \sin \theta. \tag{6-31}$$

To find v, we use another **Key Idea:** A centripetal force must act on the bob to make it travel in the circle and to give it a centripetal acceleration $\vec{a}$ of magnitude v^2/R. The free-body diagram of Fig. 6-42b shows the forces on the bob (the gravitational force $\vec{F}_g$ and the force $\vec{T}$ from the cord). It also shows the centripetal acceleration $\vec{a}$ along a radial axis r that extends from the center of

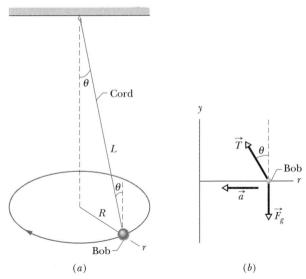

Fig. 6-42 Sample Problem 6-12. (*a*) A conical pendulum—a bob of mass *m* that traces out a horizontal circle at the end of a cord of length *L*, which makes an angle θ with the vertical. (*b*) A free-body diagram for the pendulum bob. The net force (and thus the acceleration) is directed inward along radial axis *r*, toward the center of the circle.

the circle through the bob. We see that the only radial force on the bob is the component $T \sin \theta$ of force $\vec{T}$. Thus, for the bob, we can write Newton's second law for components along the *r* axis

$(F_{\text{net},r} = ma_r)$ as

$$-T \sin \theta = m\left(-\frac{v^2}{R}\right), \qquad (6\text{-}32)$$

where *m* is the bob's mass. We need such an expression for *v*, but this one contains the extra unknowns *T* and *m*.

When we get stuck on one axis, we try the other axis—here the *y* axis in Fig. 6-42*b*. Along that axis, the upward force on the bob is the component $T \cos \theta$ and the downward force is $\vec{F}_g$, which has the magnitude *mg*. The vertical acceleration of the bob is 0, so we can write Newton's second law for components along the vertical *y* axis $(F_{\text{net},y} = ma_y)$ as

$$T \cos \theta - mg = m(0)$$

or $\qquad\qquad T \cos \theta = mg. \qquad (6\text{-}33)$

This too contains the extra unknowns *T* and *m*. However, note that dividing Eq. 6-32 by Eq. 6-33 neatly eliminates both those unknowns. Doing so and then solving for *v* give us

$$v = \sqrt{\frac{gR \sin \theta}{\cos \theta}}.$$

We next substitute into Eq. 6-30 with this expression for *v* and with Eq. 6-31 for *R*. After some algebraic manipulation, we find

$$\tau = 2\pi \sqrt{\frac{L \cos \theta}{g}}$$

$$= 2\pi \sqrt{\frac{(1.7 \text{ m})(\cos 37°)}{9.8 \text{ m/s}^2}} = 2.3 \text{ s.} \qquad \text{(Answer)}$$

QUESTIONS

11. *Organizing question:* Figure 6-43 shows six situations in which one or two forces are applied to a 2 kg box on a floor. Assume that the box remains stationary due to the friction that also acts on it. The magnitudes of the applied forces are $F_1 = 10$ N and $F_2 = 2$ N, and the coefficient of static friction between the box and the floor is 0.6. For each situation, set up equations, complete with known data, to find the magnitude f_s of the static frictional force and the magnitude *N* of the normal force on the box; use the first two equations of Eq. 5-2 ($F_{\text{net},x} = ma_x$ and $F_{\text{net},y} = ma_y$).

12. A box is on a ramp that is at angle θ to the horizontal. As θ is increased from zero, and before the box slips, do the following increase, decrease, or remain the same: (a) the component of the gravitational force on the box, along the ramp, (b) the magnitude of the static frictional force on the box from the ramp, (c) the component of the gravitational force on the box, perpendicular to the ramp, (d) the magnitude of the normal force on the box from the ramp, and (e) the maximum value $f_{s,\text{max}}$ of the static frictional force?

13. *Organizing question:* For the six situations of Question 11 and Fig. 6-43, assume that the box is sliding to the right and the coefficient of kinetic friction between the box and the floor is 0.3. For each situation, set up equations, complete with known data, to find

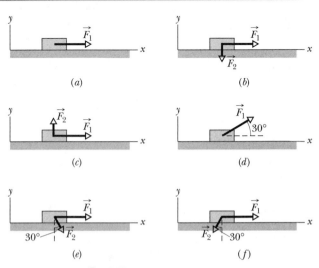

Fig. 6-43 Questions 11 and 13.

the magnitude *a* of the box's acceleration and the magnitude *N* of the normal force on the box; use the first two component equations of Eq. 5-2 ($F_{\text{net},x} = ma_x$ and $F_{\text{net},y} = ma_y$).

14. Figure 6-44 shows a section of a circular space station that rotates about its center so as to give an apparent weight to the crew. One of the crew is shown at the outer wall of the station, which has velocity $\vec{v}_s$. (a) If the astronaut moves to a point closer to the center of the station (say by taking an elevator), does his apparent weight increase, decrease, or remain the same? (b) If, instead, the astronaut runs along the outer wall in the direction opposite $\vec{v}_s$ (with a speed less than the magnitude of $\vec{v}_s$), does his apparent weight increase, decrease, or remain the same?

Fig. 6-44 Question 14.

15. Figure 6-45 is a plot of force magnitude versus angle θ for the coin in Sample Problem 6-3; 17 points are labeled with numbers. The coin slips at a certain angle θ_c. The coefficient of static friction is 0.6, and the coefficient of kinetic friction is 0.4. First consider the range of θ for which the coin is stationary. Which segments of the curves (from which number to which number) best give (a) the magnitude of the normal force on the coin from the book, (b) the magnitude of the static frictional force on the coin from the book, and (c) the maximum $f_{s,\text{max}}$ of that force? (d) Which point on the θ axis represents θ_c?

Next, consider the range of θ for which the coin slides. Which segments best give (e) the magnitude of the normal force on the coin from the book and (f) the magnitude of the kinetic frictional force on the coin from the book?

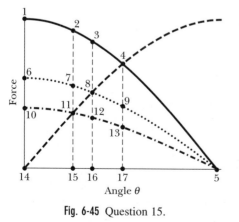

Fig. 6-45 Question 15.

16. In Fig. 6-46, a block is held stationary on a ramp by the frictional force on it from the ramp. A force $\vec{F}$, directed up the ramp, is then applied to the block and gradually increased in magnitude from zero. During the increase, what happens to the direction and magnitude of the frictional force on the block?

Fig. 6-46 Question 16.

17. Reconsider Question 16 but with the force $\vec{F}$ now directed down the ramp. As the magnitude of $\vec{F}$ is increased from zero, what happens to the direction and magnitude of the frictional force on the block?

18. During a routine flight on September 21, 1956, test pilot Tom Attridge put his Grumman F11F-1 jet fighter into a 20° dive for a test of the aircraft's 20 mm machine cannons. While traveling faster than sound at 4000 m altitude, he shot a burst of rounds. Then, after allowing the cannons to cool, he shot another burst at 2000 m; his speed was then 344 m/s, the speed of the rounds relative to him was 730 m/s, and he was still in a dive.

Almost immediately the canopy around Attridge was shredded and his right air intake was damaged. With little flying capability left, the jet crashed into a wooded area, but Attridge managed to escape the resulting explosion by crawling from the fuselage (in spite of four fractured vertebrae). Explain what apparently happened just after the second burst of cannon rounds.

19. *Organizing question:* In the two situations of Fig. 6-47, forces are applied to a stationary 2 kg tin of tuna that can slide along a plane tilted at 30°. The applied forces have magnitudes of $F_1 = 20$ N and $F_2 = 2$ N, and the coefficient of static friction is 0.8. In Fig. 6-47a, the tin is close to sliding up the plane; in Fig. 6-47b, it is close to sliding down the plane. For each situation, set up equations, complete with known data, to find the magnitude f_s of the frictional force on the tin and the magnitude N of the normal force on the tin; use the tilted xy coordinate system shown in Fig. 5-18 and the first two component equations of Eq. 5-2 ($F_{\text{net},x} = ma_x$ and $F_{\text{net},y} = ma_y$).

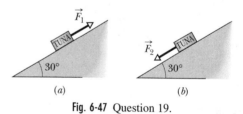

Fig. 6-47 Question 19.

20. In Fig. 6-48, a horizontal force of 100 N is to be applied to a 10 kg slab that is initially stationary on a frictionless floor, to accelerate the slab. A 10 kg block lies on top of the slab; the coefficient of friction μ between the block and the slab is not known, and the block might slip. (a) Considering that possibility, what is the possible range of values for the magnitude of the slab's acceleration a_{slab}? (*Hint:* You don't need written calculations; just consider extreme values for μ.) (b) What is the possible range for the magnitude a_{block} of the block's acceleration?

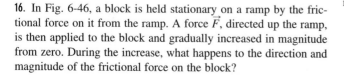

Fig. 6-48 Question 20.

21. Figure 6-49 shows five drag-racing boxes (a new sport) in which an applied force is directed rightward and a kinetic frictional force is directed leftward. The boxes, which have identical masses, cross the starting line with the same speed. Rank the boxes according to their arrival at the finish line, first box first.

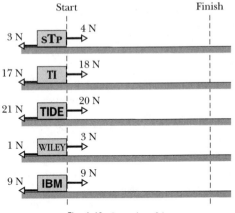

Fig. 6-49 Question 21.

22. In 1987, as a Halloween stunt, two sky divers passed a pumpkin back and forth between them while they were in free fall just west of Chicago. The stunt was great fun until the last sky diver with the pumpkin opened his parachute. The pumpkin broke free from his grip, plummeted about 0.5 km, ripped through the roof of a house, slammed into the kitchen floor, and splattered all over the newly remodeled kitchen. From the sky diver's viewpoint and from the pumpkin's viewpoint, why did the sky diver lose control of the pumpkin?

DISCUSSION QUESTIONS

23. There is a limit beyond which further polishing of a surface *increases* rather than decreases frictional resistance. Explain why.

24. Can the coefficient of static friction have a value greater than 1? What about the coefficient of kinetic friction?

25. A crate, heavier than you are, rests on a rough floor. The coefficient of static friction between the crate and the floor is the same as that between the soles of your shoes and the floor. Can you push the crate across the floor?

26. How could a person who is at rest on completely frictionless ice covering a pond reach shore? Could she do this by walking, rolling, swinging her arms, or kicking her feet? How could a person be placed in such a location in the first place?

27. Why do tires grip the road better on level ground than they do when going uphill or downhill?

28. What is the purpose of the curved surfaces, called spoilers, that are placed on the rear of race cars? They are designed so that air flowing past them exerts a downward force.

29. Which raindrops, if either, fall faster: small ones or large ones?

30. The terminal speed of a baseball is 95 mi/h. However, the measured speeds of pitched balls often exceed this, sometimes exceeding 100 mi/h. How can this be?

31. A log is floating downstream. How would you calculate the drag force acting on it?

32. What happens to a baseball that is fired downward through air at twice its terminal speed—does it speed up, slow down, or continue to move with its initial speed?

33. Consider a ball thrown vertically up. Taking air resistance into account, would you expect the time during which the ball rises to be longer or shorter than the time during which it falls? Why? Make a qualitative graph of speed v versus time t for the ball.

34. Why are train roadbeds and highways banked on curves?

35. You are flying a plane at constant altitude and wish to make a 90° turn. Why do you bank the plane?

36. In the conical pendulum of Sample Problem 6-12, what happens to the period τ and the speed v when $\theta = 90°$? Why is this angle not achievable physically? Discuss the case for $\theta = 0°$.

37. A coin is put on a phonograph turntable. The motor is started but, before the final speed of rotation is reached, the coin flies off. Explain why.

38. A car is moving at constant speed along a country road that resembles a roller coaster track. Compare the force the car exerts on a horizontal section of the road to the force it exerts on the road at the top of a hill and at the bottom of a valley. Explain.

EXERCISES & PROBLEMS

49. A small coin is placed on a flat, horizontal turntable. The turntable is observed to make three revolutions in 3.14 s. (a) What is the speed of the coin when it rides without slipping at a distance of 5.0 cm from the center of the turntable? (b) What is the acceleration (magnitude and direction) of the coin? (c) What are the magnitude and direction of the frictional force acting on the coin if the coin has a mass of 2.0 g? (d) What is the coefficient of static friction between the coin and the turntable if the coin is observed to slide off the turntable when it is more than 10 cm from the center of the turntable?

50. A model airplane of mass 0.75 kg is flying at constant speed in a horizontal circle at one end of a 30 m cord and at a height of 18 m. The other end of the cord is tethered to the ground. The airplane circles 4.4 times per minute and has its wings horizontal so that the air is pushing vertically upward. (a) What is the acceleration of the plane? (b) What is the tension in the cord? (c) What is the total upward force (lift) on the plane's wings?

51. Imagine the standard kilogram is located on Earth's equator, where it moves in a circle of radius 6.40×10^6 m (Earth's radius) at a constant speed of 465 m/s due to Earth's rotation. (a) What is the magnitude of the centripetal force on the standard kilogram during the rotation? Imagine that the standard kilogram hangs from a spring balance at that location and assume that it would weigh exactly 9.80 N if Earth did not rotate. (b) What is the reading on the spring balance; that is, what is the magnitude of the force on the spring balance from the standard kilogram?

52. In about 1915, Henry Sincosky of Philadelphia suspended himself from a rafter by gripping the rafter with the thumb of each hand on one side and the fingers on the other side (Fig. 6-50). Sincosky's mass was 79 kg. If the coefficient of static friction between hand and rafter was 0.70, what was the least magnitude of the normal force on the rafter from each thumb or opposite fingers? (After suspending himself, Sincosky chinned himself on the rafter and then moved hand-over-hand along the rafter. If you do not think Sincosky's grip was remarkable, try to repeat his stunt.)

Fig. 6-50 Problem 52.

53. Assume that Eq. 6-14 gives the drag force on a pilot and an ejection seat just after they are ejected from a jet airplane traveling horizontally at 1300 km/h. Assume also that the mass of the seat is equal to the mass of the pilot and that the drag coefficient is that of a sky diver (use Table 6-1 to get the coefficient). Making a reasonable guess of the pilot's mass, estimate the magnitudes of (a) the drag force on the pilot and seat and (b) their horizontal deceleration (in terms of g-units), both just after ejection. (The result of (a) should indicate an engineering requirement: The seat must include a protective barrier to deflect the initial wind blast away from the pilot's head.)

54. From the data in Table 6-1, deduce the diameter of the 16 lb shot. Assume that $C = 0.49$ and the density of air is 1.2 kg/m^3.

55. The three blocks in Fig. 6-51 are released from rest and accelerate at the rate of 1.5 m/s^2. If $M = 2.0$ kg, what is the magnitude of the frictional force on the block that slides horizontally?

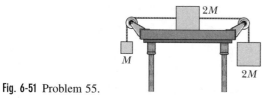

Fig. 6-51 Problem 55.

56. A four-person bobsled (total mass = 630 kg) comes down a straightaway at the start of a bobsled run. The straightaway is 80.0 m long and is inclined at a constant angle of 10.2° with the horizontal. Assume that the combined effects of friction and air resistance produce a constant force on the bobsled of 62.0 N that

acts parallel to the incline and up the incline. Answer the following questions to three significant digits. (a) If the magnitude of the velocity of the bobsled at the start of the run is 6.20 m/s, how long does the bobsled take to come down the straightaway? (b) Suppose through practice the crew is able to reduce the effects of friction and air resistance to a constant force of 42.0 N on the bobsled. For the same initial velocity, how long does the bobsled now take to come down the straightaway?

57. A car is rounding a flat curve of radius $R = 220$ m at the curve's maximum design speed $v = 94.0$ km/h. What is the magnitude of the *net* force on the seat cushion from a passenger with mass $m = 85.0$ kg?

58. Some years ago, the bookstore at Georgia Tech was in the basement of a building. As boxes were unloaded from delivery trucks at ground level, they were allowed to slide down a wooden ramp to the basement area. A strobe photograph of a box sliding down the 2.5-m-high ramp is represented in Fig. 6-52. Assume that the box is released from rest at the top of the ramp at $t = 0$ and slides with constant acceleration. The flashes on the strobe camera record the position of the box at 0.5 s intervals. (a) What is the magnitude of the acceleration of the box? (b) If the box weighs 240 N, what is the magnitude of the force of kinetic friction on the box from the ramp?

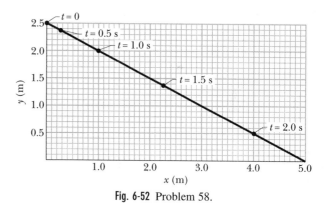

Fig. 6-52 Problem 58.

59. A ski that is placed on snow will stick to the snow. However, when the ski is moved along the snow, the rubbing warms and partially melts the snow, reducing the coefficient of kinetic friction and promoting sliding. Waxing the ski makes it water repellent and reduces friction with the resulting layer of water. A magazine reports that a new type of plastic ski is especially water repellent and that, on a gentle 200 m slope in the Alps, a skier reduced his top-to-bottom time from 61 s with standard skis to 42 s with the new skis. (a) Determine the magnitude of his average acceleration with each pair of skis. (b) Assuming a 3.0° slope, compute the coefficient of kinetic friction for each case.

60. A 1.5 kg box is initially at rest on a horizontal surface when at $t = 0$ a horizontal force, $\vec{F} = (1.8t)\hat{\imath}$ N (with t in seconds), is applied to the box. The acceleration of the box as a function of time t is given by $\vec{a} = 0$ for $0 \le t \le 2.8$ s and $\vec{a} = (1.2t - 2.4)\hat{\imath}$ m/s^2 for $t > 2.8$ s. (a) What is the coefficient of static friction between the box and the surface? (b) What is the coefficient of kinetic friction between the box and the surface?

61. A 26 kg child on a steadily rotating Ferris wheel moves at a constant speed of 5.5 m/s around a vertical circular path of radius 12 m. What are the magnitude and direction of the force of the seat on the child (a) at the highest point of the path and (b) at the lowest point of the path?

62. A car weighing 10.7 kN and traveling at 13.4 m/s attempts to round an unbanked curve with a radius of 61.0 m. (a) What force of friction is required to keep the car on its circular path? (b) If the coefficient of static friction between the tires and the road is 0.35, is the attempt at taking the curve successful?

63. A child places a picnic basket on the outer rim of a merry-go-round that has a radius of 4.6 m and revolves once every 30 s. (a) What is the speed of a point on that rim? (b) What least coefficient of static friction between the basket and the merry-go-round will allow the basket to stay on the ride?

64. A stuntman drives a car over the top of a hill, the cross section of which can be approximated by a circle of radius 250 m, as in Fig. 6-53. What is the greatest speed at which he can drive without the car leaving the road at the top of the hill?

250 m

Fig. 6-53 Problem 64.

65. A student, crazed by final exams, uses a force $\vec{P}$ of magnitude 80 N to push a 5.0 kg block across the ceiling of his room, as shown in Fig. 6-54. If the coefficient of kinetic friction between the block and the ceiling is 0.40, what is the magnitude of the acceleration of the block?

$\vec{P}$

70°

Fig. 6-54 Problem 65.

66. A warehouse worker exerts a constant horizontal force of magnitude 85 N on a 40 kg box that is initially at rest on the horizontal floor of the warehouse. When the box has moved a distance of 1.4 m, its speed is 1.0 m/s. What is the coefficient of kinetic friction between the box and the floor?

67. A child weighing 140 N sits at rest at the top of a playground slide that makes an angle of 25° with the horizontal. The child keeps from sliding by holding onto the sides of the slide. After letting go of the sides, the child has a constant acceleration of 0.86 m/s² (down the slide, of course). (a) What is the coefficient of kinetic friction between the child and the slide? (b) What maximum and minimum values for the coefficient of static friction between the child and the slide are consistent with the information given here?

68. A certain string can withstand a maximum tension of 40 N without breaking. A child ties a 0.37 kg stone to one end and, holding the other end, whirls the stone in a vertical circle of radius 0.91 m, slowly increasing the speed until the string breaks. (a) Where is the stone on its path when the string breaks? (b) What is the speed of the stone as the string breaks?

69. Suppose that the space station in Question 14 has a radius of 500 m. (a) If a crew member weighs 600 N on Earth, what must be the speed v_s of the outer wall of the station for that crew member to have an apparent weight of 300 N when standing near the outer wall? (b) What is the apparent weight of the crew member if she sprints along the outer wall at 10 m/s (relative to the outer wall) in the same direction as $\vec{v}_s$?

70. A banked circular highway curve is designed for traffic moving at 60 km/h. The radius of the curve is 200 m. Traffic is moving along the highway at 40 km/h on a rainy day. What is the minimum coefficient of friction between tires and road that will allow cars to take the turn without sliding off the road?

71. A conical pendulum (see Fig. 6-42) is formed by attaching a 50 g bob to a 1.2 m string of neligible mass. The bob swings around a horizontal circle of radius 25 cm. What are its (a) speed and (b) acceleration? (c) What is the tension in the string?

72. A 100 N force, directed at an angle θ above a horizontal floor, is applied to a 25.0 kg chair sitting on the floor. If $\theta = 0°$, what are (a) the horizontal component F_h of the applied force and (b) the magnitude N of the normal force of the floor on the chair? If $\theta = 30.0°$, what are (c) F_h and (d) N? If $\theta = 60.0°$, what are (e) F_h and (f) N? Now assume that the coefficient of static friction between chair and floor is 0.420. Does the chair slide or remain at rest if θ is (g) 0°, (h) 30.0° and (i) 60.0°?

73. A trunk with a weight of 220 N rests on the floor. The coefficient of static friction between the trunk and the floor is 0.41, and the coefficient of kinetic friction is 0.32. (a) What is the magnitude of the minimum horizontal force with which a person must push on the trunk to start it moving? (b) Once the trunk is moving, what magnitude of horizontal force must the person apply to keep it moving with constant velocity? (c) If the person continued to push with the force used to start the motion, what would be the magnitude of the trunk's acceleration?

74. As a 40 N block slides down a plane that is inclined at 25° to the horizontal, its acceleration is 0.80 m/s², directed up the plane. What is the coefficient of kinetic friction between the block and the plane?

75. A toy chest and its contents weigh 180 N. The coefficient of static friction between the toy chest and the floor is 0.42. The child in Fig. 6-55 attempts to move the chest across the floor by pulling on an attached rope. (a) If the angle θ that the rope makes with the horizontal is 42°, what is the magnitude of the force $\vec{F}$ that the child must exert on the rope to put the chest on the verge of moving? (b) Write an expression for the magnitude of the force $\vec{F}$ required to put the chest on the verge of moving as a function of the angle

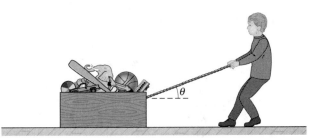

θ

Fig. 6-55 Problem 75.

θ. Determine (c) the value of θ for which the magnitude of $\vec{F}$ is a minimum and (d) that minimum magnitude.

76. An 11 kg block of steel is at rest on a horizontal table. The coefficient of static friction between block and table is 0.52. (a) What is the magnitude of the horizontal force that will just start the block moving? (b) What is the magnitude of a force acting upward 60° from the horizontal that will just start the block moving? (c) If the force acts down at 60° from the horizontal, how large can its magnitude be without causing the block to move?

77. A 5.0 kg block on an inclined plane is acted on by a horizontal force $\vec{F}$ with magnitude 50 N (Fig. 6-56). The coefficient of kinetic friction between block and plane is 0.30. The coefficient of static friction is not given (but you might still know something about it). (a) What is the acceleration of the block if it is moving up the plane? (b) With the horizontal force still acting, how far up the plane does the block go if it has an initial upward speed of 4.0 m/s? (c) What happens to the block after it reaches the highest point? Give an argument to back your answer.

Fig. 6-56 Problem 77.

78. *Left alone, write your own.* For one or more of the following situations, write a problem involving the physics in this chapter, using the style of the sample problems, providing realistic data, graphs of the variables, and explained solutions: (a) sliding large objects over a surface, (b) cars sliding with locked brakes, (c) traction in car or dragster racing, (d) air drag on a blunt object moving through the air, (e) objects moving in uniform circular motion.

CLUSTERED PROBLEMS

79. In Fig. 6-57a, a 10 kg block remains stationary on a horizontal surface when a horizontal force of magnitude 25 N is applied to it. The coefficient of static friction between the block and the surface is 0.4. What are the magnitude and direction of the frictional force on the block due to the surface?

80. In Fig. 6-57b, a 50 N force is to be applied horizontally to a 10 kg block on a horizontal surface while a second force, of magnitude F, is to be applied directly downward on the block. If the

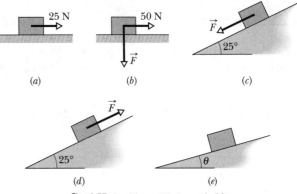

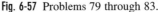

Fig. 6-57 Problems 79 through 83.

coefficient of static friction between the block and the surface is 0.4, what is the least value of F that will keep the block stationary?

81. In Fig. 6-57c, a 10 kg block is stationary on a plane inclined at 25° to the horizontal. The coefficients of static friction and kinetic friction between the block and the plane are 0.60 and 0.20, respectively. A force of magnitude F is to be applied to the block downward along the plane just long enough to start the block moving in that direction. What are (a) the maximum magnitude of the static frictional force on the block, (b) the magnitude of the component of the gravitational force on the block, along the plane, and (c) the least value of F required to start the motion? (d) What are the magnitude and direction of the block's acceleration after the applied force is removed?

82. In Fig. 6-57d, a 10 kg block is stationary on a plane inclined at 25° to the horizontal. The coefficients of static friction and kinetic friction between the block and the plane are 0.60 and 0.20, respectively. A force of magnitude F is to be applied to the block upward along the plane. (a) What is the least value of F required to make the block move up along the plane? What are the magnitude and direction of the block's acceleration if (b) the applied force is maintained on the block and (c) it is removed?

83. In Fig. 6-57e, a 10 kg block is stationary on a plane inclined at angle θ to the horizontal. The coefficients of static friction and kinetic friction between the block and the plane are 0.60 and 0.20, respectively. (a) With $\theta = 15°$ and the block stationary, what are the magnitude and direction of the frictional force acting on the block? (b) If θ is then increased, at what angle does the block begin to slide? (c) Should θ then be increased or decreased if the sliding is to be at constant speed? (d) What θ is required for that constant-speed sliding?

Kinetic Energy and Work

Sample Problem 7-11

Figure 7-34 shows an overhead view of a puck on a frictionless horizontal surface. Three constant horizontal forces act on the puck in the directions indicated. The magnitude of $\vec{F}_1$ is 10.0 N, that of $\vec{F}_2$ is 15.0 N, and that of $\vec{F}_3$ is 12.0 N. The puck starts from rest. What is the net work W done on the puck by the three forces when the puck has gone through a displacement of magnitude $d = 0.400$ m?

SOLUTION: We have three **Key Ideas** here. First, the net work W done by these three forces is equal to the work done by their net force $\vec{F}_{net}$. Second, because the forces are constant, so is $\vec{F}_{net}$. Third, we can treat the puck as a particle. With these three ideas, we can use Eq. 7-7 to write the net work as

$$W = F_{net}d \cos \phi, \qquad (7\text{-}49)$$

where ϕ is the angle between the directions of the net force and the displacement. The net force is given by

$$\vec{F}_{net} = \vec{F}_1 + \vec{F}_2 + \vec{F}_3. \qquad (7\text{-}50)$$

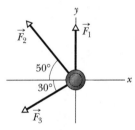

Fig. 7-34 Sample Problem 7-11. Overhead view of a puck, with three constant forces accelerating it across a frictionless horizontal surface.

Because we know the magnitudes and directions of all three applied forces, we could find $\vec{F}_{net}$ by evaluating the right side of this equation directly on a vector-capable calculator. Instead, here we shall rewrite Eq. 7-50 in component form and combine components to find $\vec{F}_{net}$. Along the x axis we have

$$F_{net,x} = 0 - (15.0 \text{ N}) \cos 50.0° - (12.0 \text{ N}) \cos 30.0°$$
$$= -20.0 \text{ N},$$

and along the y axis we have

$$F_{net,y} = (10.0 \text{ N}) + (15.0 \text{ N}) \sin 50.0° - (12.0 \text{ N}) \sin 30.0°$$
$$= 15.5 \text{ N}.$$

Combining these components, we find that $\vec{F}_{net}$ has the following magnitude and angle (from the positive direction of the x axis):

$$F_{net} = \sqrt{F_{net,x}^2 + F_{net,y}^2} = 25.3 \text{ N}$$

and

$$\theta = \tan^{-1} \frac{F_{net,y}}{F_{net,x}} = -38°.$$

However, from Fig. 7-34 we see that this value of $\theta = -38°$ is unreasonable; to get the correct value, we add 180° to $-38°$. Thus, we have $\theta = 142°$.

To evaluate the angle ϕ in Eq. 7-49, we use the following **Key Idea**: Because the puck starts from rest, it must move in the direction of the net force, namely at the angle $\theta = 142°$. Therefore, the angle ϕ between the directions of the net force and the displacement is 0°. We can now evaluate Eq. 7-49 as

$$W = (25.3 \text{ N})(0.400 \text{ m}) \cos 0°$$
$$= 10.1 \text{ J}. \qquad \text{(Answer)}$$

Tactic 1: *Integrals and Areas*
If you know a function $y = F(x)$, you can find the value of its integral between any two values of x from the rules of calculus. If you do not know the function analytically but have a plot of it, you can find the value of its integral graphically.

As an example, consider the plot of a particular force function $F(x)$ in Fig. 7-35. Let us calculate graphically the work W done by this force as the particle on which it acts moves from $x_i = 2.0$ cm

to $x_f = 5.0$ cm. According to Eq. 7-32, the work is

$$W = \int_{x_i}^{x_f} F(x) \, dx,$$

which is equal to the (total) shaded area shown under the curve between the two points.

You can approximate this area with a rectangle formed by drawing a horizontal line across Fig. 7-35. Draw it at a level such

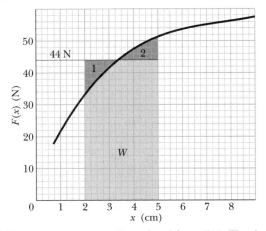

Fig. 7-35 The graph of a one-dimensional force $F(x)$. The shaded area under the curve (which represents the work done by F) is approximated by a rectangle formed by excluding area 2 and including (approximately equal) area 1.

that the areas marked "1" and "2" appear to be equal. A line at $F = 44$ N is about right, and the area of the equivalent rectangle ($= W$) is then

$$W = \text{height} \times \text{base} = (44\text{ N})(5.0\text{ cm} - 2.0\text{ cm})$$
$$= 132\text{ N} \cdot \text{cm} \approx 1.3\text{ N} \cdot \text{m} = 1.3\text{ J}.$$

You can also find the area by counting the small squares underneath the curve. The shaded area contains about 260 squares, and each square represents $(2\text{ N})(0.25\text{ cm}) = 0.5\text{ N} \cdot \text{cm}$. The work is then

$$W = (260\text{ squares})\left(\frac{0.5\text{ N} \cdot \text{cm}}{1\text{ square}}\right) = 130\text{ N} \cdot \text{cm}$$
$$= 1.3\text{ J},$$

just as above.

Sample Problem 7-12

A 2.0 kg pebble moves along an x axis on a horizontal frictionless surface, accelerated by a force $F(x)$ that varies with the pebble's position as shown in Fig. 7-36.

(a) How much work is done on the pebble by the force as the pebble moves from its initial point at $x = 0$ to $x = 5$ m?

SOLUTION: A **Key Idea** is that the work done by a one-dimensional force is given by Eq. 7-32:

$$W = \int_{x_i}^{x_f} F(x)\, dx.$$

Here the limits are $x_i = 0$ and $x_f = 5$ m, and $F(x)$ is given by Fig. 7-36. A second **Key Idea** is that we can easily evaluate the integral graphically from Fig. 7-36. To do so, we find the area between the plot of $F(x)$ and the x axis, between the limits $x_i = 0$ and $x_f = 5$ m. Note that we can split that area into three parts: a right triangle at the left (from $x = 0$ to $x = 2$ m), a central rectangle (from $x = 2$ m to $x = 4$ m), and a triangle at the right (from $x = 4$ m to $x = 5$ m).

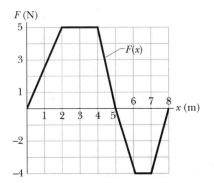

Fig. 7-36 Sample Problem 7-12. The graph of a one-dimensional force $F(x)$ that accelerates a pebble along an x axis.

Recall that the area of a triangle is $\frac{1}{2}(\text{base})(\text{height})$. The work $W_{0,5}$ that is done on the pebble from $x = 0$ and $x = 5$ m is then

$$W_{0,5} = \frac{1}{2}(2\text{ m})(5\text{ N}) + (2\text{ m})(5\text{ N}) + \frac{1}{2}(1\text{ m})(5\text{ m})$$
$$= 17.5\text{ J}. \qquad \text{(Answer)}$$

(b) The pebble starts from rest at $x = 0$. What is its speed at $x = 8$ m?

SOLUTION: A **Key Idea** here is that the pebble's speed is related to its kinetic energy, and its kinetic energy is changed because of the work done on the pebble by the force. Because the pebble is initially at rest, its initial kinetic energy K_i is 0. If we write its final kinetic energy at $x = 8$ m as $K_f = \frac{1}{2}mv^2$, then we can write the work–kinetic energy theorem of Eq. 7-11 ($K_f = K_i + W$) as

$$\frac{1}{2}mv^2 = 0 + W_{0,8}, \qquad (7\text{-}51)$$

where $W_{0,8}$ is the work done on the pebble from $x = 0$ to $x = 8$ m.

A second **Key Idea** is that, as in part (a), we can find the work graphically from Fig. 7-36 by finding the area between the plotted curve and the x axis. However, we must be careful about signs: We must take an area to be positive when the plotted curve is above the x axis and negative when it is below the x axis. We already know that work $W_{0,5} = 17.5$ J, so completing the calculation of area gives us

$$W_{0,8} = W_{0,5} + W_{5,8}$$
$$= 17.5\text{ J} - \frac{1}{2}(1\text{ m})(4\text{ N}) - (1\text{ m})(4\text{ N}) - \frac{1}{2}(1\text{ m})(4\text{ N})$$
$$= 9.5\text{ J}.$$

Substituting this and $m = 2.0$ kg into Eq. 7-51 and solving for v, we find

$$v = 3.1\text{ m/s}. \qquad \text{(Answer)}$$

Sample Problem 7-13

A horizontal cable accelerates a suspicious package across a frictionless horizontal floor. The work $W(t)$ done by the cable's force on the package is given by $W(t) = 0.20t^2$, with W in joules and time t in seconds.

(a) What is the average power P_{avg} due to the cable's force in the time interval $t = 0$ to $t = 10$ s?

SOLUTION: The **Key Idea** here is that the average power P_{avg} is the ratio of the amount of work W done in the given time interval to that time interval (Eq. 7-42). To find the work W, we evaluate $W(t)$ at $t = 0$ and $t = 10$ s. At those times, the cable has done work W_0 and W_{10}, respectively:

$$W_0 = 0.20(0)^2 = 0 \text{ J} \quad \text{and} \quad W_{10} = 0.20(10)^2 = 20 \text{ J}.$$

Therefore, in the 10 s interval, the work done is $W_{10} - W_0 = 20$ J. Equation 7-42 then gives us

$$P_{avg} = \frac{W}{\Delta t} = \frac{20 \text{ J}}{10 \text{ s}} = 2.0 \text{ W}. \qquad \text{(Answer)}$$

Thus, during the 10 s interval, the cable does work at the average rate of 2.0 joules per second.

(b) What is the instantaneous power P due to the cable's force at $t = 3.0$ s, and is P then increasing or decreasing?

SOLUTION: The **Key Idea** here is that the instantaneous power P at $t = 3.0$ s is the time derivative of the work dW/dt evaluated at $t = 3.0$ s (Eq. 7-43). Taking the derivative of $W(t)$ gives us

$$P = \frac{dW}{dt} = \frac{d}{dt}(0.20t^2) = 0.40t,$$

with P in watts. This result tells us that as time t increases, so does P. Evaluating P for $t = 3.0$ s, we find

$$P = (0.40)(3.0) = 1.20 \text{ W}. \qquad \text{(Answer)}$$

Thus, at $t = 3.0$ s, the cable is doing work at the rate of 1.20 joules per second, and that rate is increasing.

QUESTIONS

14. Figure 7-37 gives four plots of position versus time of a cat carrier set in motion along an x axis on a frictionless floor by an applied force. Plots 1 and 2 are symmetric about the time axis with plots 4 and 3, respectively. Rank the plots according to the kinetic energy of the carrier, greatest first.

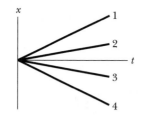

Fig. 7-37 Question 14.

15. In four situations, a briefly applied horizontal force changes the velocity of a hockey puck that slides over frictionless ice. The overhead views of Fig. 7-38 indicate, for each situation, the puck's initial speed v_i, its final speed v_f, and the directions of the corresponding velocity vectors. Rank the situations according to the work done on the puck by the applied force, most positive first and most negative last.

16. Here, for three situations, is the force acting on a particle and the (instantaneous) velocity of the particle: (a) $\vec{F} = -2\hat{i}$, $\vec{v} = -3\hat{i}$; (b) $\vec{F} = 3\hat{i}$, $\vec{v} = 2\hat{i}$; (c) $\vec{F} = 4\hat{i}$, $\vec{v} = -2\hat{i}$; (d) $\vec{F} = 3\hat{i}$, $\vec{v} = 4\hat{j}$. Rank the situations according to the rate at which the force is

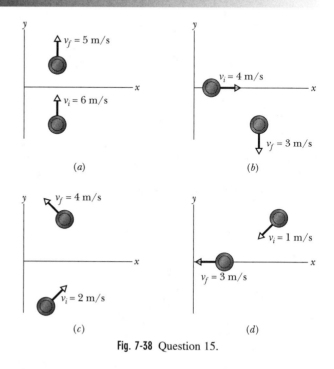

(a)   (b)

(c)   (d)

Fig. 7-38 Question 15.

transferring energy to or from the particle at that instant, with the greatest rate *to* the particle first and the greatest rate *from* the particle last.

17. *Organizing question:* Figure 7-39 shows three situations in which a 2 kg box moves downward through a distance of 4 m while forces $\vec{F}_1$ and $\vec{F}_2$ are applied to it. The magnitudes of those forces are $F_1 = 10$ N and $F_2 = 2$ N. In each situation, the box begins with a speed $v_i = 3$ m/s. For each situation, set up an equation, complete with known data, for the kinetic energy K_f of the box at the end of the 4 m.

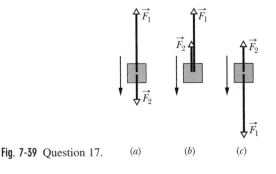

Fig. 7-39 Question 17.   (a)   (b)   (c)

18. Figure 7-40 shows five situations in which this book is pulled by a force along a frictionless surface through a distance of 4 m. In Fig. 7-40a the force is directed along the surface and its magnitude is a constant 3 N. In Fig. 7-40b the force is directed along the surface and its magnitude varies with time t as (3 N/s)t. In Fig. 7-40c the force is directed along the surface and its magnitude varies with the distance x along the surface as (3 N/m)x. In Fig. 7-40d the force is directed at a constant angle $\phi = 0.3$ rad relative to the surface and its magnitude is a constant 3 N. In Fig. 7-40e the force has a constant magnitude of 3 N but its angle ϕ relative to the surface varies with time as (0.3 rad/s)t. In which of the situations can we find the work done on the book by the force with (a) Eq. 7-7 and (b) Eq. 7-32?

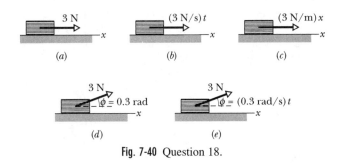

Fig. 7-40 Question 18.

19. *Organizing question:* Figure 7-41 shows four situations in which a 2 kg box slides over a frictionless floor while forces $\vec{F}_1$ and $\vec{F}_2$ are applied to it. The magnitudes of those forces are $F_1 = 10$ N and $F_2 = 2$ N. In each situation, the box begins with speed $v_i = 3$ m/s and slides through a distance of 4 m in the positive direction along an x axis. For each situation, set up an equation, complete with known data, for the kinetic energy K_f of the box at the end of the 4 m.

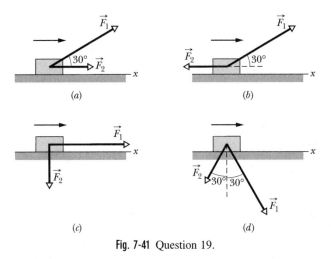

Fig. 7-41 Question 19.

20. A glob of slime is launched or dropped from the edge of a cliff. Which of the graphs in Fig. 7-42 could possibly show how the kinetic energy of the glob changes during its flight?

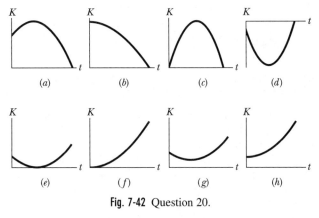

Fig. 7-42 Question 20.

21. A force $\vec{F}$ that is directed along an x axis moves a particle 5 m along that axis. Can we use Eqs. 7-7 and 7-8 to find the work done if the magnitude (in newtons) of $\vec{F}$ is given by (a) $F = 3$, (b) $F = 2x$, and (c) $F = 2t$?

22. If you cut a spring in half, the spring constant of either half is what multiple of the spring constant of the original spring? (*Hint:* Consider the amount by which each half stretches for a given applied force.)

23. In Fig. 7-43, an initially stationary load of *Rolling Stone* magazines is lifted through (and then past) a height h via a cable in three ways. In Fig. 7-43a, the load is accelerated to a speed of 3 m/s through the first $h/3$ and then moved at that speed through the rest of the distance. In Fig. 7-43b it is accelerated to 3 m/s through the first $h/2$ and then moved at 3 m/s through the rest of the distance. In Fig. 7-43c the acceleration to 3 m/s takes the full height h. Rank the three ways of lifting according to (a) the work done on the load by the cable's force during the lift of height h and (b) the rate at which that work is done, greatest first.

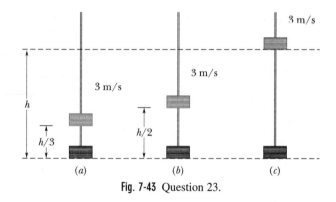

Fig. 7-43 Question 23.

24. Figure 7-44 shows three arrangements of a block attached to identical springs that are in their relaxed state when the block is centered as shown. Rank the arrangements according to the magnitude of the net force on the block, largest first, when the block is displaced by distance d (a) to the right and (b) to the left. Rank the arrangements according to the work done on the block by the spring forces, greatest first, when the block is displaced by d (c) to the right and (d) to the left.

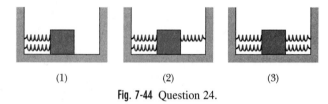

Fig. 7-44 Question 24.

25. Figure 7-45 shows four situations in which the same can of corned beef hash is pulled by an applied force $\vec{F}$ up a frictionless ramp through (and then past) the same vertical distance h. In each situation that force has a magnitude of 10 N. In situations 2 and 4, the force is directed along the plane; in situations 1 and 3, it is directed at an angle $\theta = 20°$ to the plane, as drawn. Rank the situations according to the work done on the can in the vertical distance h by (a) the applied force and (b) the gravitational force on the can, greatest first.

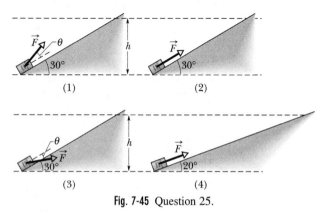

Fig. 7-45 Question 25.

26. Figure 7-46 gives the velocity v versus time t of a carriage being moved along an axis by an applied force. The time axis shows five time periods, with $\Delta t_2 = \Delta t_3 = \Delta t_4 = \Delta t_5 = 2\Delta t_1$. Rank the time periods according to (a) the work done by that applied force during

that time period and (b) the rate at which the work is done, greatest first.

Fig. 7-46 Question 26.

DISCUSSION QUESTIONS

27. What other words are like "work" in that their colloquial meanings differ from their scientific meanings?

28. Why is it tiring to hold a heavy object even though no work is done?

29. The inclined plane is a simple "machine" that enables us to do work with the application of a smaller force than is otherwise necessary. The same statement applies to a wedge, a lever, a screw, a gear wheel, and a set of pulleys. Far from saving us work, such machines in practice require that we do a little more work with them than without them. Why is this so? Why do we use such machines?

30. In a tug-of-war, one team is slowly giving way to the other. Is work being done on the losing team? How about on the winning team?

31. Give a situation in which positive work is done by a static frictional force.

32. Suppose that Earth orbits the Sun in a perfect circle. Does the Sun do any work on Earth?

33. If you slowly lift a bowling ball from the floor, two forces act on the ball: the gravitational force $\vec{F}_g = m\vec{g}$ and your upward force $\vec{F} = -m\vec{g}$. These two forces cancel each other so that it would seem that no work is done. On the other hand, you know that you have done some work. What is wrong?

34. An ant must carry a food morsel to the top of a cone (Fig. 7-47). Compare the work it does on the food if it follows a spiraling path around the cone to the work it does by climbing directly up the side of the cone.

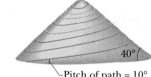

Fig. 7-47 Question 34.   —Pitch of path = 10°

35. In picking up a book from the floor and putting it on a table, you do work. However, the initial and final values of the book's kinetic energy are zero. Is there a violation of the work–kinetic energy theorem here? Explain why or why not.

36. You throw a ball vertically into the air and catch it when it returns. What happens to the ball's kinetic energy during the flight? First ignore air resistance, and then take it into account.

37. Does the power needed to raise a box onto a platform depend on how fast it is raised?

38. We have heard a lot about the "energy crisis." Would it be more accurate to speak of a "power crisis"?

39. The displacement of a body depends on the reference frame of the observer who measures it. It follows that the work done on a body should also depend on the observer's reference frame. Suppose you drag a crate across a rough floor by pulling on it with a rope. Identify reference frames in which the work done on the crate by the rope would be (a) positive, (b) zero, and (c) negative.

40. Sally and Yuri are flying jet planes at the same speed on parallel low-altitude paths. Suddenly, Sally lowers her flaps and slows to a new speed. Consider how this looks from the reference frame of Yuri, who keeps on flying at the original speed. (a) Would he say that Sally's plane had gained or lost kinetic energy? (b) Would he say that the work done *on* her plane is positive or negative? (c) Would he conclude that the work–kinetic energy theorem holds? (d) Answer these same questions from the reference frame of Chang, who watches from the ground.

EXERCISES & PROBLEMS

41. If a ski lift raises 100 passengers averaging 660 N in weight to a height of 150 m in 60 s, at constant speed, what average power is required of the force making the lift?

42. A can of sardines is made to move along an x axis from $x = 0.25$ m to $x = 1.25$ m by a force with a magnitude given by $F = \exp(-4x^2)$, with x in meters and F in newtons. (Here exp is the exponential function.) How much work is done on the can by the force?

43. In Fig. 7-10a, a block of mass m lies on a horizontal frictionless surface and is attached to one end of a horizontal spring (with spring constant k) whose other end is fixed. The block is initially at rest at the position where the spring is unstretched ($x = 0$) when a constant horizontal force $\vec{F}$ in the positive direction of the x axis is applied to it. A plot of the resulting kinetic energy of the block versus its position x is shown in Fig. 7-48. (a) What is the magnitude of $\vec{F}$? (b) What is the value of the spring constant k?

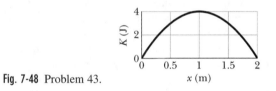

Fig. 7-48 Problem 43.

44. A constant force of magnitude 10 N makes an angle of 150° (measured counterclockwise) with the positive x direction as it acts on a 2.0 kg object moving in the xy plane. How much work is done on the object by the force as the object moves from the origin to the point with position vector $(2.0\ \text{m})\hat{i} - (4.0\ \text{m})\hat{j}$?

45. A 50 kg ice-skater is moving toward the east across a frictionless frozen lake with a constant velocity when the constant force from a sudden wind pushes the skater toward the west. A plot of position x versus time t for the skater's motion is shown in Fig. 7-49, where $t = 0$ is taken to be the instant the wind starts to blow and the positive direction of the x axis is toward the east. (a) What is the kinetic energy of the ice-skater at $t = 0$? (b) How much work does the force of the wind do on the ice-skater between $t = 0$ and $t = 3.0$ s?

46. A particle moves along a straight path through displacement $\vec{d} = (8\ \text{m})\hat{i} + c\hat{j}$ while force $\vec{F} = (2\ \text{N})\hat{i} - (4\ \text{N})\hat{j}$ acts on it. (Other forces also act on the particle.) What is the value of c if the work done by $\vec{F}$ on the particle is (a) zero, (b) positive, and (c) negative?

47. If a Saturn V rocket with an Apollo spacecraft attached

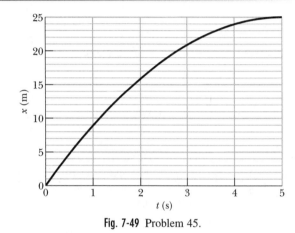

Fig. 7-49 Problem 45.

had a combined mass of 2.9×10^5 kg and reached a speed of 11.2 km/s, how much kinetic energy would it then have?

48. A spring with a pointer attached is hanging next to a scale marked in millimeters. Three different packages are hung from the spring, in turn, as shown in Fig. 7-50. (a) Which mark on the scale will the pointer indicate when no package is hung from the spring? (b) What is the weight W of the third package?

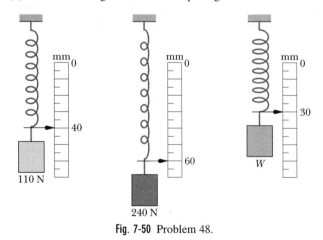

Fig. 7-50 Problem 48.

49. A 230 kg crate hangs from the end of a 12.0 m rope. You push horizontally on the crate with a varying force $\vec{F}$ to move it 4.00 m to the side (Fig. 7-51). (a) What is the magnitude of $\vec{F}$ when the crate is in this final position? During the crate's displacement, what are (b) the total work done on it, (c) the work done by the gravitational force on the crate, and (d) the work done by the pull on the crate from the rope? (e) Knowing that the crate is motionless before and after its displacement, use the answers to (b), (c), and (d) to find the work your force $\vec{F}$ does on the crate. (f) Why is the work of your force not equal to the product of the horizontal displacement and the answer to (a)?

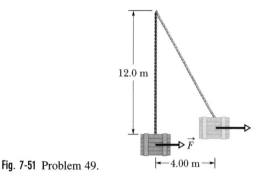

Fig. 7-51 Problem 49.

50. A frightened child is restrained by her mother as the child slides down a frictionless playground slide. If the mother exerts a force of 100 N up the slide on the child, the child's kinetic energy increases by 30 J as she moves down the slide a distance of 1.8 m. (a) How much work is done on the child by the gravitational force during the 1.8 m descent? (b) If the child is not restrained by her mother, how much will the child's kinetic energy increase as she comes down the slide that same distance of 1.8 m?

51. An iceboat is at rest on a frictionless frozen lake when a sudden wind exerts a constant force of 200 N, toward the east, on the boat. Due to the angle of the sail, the wind causes the boat to slide in a straight line for a distance of 8.0 m in a direction 20° north of east. What is the kinetic energy of the iceboat at the end of that 8.0 m?

52. In Fig. 7-52, two identical springs, each with a relaxed length of 50 cm and a spring constant of 500 N/m, are connected by a short cord of length 10 cm. The upper spring is attached to the ceiling; a box that weighs 100 N hangs from the lower spring. Two additional cords, each 85 cm long, are also tied to the assembly; they are limp. (a) If the short cord is cut, so that the box then hangs from the springs and the two longer cords, does the box move up or down from its initial location? (b) How far does the box move,

and (c) how much total work do the two spring forces (one directly, the other via a cord) do on the box during that move?

Fig. 7-52 Problem 52.

53. An explosion at ground level leaves a crater with a diameter that is proportional to the energy of the explosion raised to the $\frac{1}{3}$ power; an explosion of 1 megaton of TNT leaves a crater with a 1 km diameter. Below Lake Huron in Michigan there appears to be an ancient impact crater with a 50 km diameter. What was the kinetic energy associated with that impact, in terms of (a) megatons of TNT (1 megaton yields 4.2×10^{15} J) and (b) Hiroshima bomb equivalents (13 kilotons of TNT each)? (Ancient meteorite or comet impacts may have significantly altered Earth's climate and contributed to the extinction of the dinosaurs and other life-forms.)

54. A force $\vec{F} = (3.00 \text{ N})\hat{i} + (7.00 \text{ N})\hat{j} + (7.00 \text{ N})\hat{k}$ acts on a 2.00 kg mobile object that moves from an initial position of $\vec{d}_i = (3.00 \text{ m})\hat{i} - (2.00 \text{ m})\hat{j} + (5.00 \text{ m})\hat{k}$ to a final position of $\vec{d}_f = -(5.00 \text{ m})\hat{i} + (4.00 \text{ m})\hat{j} + (7.00 \text{ m})\hat{k}$ in 4.00 s. Find (a) the work done on the object by the force in the 4.00 s interval, (b) the average power due to the force during that interval, and (c) the angle between vectors $\vec{d}_i$ and $\vec{d}_f$.

55. *Left alone, write your own.* For one or more of the following situations, write a problem involving the physics in this chapter, using the style of the sample problems and providing realistic data, graphs of the variables, and explained solutions: (a) lifting or lowering a victim or a large object, (b) multiple forces moving an object over a frictionless surface or in space, (c) animals racing or chasing one another, (d) a system that includes a spring, such as in an inner-spring mattress.

8
Potential Energy and Conservation of Energy

Sample Problem 8-9

The spring of a spring gun is compressed a distance $d = 3.2$ cm from its relaxed state, and a ball of mass $m = 12$ g is put in the barrel. With what speed will the ball leave the barrel when the gun is fired while fixed in place? The spring constant k is 7.5 N/cm. Assume no friction and a horizontal gun barrel. Also assume that the ball leaves the spring and the spring stops when the spring reaches its relaxed length.

SOLUTION: A starting **Key Idea** is to examine all the forces acting on the ball as it is being propelled in the barrel, and then to determine whether we have an isolated system or a system on which an external force is doing work.

Forces: The spring pushes on the ball and on the opposite end of the barrel with a spring force, which transfers energy from elastic potential energy of the spring to kinetic energy of the ball. That is the only force doing work on the ball.

System: The ball–spring–barrel system includes the energy transfers in one isolated system. Therefore, a second **Key Idea** is that, because the system is isolated, its total energy cannot change. We can then apply the law of conservation of energy to the system.

The mechanical energy of the system is the sum of the ball's kinetic energy ($K = \frac{1}{2}mv^2$) and the spring's potential energy ($U = \frac{1}{2}kx^2$). Let the mechanical energy of the system be $E_{mec,1}$ before the gun is fired, and $E_{mec,2}$ as the ball leaves the barrel. Then we can write the principle of conservation of mechanical energy as

$$E_{mec,1} = E_{mec,2}$$

or as
$$U_1 + K_1 = U_2 + K_2. \qquad (8\text{-}40)$$

In state 1, the spring is compressed by distance d and the ball has a speed of zero. In state 2, the spring is not compressed and the ball has speed v. Thus, we may rewrite Eq. 8-40 as

$$\tfrac{1}{2}kd^2 + 0 = 0 + \tfrac{1}{2}mv^2.$$

Solving for v yields

$$v = d\sqrt{\frac{k}{m}} = (0.032 \text{ m})\sqrt{\frac{750 \text{ N/m}}{12 \times 10^{-3} \text{ kg}}}$$
$$= 8.0 \text{ m/s}. \qquad \text{(Answer)}$$

Sample Problem 8-10

Figure 8-55 shows a disabled robot of mass $m = 40$ kg being dragged by a cable up the 30° inclined wall inside a volcano crater. The force $\vec{F}_{cab}$ on the robot from the cable has a magnitude of 332 N. The kinetic frictional force $\vec{f}_k$ on the robot from the crater wall has a magnitude of 136 N. The robot moves through a displacement $\vec{d}$ of magnitude 0.50 m along the wall.

(a) What is the work W done by the cable's force in producing displacement $\vec{d}$, and on what system is that work done?

SOLUTION: Because the cable's force $\vec{F}_{cab}$ acts directly on the robot, it can change the robot's energy—so the robot is certainly part of the system. A **Key Idea** at this point is to identify all other forces acting on the robot, to help us identify the rest of the system.

Forces: Besides $\vec{F}_{cab}$, the gravitational force acts on the robot, with a resulting change in gravitational potential energy. There is also friction between the robot and the crater wall, so the sliding increases their thermal energy.

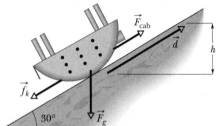

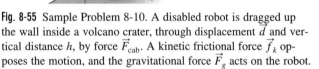

Fig. 8-55 Sample Problem 8-10. A disabled robot is dragged up the wall inside a volcano crater, through displacement $\vec{d}$ and vertical distance h, by force $\vec{F}_{cab}$. A kinetic frictional force $\vec{f}_k$ opposes the motion, and the gravitational force $\vec{F}_g$ acts on the robot.

System: The robot–wall–Earth system contains these energy changes, and force $\vec{F}_{cab}$ is an external force doing work W on the system. We can calculate W with Eq. 7-7 ($W = Fd \cos \phi$), where the angle ϕ between the directions of $\vec{F}_{cab}$ and $\vec{d}$ is 0°:

$$W = F_{cab}d \cos \phi = (332 \text{ N})(0.50 \text{ m}) \cos 0°$$
$$= 166 \text{ J}. \hspace{3cm} \text{(Answer)}$$

(b) What is the change ΔU in the gravitational potential energy of the system?

SOLUTION: The **Key Idea** here is that the change ΔU depends only on the change h in the elevation of the robot, not on the displacement $\vec{d}$. From Fig. 8-55, we see that $h = d \sin 30°$. Then, from Eq. 8-7 ($\Delta U = mg \, \Delta y$), we have

$$\Delta U = mgh = mgd \sin 30°$$
$$= (40 \text{ kg})(9.8 \text{ m/s}^2)(0.50 \text{ m}) \sin 30°$$
$$= 98 \text{ J}. \hspace{3cm} \text{(Answer)}$$

(c) What is the increase ΔE_{th} in thermal energy of the robot and crater wall because of the friction?

SOLUTION: The **Key Idea** here is that the increase ΔE_{th} is related to the magnitude f_k of the frictional force and the magnitude d of the robot's displacement (according to Eq. 8-29):

$$\Delta E_{th} = f_k d = (136 \text{ N})(0.50 \text{ m}) = 68 \text{ J}. \hspace{0.5cm} \text{(Answer)}$$

(d) What is the change ΔK in the kinetic energy of the robot?

SOLUTION: The **Key Idea** here is that we can apply the law of conservation of energy to our robot–wall–Earth system. In particular, we can relate the work W done on the system to the energy changes inside the system with Eq. 8-33:

$$W = \Delta E_{mec} + \Delta E_{th}.$$

Substituting $\Delta E_{mec} = \Delta K + \Delta U$ and solving for ΔK, we find

$$\Delta K = W - \Delta U - \Delta E_{th}$$
$$= 166 \text{ J} - 98 \text{ J} - 68 \text{ J} = 0. \hspace{1cm} \text{(Answer)}$$

The result tells us that the kinetic energy of the robot does not change; that is, the robot moves along the wall at constant speed. Force $\vec{F}_{cab}$ is transferring energy into the system but not to kinetic energy of the robot.

Sample Problem 8-11

In Fig. 8-56, a steel ball of mass $m = 5.2$ g that is fired vertically downward from height $H = 18$ m with initial speed $v_1 = 14$ m/s penetrates the sand below it to depth $D = 21$ cm. Assume that the force from the sand that slows and stops the ball is a constant kinetic frictional force.

(a) What is the change ΔE_{th} in the thermal energy of the ball and sand?

SOLUTION: A starting **Key Idea** is to examine all the forces on the ball, and from them to see if we have an isolated system or a system on which an external force is doing work.

Forces: Throughout the ball's downward displacement, the gravitational force does work on the ball. As the ball travels through the sand there is friction between the ball and sand, so the rubbing between them increases their thermal energies.

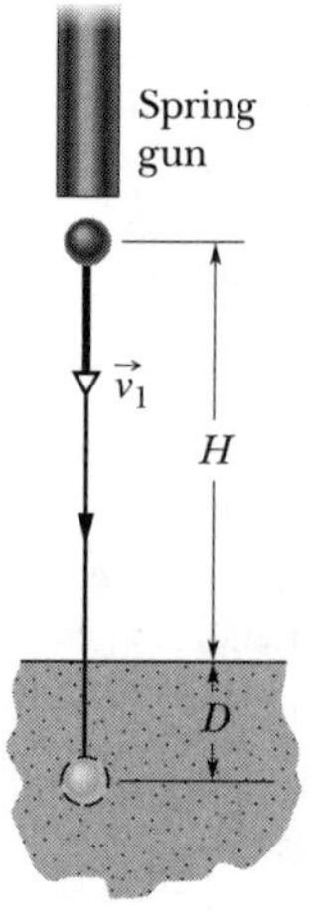

System: The ball–Earth–sand system contains these forces and energy changes, and that system is isolated. Thus, a second **Key Idea** is that, because the system is isolated, its total energy cannot change. We can then apply the law of conservation of energy in the form of Eq. 8-35 to this isolated system:

$$E_{mec,2} = E_{mec,1} - \Delta E_{th}, \hspace{1cm} (8\text{-}41)$$

where subscript 1 corresponds to the initial state of the system, when the ball is at height H, and subscript 2 corresponds to the final state with the ball buried at depth D. The mechanical energy of the system is the sum of the ball's kinetic energy ($K = \frac{1}{2}mv^2$) and the system's gravitational potential energy ($U = mgy$). Let us take the reference point $y = 0$ to be at the ball's lowest point, in the sand. Then for state 1, $K_1 = \frac{1}{2}mv_1^2$ and $U_1 = mg(H + D)$. Thus, we have

$$E_{mec,1} = K_1 + U_1 = \frac{1}{2}mv_1^2 + mg(H + D).$$

For state 2, $K_2 = 0$ (because the ball has stopped) and $U_2 = 0$ (because of our choice of reference point), so we have $E_{mec,2} = 0$. Substituting these expressions for mechanical energy into Eq. 8-41 then gives us

$$0 = \frac{1}{2}mv_1^2 + mg(H + D) - \Delta E_{th}.$$

Solving for ΔE_{th}, we find

$$\Delta E_{th} = \frac{1}{2}(5.2 \times 10^{-3} \text{ kg})(14 \text{ m/s})^2$$
$$+ (5.2 \times 10^{-3} \text{ kg})(9.8 \text{ m/s}^2)(18 \text{ m} + 0.21 \text{ m})$$
$$= 1.437 \text{ J} \approx 1.4 \text{ J}. \hspace{1cm} \text{(Answer)}$$

(b) What is the magnitude f_k of the force on the ball from the sand as the ball is being slowed to a stop?

Fig. 8-56 Sample Problem 8-11. A ball is fired downward with velocity $\vec{v}_1$ from height H and stops in sand at depth D.

SOLUTION: The **Key Idea** here is that the change ΔE_{th} in thermal energy is related to f_k by Eq. 8-29 ($\Delta E_{th} = f_k d$). Here the distance over which the frictional force acts is D. Solving Eq. 8-29 for f_k and substituting known values give us

$$f_k = \frac{\Delta E_{th}}{D} = \frac{1.437 \text{ J}}{0.21 \text{ m}} = 6.8 \text{ N.} \qquad \text{(Answer)}$$

QUESTIONS

10. In Question 6 and Fig. 8-20, what is the least mechanical energy the particle can have in regions (a) *BC*, (b) *DE*, and (c) *FG*?

11. *Organizing question:* Figure 8-57 shows 10 situations in which a block of mass m with initial speed v_i is brought to a stop. For each situation, what are the initial total mechanical energy $E_{mec,i}$ of the block, the change $\Delta E_{mec,d}$ in the block's mechanical energy due to a transfer to thermal energy, and the final total mechanical energy $E_{mec,f}$ of the block or, in some situations, of the block–spring system? In the 10 situations of Fig. 8-57, the block slides

(a) into a region of frictional force f, where it stops in distance d;

(b) up a frictionless ramp at angle θ until it (momentarily) stops at height h;

(c) up a ramp at angle θ against a frictional force f until it stops at height h;

(d) across a frictionless floor and against a spring of spring constant k, stopping (momentarily) when the spring is compressed by distance d;

(e) against a spring of spring constant k and simultaneously into a region of frictional force f, stopping when the spring is compressed by distance d;

(f) downward by height h onto a lower floor and into a region of frictional force f, stopping in that region in distance d;

(g) downward by height h onto a lower floor and then simultaneously against a spring of spring constant k and into a region of frictional force f, stopping when the spring is compressed by distance d;

(h) upward by height h onto a higher floor and then into a region of frictional force f, stopping in distance d in that region;

(i) upward by height h onto a higher floor and then against a spring of spring constant k, stopping when the spring is compressed by distance d; and

(j) upward by height h onto a higher floor and then simultaneously against a spring of spring constant k and into a region of frictional force f, stopping when the spring is compressed by distance d.

12. Figure 8-58 shows three plums that are launched from the same level with the same speed. One moves straight upward, one is launched at a small angle to the vertical, and one is launched along a frictionless incline. Rank the plums according to their speed when they reach the level of the dashed line, greatest first.

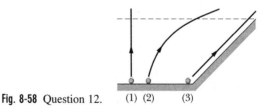

Fig. 8-58 Question 12.   (1) (2)   (3)

13. In Fig. 8-59, a horizontally moving block can take three frictionless routes, differing only in elevation, to reach the dashed finish line. Rank the routes according to (a) the speed of the block at the finish line and (b) the travel time of the block to the finish line, greatest first.

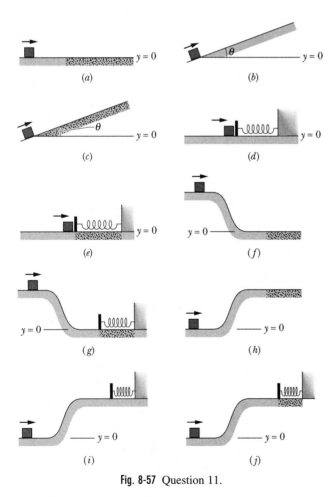

Fig. 8-57 Question 11.

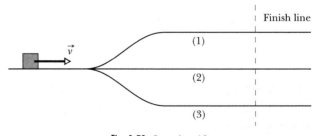

Fig. 8-59 Question 13.

14. A spring that lies along an x axis is attached to a wall at one end and a block at the other end. The block rests on a frictionless surface at $x = 0$, as in Fig. 7-10a. Then an applied force of constant magnitude begins to compress the spring, displacing the block by a distance x, until the block comes to a maximum displacement x_{max}. During this displacement, which of the curves in Fig. 8-60 best represents (a) the elastic potential energy of the spring, (b) the kinetic energy of the block, and (c) the work done on the spring–block system by the applied force? (d) What is the sum of curves 2 and 3? In what range of displacement does the block's kinetic energy (e) increase and (f) decrease? (g) At what displacement is the block's kinetic energy maximum? (h) What is the block's kinetic energy when the block reaches x_{max}? Does the applied force transfer more energy to the block's kinetic energy or the spring's potential energy during (i) the first half of the compression and (j) the second half?

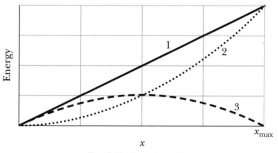

Fig. 8-60 Question 14.

15. When a particle moves from f to i and from j to i along the paths shown in Fig. 8-61, and in the indicated directions, a conservative force $\vec{F}$ does the indicated amounts of work on it. How much work is done on the particle by $\vec{F}$ when the particle moves directly from f to j?

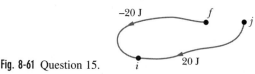

Fig. 8-61 Question 15.

16. Figure 8-62 shows three situations involving a plane that is not frictionless and a block sliding along the plane. The block begins with the same speed in all three situations and slides until the kinetic frictional force has stopped it. Rank the situations according to the increase in thermal energy due to the sliding, greatest first.

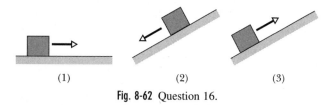

Fig. 8-62 Question 16.

DISCUSSION QUESTIONS

17. An automobile is moving along a highway. The driver jams on the brakes and the car skids to a halt, its kinetic energy decreasing to zero. What type of energy increases as a result of the action?

18. In Question 17, assume that the driver operates the brakes in such a way that there is no skidding or sliding. In this case, what type of energy increases?

19. You drop an object and observe that it bounces to half its original height. What conclusions can you draw? What if it bounces to 1.5 times its original height?

20. When an elevator descends from the top of a building and stops at the ground floor, what becomes of the energy that had been potential energy?

21. Why do mountain roads rarely go straight up the slope but instead wind up gradually?

22. Air bags greatly reduce the chance of injury in a car accident. Explain how they do so, in terms of energy transfers.

23. You see a duck flying by and declare it to have a certain amount of kinetic energy. However, another duck, who flies alongside the first one and who knows a bit of physics, declares it to have no kinetic energy at all. Who is right, you or the second duck? How does the law of conservation of energy fit into this situation?

24. An earthquake can release enough energy to devastate a city. Where does this energy "reside" an instant before the earthquake takes place?

25. Figure 8-63 shows a circular glass tube that is fastened to a vertical wall. The tube is filled with water except for an air bubble that is temporarily at rest at the bottom of the tube. Discuss the subsequent motion of the bubble in terms of energy transfers. Do so first while neglecting retarding forces and then while taking them into account.

Fig. 8-63 Question 25. ⎯Bubble

26. Give physical examples of unstable equilibrium, neutral equilibrium, and stable equilibrium.

27. In an article "Energy and the Automobile," which appeared in the October 1980 issue of *The Physics Teacher,* author Gene Waring states: "It is interesting to note that *all* the fuel input energy is eventually transformed to thermal energy and strung out along the car's path." Analyze the various mechanisms by which this might come about. Consider, for example, road friction, air resistance, braking, the car radio, the headlamps, the battery, internal engine and drive train losses, the horn, and so on. Assume a straight and level roadway.

28. Trace back to the Sun as many of our present energy sources as you can. Can you think of any that cannot be so traced?

29. Explain, using work and energy ideas, how you can pump a swing to make it go higher. If the swing is initially at rest, can you get it going by pumping it?

30. Two disks are connected by a stiff spring (Fig. 8-64). Can you press the upper disk down far enough so that when it is released it will spring back and raise the lower disk off the table? Can mechanical energy be conserved in such a case?

Fig. 8-64 Question 30.

31. Discuss the words "conservation of energy" as used (a) in this chapter and (b) in connection with an "energy crisis." How do these two usages differ?

32. The electric power for a small town is provided by a hydroelectric plant at a nearby river. If you turn off a lightbulb in this system, conservation of energy requires that an equal amount of energy, perhaps in another form, appears somewhere else in the system. Where and in what form does this energy appear?

33. A spring is compressed by tying its ends together tightly. It is then placed in acid and dissolves. What happens to its stored potential energy?

EXERCISES & PROBLEMS

68. A 20 kg block on a horizontal surface is attached to a horizontal spring of spring constant $k = 4.0$ kN/m. The block is pulled to the right so that the spring is extended 10 cm beyond its unstretched length, and the block is then released from rest. The frictional force between the sliding block and the surface has a magnitude of 80 N. (a) What is the kinetic energy of the block when it has moved 2.0 cm from its point of release? (b) What is the kinetic energy of the block when it first slides back through the point at which the spring is unstretched? (c) What is the maximum kinetic energy attained by the block as it slides from its point of release to the point at which the spring is unstretched?

69. Fasten one end of a vertical spring to a ceiling, attach a cabbage to the other end, and then slowly lower the cabbage until the upward force on it from the spring balances the gravitational force on it. Show that the loss of gravitational potential energy of the cabbage–Earth system equals twice the gain in the spring's potential energy. Why are these two quantities not equal?

70. A 68 kg sky diver falls at a constant terminal speed of 59 m/s. (a) At what rate is the gravitational potential energy of the Earth–sky diver system being reduced? (b) At what rate is the system's mechanical energy being reduced?

71. Each second, 1200 m³ of water passes over a waterfall 100 m high. Three-fourths of the kinetic energy gained by the water in falling is transferred to electrical energy by a hydroelectric generator. At what rate does the generator produce electrical energy? (The mass of 1 m³ of water is 1000 kg.)

72. A 50 kg trunk is pushed 6.0 m up along a 30° incline at constant speed by a constant horizontal force. The coefficient of kinetic friction between the trunk and the incline is 0.20. What are (a) the work done by the applied force and (b) the increase in the thermal energy of the trunk and incline?

73. Figure 8-65a shows a molecule consisting of two atoms of masses m and M (with $m \ll M$) and separation r. Figure 8-65b shows the potential energy $U(r)$ of the molecule as a function of r. Describe the motion of the atoms (a) if the total mechanical energy E of the two-atom system is greater than zero (as is E_1), and (b) if E is less than zero (as is E_2). For $E_1 = 1 \times 10^{-19}$ J and $r = 0.3$ nm, find (c) the potential energy of the system, (d) the total kinetic energy of the atoms, and (e) the force (magnitude and direction) acting on each atom. For what values of r is the force (f) repulsive, (g) attractive, and (h) zero?

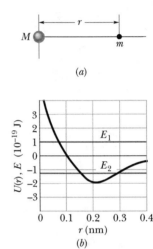

Fig. 8-65 Problem 73.

74. A skier weighing 600 N goes over a frictionless circular hill of radius 20 m (Fig. 8-66). Assume that the effects of air resistance on the skier are negligible. As she comes up the hill, her speed is 8.0 m/s at point B. (a) What is the skier's speed at the top of the hill (point A) if she coasts over the hill without using her poles? (b) What minimum speed can the skier have at point B and still coast to the top of the hill? (c) Do the answers to these two questions increase, decrease, or remain the same if the skier weighs, instead, 700 N?

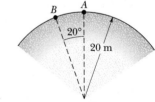

Fig. 8-66 Problem 74.

75. A spring with spring constant $k = 200$ N/m is suspended vertically with its upper end fixed to the ceiling and its lower end at position $y = 0$ (Fig. 8-67). A block of weight 20 N is attached to the lower end, held still for a moment, and then released. What are the kinetic energy K and the changes (from the initial values) in the gravitational potential energy ΔU_g and the elastic potential energy ΔU_e of the spring–block system when the block is at y values of (a) -5.0 cm, (b) -10 cm, (c) -15 cm, and (d) -20 cm?

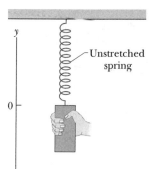

Fig. 8-67 Problem 75.

76. A 20 kg object is acted on by a conservative force given by $F = -3.0x - 5.0x^2$, with F in newtons and x in meters. Take the potential energy associated with the force to be zero when the object is at $x = 0$. (a) What is the potential energy of the system associated with the force when the object is at $x = 2.0$ m? (b) If the object has a velocity of 4.0 m/s in the negative direction of the x axis when it is at $x = 5.0$ m, what is its speed when it passes through the origin? (c) What are the answers to (a) and (b) if the potential energy of the system is taken to be -8.0 J when the object is at $x = 0$?

77. A 50 g ball is thrown from a window with an initial velocity of 8.0 m/s at an angle of 30° above the horizontal. Using energy methods, determine (a) the kinetic energy of the ball at the top of its flight and (b) its speed when it is 3.0 m below the window. Does the answer to (b) depend on either (c) the mass of the ball or (d) the initial angle?

78. Figure 8-68a applies to the spring in a cork gun (Fig. 8-68b); it shows the spring force as a function of the stretch or compression of the spring. The spring is compressed by 5.5 cm and used to propel a 3.8 g cork from the gun. (a) What is the speed of the cork if it is released as the spring passes through its relaxed position? (b) Suppose, instead, that the cork sticks to the spring and stretches it 1.5 cm before separation occurs. What now is the speed of the cork at the time of release?

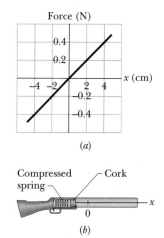

Fig. 8-68 Problem 78.

79. A 70.0 kg man jumping from a window lands in an elevated fire rescue net 11.0 m below the window. He momentarily stops when he has stretched the net by 1.50 m. Assuming that mechanical energy is conserved during this process and that the net functions like an ideal spring, find the elastic potential energy of the net when it is stretched by 1.50 m.

80. The only force acting on a particle is conservative force $\vec{F}$. If the particle is at point A, the potential energy of the system associated with $\vec{F}$ and the particle is 40 J. If the particle moves from point A to point B, the work done on the particle by $\vec{F}$ is $+25$ J. What is the potential energy of the system with the particle at B?

81. A 0.42 kg shuffleboard disk is initially at rest when a player uses a cue to increase its speed to 4.2 m/s at constant acceleration. The acceleration takes place over a 2.0 m distance, at the end of which the cue loses contact with the disk. Then the disk slides an additional 12 m before stopping. Assume that the shuffleboard court is level and that the force of friction on the disk is constant. What is the increase in the thermal energy of the disk–court system (a) for that additional 12 m and (b) for the entire 14 m distance? (c) How much work is done on the disk by the cue?

82. A 70 kg firefighter slides, from rest, 4.3 m down a vertical pole. (a) If the firefighter holds onto the pole lightly, so that the frictional force of the pole on her is negligible, what is her speed just before reaching the ground? (b) If the firefighter grasps the pole more firmly as she slides, so that the average frictional force of the pole on her is 500 N upward, what is her speed just before reaching the ground floor?

83. The surface of the continental United States has an area of about 8×10^6 km^2 and an average elevation of about 500 m (above sea level). The average yearly rainfall is 75 cm. The fraction of this rainwater that returns to the atmosphere by evaporation is $\frac{2}{3}$; the rest eventually flows into the ocean. If the decrease in gravitational potential energy of the water–Earth system associated with that flow could be fully converted to electrical energy, what would be the average power? (The mass of 1 m^3 of water is 1000 kg.)

84. A 0.63 kg ball, thrown directly upward with an initial speed of 14 m/s, reaches a maximum height of 8.1 m. What is the change in the mechanical energy of the ball–Earth system during the ascent of the ball to that maximum height?

85. A 1400 kg block of granite is pulled up an incline at a constant speed of 1.34 m/s by a cable and winch (Fig. 8-69). The coefficient of kinetic friction between the block and the incline is 0.40. What is the power due to the force applied to the block by the cable?

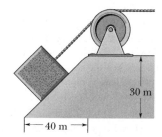

Fig. 8-69 Problem 85.

86. A 2.0 kg bread box on a frictionless 40° incline is connected, by a cord that runs over a pulley, to a light spring of spring constant $k = 120$ N/m, as shown in Fig. 8-70. The box is released from rest when the spring is unstretched. Assume that the pulley is massless and frictionless. (a) What is the speed of the box when it has moved 10 cm down the incline? (b) How far down the incline from its point of release does the box slide before momentarily stopping, and (c) what are the magnitude and direction of the acceleration of the box at the instant it momentarily stops?

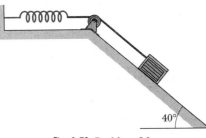

Fig. 8-70 Problem 86.

87. In a circus act, a 60 kg clown is shot from a cannon with an initial velocity of 16 m/s at some unknown angle above the horizontal. A short time later the clown lands in a net that is 3.9 m vertically above the clown's initial position. Disregard air drag. What is the kinetic energy of the clown as he lands in the net?

88. A 1.50 kg snowball is shot upward at an angle of 34.0° to the horizontal with an initial speed of 20.0 m/s. (a) What is its initial kinetic energy? (b) By how much does the gravitational potential energy of the snowball–Earth system change as the snowball moves from the launch point to the point of maximum height? (c) What is that maximum height?

89. In Fig. 8-71, the pulley is massless, and both it and the inclined plane are frictionless. If the blocks are released from rest with the connecting cord taut, what is their total kinetic energy when the 2.0 kg block has fallen 25 cm?

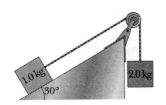

Fig. 8-71 Problem 89.

90. A 0.55 kg projectile is launched from the edge of a cliff with an initial kinetic energy of 1550 J and at its highest point is 140 m above the launch point. (a) What is the horizontal component of its velocity? (b) What was the vertical component of its velocity just after launch? (c) At one instant during its flight the vertical component of its velocity is 65 m/s. At that time, how far is it above or below the launch point?

91. A 1.50 kg water balloon is shot straight up with an initial speed of 3.00 m/s. (a) What is the kinetic energy of the balloon just as it is launched? (b) How much work does the gravitational force do on the balloon during the balloon's full ascent? (c) What is the change in the gravitational potential energy of the balloon–Earth system during the full ascent? (d) If the gravitational potential energy is taken to be zero at the launch point, what is its value when the balloon reaches its maximum height? (e) If, instead, the grav-

itational potential energy is taken to be zero at the maximum height, what is its value at the launch point? (f) What is the maximum height of the balloon?

92. A single conservative force $\vec{F} = (6.0x - 12)\hat{i}$ N, where x is in meters, acts on a particle moving along an x axis. The potential energy associated with this force is assigned a value of 27 J at $x = 0$. (a) Write an expression for the potential energy U as a function of x. (b) What is the maximum positive potential energy? (c) At what values of x is the potential energy equal to zero?

93. *Left alone, write your own.* For one or more of the following situations, write problems involving the physics in this chapter, using the style of the sample problems and providing realistic data, graphs of the variables, and explained solutions: (a) a roller coaster moves along track with sections of friction and sections without friction, (b) a spring launches an object into projectile motion, (c) a spring stops a free-fall amusement park ride, (d) a person slides down a rope, with and without acceleration.

CLUSTERED PROBLEMS

In problems involving friction, the coefficient of friction is usually given with only two significant figures, so answers are usually rounded off to two significant figures. However, such rounding off should be done only as a last step in a solution. Also, if the answer to, say, part (a) of a problem is needed in part (c), you should use the unrounded answer.

94. A 5.0 kg block is projected at 5.0 m/s up a plane that is inclined at 30° with the horizontal. How far up along the plane does the block go (a) if the plane is frictionless and (b) if the coefficient of kinetic friction between the block and the plane is 0.40? (c) In the latter case, what is the increase in thermal energy of block and plane during the block's ascent? (d) If the block then slides back down against the frictional force, what is the block's speed when it reaches the original projection point?

95. A 1500 kg car begins sliding down a 5.0° inclined road with a speed of 30 km/h. The engine is turned off, and the only forces acting on the car are a net frictional force from the road and the gravitational force. After traveling 50 m along the road, the car's speed is 40 km/h. (a) How much is the mechanical energy of the car reduced because of the net frictional force? (b) What is the magnitude of that net frictional force?

96. A 15 kg block is accelerated at 2.0 m/s² along a horizontal frictionless surface, with the speed increasing from 10 m/s to 30 m/s. What are (a) the change in the block's mechanical energy and (b) the average rate at which energy is transferred to the block? What is the instantaneous rate of that transfer when the block's speed is (c) 10 m/s and (d) 30 m/s?

97. Repeat Problem 96, but now with the block accelerated up along a frictionless plane inclined at 5.0° to the horizontal.

9

Systems of Particles

Sample Problem 9-10

Find the center of mass of the uniform triangular plate that is shown in Fig. 9-37.

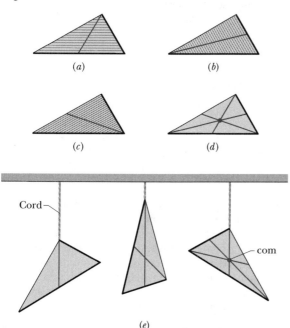

(a)

(b)

(c)

(d)

Cord

com

(e)

Fig. 9-37 Sample Problem 9-10. In (a), (b), and (c), the triangular plate is divided into thin slats, parallel to one side. The center of mass must lie along the bisecting line shown. (d) The dot, the only point common to all three lines, is the position of the center of mass. (e) Finding the center of mass by suspending the triangle from each vertex in turn.

SOLUTION: Figure 9-37a shows the plate divided into thin slats, parallel to one side of the triangle. The **Key Idea** here is that, from symmetry, the center of mass of a thin, uniform slat is at its midpoint. Thus, the center of mass of the triangular plate must lie somewhere along the line that connects the midpoints of all slats. That bisecting line also connects the upper vertex with the midpoint of the opposite side. The plate would balance if it were placed on a knife-edge coinciding with this line of symmetry.

In Figs. 9-37b and 9-37c, we subdivide the plate into slats parallel to the other two sides of the triangle. Again, the center of mass must lie somewhere along each of the bisecting lines shown. Hence the center of mass of the plate must lie at the intersection of these three symmetry lines, as Fig. 9-37d shows. That intersection is the only point that the three lines have in common.

You can check this conclusion experimentally by taking advantage of the (correct) intuitive notion that an object suspended from one of its points will orient itself so that its center of mass lies vertically below that point. Suspend the triangle from each vertex in turn, and draw a line vertically downward from the suspension point, as in Fig. 9-37e. The center of mass of the triangle will be at the intersection of the three lines.

Sample Problem 9-11

Silbury Hill (Fig. 9-38a), a mound on the plains near Stonehenge, was built 4600 years ago for unknown reasons, perhaps as a burial site. It is an incomplete right circular cone (see Fig. 9-38b), with a flattened top of radius $r_2 = 16$ m, a base of radius $r_1 = 88$ m, a height $h = 40$ m, and a volume V of 4.09×10^5 m^3. The sides of the cone make an angle $\theta = 30°$ with the horizontal.

(a) Where is the center of mass of the mound?

SOLUTION: One **Key Idea** here is that, because the mound is a solid body, we must integrate with Eqs. 9-11 to locate its center mass.

However, a second **Key Idea** simplifies our effort: Because of the circular symmetry of the mound, the center of mass must lie on the central axis of the cone, at some height z_{com} above the base. Now we need only the last equation of Eqs. 9-11 to find z_{com}.

We also use symmetry to evaluate that equation. Specifically, we consider the mound as a stack of thin, horizontal, circular wafers, each of smaller radius than the one below. Figure 9-38b shows a typical wafer, of radius r, thickness dz, and horizontal area πr^2, at height z from the base. Its volume dV is

$$dV = \pi r^2 \, dz. \qquad (9\text{-}56)$$

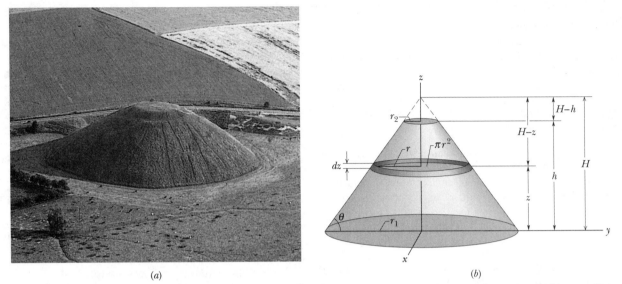

(a)   (b)

Fig. 9-38 Sample Problem 9-11. (a) Silbury Hill in England, built by Neolithic people, required an estimated 1.8×10^7 hours of labor to construct. (b) An incomplete right circular cone that resembles Silbury Hill. A "wafer" of radius r and thickness dz is shown at height z from the base.

The radii of the wafers range from r_1 at the bottom of the stack to r_2 at the top. If the cone were complete, it would have a height that we call H in Fig. 9-38b. We can relate the radius r of our typical wafer to H and z by writing

$$\tan \theta = \frac{H}{r_1} = \frac{H - z}{r},$$

which gives us

$$r = (H - z)\frac{r_1}{H}. \qquad (9\text{-}57)$$

Substituting Eqs. 9-56 and 9-57 into the last equation of Eqs. 9-11, we have

$$z_{\text{com}} = \frac{1}{V} \int z\, dV = \frac{\pi r_1^2}{VH^2} \int_0^h z(H - z)^2\, dz$$

$$= \frac{\pi r_1^2}{VH^2} \int_0^h (z^3 - 2z^2H + zH^2)\, dz$$

$$= \frac{\pi r_1^2}{VH^2} \left[\frac{z^4}{4} - \frac{2z^3H}{3} + \frac{z^2H^2}{2} \right]_0^h$$

$$= \frac{\pi r_1^2 h^4}{VH^2} \left[\frac{1}{4} - \frac{2H}{3h} + \frac{H^2}{2h^2} \right].$$

Finally, computing $H = r_1 \tan 30° = 50.8$ m and substituting known values, we find

$$z_{\text{com}} = \frac{\pi(88 \text{ m})^2(40 \text{ m})^4}{(4.09 \times 10^5 \text{ m}^3)(50.8 \text{ m})^2}$$

$$\times \left[\frac{1}{4} - \frac{2(50.8 \text{ m})}{3(40 \text{ m})} + \frac{(50.8 \text{ m})^2}{2(40 \text{ m})^2} \right]$$

$$= 12.38 \text{ m} \approx 12 \text{ m}. \qquad \text{(Answer)}$$

(b) If Silbury Hill has density $\rho = 1.5 \times 10^3$ kg/m³, then how much work was required to lift the dirt from the level of the base to build the mound?

SOLUTION: A Key Idea here is to first find an expression for the work dW needed to lift a mass element dm into place, and then integrate that expression over the full mound. From Eq. 7-17 with $\phi = 180°$ the work dW needed to lift element dm to height z is

$$dW = -dm\, gz \cos 180° = dm\, gz.$$

With Eq. 9-10, we substitute $\rho\, dV$ for dm, finding

$$dW = \rho gz\, dV.$$

To find the total work required to lift all the mass of Silbury Hill into place, we sum, via integration, the work dW associated with each volume element dV:

$$W = \int dW = \int \rho gz\, dV = \rho g \int z\, dV.$$

The last equation of Eqs. 9-11 tells us we can replace the integral with Vz_{com}, so we get

$$W = \rho V g z_{\text{com}}. \qquad (9\text{-}58)$$

Equation 7-17 tells us that the work needed to lift a mass m by a height h is mgh. Thus Eq. 9-58 tells us that the work required to lift all the mass of Silbury Hill into place is the same as if all the mass (ρV) were lifted to (and somehow concentrated at) the center of mass of the hill, at height z_{com}.

Substituting known data into Eq. 9-58 now yields

$$W = (1.5 \times 10^3 \text{ kg/m}^3)(4.09 \times 10^5 \text{ m}^3)$$

$$\times (9.8 \text{ m/s}^2)(12.37 \text{ m})$$

$$= 7.4 \times 10^{10} \text{ J}. \qquad \text{(Answer)}$$

Sample Problem 9-12

In Fig. 9-39, a mobile cannon of mass $M = 1300$ kg fires a ball of mass $m = 72$ kg in the positive direction of an x axis, at velocity $\vec{v}_{rel}$ relative to the cannon. Because of the firing, the cannon recoils (moves) with velocity $\vec{v}_{CG}$ relative to the ground. The magnitude of $\vec{v}_{rel}$ is 55 m/s. What are the magnitude and direction of $\vec{v}_{CG}$? (Although $\vec{v}_{CG}$ is drawn leftward in Fig. 9-39, we actually do not yet know its direction.)

SOLUTION: A **Key Idea** here is to check whether we can apply the law of conservation of linear momentum to the situation. If so, maybe we can find $\vec{v}_{CG}$ from the linear momentum of the recoiling cannon. To make this check, we let our system include the cannon and the ball so that the forces propelling them are internal to the system. That way, those forces cannot change the total linear momentum $\vec{P}$ of the system and we need not consider them. What *can* change $\vec{P}$ is an external force. Since the only *external* forces acting on the ball–cannon system are vertical, the *horizontal* component of $\vec{P}$ cannot change as the cannon is fired. Therefore, we can apply the law of conservation of momentum horizontally.

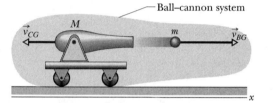

Fig. 9-39 Sample Problem 9-12. A cannon of mass M fires a ball of mass m. Relative to the ground, the cannon recoils with velocity $\vec{v}_{CG}$ and the ball has velocity $\vec{v}_{BG}$. The ball–cannon system is indicated.

First we choose the ground as our reference frame. Then, because the motion here is along only the x axis, we can simplify the notation by using x components of the velocities and momenta. Finally, we let $\vec{v}_{BG}$ be the velocity of the ball relative to the ground.

We shall apply the law of conservation of momentum in the form $P_i = P_f$, for horizontal momenta. We choose i to be an instant before the cannon is fired, and f to be an instant while the ball is in flight and the cannon is recoiling. Because the system is motionless before the cannon is fired, $P_i = 0$. We can now write the conservation law in component form as

$$0 = mv_{BG} + Mv_{CG}. \tag{9-59}$$

We cannot yet solve Eq. 9-59, because we do not know v_{BG}. However, another **Key Idea** helps: We can relate v_{BG} to the given velocities by writing

$$\left(\begin{array}{c}\text{velocity of ball}\\\text{relative to ground}\end{array}\right) = \left(\begin{array}{c}\text{velocity of ball}\\\text{relative to cannon}\end{array}\right)$$
$$+ \left(\begin{array}{c}\text{velocity of cannon}\\\text{relative to ground}\end{array}\right).$$

In symbols, this gives us

$$v_{BG} = v_{rel} + v_{CG}.$$

Substituting this for v_{BG} in Eq. 9-59 and solving for v_{CG}, we find

$$v_{CG} = -\frac{mv_{rel}}{m + M} = -\frac{(72 \text{ kg})(55 \text{ m/s})}{72 \text{ kg} + 1300 \text{ kg}}$$
$$= -2.9 \text{ m/s}. \tag{Answer}$$

The sign tells us that $\vec{v}_{CG}$ is directed leftward in Fig. 9-39, as drawn.

Sample Problem 9-13

Figure 9-40 shows two blocks that are connected by an ideal spring and are free to slide along an x axis on a frictionless horizontal surface. Block 1 has mass m_1 and block 2 has mass m_2. The blocks are pulled in opposite directions (stretching the spring) and then released from rest.

(a) What is the ratio v_1/v_2 of the velocity of block 1 to the velocity of block 2 as the separation between the blocks decreases?

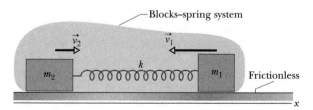

Fig. 9-40 Sample Problem 9-13. Two blocks, which are on a frictionless surface and connected by a spring, are pulled away from each other and then released from rest. The blocks–spring system is indicated.

SOLUTION: A **Key Idea** is to check whether we can use the law of conservation of linear momentum. If so, maybe we can relate the velocities of the blocks by relating their momenta. To make this check, we consider the system consisting of the two blocks and the spring. Then the spring force that accelerates the blocks is internal to the system and so cannot change the total linear momentum $\vec{P}$ of the system; thus, we need not consider that force. What *can* change $\vec{P}$ is an external force. The only *external* forces acting on the system are vertical, so the horizontal component of $\vec{P}$ cannot change as the blocks move.

As in Sample Problem 9-12, we will apply the law of conservation of momentum in the form $P_i = P_f$, for x components of the momenta. We let i correspond to the release of the blocks from rest. Each block has a momentum of zero just then, so $P_i = 0$. We also let f correspond to any instant as the blocks approach each other. Then the total momentum is the sum of momenta m_1v_1 for block 1 and m_2v_2 for block 2. For our system, we can now write the conservation law as

$$0 = m_1v_1 + m_2v_2, \tag{9-60}$$

which gives us

$$\frac{v_1}{v_2} = -\frac{m_2}{m_1}. \qquad (9\text{-}61)$$

The minus sign tells us that the two velocities always have opposite signs. Equation 9-61 holds at every instant after release, no matter what the actual speeds of the blocks are.

(b) What is the ratio K_1/K_2 of the kinetic energies of the blocks as their separation decreases?

SOLUTION: The Key Idea here is that we can find the ratio of the kinetic energies by using the definition of kinetic energy; that is,

$$\frac{K_1}{K_2} = \frac{\frac{1}{2}m_1 v_1^2}{\frac{1}{2}m_2 v_2^2} = \frac{m_1}{m_2}\left(\frac{v_1}{v_2}\right)^2.$$

Substituting for v_1/v_2 from Eq. 9-61 and simplifying, we find

$$\frac{K_1}{K_2} = \frac{m_2}{m_1}. \qquad (9\text{-}62)$$

As the blocks move toward each other and the stretch of the spring decreases, energy is transferred from the elastic potential energy of the spring to the kinetic energies K_1 and K_2 of the blocks. Although K_1 and K_2 then increase, Eq. 9-62 tells us that their ratio does not change but is preset by the ratio of the masses. After the spring reaches its rest length and begins to be compressed by the blocks, the energy transfer is reversed, but Eq. 9-62 still holds.

Equations 9-60 through 9-62 apply to other situations in which two bodies attract (or repel) each other. For example, they apply to a stone falling toward Earth. The stone corresponds to, say, block 1 in Fig. 9-40, and Earth to block 2. The gravitational force between the stone and Earth corresponds to the mutual force between the blocks that is provided by the spring in Fig. 9-40. Our reference frame is the frame in which the center of mass of the stone–Earth system is stationary. *In this frame,* Eq. 9-60 tells us that the magnitudes of the linear momenta of the stone and Earth remain equal to each other throughout the fall. Equations 9-61 and 9-62 tell us that, because $m_2 \gg m_1$, the stone has much greater speed and kinetic energy than Earth during the fall, again *in the given frame.*

QUESTIONS

10. (a) If you drop two watermelons side by side from a bridge, what is the acceleration of the center of mass of the two-melon system? (b) If you delay dropping one of the melons, what is the acceleration of the two-melon system while both are falling?

11. *Organizing question:* A 5 kg block is sliding with a velocity of $+4$ m/s along an x axis on a frictionless surface when it explodes into two pieces. Piece A, with a mass of 3 kg, and piece B are sent sliding over that surface. Figure 9-41 shows the speed and direction of travel of piece A for three situations. For each situation, set up an equation to find velocity $\vec{v}_{Bf}$ of piece B, complete with known data.

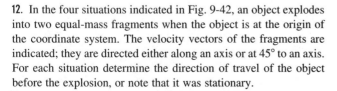

Fig. 9-41 Question 11.

12. In the four situations indicated in Fig. 9-42, an object explodes into two equal-mass fragments when the object is at the origin of the coordinate system. The velocity vectors of the fragments are indicated; they are directed either along an axis or at 45° to an axis. For each situation determine the direction of travel of the object before the explosion, or note that it was stationary.

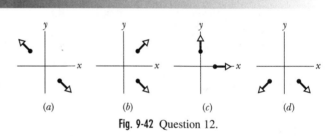

Fig. 9-42 Question 12.

13. Figure 9-43 shows overhead views of three two-dimensional explosions in which a stationary grapefruit is blown by a firecracker into three pieces, seven pieces, and nine pieces. The pieces then slide over a frictionless floor. For each situation, Fig. 9-43 also shows the momentum vectors of all but one piece; that piece has momentum $\vec{P}'$. The numbers next to the vectors are the magnitudes of the momenta (in kilogram-meters per second). Rank the three situations according to the magnitudes of (a) P'_x, (b) P'_y, and (c) $\vec{P}'$, greatest first.

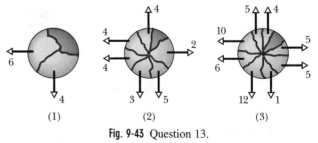

Fig. 9-43 Question 13.

14. Figure 9-44 is an overhead view showing six particles that have emerged from a region in which a two-dimensional explosion took place. The explosion was of an object that had been stationary on a frictionless floor. The directions of the momenta of the particles

are indicated by the vectors; the numbers give the magnitudes of the momenta (in kilogram-meters per second). (a) Will more particles be emerging from the explosion region? If so, give (b) their net momentum and (c) their direction of travel.

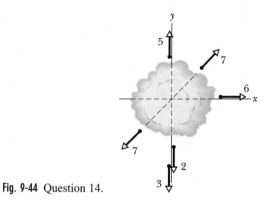

Fig. 9-44 Question 14.

15. Figure 9-45 shows, from overhead, the path taken by a toy car moving at constant speed; the straight sections are either parallel to the x axis, parallel to the y axis, or at 45° to the axes. (a) Rank the curved sections according to the magnitude of the change $\Delta \vec{p}$ in linear momentum of the car due to them, greatest first. For which curved sections does $\Delta \vec{p}$ have a component in (b) the negative direction of the y axis and (c) the positive direction of the x axis?

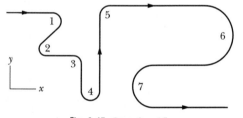

Fig. 9-45 Question 15.

16. The free-body diagrams in Fig. 9-46 give, from overhead views, the horizontal forces acting on three boxes of chocolates as the boxes move over a frictionless confectioner's counter. For each box, is its linear momentum conserved along the x axis and the y axis?

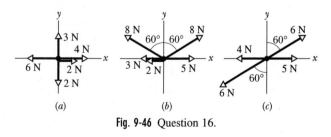

Fig. 9-46 Question 16.

17. Three rockets, all with the same uniformly distributed mass, race along a deep-space raceway. At a certain instant they are even with one another and are traveling at the same speed. However, just then they each jettison a rear section, as shown in Fig. 9-47. The resulting relative speeds between the front and rear sections happen to be identical for the rockets. Rank the rockets according to their travel time to the end of the raceway, greatest first.

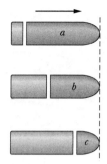

Fig. 9-47 Question 17.

18. An initially stationary box on a frictionless floor explodes into two pieces: piece A with mass m_A and piece B with mass m_B. These pieces then move across the floor along an x axis. Graphs of position versus time for the two pieces are given in Fig. 9-48. (a) Which graphs pertain to physically possible explosions? Of those graphs, which best corresponds to the situation in which (b) $m_A = m_B$, (c) $m_A > m_B$, and (d) $m_A < m_B$?

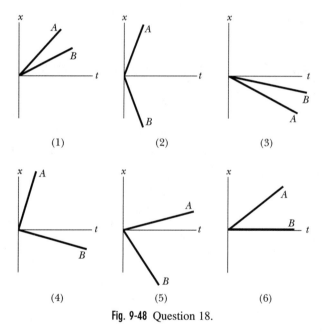

Fig. 9-48 Question 18.

DISCUSSION QUESTIONS

19. Figure 9-49 shows (a) an isosceles triangle and (b) a right-circular cone whose diameter has the same length as the base of the triangle. The center of mass of the triangle is one-third of the way up from the base but that of the cone is only one-fourth of the way up. Can you explain this difference?

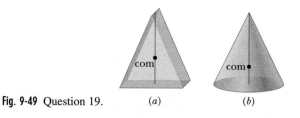

Fig. 9-49 Question 19.

20. Where is the center of mass of Earth's atmosphere?

21. Someone claims that when a skillful high jumper clears the bar her or his center of mass actually goes *under* the bar. Is this possible?

22. A bird is in a wire cage that hangs from a spring balance. Is the reading of the balance when the bird is flying greater than, less than, or the same as that when the bird sits in the cage?

23. Can a sailboat be propelled by air blown at the sails from a fan attached to the boat? Explain your answer.

24. A canoeist in a still pond can reach shore by jerking sharply on a rope attached to the bow of the canoe. How do you explain this? (It really can be done.)

25. How might a person sitting at rest on a frictionless horizontal surface get altogether off it?

26. A man stands still on a large sheet of slick ice; in his hand he holds a lighted firecracker. He throws the firecracker into the air.

Describe briefly, but as exactly as you can, the motion of the center of mass of the firecracker and the motion of the center of mass of the system consisting of man and firecracker. It will be most convenient to describe each motion during each of the following periods: (a) after he throws the firecracker, but before it explodes; (b) between the explosion and the first piece of firecracker hitting the ice; (c) between the first fragment hitting the ice and the last fragment landing; and (d) during the time when all fragments have landed but none has reached the edge of the ice.

27. In 1920, a prominent newspaper editorialized as follows about the pioneering rocket experiments of Robert H. Goddard, dismissing the notion that a rocket could operate in a vacuum: "That Professor Goddard, with his 'chair' in Clark College and the countenancing of the Smithsonian Institution, does not know the relation of action to reaction, and of the need to have something better than a vacuum against which to react—to say that would be absurd. Of course, he only seems to lack the knowledge ladled out daily in high schools." What is wrong with this argument?

EXERCISES & PROBLEMS

60. A certain nucleus, at rest, transforms into three particles. Two of them are detected; their masses and velocities are as shown in Fig. 9-50. (a) In unit-vector notation, what is the linear momentum of the third particle, with a mass of 11.7×10^{-27} kg? (b) How much kinetic energy appears in this transformation process?

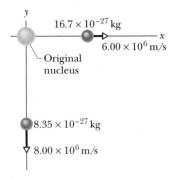

Fig. 9-50 Problem 60.

61. A rocket sled with a mass of 2900 kg moves at 250 m/s on a set of rails. At a certain point, a scoop on the sled dips into a trough of water located between the tracks and scoops water into an empty tank on the sled. By applying the principle of conservation of linear momentum, determine the speed of the sled after 920 kg of water has been scooped up. Ignore any retarding force on the scoop.

62. A 2140 kg railroad flatcar, which can move with negligible friction, is motionless next to a platform. A 242 kg sumo wrestler runs at 5.3 m/s along the platform (parallel to the track) and then jumps onto the flatcar. What is the speed of the flatcar if he then (a) stands on it, (b) runs at 5.3 m/s relative to it in his original direction, and (c) turns and runs at 5.3 m/s relative to the flatcar opposite his original direction?

63. A suspicious package is sliding on a frictionless surface when it explodes into three pieces of equal masses and with the velocities (1) 7.0 m/s, north, (2) 4.0 m/s, 30° south of west, and (3) 4.0 m/s, 30° south of east. (a) What is the velocity (magnitude and direction)

of the package before it explodes? (b) What is the displacement of the center of mass of the three-piece system (with respect to the point where the explosion occurs) 3.0 s after the explosion?

64. The Great Pyramid of Cheops at El Gizeh, Egypt (Fig. 9-51a), had height $H = 147$ m before its topmost stone fell. Its base is a square with edge length $L = 230$ m (see Fig. 9-51b). Its volume V

(a)

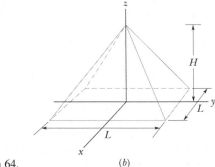

Fig. 9-51 Problem 64.   (b)

is equal to $L^2H/3$. Assuming $\rho = 1.8 \times 10^3$ kg/m^3 is its uniform density, find (a) the original height of its center of mass above the base and (b) the work required to lift the blocks into place from the base level.

65. A 40 kg child and her 75 kg father simultaneously dive from a 100 kg boat that is initially motionless. The child dives horizontally toward the east with a speed of 2.0 m/s, and the father dives toward the south with a speed of 1.5 m/s at an angle of 37° above the horizontal. (Assume the boat's vertical motion due to the father's dive does not alter its horizontal motion.) Determine the magnitude and direction of the velocity of the boat along the water's surface immediately after their dives.

66. A 50 g ball is thrown from ground level into the air with an initial speed of 16 m/s at an angle of 30° above the horizontal. (a) What are the values of the kinetic energy of the ball initially and just before it hits the ground? (b) Find the corresponding linear momenta (magnitude and direction). (c) Show that the change in the magnitude of the linear momentum is equal to the weight of the ball multiplied by the time of flight.

67. A 2.00 kg block is released from rest over the side of a very tall building at time $t = 0$. At time $t = 1.00$ s, a 3.00 kg block is released from rest at the same point. The first block hits the ground at $t = 5.00$ s. Plot, for the time interval $t = 0$ to $t = 6.00$ s, (a) the position and (b) the speed of the center of mass of the two-block system. Take $y = 0$ at the release point.

68. At the instant a 3.0 kg particle has a velocity of 6.0 m/s in the negative y direction, a 4.0 kg particle has a velocity of 7.0 m/s in the positive x direction. What is the speed of the center of mass of the two-particle system?

69. A 1500 kg car and a 4000 kg truck are moving north and east, respectively, with constant velocities. The center of mass of the car–truck system has a velocity of 11 m/s in a direction 55° north of east. (a) What is the magnitude of the car's velocity? (b) What is the magnitude of the truck's velocity?

70. While a 1710 kg automobile is moving at a constant speed of 15.0 m/s, the engine supplies 16.0 kW of power to overcome friction, air drag, and other retarding factors. (a) What is the effective retarding force associated with all these factors combined? (b) What power must the engine supply if the car is to move up an 8.00% grade (8.00 m vertically for each 100 m horizontally) at 15.0 m/s? (c) On what downgrade, expressed as a percentage, would the car coast at 15.0 m/s?

71. At $t = 0$, a 1.0 kg jelly jar is projected vertically upward from the base of a 50-m-tall building with an initial velocity of 40 m/s. At the same instant and directly overhead, a 2.0 kg peanut butter jar is dropped from rest from the top of the building. (a) How far above ground level is the center of mass of the two-jar system at $t = 3.0$ s? (b) What maximum height above ground level is reached by the center of mass?

72. A cannon and a supply of cannonballs are inside a sealed railroad car of length L, as in Fig. 9-52. The cannon fires to the right; the car recoils to the left. Fired cannonballs travel a horizontal distance L and remain in the car after hitting the far wall and landing on the floor there. (a) After all the cannonballs have been fired, what is the greatest distance the car could have moved from its

original position? (b) What is the speed of the car just after the last cannonball has hit the far wall?

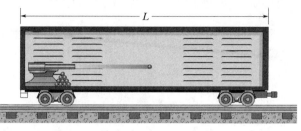

Fig. 9-52 Problem 72.

73. A 1400 kg cannon, which fires a 70.0 kg shell with a speed of 556 m/s relative to the muzzle, is set at an elevation angle of 39.0° above the horizontal. The cannon is mounted on frictionless rails so that it can recoil freely. (a) At what speed relative to the ground is the shell fired? (b) At what angle with the ground is the shell fired? (*Hint:* The horizontal component of the linear momentum of the system remains unchanged as the cannon is fired.)

74. A single-stage rocket, at rest in a certain inertial reference frame, has mass M when the rocket engine is ignited. Show that when the mass has decreased to $0.368\,M$, the gases steaming out of the rocket engine at that time will be at rest in the original reference frame.

75. You are on an iceboat on frictionless, flat ice; you and the boat have a combined mass M. Along with you are two stones with masses m_1 and m_2 such that $M = 6.00m_1 = 12.0m_2$. To get the boat moving, you throw the stones rearward, either in succession or together, but in each case with a certain speed v_{rel} relative to the boat after the stone is thrown. What is the resulting speed of the boat if you throw the stones (a) simultaneously, (b) in the order m_1 and then m_2, and (c) in the order m_2 and then m_1?

76. A jet airplane is traveling at a speed of 180 m/s through air that is stationary relative to the ground. Each second, the engine takes in 68 m^3 of air, which has a mass of 70 kg. The air is used to burn 2.9 kg of fuel each second. The fuel's energy is used to compress the products of combustion and eject them at the rear of the plane at 490 m/s relative to the plane. Take the forward direction as the positive direction of motion. What are the magnitudes of (a) the intake air velocity relative to the airplane, (b) the force on the airplane due to the intake of the air, (c) the force on the airplane due to the ejection of the air, and (d) the net thrust (the net of those two forces) produced by the engine? (e) What is the power associated with that net thrust?

77. A 4.0 kg particle-like object is located at $x = 0$, $y = 2.0$ m; a 3.0 kg particle-like object is located at $x = 3.0$ m, $y = 1.0$ m. At what (a) x and (b) y coordinates must a 2.0 kg particle-like object be placed for the center of mass of the three-particle system to be located at the origin?

78. The following table gives the masses of three objects and, at a certain instant, the coordinates (x, y) and the velocities of the objects. At that instant, what are (a) the position and (b) the velocity of the center of mass of the three-particle system, and (c) what is the net linear momentum of the system?

Object	Mass (kg)	Coordinates (m)	Velocity (m/s)
1	4.00	(0,0)	$1.50\hat{i} - 2.50\hat{j}$
2	3.00	(7.00, 3.00)	0
3	5.00	(3.00, 2.00)	$2.00\hat{i} - 1.00\hat{j}$

79. *Left alone, write your own.* For one or more of the following situations, write a problem involving physics in this chapter, using the style of the sample problems and providing realistic data, graphs of the variables, and explained solutions: (a) determining the center of mass of a large object, (b) a system separated into parts by an internal explosion, (c) someone climbing or descending a structure, (d) track and field events.

CLUSTERED PROBLEMS

In each problem of this cluster, find the requested center of mass by using the xy coordinate system shown and either symmetry or Eqs. 9-9.

80. In Fig. 9-53a, a uniform wire forms an isosceles triangle of base B and height H. (a) Find the x and y coordinates of the figure's center of mass by assuming that each side can be replaced with a particle of the same mass as that side and positioned at the center of the side. (*Be careful:* Note that the base and, say, the left-hand side do not have the same mass.) (b) Use Eqs. 9-9 to find the x and y coordinates of the center of mass of the left-hand side.

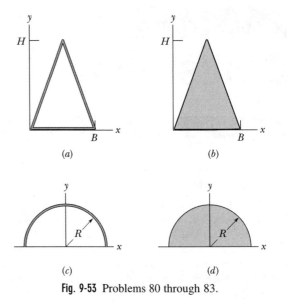

(a)　　　　　　(b)

(c)　　　　　　(d)

Fig. 9-53 Problems 80 through 83.

81. Figure 9-53b shows a uniform, solid plate in the shape of an isosceles triangle with base B and height H. What are the x and y coordinates of the plate's center of mass?

82. In Fig. 9-53c, a uniform wire forms a semicircle of radius R. What are the x and y coordinates of the figure's center of mass?

83. Figure 9-53d shows a uniform, solid plate in the shape of a semicircle with radius R. What are the x and y coordinates of the plate's center of mass?

10
Collisions

Sample Problem 10-6

An engineer has been hired to design a wall to protect a hillside home from a possible snow avalanche. The engineer makes these assumptions: Snow in an avalanche has a density ρ of 500 kg/m^3, would hit the wall perpendicularly at a speed of 20 m/s and then stop, and would hit it at the rate of 1000 m^3/s. For these assumptions, what would be the magnitude of the average force on the wall from the snow?

SOLUTION: A **Key Idea** here is that the snow hits the wall as a steady stream of bodies, as in Fig. 10-5. Thus, we can use Eq. 10-13 to find the magnitude of the average force on the wall. However, because the snow hits in a continuous flow, we take a differential limit of $\Delta m/\Delta t$ and write

$$F_{avg} = -\frac{dm}{dt}\,\Delta v. \qquad (10\text{-}48)$$

Because the snow stops on hitting the wall, we use Eq. 10-11

($\Delta v = -v$) for the change Δv in the velocity of the snow. We are given the volume rate $dV/dt = 1000$ m^3/s at which snow collides with the wall. We can find the mass rate dm/dt from

$$\frac{\text{mass}}{\text{second}} = \frac{\text{mass}}{\text{volume}} \times \frac{\text{volume}}{\text{second}}$$

or

$$\frac{dm}{dt} = \rho\,\frac{dV}{dt}.$$

Substituting $-v$ for Δv and $\rho\,dV/dt$ for dm/dt in Eq. 10-48, we find

$$F_{avg} = \rho v\,\frac{dV}{dt} = (500 \text{ kg/m}^3)(20 \text{ m/s})(1000 \text{ m}^3/\text{s})$$

$$= 1.0 \times 10^7 \text{ N}. \qquad \text{(Answer)}$$

The answer suggests why avalanches typically flatten homes and other structures in their paths.

Sample Problem 10-7

In a nuclear reactor, newly produced fast neutrons must be slowed before they can participate effectively in the chain-reaction process. This is done by allowing them to collide with the nuclei of atoms in a *moderator*.

(a) By what fraction is the kinetic energy of a neutron (of mass m_1) reduced in a head-on elastic collision with a nucleus of mass m_2, initially at rest?

SOLUTION: The initial and final kinetic energies of the neutron are

$$K_i = \tfrac{1}{2}m_1 v_{1i}^2 \quad \text{and} \quad K_f = \tfrac{1}{2}m_1 v_{1f}^2.$$

The fraction we seek (call it frac) is then

$$\text{frac} = \frac{K_i - K_f}{K_i} = \frac{v_{1i}^2 - v_{1f}^2}{v_{1i}^2} = 1 - \frac{v_{1f}^2}{v_{1i}^2}. \qquad (10\text{-}49)$$

To evaluate Eq. 10-49, we need two **Key Ideas:** Assuming the neutron–nucleus system is closed and isolated, the total linear momentum of the system is conserved during the collision. Because we know the collision is elastic, the total kinetic energy of the system is also conserved. Thus, we can use Eq. 10-30 to write the following:

$$\frac{v_{1f}}{v_{1i}} = \frac{m_1 - m_2}{m_1 + m_2}. \qquad (10\text{-}50)$$

Substituting Eq. 10-50 into Eq. 10-49 yields, after a little algebra,

$$\text{frac} = \frac{4m_1 m_2}{(m_1 + m_2)^2}. \qquad \text{(Answer)} \quad (10\text{-}51)$$

(b) Evaluate the fraction for lead, carbon, and hydrogen. The ratio m_2/m_1 of the mass of a nucleus to the mass of a neutron for these nuclei is 206 for lead, 12 for carbon, and about 1 for hydrogen.

SOLUTION: The following values of the fraction can be calculated with Eq. 10-51: for lead ($m_2 = 206m_1$),

$$\text{frac} = \frac{(4)(206)}{(1 + 206)^2} = 0.019 \text{ or } 1.9\%; \qquad \text{(Answer)}$$

for carbon ($m_2 = 12m_1$),

$$\text{frac} = \frac{(4)(12)}{(1 + 12)^2} = 0.28 \text{ or } 28\%; \qquad \text{(Answer)}$$

and for hydrogen ($m_2 \approx m_1$),

$$\text{frac} = \frac{(4)(1)}{(1 + 1)^2} = 1 \text{ or } 100\%. \qquad \text{(Answer)}$$

These results show that the kinetic energy of the projectile neutron is reduced more when there is a closer match of the target mass with the neutron's mass. They partially explain why water, which contains lots of hydrogen, is a much better moderator of neutrons than lead.

Sample Problem 10-8

Figure 10-43a shows two particles of equal masses before and after they have an elastic collision. The particles form a closed, isolated system. Show that when the collision is not head-on the two particles must leave the collision along perpendicular paths.

SOLUTION: There are two **Key Ideas** here. Because the system is closed and isolated, the total linear momentum of the two particle system is conserved. Also, because the collision is elastic, the total kinetic energy of the system is conserved. Therefore, Eqs. 10-40, 10-41, and 10-44 apply to this collision. Since the arrangement shown in Fig. 10-43a is like that in Fig. 10-16, we could actually adapt Eqs. 10-42 through 10-44 to fit this collision. Instead, we can do a graphical solution that is much neater than using those equations.

Figure 10-43a shows the linear momenta of the particles before and after the collision. Because linear momentum is conserved in the collision, these vectors must form a closed triangle, as Fig. 10-43b shows. (The vector $m\vec{v}_{1i}$ must be the vector sum of $m\vec{v}_{1f}$ and $m\vec{v}_{2f}$.) Because the masses of the particles are equal, the closed linear-momentum triangle of Fig. 10-43b is also a closed velocity triangle, because dividing by the scalar m does not change the relation of the vectors. Thus, we may draw the velocity vectors as in Fig. 10-43c, because

$$\vec{v}_{1i} = \vec{v}_{1f} + \vec{v}_{2f}. \tag{10-52}$$

Equation 10-44, with the equal terms $\frac{1}{2}m$ canceled out, tells us that

$$v_{1i}^2 = v_{1f}^2 + v_{2f}^2. \tag{10-53}$$

This equation relates the lengths of the sides in the triangle of Fig. 10-43c. For it to hold, the triangle must be a right triangle (and Eq. 10-53 is then the Pythagorean theorem). Therefore, the angle ϕ between the directions of vectors $\vec{v}_{1f}$ and $\vec{v}_{2f}$ in Fig. 10-43c must be 90°, which is what we set out to prove.

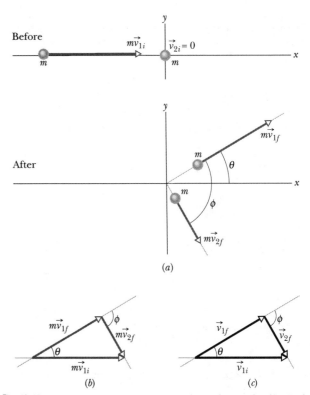

Fig. 10-43 Sample Problem 10-8. (a) An incoming projectile particle and a stationary target particle, of equal mass, have an elastic collision. Their momenta before and after the collision are shown. (b) Their momentum vectors form a closed triangle, and (c) so do their velocity vectors.

QUESTIONS

11. *Organizing question:* Figure 10-44 shows four situations in which there is a one-dimensional collision between two blocks of masses $m_1 = 2$ kg and $m_2 = 3$ kg on a frictionless floor. Information about the speeds and directions of travel are given for before and after each collision, except for the final velocity v_{2f} of block 2. For each situation, set up an equation, complete with known data, to find v_{2f}. (*Hint:* Do not assume that the collision is elastic.)

12. One body catches up with a second body and they then undergo a one-dimensional collision. The bodies form a closed, isolated system. Figure 10-45 is a graph of position versus time for the bodies and for their center of mass. Which line segment corresponds to (a) the faster body before the collision, (b) the center of mass before the collision, (c) the center of mass after the collision, and (d) the initially faster body after the collision? (e) Is the mass

of the initially faster body greater than, less than, or equal to that of the other body?

13. *Organizing question:* Set up an equation, complete with known data, to find the change ΔK in the total kinetic energy of the blocks in Question 11.

14. Figure 10-46 shows, for four situations, three identical blocks that undergo one-dimensional elastic collisions on a frictionless surface. In situations 1 and 2, two of the blocks are glued together. In all four situations, the initially moving blocks have the same velocity $\vec{v}$. Rank the situations according to (a) the total linear momentum of the blocks after the collisions and (b) the speed of the rightmost block after the collisions, greatest first.

15. *Organizing question:* Figure 10-47 shows two situations in which a projectile box 1 collides with a stationary target box 2

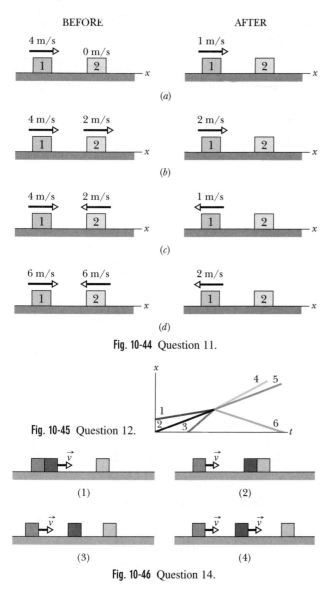

BEFORE   AFTER

4 m/s   0 m/s   1 m/s
⊳
1   2   1   2
x   x
(a)

4 m/s   2 m/s   2 m/s
1   2   1   2
x   x
(b)

4 m/s   2 m/s   1 m/s
1   2   1   2
x   x
(c)

6 m/s   6 m/s   2 m/s
1   2   1   2
x   x
(d)

Fig. 10-44 Question 11.

Fig. 10-45 Question 12.

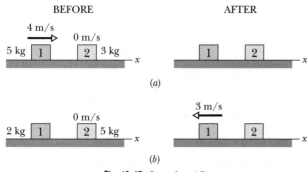

(1)   (2)

(3)   (4)

Fig. 10-46 Question 14.

on a frictionless floor. In both situations the collisions are one-dimensional and elastic; some of the speeds and directions of travel of the boxes are given. For each situation set up equations, complete with known data, to find the velocities whose values are not given.

BEFORE   AFTER

4 m/s   0 m/s
5 kg 1   2 3 kg   1   2
x   x
(a)

0 m/s   3 m/s
2 kg 1   2 5 kg   1   2
x   x
(b)

Fig. 10-47 Question 15.

16. Recall that Eqs. 10-30 and 10-31 correspond to the one-dimensional elastic collision generically depicted in Fig. 10-13. (a) Which of the curves in Fig. 10-48 is the graph of the velocity ratio v_{1f}/v_{1i} versus the mass ratio m_1/m_2? (b) What are the values of the ratios at the intercept of that curve with the vertical axis? (c) What velocity ratio does that curve approach beyond the right side of the graph? (d) What point on that curve corresponds to equal masses of the projectile and the target?

(e) Which of the curves is the graph of the velocity ratio v_{2f}/v_{1i} versus m_1/m_2? (f) What are the values of the ratios at the intercept of the curve with the vertical axis? (g) What velocity ratio does the curve approach beyond the right side of the graph? (h) What point on the curve corresponds to equal masses of the projectile and the target?

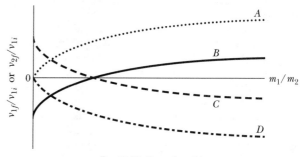

Fig. 10-48 Question 16.

17. Body Q, with linear momentum $\vec{p}_Q = (2\hat{i} - 3\hat{j})$ kg·m/s, collides with and also sticks to body R, with linear momentum $\vec{p}_R = (8\hat{i} + 3\hat{j})$ kg·m/s. The bodies form a closed, isolated system. In what direction do they move after the collision? (You do not need a calculator or written calculations.)

18. An evacuated box (of small mass) is at rest on a frictionless table. You punch a small hole in one side so that air can rush into the box (Fig. 10-49). If the box then remains at rest, explain why. If, instead, the box moves, explain why and give its direction and the duration of the motion.

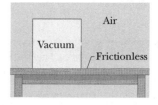

Fig. 10-49 Question 18.

19. Figure 10-50 shows seven identical blocks on a frictionless floor. Initially, blocks a and b are moving rightward and block g is moving leftward, each with speed $v = 3$ m/s. The other blocks are stationary. A series of elastic collisions occurs. After the last collision, what are the speeds and directions of motion of each of the seven blocks?

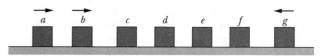

Fig. 10-50 Question 19.

DISCUSSION QUESTIONS

20. Explain how conservation of linear momentum applies to a handball bouncing off a wall.

21. Can the impulse of a force be zero, even if the force is not zero? Explain why or why not.

22. Figure 10-51 shows a popular carnival "strongarm" device, in which contestants try to see how high they can raise a weighted marker by hitting a target with a sledgehammer. What physical quantity does the device measure? Is it the average force, the maximum force, the work done, the impulse, the energy transferred, the linear momentum transferred, or something else? Discuss your answer.

Fig. 10-51 Question 22.

23. Different batters swing bats differently. What features of the swing help determine the speed and trajectory of the ball?

24. Many features of cars, such as collapsible steering wheels and padded dashboards, are meant to protect passengers during accidents. Explain their usefulness, using the impulse concept.

25. Why does the use of gloves make modern boxing somewhat safer than bare-knuckle fighting? When stuntpeople fall from buildings, why does landing on an air bag help protect them from injury or death? Why can some victims falling from great heights survive the landing if the ground is covered with soft snow, if they crash through tree branches before reaching the ground, or if they land on the side of a ravine and then slide down it? In each case, argue your point from considerations of average force.

26. It is said that, during a 50 km/h collision, a 45 kg child can exert a 1300 N force against a parent's grip. How can such a large force come about?

27. The following statement was taken from an exam paper: "The collision between two helium atoms is perfectly elastic, so that momentum is conserved." Is the statement logically correct? Explain.

28. You are driving along a highway at 80 km/h, followed by another car moving at the same speed. You slow to 60 km/h but the other driver does not and there is a collision. What are the initial velocities of the colliding cars as seen from the reference frame of (a) yourself, (b) the other driver, and (c) a state trooper, who is in a patrol car parked by the roadside? (d) A judge asks whether you bumped into the other driver or the other driver bumped into you. As a physicist, how would you answer?

29. Two identical cubical blocks, moving in the same direction with a common speed v, strike a third such block initially at rest on a horizontal frictionless surface. What is the motion of the blocks after the collision? Does it matter whether or not the two initially moving blocks were in contact? Does it matter whether these two blocks were glued together?

30. Two clay balls of equal mass and speed strike each other head-on, stick together, and come to rest. Kinetic energy is certainly not conserved. What happened to the energy? How is linear momentum conserved?

31. A football player, momentarily at rest on the field, catches a football as he is tackled by a running player on the other team. This is certainly a collision (inelastic!), and linear momentum must be conserved. In the reference frame of the football field, there is linear momentum before the collision but there seems to be none after the collision. Is linear momentum really conserved? If so, explain how. If not, explain why not.

32. Consider a one-dimensional elastic collision between a moving object A and an object B initially at rest. How would you choose the mass of B, in comparison to the mass of A, in order for B to recoil with (a) the greatest speed, (b) the greatest linear momentum, and (c) the greatest kinetic energy?

33. An inverted hourglass is weighed on a sensitive balance from the time the first grain moves to after the last grain has landed. How does the weight vary during that time? Why?

EXERCISES & PROBLEMS

58. *Domino-effect energy amplifier.* Figure 10-52*a* shows two dominoes in a series of dominoes in which each domino (except for the first one) is scaled up in each dimension from the previous domino by a scale factor of 1.5. If the first (smallest) domino is

pushed slightly so that its center of mass barely passes over one edge, it then topples against the second domino and a wave of collisions sweeps along the series of dominoes in Fig. 10-52*b*. (a) Assume that the first domino has a height of 1.0 cm. What is

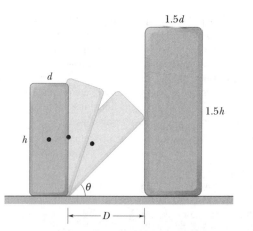

(a)

(b)

Fig. 10-52 Problem 58.

(and larger) domino receive during this collision in order to be toppled? (e) What is the ratio $\Delta E_{1,\text{out}}/\Delta E_{2,\text{in}}$? From this result and the fact that the next domino is indeed toppled, we can see that the domino series is an "energy amplifier," triggered by a small push and drawing on the energy stored as gravitational potential energy in the upright dominoes. (Adapted from Lorne Whitehead, "Domino 'Chain Reaction'," *American Journal of Physics*, 1983, Vol. 51, p. 182.)

59. *Body armor.* The textile engineering behind modern body armor has saved countless lives of police and military personnel. When a high-speed projectile such as a bullet or a bomb fragment strikes the armor, the projectile is snared by the armor's woven fabric. Figure 10-53 is a graph of the speed v versus time t for a 10.2 g bullet fired from a .38 Special revolver directly into body armor. In the bullet–armor collision, what are (a) the magnitude a of the bullet's deceleration, (b) the magnitude Δp of the bullet's change in momentum, (c) the change ΔK in the bullet's kinetic energy, and (d) the stopping distance d of the bullet? What are (e) the magnitude J of the impulse on the armor from the bullet and (f) the magnitude F of the corresponding force?

Suppose that the total mass of the body armor and the person wearing it is 65 kg and the person is on a frictionless surface. What are (g) the magnitude a_p of the person's acceleration during the collision and (h) the person's speed v_p just after the collision? If a textile engineer decreases the bullet-stopping distance d by changing the mesh and layering of the fabric, the snared bullet produces less of a dent in the chest. (i) If distance d is decreased in this way, do the quantities in parts (a) through (c) and (e) through (h) increase, decrease, or remain the same?

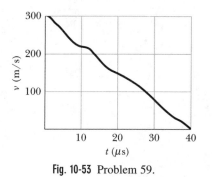

Fig. 10-53 Problem 59.

60. A 3000 kg block falls vertically through 6.0 m and then collides with a 500 kg pile, driving it 3.0 cm into bedrock. Assuming that the block–pile collision is completely inelastic, find the magnitude of the average force on the pile from the bedrock during the 3.0 cm descent.

61. A 3.0 kg object moving at 8.0 m/s in the positive direction of x has a one-dimensional, completely elastic collision with an object of mass M, initially at rest. After the collision the object of mass M has a velocity of 6.0 m/s in the positive direction of x. What is mass M?

62. A railroad freight car weighing 280 kN and traveling at 1.52 m/s overtakes one weighing 210 kN and traveling at 0.914 m/s in the same direction. If the cars couple together, find (a) the speed of the cars after the collision and (b) the loss of kinetic energy during the collision. If instead, as is very unlikely, the col-

the height of the 32nd domino? (b) For each domino, take the ratio of height h to thickness d to be $h/d = 10$. Let U_1 be the gravitational potential energy of the first domino's center of mass before that domino is pushed. In terms of U_1, how much energy $\Delta E_{1,\text{in}}$ must the domino receive if its center of mass is to barely pass over its edge?

Once the domino's center of mass passes over its edge, the center of mass falls until the domino strikes the next domino, when it makes an angle θ with the floor. (c) Take $\theta = 45°$ and approximate the height of the center of mass to then be $(h/2) \sin \theta$. In terms of U_1, how much energy $\Delta E_{1,\text{out}}$ is transferred to kinetic energy by the fall? (d) In terms of U_1, how much energy $\Delta E_{2,\text{in}}$ must the next lision is elastic, find the speeds of (c) the lighter car and (d) the heavier car after the collision.

63. A 60 kg man is ice-skating due north with a velocity of 6.0 m/s when he collides with a 38 kg child. The man and child stay together and have a velocity of 3.0 m/s at an angle of 35° north of east immediately after the collision. What are the magnitude and direction of the velocity of the child just before the collision?

64. A barge with mass 1.50×10^5 kg is proceeding downriver at 6.2 m/s in heavy fog when it collides with a barge heading directly across the river (see Fig. 10-54). The second barge has mass 2.78×10^5 kg and before the collision is moving at 4.3 m/s. Immediately after impact, the second barge finds its course deflected by 18° in the downriver direction and its speed increased to 5.1 m/s. The river current is approximately zero at the time of the accident. (a) What are the speed and direction of motion of the first barge immediately after the collision? (b) How much kinetic energy is lost in the collision?

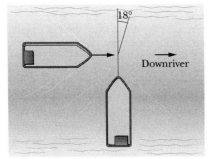

Fig. 10-54 Problem 64.

65. In a game of billiards, the cue ball is given an initial speed V and strikes the pack of 15 stationary balls. All 16 balls then engage in numerous ball–ball and ball–cushion collisions. Some time later, all 16 balls have the same speed v (as improbable as that is). Assuming that all collisions are elastic and ignoring the rotational aspect of the balls' motion, calculate v in terms of V.

66. A platform scale is calibrated to indicate the mass in kilograms of an object placed on it. Particles, initially at rest, fall from a height of 3.5 m and collide with the platform of the scale. The collisions are elastic; the particles rebound upward with the same speed they had before hitting the platform. If each particle has a mass of 110 g and collisions occur at the rate of 42 s^{-1}, what is the average scale reading?

67. In Fig. 10-55, a 6.0 kg block and a 4.0 kg block are moving on a frictionless surface. A spring of spring constant $k = 8000$ N/m is fixed to the 4.0 kg block. The 6.0 kg block has an initial velocity of 8.0 m/s toward the right, and the 4.0 kg block has an initial velocity of 2.0 m/s toward the right. Eventually, the larger block overtakes the smaller block. (a) What is the velocity of the 4.0 kg block at the instant the 6.0 kg block has a velocity of

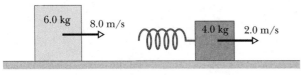

Fig. 10-55 Problem 67.

6.4 m/s toward the right? (b) What is the elastic potential energy of the spring just then?

68. In Fig. 10-56, a 3.2 kg box of running shoes slides on a horizontal frictionless table and collides with a 2.0 kg box of ballet slippers initially at rest on the edge of the table. The speed of the 3.2 kg box is 3.0 m/s just before the collision. If the two boxes stick together because of packing tape on their sides, what is their kinetic energy just before they strike the floor 0.40 m below the table's surface?

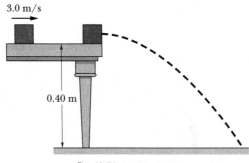

Fig. 10-56 Problem 68.

69. The only force acting on a 1.6 kg stone moving along an x axis is given by $\vec{F} = (16 - t^2)\hat{i}$ N for $0 \le t \le 8.0$ s. The velocity of the stone at $t = 0$ is zero. (a) What is the velocity of the stone at $t = 3.0$ s? (b) At what value of t will the velocity of the stone again be equal to zero? (c) What maximum velocity in the positive x direction is reached by the stone?

70. A 60.0 kg diver has a downward velocity of 3.00 m/s just before making contact with a diving board. The diver leaves the board 1.20 s later with a velocity of 5.00 m/s at an angle of 40.0° with the vertical, as shown in Fig. 10-57. (a) What is the magnitude of the average *net* force on the diver while she is in contact with the board? (b) What is the magnitude of the average force of the board on the diver while she is in contact with the board? (*Hint:* The *net* force includes this force from the board *and* the gravitational force on her.)

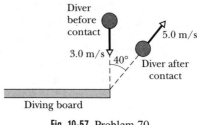

Fig. 10-57 Problem 70.

71. A movie-set machine gun fires 50 g bullets, each at a speed of 1000 m/s. An actor, holding the machine gun in his hands, can exert an average horizontal force of magnitude 180 N against the gun. Determine the maximum number of bullets he can fire per minute while still holding the gun steady.

72. The linear momentum of a 1500 kg car increased in magnitude by 9.0×10^3 kg · m/s in 12 s. (a) What is the magnitude of the constant force that accelerated the car? (b) By how much did the speed of the car increase?

73. Two vehicles A and B are traveling west and south, respectively, toward the same intersection, where they collide and lock together. Before the collision, A (total weight 12.0 kN) has a speed of 64.4 km/h, and B (total weight 16.0 kN) has a speed of 96.6 km/h. Find (a) the magnitude and (b) the direction of the velocity of the (interlocked) vehicles immediately after the collision, assuming the collision is isolated.

74. A 2.0 kg tin cookie, with an initial velocity of 8.0 m/s to the east, collides with a stationary 4.0 kg cookie tin. Just after the collision, the cookie has a velocity of 4.0 m/s at an angle of 37° north of east. Just then, what are (a) the magnitude and (b) the direction of the velocity of the cookie tin?

75. A 5.0 kg ball moving due east at 4.0 m/s collides with a 4.0 kg ball moving due west at 3.0 m/s. Just after the collision, the 5.0 kg ball has a velocity of 1.2 m/s, due south. (a) What is the magnitude of the velocity of the 4.0 kg ball just after the collision? (b) By how much is the mechanical energy of the two-ball system reduced by the collision?

76. Two pendulums, both of length l, are initially situated as in Fig. 10-58, with the center of one bob held a distance d above the center of the other bob. The left pendulum is released and strikes the other. Assume that the collision is completely inelastic, and neglect the mass of the strings and any frictional effects. How high does the center of mass of the pendulum system rise after the collision?

79. The bumper of a 1200 kg car is designed so that it can just absorb all the car's energy when the car runs head-on into a solid wall at 5.00 km/h. The car is involved in a collision in which it runs at 70.0 km/h into the rear of a 900 kg car moving at 60.0 km/h in the same direction. The 900 kg car is accelerated to 70.0 km/h as a result of the collision. (a) What is the speed of the 1200 kg car immediately after impact? (b) What is the ratio of the kinetic energy absorbed in the collision to that which can be absorbed by the bumper of the 1200 kg car?

80. In the ballistic pendulum of Sample Problem 10-2, assume the bullet's mass m is 8.00 g, the block's mass M is 7.00 kg, and the vertical distance h the block rises is 5.00 cm. (a) When the bullet is fired into the block, what fraction of the bullet's initial kinetic energy remains as mechanical energy of the bullet–block pendulum after the collision? (b) If we increase the initial speed of the bullet, does that fraction increase, decrease, or remain the same? Why? (c) What is that fraction in terms of the symbols used in the sample problem?

81. A remote-controlled toy car of mass 2.0 kg starts from rest at the origin at $t = 0$ and moves in the positive direction of an x axis. The net force on the car as a function of time is given by Fig. 10-60. (a) What is the time rate of change of the momentum of the car at $t = 3.0$ s? (b) What is the momentum of the car at $t = 3.0$ s?

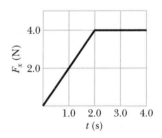

Fig. 10-60 Problem 81.

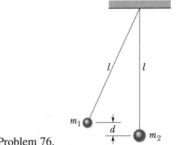

Fig. 10-58 Problem 76.

77. In Fig. 10-59, a target glider, whose mass m_2 is 350 g, is at rest on an air track, a distance $d = 53$ cm from the left end of the track. A projectile glider, whose mass m_1 is 590 g, approaches the target glider with velocity $v_{1i} = -75$ cm/s and collides elastically with it. Then the target glider rebounds elastically from a spring at the left end of the track and meets the projectile glider for a second time. (Assume the spring has a negligible length when compressed by the glider.) How far from the left end of the track does this second collision occur?

82. An initially stationary croquet ball with mass 0.50 kg is struck by a mallet. A graph of the force magnitude versus time is shown in Fig. 10-61. By analyzing the graph, determine the ball's speed just after the force magnitude has become zero.

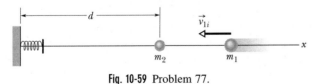

Fig. 10-59 Problem 77.

78. A hovering fly is approached by an enraged elephant charging at 2.1 m/s. Assuming that the collision is elastic, at what speed does the fly rebound? Note that the projectile (the elephant) is much more massive than the stationary target (the fly).

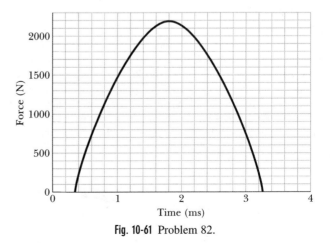

Fig. 10-61 Problem 82.

83. A golfer hits a golf ball, giving it an initial velocity of magnitude 50 m/s directed 30° above the horizontal. Assuming that the mass of the ball is 46 g and the club and ball are in contact for 1.7 ms, find (a) the impulse on the ball, (b) the impulse on the club, (c) the average force on the ball from the club, and (d) the work done on the ball by the club.

84. In Fig. 10-62, a stream of water strikes a stationary "dished" turbine blade. The speed of the water is v, both before and after it strikes the curved surface of the blade, and the rate dm/dt at which the mass strikes the blade is constant. Find the magnitude of the force on the blade from the water.

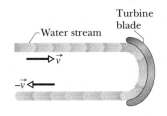

Fig. 10-62 Problem 84.

85. A stream of water from a hose is sprayed directly on a wall. If the speed of the water is 5.0 m/s and the hose sprays 300 cm³/s, what is the magnitude of the average force on the wall from the stream of water? Assume that the water does not spatter back appreciably. Each cubic centimeter of water has a mass of 1.0 g.

86. *Left alone, write your own.* For one or more of the following situations, write problems involving physics in this chapter, writing in the style of the sample problems, providing realistic data, graphs of the variables, and explained solutions: (a) pinball collisions, (b) collisions of dodge-em cars, (c) hail damaging cars, (d) professional boxing, (e) collisions in any ball sport.

CLUSTERED PROBLEMS

Cluster 1

87. A projectile body of mass m_1 and initial velocity v_{1i} collides with an initially stationary target body of mass m_2 in a one-dimensional collision. What are the velocities of the bodies after the collision if (a) the bodies stick together (as in Fig. 10-9) and (b) the collision is elastic (as in Fig. 10-13)?

88. A projectile body of mass m_1 and initial velocity $v_{1i} = 10.0$ m/s collides with an initially stationary target body of mass $m_2 = 2.00m_1$ in a one-dimensional collision. We do not know whether the collision is elastic, but we do know that there is no energy production during the collision (such as with an explosion between the bodies). (a) What is v_{2f} as a function of v_{1f}? (b) Plot this function. What are (c) the greatest possible value of v_{1f}, (d) the corresponding value of v_{2f}, and (e) the circumstance that produces this limiting value of v_{1f}? (f) What is the physical reason why v_{1f} cannot be greater than this limiting value? (g) On your plot of v_{2f} versus v_{1f}, mark the point corresponding to this limiting value.

What are (h) the least (most negative) possible value of v_{1f}, (i) the corresponding value of v_{2f}, and (j) the circumstance that produces this limiting value of v_{1f}? (k) What is the physical reason why v_{1f} cannot be smaller (even more negative) than this limiting value? (l) On your plot of v_{2f} versus v_{1f}, mark the point corresponding to this limiting value, then label the physically possible portion of the plot and indicate the physical reasons for the two impossible portions.

89. Repeat Problem 88 but with $m_2 = 0.500m_1$. Here, however, the least possible value of v_{1f} is positive.

Cluster 2

For the problems in this cluster, consider the two-dimensional collision of Fig. 10-16 with the following data: the projectile body has mass m_1, an initial velocity of magnitude $v_{1i} = 10.0$ m/s, and a final angle of $\theta_1 = 30.0°$; the initially stationary target body has mass $m_2 = 2.00m_1$. In solving Problems 91 and 92, you might make use of the trigonometric identity

$$\sin^2 \theta + \cos^2 \theta = 1.$$

90. See the setup for this cluster. We also know that the projectile body has a final velocity $\vec{v}_{1f}$ of magnitude 5.00 m/s, but we do not know whether the collision is elastic. (a) After the collision, what are the speed v_{2f} of the target body and the angle θ_2 of its travel? (b) Is the collision elastic? If not, what fraction of the initial kinetic energy of the projectile body is transformed into other types of energy?

91. See the setup for this cluster. We also know that the collision is elastic. After the collision, what are (a) the speed v_{1f} of the projectile body, (b) the speed v_{2f} of the target body, and (c) the angle θ_2 of the target body's travel?

92. See the setup for this cluster. We do not know whether the collision is elastic. (a) What is the final speed v_{2f} of the target body as a function of the final speed v_{1f} of the projectile body? (b) Plot this function. What are (c) the greatest possible value of v_{1f}, (d) the corresponding value of v_{2f}, and (e) the circumstance that produces this limiting value of v_{1f}? (f) What is the physical reason why v_{1f} cannot be greater than this limiting value? (g) On your plot of v_{2f} versus v_{1f}, mark the point corresponding to this limiting value.

(h) What is the angle θ_2 of the target body's motion as a function of the final speed v_{1f} of the projectile body? (i) Plot this function. (j) What is the value of θ_2 corresponding to the greatest value of v_{1f}? (k) On your plot of θ_2 versus v_{1f}, mark the point corresponding to this limiting value.

What are (l) the least possible value of v_{1f}, (m) the corresponding value of v_{2f}, (n) the corresponding value of θ_2, and (o) the circumstance that produces this limiting value of v_{1f}?

11

Rotation

Sample Problem 11-11

A child's top is spun with angular acceleration

$$\alpha = 5t^3 - 4t,$$

with t in seconds and α in radians per second–squared. At $t = 0$, the top has angular velocity 5 rad/s, and a reference line on it is at angular position $\theta = 2$ rad.

(a) Obtain an expression for the angular velocity $\omega(t)$ of the top.

SOLUTION: The **Key Idea** here is that, by definition, $\alpha(t)$ is the derivative of $\omega(t)$ with respect to time. Thus, we can find $\omega(t)$ by integrating $\alpha(t)$ with respect to time; that is, Eq. 11-8 tells us

$$d\omega = \alpha \, dt,$$

so

$$\int d\omega = \int \alpha \, dt.$$

From this we find

$$\omega = \int (5t^3 - 4t) \, dt = \tfrac{5}{4}t^4 - \tfrac{4}{2}t^2 + C.$$

To evaluate the constant of integration C, we note that $\omega =$

5 rads/s at $t = 0$. Substituting these values in our expression for ω yields

$$5 \text{ rad/s} = 0 - 0 + C,$$

so $C = 5$ rad/s. Then

$$\omega = \tfrac{5}{4}t^4 - 2t^2 + 5. \qquad \text{(Answer)}$$

(b) Obtain an expression for the angular position $\theta(t)$ of the top.

SOLUTION: Here the **Key Idea** is similar: By definition, $\omega(t)$ is the derivative of $\theta(t)$ with respect to time. Therefore, we can find $\theta(t)$ by integrating $\omega(t)$ with respect to time. Since Eq. 11-6 tells us that

$$d\theta = \omega \, dt,$$

we can write

$$\theta = \int \omega \, dt = \int (\tfrac{5}{4}t^4 - 2t^2 + 5) \, dt$$

$$= \tfrac{1}{4}t^5 - \tfrac{2}{3}t^3 + 5t + C'$$

$$= \tfrac{1}{4}t^5 - \tfrac{2}{3}t^3 + 5t + 2, \qquad \text{(Answer)}$$

where C' has been evaluated by noting that $\theta = 2$ rad at $t = 0$.

Sample Problem 11-12

During an analysis of a helicopter engine, you determine that the angular velocity of the rotor blades changes from $\omega_0 = 320$ rev/min to $\omega_1 = 225$ rev/min in a time interval $\Delta t_1 = 1.50$ min as the rotor slows to rest at constant acceleration.

(a) What time interval Δt_2 is required for the blades to come to rest from their initial angular velocity ω_0?

SOLUTION: One **Key Idea** here is that the angular acceleration α of the blades is constant. Therefore, we can use the angular equations of Table 11-1. Because we are given data for angular velocities and one time interval, we try Eq. 11-12 ($\omega = \omega_0 + \alpha t$). For the time interval Δt_1, the initial velocity is ω_0 and the final velocity is ω_1. So Eq. 11-12 gives us

$$\alpha = \frac{\omega_1 - \omega_0}{\Delta t_1}.$$

Similarly, for the interval Δt_2, the initial velocity is ω_0 and the final velocity is ω_2 ($= 0$). Now Eq. 11-12 gives us

$$\alpha = \frac{\omega_2 - \omega_0}{\Delta t_2}.$$

Because α is the same in both time intervals, we can write

$$\frac{\omega_1 - \omega_0}{\Delta t_1} = \frac{\omega_2 - \omega_0}{\Delta t_2}$$

or

$$\frac{225 \text{ rev/min} - 320 \text{ rev/min}}{1.50 \text{ min}} = \frac{0 - 320 \text{ rev/min}}{\Delta t_2}.$$

Thus,

$$\Delta t_2 = 5.05 \text{ min.} \qquad \text{(Answer)}$$

(b) How many revolutions do the blades make in coming to rest from the initial velocity ω_0?

SOLUTION: The **Key Idea** is again that the equations for constant angular acceleration apply. We want an equation that contains the angular displacement $\theta - \theta_0$. If we do not have the list of Table

11-1, we would solve the basic equations (Eqs. 11-12 and 11-13) simultaneously for $\theta - \theta_0$ by eliminating the unknown angular acceleration α. If we do have the list, we use Eq. 11-15 because we have data for all its variables except the angular displacement $\theta - \theta_0$. We then find

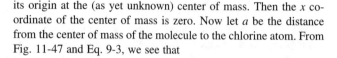

$$\theta - \theta_0 = \tfrac{1}{2}(\omega_0 + \omega_2)t$$
$$= \tfrac{1}{2}(320 \text{ rev/min} + 0)(5.05 \text{ min}) = 808 \text{ rev.} \quad \text{(Answer)}$$

Sample Problem 11-13

A hydrogen chloride molecule consists of a hydrogen atom whose mass m_H is 1.01 u (atomic mass units) and a chlorine atom whose mass m_{Cl} is 35.0 u. The centers of the two atoms are a distance $d = 1.27 \times 10^{-10}$ m = 127 pm apart (Fig. 11-47). What is the rotational inertia I_{com} of the molecule about an axis perpendicular to the line joining the two atoms and passing through the center of mass of the molecule? Work in atomic mass units and picometers.

SOLUTION: The **Key Idea** is that, because we have only two atoms, we can find the rotational inertia I_{com} with Eq. 11-26. However, first we must locate the center of mass so that we know the perpendicular distance of each atom from it.

We place an x axis along the line joining the two atoms, with

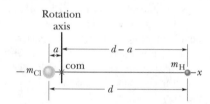

Fig. 11-47 Sample Problem 11-13. A hydrogen chloride molecule, shown schematically. A rotation axis extends through its center of mass and is perpendicular to the line joining the two atomic centers.

its origin at the (as yet unknown) center of mass. Then the x coordinate of the center of mass is zero. Now let a be the distance from the center of mass of the molecule to the chlorine atom. From Fig. 11-47 and Eq. 9-3, we see that

$$0 = \frac{-m_{Cl}a + m_H(d - a)}{m_{Cl} + m_H}$$

or
$$m_{Cl}a = m_H(d - a).$$

This yields

$$a = \frac{m_H}{m_{Cl} + m_H}\, d. \qquad (11\text{-}57)$$

Now, from Eq. 11-26, the rotational inertia about an axis through the center of mass is

$$I_{com} = \sum m_i r_i^2 = m_H(d - a)^2 + m_{Cl}a^2.$$

Substituting for a from Eq. 11-57 leads, after some algebra, to

$$I_{com} = d^2 \frac{m_H m_{Cl}}{m_{Cl} + m_H} = (127 \text{ pm})^2 \frac{(1.01 \text{ u})(35.0 \text{ u})}{35.0 \text{ u} + 1.01 \text{ u}}$$
$$= 15{,}800 \text{ u} \cdot \text{pm}^2. \qquad \text{(Answer)}$$

These units are commonly used for the rotational inertias of molecules.

Sample Problem 11-14

Figure 11-48 shows a thin, uniform rod of mass M and length L, on an x axis with the origin at the rod's center.

(a) What is the rotational inertia of the rod about the rotation axis shown, which is perpendicular to the rod and through its center?

SOLUTION: One **Key Idea** is that, because the rod is uniform, the rod's center of mass is at the rod's center. Therefore, we are looking for I_{com}. A second **Key Idea** is that, because the rod is a continuous object, we must use the integral of Eq. 11-28,

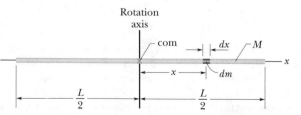

Fig. 11-48 Sample Problem 11-14. A uniform rod of length L and mass M. An element of mass dm and length dx is represented.

$$I = \int r^2 \, dm, \qquad (11\text{-}58)$$

to find the rotational inertia. However, we want to integrate with respect to coordinate x (not mass m as indicated in the integral), so we must relate the mass dm of an element of the rod to its length dx along the rod. (Such an element is shown in Fig. 11-48.) Because the rod is uniform, the ratio of mass to length is the same for all the elements and for the rod as a whole. Thus, we can write

$$\frac{\text{element's mass } dm}{\text{element's length } dx} = \frac{\text{rod's mass } M}{\text{rod's length } L}$$

or
$$dm = \frac{M}{L}\, dx.$$

We can now substitute this for dm and x for r in Eq. 11-58. Then we integrate from end to end of the rod (from $x = -L/2$ to $x = L/2$) to include all the elements. We find

$$I = \int_{x=-L/2}^{x=+L/2} x^2 \left(\frac{M}{L}\right) dx$$

$$= \frac{M}{3L} \left[x^3\right]_{-L/2}^{+L/2} = \frac{M}{3L}\left[\left(\frac{L}{2}\right)^3 - \left(-\frac{L}{2}\right)^3\right]$$

$$= \tfrac{1}{12}ML^2. \qquad \text{(Answer)}$$

This agrees with the result given in Table 11-2e.

(b) What is the rod's rotational inertia I about a new rotation axis that is perpendicular to the rod and through the left end?

SOLUTION: Here is the more difficult of two ways to solve for I: We use the same **Key Ideas** as previously but this time shift the origin

of the x axis to the left end of the rod and then integrate from $x = 0$ to $x = L$. Doing so, we find

$$I = \tfrac{1}{3}ML^2. \qquad \text{(Answer)}$$

The easier way is to use another **Key Idea:** If we place the axis at the rod's end so that it is parallel to the axis through the center of mass, then we can use the parallel-axis theorem (Eq. 11-29). We know from part (a) that I_{com} is $\tfrac{1}{12}ML^2$. From Fig. 11-48, the perpendicular distance h between the new rotation axis and the center of mass is $\tfrac{1}{2}L$. Equation 11-29 then gives us

$$I = I_{\text{com}} + Mh^2 = \tfrac{1}{12}ML^2 + (M)(\tfrac{1}{2}L)^2 = \tfrac{1}{3}ML^2. \quad \text{(Answer)}$$

Actually, this result holds for any axis through the left or right end that is perpendicular to the rod, whether it is parallel to the axis shown in Fig. 11-48 or not.

QUESTIONS

13. *Organizing question:* A flywheel is spun about its central axis in the four situations given in the table. (a) For situations A and B, set up equations, complete with known data, to find the time t the flywheel takes to rotate to the angular position of $+1.0$ rad. (b) Without solving those equations, determine whether the time to reach $+1.0$ rad in situation A is greater than, less than, or the same as that in situation B.

(c) Next, for situations C and D, set up equations, complete with known data, to find the angular position θ of the flywheel when it has an angular velocity of 8 rad/s. (d) Without solving those equations, determine whether the magnitude of the angle θ in situation C is greater than, less than, or the same as that in situation D.

Situation	A	B	C	D
Initial angular position (rad)	0	0	0	0
Initial angular velocity (rad/s)	$+5$	-5	-5	$+5$
Constant angular acceleration (rad/s²)	$+2$	$+2$	$+2$	-2

14. Figure 11-49 gives the angular acceleration $\alpha(t)$ of a disk that rotates like a merry-go-round. In which of the time periods indicated, if any, does the disk move at constant angular speed?

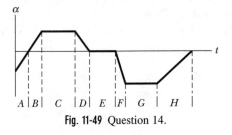

Fig. 11-49 Question 14.

15. *Organizing question:* Figure 11-50 shows, from overhead, four situations in which a lazy Susan turns like a merry-go-round without friction while one or two forces are applied to its rim. The

magnitudes of those forces are $F_1 = 1$ N and $F_2 = 2$ N; during the rotation the forces maintain the angles shown; the radius of the lazy Susan is 0.5 m; the rotational inertia of the lazy Susan is 2 kg · m²; initially the angular velocity is 3 rad/s counterclockwise. (a) For each situation, set up an equation, complete with known data, to find the angular acceleration α of the lazy Susan. (b) Assuming that we then know α, set up an equation, complete with known data, to find the angular displacement $\theta - \theta_0$ of the lazy Susan during the first 2 s of rotation.

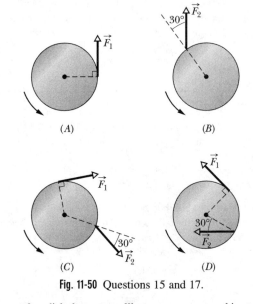

Fig. 11-50 Questions 15 and 17.

16. At $t = 0$, a disk that rotates like a merry-go-round is at angular position $\theta_0 = -2$ rad. Here are the signs of the disk's initial angular velocity and constant angular acceleration, respectively, for four situations: (1) +, +; (2) +, −; (3) −, +; (4) −, −. For which situations will the disk (a) undergo a momentary stop, (b) definitely pass through angular position $\theta = 0$ (given enough time), and (c) definitely not pass through that angular position?

17. *Organizing question:* Figure 11-50 shows, from overhead, four situations in which a lazy Susan turns like a merry-go-round without friction while one or two forces are applied to its rim. The magnitudes of those forces are $F_1 = 1$ N and $F_2 = 2$ N; during the rotation the forces maintain the angles shown; the radius of the lazy Susan is 0.5 m; the rotational inertia of the lazy Susan is 2 kg $\cdot$ m^2. In each situation, the lazy Susan begins with angular velocity $\omega_i = 3$ rad/s counterclockwise and rotates through an angular displacement of 1.2 rad counterclockwise.

(a) For each situation, determine whether the torques associated with the one or two forces tend to transfer energy *to* or *from* the merry-go-round during the 1.2 rad rotation. (b) For each situation, set up an equation, complete with known data, to find the rotational kinetic energy K_f of the lazy Susan at the end of the 1.2 rad rotation.

18. Figure 11-51a shows a meter stick, half wood and half steel, that is pivoted at the wood end at O. A force $\vec{F}$ is applied to the steel end at a. In Fig. 11-51b, the stick is reversed and pivoted at the steel end at O', and the same force is applied at the wood end at a'. Is the resulting angular acceleration of Fig. 11-51a greater than, less than, or the same as that of Fig. 11-51b?

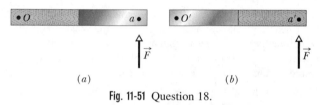

(a)   (b)

Fig. 11-51 Question 18.

19. Figure 11-52 shows three uniform disks, along with their radii R and masses M. Rank the disks according to their rotational inertias about their central axes, greatest first.

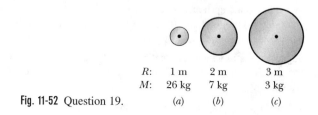

R:	1 m	2 m	3 m
M:	26 kg	7 kg	3 kg
	(a)	(b)	(c)

Fig. 11-52 Question 19.

20. Five solids with identical masses are shown in cross section in Fig. 11-53. The cross sections have equal widths at the widest part and equal heights (but not necessarily equal thicknesses). (a) Which one has the greatest rotational inertia about an axis through its center of mass and perpendicular to its cross section? (b) Which has the least?

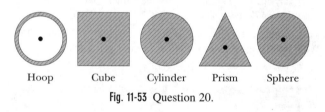

Hoop   Cube   Cylinder   Prism   Sphere

Fig. 11-53 Question 20.

21. For a disk that rotates like a merry-go-round, here are the initial and final angular velocities, respectively, in four situations: (a) 2 rad/s, 3 rad/s; (b) −2 rad/s, 3 rad/s; (c) −2 rad/s, −3 rad/s;

(d) 2 rad/s, −3 rad/s. The magnitude of the angular acceleration of the disk has the same constant value in all four situations. Rank the situations according to the magnitude of the disk's displacement, greatest first, during the change from the initial to the final angular velocity.

22. In Fig. 11-54, assume that the vertical axis gives the angular velocity ω of a lazy Susan turning like a merry-go-round. Then determine which of the 10 plots of ω versus time t best describes the motion for the following four situations. (Plots 2, 3, 7, and 9 are straight; the others are curved.)

Situation	(a)	(b)	(c)	(d)
Initial θ (rad)	+10	−10	+10	−10
Initial ω (rad/s)	+5	−5	−5	+5
Constant α (rad/s^2)	+2	−2	+2	−2

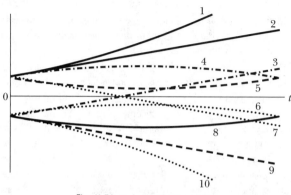

Fig. 11-54 Questions 22 and 23.

23. Question 22 continued: If, instead, the vertical axis gives the angular position θ of the lazy Susan, then which of the 10 plots best describes the motion for the given four sets of values?

24. In Fig. 11-55 forces $\vec{L}$ and $\vec{R}$ are applied to an oddly shaped plate of iron that can rotate about a perpendicular axis through point O. The tangential and radial vector components of those applied

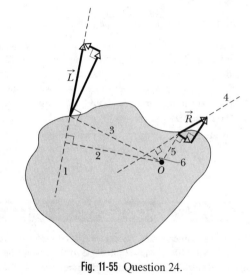

Fig. 11-55 Question 24.

forces are also shown, all drawn to the same scale. With respect to *O*, which applied force has (a) the greater tangential component and (b) the greater radial component? Which of the dashed lines are (c) lines of actions of the applied forces, (d) moment arms to the applied forces, and (e) moment arms to the tangential components? Do forces (f) $\vec{L}$ and (g) $\vec{R}$ tend to rotate the iron plate clockwise or counterclockwise?

25. A force is applied to the rim of a disk that can rotate like a merry-go-round, so as to change its angular velocity. Its initial and final angular velocities, respectively, for four situations are: (a) −2 rad/s, 5 rad/s; (b) 2 rad/s; 5 rad/s; (c) −2 rad/s, −5 rad/s; and (d) 2 rad/s, −5 rad/s. Rank the situations according to the work done by the torque due to the force, greatest first.

DISCUSSION QUESTIONS

26. Could the angular quantities θ, ω, and α be expressed in degrees instead of radians in the rotational equations of Table 11-1? Explain.

27. In what sense is the radian a "natural" measure of angle and the degree an "arbitrary" measure of that same quantity? (Consider the definitions of the two measures.)

28. Does the vector representing the angular velocity of a wheel rotating about a fixed axis necessarily have to lie along that axis?

29. Experiment by rotating a book after the fashion of Fig. 11-7, but this time using angular displacements of 180° rather than 90°. What do you conclude about the final positions of the book? Does this change your mind about whether (finite) angular displacements can be treated as vectors?

30. Suppose you hold a coin flat against a table while you rotate a second coin around it without slippage (Fig. 11-56). When the second coin returns to its original position, through what angle has it turned?

Fig. 11-56   Question 30.

31. The rotation of the Sun can be monitored by tracking *sunspots,* magnetic storms on the Sun that appear dark against the otherwise bright solar disk. Figure 11-57*a* shows the initial positions of five spots, and Fig. 11-57*b* the positions of these same spots one solar rotation later. What can we conclude about the structure of the Sun from these two observations?

32. Why is it suitable to express α in revolutions per second-squared in the expression $\theta = \omega_0 t + \frac{1}{2}\alpha t^2$ but not in the expression $a_t = \alpha r$?

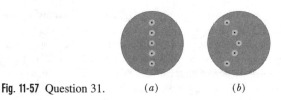

Fig. 11-57   Question 31.   (*a*)   (*b*)

33. A rigid body is free to rotate about a fixed axis. Can the body have nonzero angular acceleration even if the angular velocity of the body is (perhaps instantaneously) zero? What is the linear equivalent of this question? Give examples to illustrate both the angular and linear situations.

34. A golfer swings a golf club, making a long drive from the tee. Do all points on the club have the same angular velocity ω at every instant, as long as the club is in motion?

35. When we say that a point on the equator has an angular speed of 2π rad/day, what reference frame do we have in mind?

36. Taking into account Earth's rotation and orbiting (which are in the same direction), explain whether a tree moves faster during the day or during the night. With respect to what reference frame is your answer given?

37. Suppose that you were asked to determine the distance traveled by a needle in playing a phonograph record. What information would you need? Discuss the situation from the point of view of reference frames (a) fixed in the room, (b) fixed on the rotating record, and (c) fixed on the arm of the record player.

38. What is the relation between the angular velocities of a pair of coupled gears of different radii?

39. Can the mass of a body be considered as concentrated at the center of mass of the body for purposes of computing its rotational inertia? If yes, explain why. If no, offer a counterexample.

40. When you stand upright, about what axis is the rotational inertia of your body (a) the least and (b) the greatest? (c) For (a) and (b), how can you change the value of your rotational inertia?

41. Suggest ways in which the rotational inertia about a particular axis of a body with a very complicated shape could be measured experimentally.

42. Comment on each of these assertions about skiing. (a) In downhill racing, one wants skis that do not turn easily. (b) In slalom racing, one wants skis that turn easily. (c) Therefore, the rotational inertia of downhill skis should be larger than that of slalom skis. (d) Considering that there is low friction between skis and snow and that the skier's center of mass is about over the center of the skis, how does a skier exert torques to turn or to stop a turn? (From "The Physics of Ski Turns," by J. I. Shonie and D. L. Mordick, *The Physics Teacher,* December 1972.)

43. Consider a straight stick standing on end on (frictionless) ice. What will be the path of its center of mass if it falls?

EXERCISES & PROBLEMS

71. A uniform helicopter rotor blade (see Fig. 11-35) is 7.80 m long, has a mass of 110 kg, and is attached to the rotor axle by a single bolt. (a) What is the magnitude of the force on the bolt from

the axle when the rotor is turning at 320 rev/min? (*Hint:* For this calculation the blade can be considered to be a point mass at its center of mass. Why?) (b) Calculate the torque that must be applied to the rotor to bring it to full speed from rest in 6.7 s. Ignore air resistance. (The blade cannot be considered to be a point mass for this calculation. Why not? Assume the mass distribution of a uniform thin rod.) (c) How much work does the torque do on the blade in order for the blade to reach a speed of 320 rev/min?

72. Attached to each end of a thin steel rod of length 1.20 m and mass 6.40 kg is a small ball of mass 1.06 kg. The rod is constrained to rotate in a horizontal plane about a vertical axis through its midpoint. At a certain instant, it is rotating at 39.0 rev/s. Because of friction, it slows to a stop in 32.0 s. Assuming a constant frictional torque, compute (a) the angular acceleration, (b) the retarding torque due to friction, (c) the total energy transferred from mechanical energy to thermal energy by friction, and (d) the number of revolutions rotated during the 32.0 s. (e) Now suppose that the frictional torque is known not to be constant. Which, if any, of the quantities (a), (b), (c), or (d) can still be computed without additional information? If such quantities exist, give their value.

73. Between 1911 and 1990, the top of the leaning bell tower at Pisa, Italy, moved toward the south at an average rate of 1.2 mm/y. The tower is 55 m tall. In radians per second, what is the average angular speed of the tower's top about its base?

74. George Washington Gale Ferris, Jr., a civil engineering graduate from Rensselaer Polytechnic Institute, built the original Ferris wheel for the 1893 World's Columbian Exposition in Chicago. The wheel, an astounding engineering construction at the time, carried 36 wooden cars, each holding up to 60 passengers, around a circle 76 m in diameter. The cars were loaded 6 at a time, and once all 36 cars were full, the wheel made a complete rotation at constant angular speed in about 2 min. Estimate the amount of work that was required of the machinery to rotate the passengers alone.

75. You are hired as an expert witness on physics and engineering to give a deposition regarding a personal injury case that has been brought against the manufacturer of a certain type of picnic table. Each table consists of a tabletop and two benches that are fastened at each end by cross pieces to a solid rod running along the ground (Fig. 11-58). The injury reportedly occurred at a picnic of retired NFL linebackers when a linebacker was the first to sit down at one of these picnic tables. He claims that he squeezed his legs and abdomen into the space between the tabletop and the bench while bending over the tabletop. Then, when he sat down on the bench and straightened up, the table rotated as a whole around a point below him, throwing him backward onto the ground where he struck his head.

The dimensions of the table are d_1 = 60 cm, d_2 = 13 cm, d_3 = 5.0 cm, and d_4 = 18 cm; the mass of the table is 90 kg.

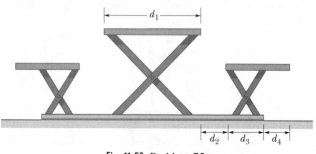

Fig. 11-58 Problem 75.

According to a previous deposition, the linebacker's mass that day was 130 kg and, when seated upright on the bench, his center of mass was d_5 = 20 cm beyond the outside edge of the bench. Assume that the gravitational force on him acts at his center of mass. (a) Was the claimed rotation physically possible? (b) Approximately what minimum mass is needed to cause such a rotation?

76. In Fig. 11-59, two blocks, of mass m_1 = 400 g and m_2 = 600 g, are connected by a massless cord that is wrapped around a uniform disk of mass M = 500 g and radius R = 12.0 cm. The disk can rotate without friction about a fixed horizontal axis through its center; the cord cannot slip on the disk. The system is released from rest. Find (a) the magnitude of the acceleration of the blocks, (b) the tension T_1 in the cord at the left, and (c) the tension T_2 in the cord at the right.

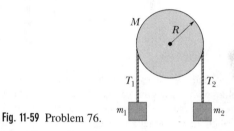

Fig. 11-59 Problem 76.

77. In Fig. 11-60, a wheel of radius 0.20 m is mounted on a frictionless horizontal axis. The rotational inertia of the wheel about the axis is 0.40 kg · m². A massless cord wrapped around the wheel's circumference is attached to a 6.0 kg box. A short time after being released from rest, the box has a kinetic energy of 6.0 J. Just then, what are (a) the rotational kinetic energy of the wheel and (b) the distance the box has fallen from its initial position?

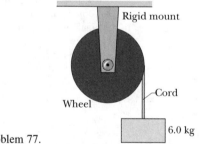
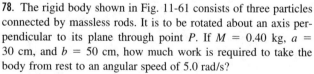
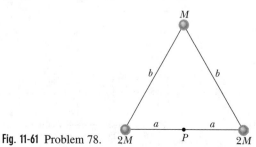

Fig. 11-60 Problem 77.

78. The rigid body shown in Fig. 11-61 consists of three particles connected by massless rods. It is to be rotated about an axis perpendicular to its plane through point P. If M = 0.40 kg, a = 30 cm, and b = 50 cm, how much work is required to take the body from rest to an angular speed of 5.0 rad/s?

Fig. 11-61 Problem 78.

79. Two identical blocks, each of mass M, are connected by a massless string over a pulley of radius R and rotational inertia I (Fig. 11-62). The string does not slip on the pulley; it is not known whether there is friction between the table and the sliding block; the pulley's axis is frictionless. When this system is released from rest, the pulley turns through an angle θ in time t and the acceleration of the blocks is constant. In terms of these symbols and g, what are (a) the magnitude of the pulley's angular acceleration, (b) the magnitude of either block's acceleration, (c) string tension T_1, and (d) string tension T_2?

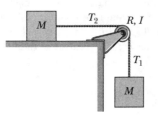

Fig. 11-62 Problem 79.

80. A small ball with mass 1.30 kg is mounted on one end of a rod 0.780 m long and of negligible mass. The system rotates in a horizontal circle about the other end of the rod at 5010 rev/min. (a) Calculate the rotational inertia of the system about the axis of rotation. (b) There is an air drag of 2.30×10^{-2} N on the ball, directed opposite its motion. What torque must be applied to the system to keep it rotating at constant speed?

81. The rigid object shown in Fig. 11-63 consists of three balls and three connecting rods, with $M = 1.6$ kg and $L = 0.60$ m. The balls may be treated as particles, and the connecting rods have negligible mass. Determine the rotational kinetic energy of the object if it has an angular speed of 1.2 rad/s about (a) an axis that passes through point P and is perpendicular to the plane of the figure and (b) an axis that passes through point P, is perpendicular to the rod of length $2L$, and lies in the plane of the figure.

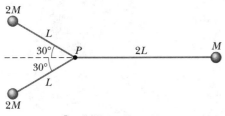

Fig. 11-63 Problem 81.

82. Two thin rods (each of mass 0.20 kg) are joined together to form a rigid body as shown in Fig. 11-64. One of the rods is 0.40 m long and the other is 0.50 m long. What is the rotational inertia of this rigid body about (a) an axis that is perpendicular to the plane of the paper and passes through the center of the shorter rod and (b) an axis that is perpendicular to the plane of the paper and passes through the center of the longer rod?

83. Derive the expression for the rotational inertia of a hoop of mass M and radius R about its central axis; see Table 11-2a.

84. The turntable of a record player has an angular speed of 3.5 rad/s at the instant it is turned off. The turntable stops 1.6 s after being turned off. The radius of the turntable is 15 cm. (a) If the angular acceleration is constant, through how many radians does the turntable turn after being turned off? (b) What is the magnitude of the linear acceleration of a point on the rim of the turntable 1.0 s after it is turned off? (*Hint:* The linear acceleration has both a radial and a tangential component.)

85. Assume that Earth's orbit about the Sun is a circle. With respect to the Sun, what are Earth's (a) angular speed, (b) linear speed, and (c) acceleration?

86. The turntable of a record player has an angular speed of 8.0 rad/s at the instant it is switched off. Three seconds later, the turntable has an angular speed of 2.6 rad/s. Through how many radians does the turntable rotate from the time it is turned off until it stops? (Assume constant angular acceleration.)

87. Starting from rest at $t = 0$, a wheel undergoes a constant angular acceleration. When $t = 2.0$ s, the angular velocity of the wheel is 5.0 rad/s. The acceleration continues until $t = 20$ s, when it abruptly ceases. Through what angle does the wheel rotate in the interval $t = 0$ to $t = 40$ s?

88. The flywheel of an engine is rotating at 25.0 rad/s. When the engine is turned off, the flywheel slows at a constant rate and stops in 20.0 s. Calculate (a) the angular acceleration of the flywheel, (b) the angle through which the flywheel rotates in stopping, and (c) the number of revolutions made by the flywheel in stopping.

89. A wheel rotates with angular acceleration $\alpha = 4at^3 - 3bt^2$, where t is the time and a and b are constants. If the wheel has initial angular velocity ω_0, write expressions for (a) the angular velocity and (b) the angular position of the wheel as functions of time.

90. Four particles, each of mass 0.20 kg, are placed at the vertices of a square with sides of length 0.50 m. The particles are connected by rods of negligible mass. This rigid body can rotate in a vertical plane about a horizontal axis A that passes through one of the particles. The body is released from rest with rod AB horizontal, as shown in Fig. 11-65. (a) What is the rotational inertia of the body about axis A? (b) What is the angular speed of the body about axis A at the instant rod AB swings through the vertical position?

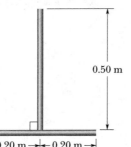

Fig. 11-64 Problem 82.

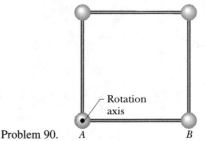

Fig. 11-65 Problem 90.

91. The thin uniform rod in Fig. 11-66 has mass 1.5 kg and length 2.0 m and can pivot about a horizontal, frictionless pin through one end. It is released from rest at an angle of 40° above the horizontal. Use the principle of conservation of energy to determine the angular speed of the rod as it passes through the horizontal position.

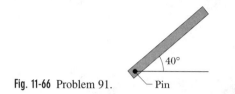

Fig. 11-66 Problem 91.

92. Two uniform solid spheres have the same mass of 1.65 kg but one has a radius of 0.226 m whereas the other has a radius of 0.854 m. Each can rotate about an axis through its center. (a) What is the magnitude τ of the torque required to bring the smaller sphere from rest to an angular speed of 317 rad/s in 15.5 s? (b) What is the magnitude F of the force that must be applied tangentially at the sphere's equator to give that torque? What are the corresponding values of (c) τ and (d) F for the larger sphere?

93. In Fig. 11-67, a wheel of radius 0.20 m is mounted on a frictionless horizontal axle. A massless cord is wrapped around the wheel and attached to a 2.0 kg box that slides on a frictionless surface inclined at an angle of 20° with the horizontal. The box accelerates down the incline at 2.0 m/s². What is the rotational inertia of the wheel about the axle?

Fig. 11-67 Problem 93.

94. A bicyclist of mass 70 kg puts all his mass on each downward-moving pedal as he pedals up a steep road. Take the diameter of the circle in which the pedals rotate to be 0.40 m, and determine the magnitude of the maximum torque he exerts.

95. Two thin disks, each of mass 4.0 kg and radius 0.40 m, are attached as shown in Fig. 11-68 to form a rigid body. What is the rotational inertia of this body about an axis A that is perpendicular

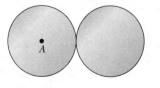

Fig. 11-68 Problem 95.

to the plane of the disks and passes through the center of one of the disks?

96. Three particles, each of mass 0.50 kg, are positioned at the vertices of an equilateral triangle with sides of length 0.60 m. The particles are connected by rods of negligible mass. What is the rotational inertia of this rigid body about (a) an axis that passes through one of the particles and is parallel to the rod connecting the other two, (b) an axis that passes through the midpoint of one of the sides and is perpendicular to the plane of the triangle, and (c) an axis that is parallel to one side of the triangle and passes through the midpoints of the other two sides?

97. Show that the axis about which a given rigid body has its smallest rotational inertia must pass through its center of mass.

98. A car starts from rest and moves around a circular track of radius 30.0 m. Its speed increases at the constant rate of 0.500 m/s². (a) What is the magnitude of its *net* linear acceleration 15.0 s later? (b) What angle does this net acceleration vector make with the car's velocity at this time?

99. If an airplane propeller rotates at 2000 rev/min while the airplane flies at a speed of 480 km/h relative to the ground, what is the speed of a point on the tip of the propeller, at radius 1.5 m, as seen by (a) the pilot and (b) an observer on the ground? The plane's velocity is parallel to the propeller's axis of rotation.

100. A point on the rim of a 0.75-m-diameter grinding wheel changes speed at a constant rate from 12 m/s to 25 m/s in 6.2 s. What is the average angular acceleration of the wheel?

101. A wheel turns at constant angular acceleration through 90 rev in 15 s, reaching an angular speed of 10 rev/s. (a) What is its angular speed at the beginning of the 15 s interval? (b) How much time elapses between the time the wheel is at rest and the beginning of the 15 s interval?

102. A pulley wheel that is 8.0 cm in diameter has a 5.6-m-long cord wrapped around its periphery. Starting from rest, the wheel is given a constant angular acceleration of 1.5 rad/s². (a) Through what angle must the wheel turn for the cord to unwind? (b) How long does the unwinding take?

103. A good baseball pitcher can throw a baseball toward home plate at 85 mi/h with a spin of 1800 rev/min. How many revolutions does the baseball make on its way to home plate? For simplicity, assume that the 60 ft trajectory is a straight line.

104. *Left alone, write your own.* For one or more of the following situations, write problems involving physics in this chapter, writing in the style of the sample problems and providing realistic data, graphs of the variables, and explained solutions: (a) upright spinning tops, (b) multiple forces causing the rotation of an abstract sculpture about a fixed axis, (c) several objects in a rotation race, (d) judo throws other than the basic hip throw, (e) aikido techniques, (f) arm-wrestling.

12

Rolling, Torque, and Angular Momentum

ADDITIONAL SAMPLE PROBLEMS

Sample Problem 12-10

Consider a hoop, a disk, and a sphere, each of mass M and radius R, that roll smoothly along a horizontal table. For each, what fraction of its kinetic energy is associated with the translation of its center of mass?

SOLUTION: The **Key Idea** here is that the kinetic energy of a smoothly rolling body is the sum of its translational kinetic energy ($\frac{1}{2}Mv_{com}^2$) and its rotational kinetic energy ($\frac{1}{2}I_{com}\omega^2$), as stated by Eq. 12-5. Therefore, the fraction of the kinetic energy associated with translation is

$$\text{frac} = \frac{\frac{1}{2}Mv_{com}^2}{\frac{1}{2}Mv_{com}^2 + \frac{1}{2}I_{com}\omega^2}. \qquad (12\text{-}43)$$

We can greatly simplify the right side of Eq. 12-43 by substituting v_{com}/R for ω (Eq. 12-2) and realizing that the expressions for rotational inertia in Table 11-2 are all of the form βMR^2, where β is a numerical coefficient (the "front number"). Here β is 1 for a hoop, $\frac{1}{2}$ for a disk, and $\frac{2}{5}$ for a sphere. Thus, we can substitute βMR^2 for I_{com} in Eq. 12-43.

After these substitutions and some cancellations, Eq. 12-43 becomes

$$\text{frac} = \frac{1}{1 + \beta}. \qquad (12\text{-}44)$$

Now, substituting the β values for the hoop, disk, and sphere, we can generate Table 12-2 to show the fractional splits of translational

and rotational kinetic energy. For example, 0.67 of the kinetic energy of the disk is associated with the translation.

The relative split between translational and rotational energy depends on the relative size of the rotational inertia of the rolling object. As Table 12-2 shows, the rolling object (the hoop) that has its mass farthest from the central axis of rotation (and so has the largest rotational inertia) has the largest share of its kinetic energy in rotational motion. The object (the sphere) that has its mass closest to the central axis of rotation (and so has the smallest rotational inertia) has the smallest share in rotational motion.

TABLE 12-2 The Relative Splits Between Rotational and Translational Energy for Rolling Objects

Object	Rotational Inertia I_{com}	Fraction of Energy in	
		Translation	Rotation
Hoop	$1MR^2$	0.50	0.50
Disk	$\frac{1}{2}MR^2$	0.67	0.33
Sphere	$\frac{2}{5}MR^2$	0.71	0.29
General[a]	βMR^2	$\dfrac{1}{1 + \beta}$	$\dfrac{\beta}{1 + \beta}$

[a]β may be computed for any rolling object as I_{com}/MR^2.

Sample Problem 12-11

A uniform hoop, disk, and sphere, with the same mass M and same radius R, are released simultaneously from rest at the top of a ramp of length $L = 2.5$ m and angle $\theta = 12°$ with the horizontal. The objects roll smoothly down the ramp.

(a) Which object wins the race to the bottom of the ramp?

SOLUTION: Two **Key Ideas** are these: First, the objects begin with the same mechanical energy E, because they start from rest and the same height. Second, E is conserved during the race to the bottom, because the only force doing work on the objects is the gravitational force. (The normal force on them from the ramp and the frictional

force at their point of contact with the ramp do not cause energy transfers.) Further, at any given point along the ramp, the objects must have the same kinetic energy K because the same amount of energy has been transferred from gravitational potential energy to kinetic energy.

If the objects were sliding down the ramp, this means they would have the same speed. However, another **Key Idea** is that they do not have the same speed v_{com} because each object shares its kinetic energy between its translational motion down the ramp and its rotational motion around its center of mass. As we saw in Sample Problem 12-10 and Table 12-2, the sphere has the greatest fraction (0.71) as translational energy, so it has the greatest v_{com} and wins the race. Figure 12-48 shows the order of the objects during the race.

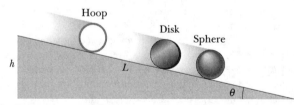

Fig. 12-48 Sample Problem 12-11. A hoop, a disk, and a sphere roll smoothly from rest down a ramp of angle θ.

(b) What is v_{com} for each object at the bottom of the ramp?

SOLUTION: Again, the **Key Idea** here is that mechanical energy is conserved. Let us choose the bottom of the ramp as our reference height for zero gravitational potential energy, so at the finish each object–Earth system has $U_f = 0$. The final kinetic energy at the finish is given by Eq. 12-5 for each object. The initial kinetic energy for all three objects is $K_i = 0$. The initial potential energy is $U_i = Mgh = Mg(L \sin \theta)$. Now we can write the conservation of mechanical energy ($E_f = E_i$) as

$$K_f + U_f = K_i + U_i$$

or

$$(\tfrac{1}{2}I_{com}\omega^2 + \tfrac{1}{2}Mv_{com}^2) + 0 = 0 + Mg(L \sin \theta).$$

Substituting $\omega = v_{com}/R$ and solving for v_{com} give us

$$v_{com} = \sqrt{\frac{2gL \sin \theta}{1 + I_{com}/MR^2}}, \quad \text{(Answer)} \quad (12\text{-}45)$$

which is the symbolic answer to the question.

Note that the speed depends not on the mass or the radius of the rolling object, but only on the distribution of its mass about its central axis, which enters through the term I_{com}/MR^2. A marble and a bowling ball will have the same speed at the bottom of the ramp and will thus roll down the ramp in the same time. A bowling ball will beat a disk of any mass or radius, and almost anything that rolls will beat a hoop.

For the rolling hoop (see the hoop listing in Table 12-2) we have $I_{com}/MR^2 = 1$, so Eq. 12-45 yields

$$v_{com} = \sqrt{\frac{2gL \sin \theta}{1 + I_{com}/MR^2}}$$

$$= \sqrt{\frac{(2)(9.8 \text{ m/s}^2)(2.5 \text{ m})(\sin 12°)}{1 + 1}}$$

$$= 2.3 \text{ m/s.} \quad \text{(Answer)}$$

From a similar calculation, we obtain $v_{com} = 2.6$ m/s for the disk ($I_{com}/MR^2 = \tfrac{1}{2}$) and 2.7 m/s for the sphere ($I_{com}/MR^2 = \tfrac{2}{5}$).

Sample Problem 12-12

A yo-yo has mass $M = 0.550$ kg, axle radius $R_0 = 3.2$ mm, and rotational inertia $I = 3.4 \times 10^{-4}$ kg·m² about its rotation axis. Assume that the string has negligible mass and thickness.

(a) What is the linear acceleration of the yo-yo as it rolls down the string from rest?

SOLUTION: The **Key Idea** here is that the linear acceleration a_{com} is determined by Newton's second law in its linear and angular forms. When those two equations are solved simultaneously, as suggested in Section 12-4, they yield Eq. 12-13. Substituting known data into that equation, we find

$$a_{com} = -\frac{g}{1 + I_{com}/MR_0^2}$$

$$= -\frac{9.8 \text{ m/s}^2}{1 + \dfrac{3.4 \times 10^{-4} \text{ kg·m}^2}{(0.550 \text{ kg})(0.0032 \text{ m})^2}}$$

$$= -0.16 \text{ m/s}^2. \quad \text{(Answer)}$$

The acceleration is directed downward and has this value whether the yo-yo is rolling down the string or climbing it.

(b) What is the tension in the string as the yo-yo descends?

SOLUTION: Note that the tension in the string is equal to the magnitude T of the force on the yo-yo from the string. A **Key Idea** here is that this force and the gravitational force are the only forces acting on the yo-yo, so they determine the acceleration we just calculated (see the free-body diagram of Fig. 12-8b). Thus, we can write Newton's second law for components along a vertical y axis ($F_{net,y} = ma_y$) as

$$T - Mg = Ma_{com}.$$

Solving for T and substituting $a_{com} = -0.16$ m/s², we have

$$T = M(a_{com} + g) = (0.550 \text{ kg})(-0.16 \text{ m/s}^2 + 9.8 \text{ m/s}^2)$$

$$= 5.3 \text{ N.} \quad \text{(Answer)}$$

This is the magnitude of the force on the yo-yo from the string, the tension in the string, and the magnitude of the force on your finger from the string, whether the yo-yo is rolling up or down the string.

QUESTIONS

14. A solid brass cylinder and a solid wood cylinder have the same radius and mass (the wood cylinder is longer). Released together from rest, they roll down an incline. (a) Which cylinder reaches the bottom first, or do they tie? (b) The wood cylinder is then shortened to match the length of the brass cylinder, and the brass cylinder is drilled out along its long axis to match the mass of the wood cylinder. Which cylinder now wins the race?

15. *Organizing question:* Figure 12-49 shows three situations in which an object of mass m, rotational inertia I about its center, and radius r rolls smoothly (and without slipping) up or down a track. For each situation, what are the object's initial total mechanical energy E_i and its final total mechanical energy E_f? In (a) the object has initial speed v_i, rolls up a slope and onto a plateau at vertical height h, and there has final speed v_f; in (b) the object has initial speed v_i and rolls up a slope to some maximum vertical height h (take this state as the final state); in (c) the object starts from rest at vertical height h and rolls down into a circular track of radius R and up to the top of that circular track, where it has speed v_f (take this state as the final state).

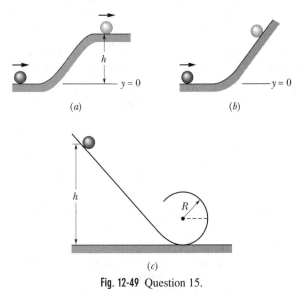

(a)   (b)

(c)

Fig. 12-49 Question 15.

16. What happens to the initially stationary yo-yo in Fig. 12-50 if you pull it via its string with (a) force $\vec{F}_2$ (the line of action passes through the point of contact on the table, as indicated), (b) force $\vec{F}_1$ (the line of action passes above the point of contact), and (c) force $\vec{F}_3$ (the line of action passes to the right of the point of contact)?

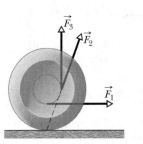

Fig. 12-50 Question 16.

17. In Fig. 12-51, a disk rolls smoothly (and without slipping) up an incline at angle θ. (a) What is the direction of the frictional force on the disk at P, the point of contact of the disk with the incline? (b) Which is greater in magnitude, that frictional force or the component of the gravitational force along the plane? (c) About the center, which is greater, the torque due to the gravitational force or the torque due to the frictional force on the disk? (d) Repeat part (c), but now for torques about point P. (e) If we decrease the angle θ, does the torque about P due to the gravitational force increase, decrease, or remain the same?

Fig. 12-51 Question 17.

18. A cannonball and a marble roll from rest down an incline without sliding. (a) Does the cannonball take more, less, or the same time as the marble to reach the bottom? (b) Is the fraction of the cannonball's kinetic energy that is associated with translation more than, less than, or the same as that of the marble?

19. The angular momenta $\ell(t)$ of a particle in four situations are (1) $\ell = 3t + 4$; (2) $\ell = -6t^2$; (3) $\ell = 2$; (4) $\ell = 4/t$. In which situation is the net torque on the particle (a) zero, (b) positive and constant, (c) negative and increasing in magnitude ($t > 0$), and (d) negative and decreasing in magnitude ($t > 0$)?

20. The table gives, for four situations and at the same instant, the sign of the angular momentum L and the sign of the change dL/dt of a merry-go-round about its center. For each situation, are (a) the angular speed and (b) the rotational kinetic energy of the merry-go-round increasing, decreasing, or staying the same?

	(1)	(2)	(3)	(4)
L	+	+	−	−
dL/dt	+	−	+	−

21. *Organizing question:* In the overhead view of Fig. 11-23, a 0.2 kg Texas cockroach rides the rim of a uniform 3.0 kg disk that is rotating like a merry-go-round. The angular speed is 1.5 rad/s; assume the cockroach is small enough to be a particle. The cockroach is to crawl to either (a) a point halfway to the center of the disk or (b) the center of the disk. For each situation, set up an equation, complete with known data, to find the angular speed ω_f of the disk and cockroach once the move is complete. (c) Reaching which final point, the halfway point or the center, requires more effort by the cockroach? In which situation, (a) or (b), will (d) ω_f and (e) the final rotational kinetic energy of the cockroach–disk system be greater?

22. If the rotating student in Fig. 12-16 were to drop the dumbbells, would the following, each relative to the rotation axis, increase, decrease, or remain the same: (a) the student's angular momentum, (b) the student's angular speed, (c) the angular momentum of the dumbbells, (d) the speed at which the dumbbells move vertically, (e) the speed at which they move horizontally, (f) the angular speed of the dumbbells, and (g) the angular momentum of the student–dumbbell system?

23. In ballet's *tour jeté* (or turning jump), the performer leaps from the floor with no apparent spin and then somehow turns on the spin while in midair with a simple rearrangement of arms and legs. Just before the performer lands, the spinning is somehow eliminated. Does this maneuver violate the law of conservation of angular momentum? How is it accomplished?

24. When an astronaut floats in a spacecraft while facing a wall, how can the astronaut turn to face another wall to the left or right without touching anything? Such motion is called *yaw*. How might the astronaut *pitch,* which is to rotate forward or backward around a horizontal axis that runs left and right? Is a *roll,* which is a rotation around a horizontal axis that runs forward and rearward, possible?

25. In Fig. 12-52, a disk is spinning freely at the bottom of an axle and with an angular momentum (about its center) of 50 units, counterclockwise. Four more spinning disks are to be dropped down the axle to land on and couple (via friction) to the first disk. Their angular momenta are: (1) 20 units clockwise, (2) 10 units counterclockwise, (3) 10 units clockwise, and (4) 60 units clockwise. (a) What is the final angular momentum of the system? (b) In what order should the four disks be dropped so that at one stage the disks that are then coupled on the axle are stationary?

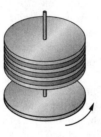

Fig. 12-52 Question 25.

DISCUSSION QUESTIONS

26. A cylindrical can filled with beef and an identical can filled with water both roll down an incline. Compare their angular and linear accelerations. Explain the difference.

27. If a car's speedometer is set to read at a speed proportional to the angular speed of its rear wheels, is it necessary to correct the reading when tires with a larger outside diameter (such as snow tires) are used?

28. Two heavy disks are connected by a short rod of much smaller radius. This system is placed on a ramp so that the disks hang over the sides as in Fig. 12-53. The system rolls down the ramp without slipping. (a) Near the bottom of the ramp the disks touch the horizontal table and the system takes off with greatly increased translational speed. Explain why. (b) If this system raced a hoop (of any radius) down the ramp, which would reach the bottom first?

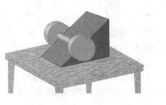

Fig. 12-53 Question 28.

(c) Show that the system has $\beta > 1$, where β is the rotational inertia parameter of Table 12-2. (See Sample Problem 12-10 above.)

29. A yo-yo falls to the bottom of its cord and then climbs back up. Does it reverse its direction of rotation at the bottom? Explain your answer.

30. A rear-wheel-drive car accelerates quickly from rest, and the driver observes that the car "noses up." Why does it do that? Would a front-wheel-drive car do that?

31. The mounting bolts that fasten the engines of jet planes to the structural framework of the plane are arranged to snap apart if the (rapidly rotating) engine suddenly seizes up because of some mishap. Why are such "structural fuses" used?

32. Is there any advantage in setting the wheels of the landing gear of an airplane in rotation just before the plane lands? If so, how would you find the optimum speed and direction of rotation?

33. A disgruntled hockey player throws a hockey stick along the ice. It rotates about its center of mass as it slides and is eventually brought to rest by friction. Why must its rotational motion stop exactly when its center of mass comes to rest?

34. When the angular velocity ω of an object increases, its angular momentum may or may not also increase. Give an example in which it does and one in which it does not.

35. A helicopter flies off, its rotor blades rotating. Why doesn't the body of the helicopter rotate in the opposite direction?

36. If the entire population of the world moved to Antarctica, would the length of the day be affected? If so, in what way?

37. A circular turntable rotates at constant angular speed about a vertical axis. There is no friction and no driving torque. A circular pan rests on the turntable and rotates with it (see Fig. 12-54). The bottom of the pan is covered with a layer of ice of uniform thickness, which is, of course, also rotating with the pan. Suppose the ice melts, but none of the water escapes from the pan. Is the angular speed now greater than, the same as, or less than the original speed? Give reasons for your answer.

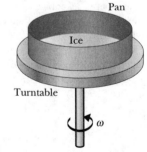

Fig. 12-54 Question 37.

38. You can distinguish between a raw egg and a hard-boiled one by spinning each egg on the table. Explain how. Also, if you stop a spinning raw egg by very briefly touching its top, why will it resume spinning?

39. Figure 12-55*a* shows an acrobat propelled upward by a trampoline with zero angular momentum. Can the acrobat, by maneuvering the body, manage to land on his back as in Fig. 12-55*b*? Interestingly, 38% of questioned diving coaches and 34% of a sample of physicists gave the wrong answer. What do *you* think? (From "Do Springboard Divers Violate Angular Momentum Conservation?" by Cliff Frohlich, *American Journal of Physics,* July 1979.)

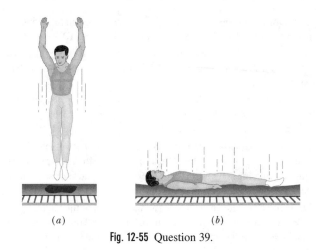

(a)   (b)

Fig. 12-55 Question 39.

40. Can you "pump" a swing so that it turns in a complete circle, moving completely around its support? Assume (if you wish) that the seat of the swing is connected to its support by a rigid rod rather than by a rope or a chain. Explain your answer.

41. A massive spinning wheel can be used for a stabilizing effect on a ship. If the wheel is mounted with its axis of rotation perpendicular to the ship deck, what is its effect when the ship tends to roll from side to side?

42. Why must a football quarterback put a lot of spin on the ball if it is to stay aligned with its path during a long pass?

EXERCISES & PROBLEMS

60. In 1978, a New Zealand motorcyclist roared off a 15° ramp at a speed of 32 m/s, flying into the air, over a parked airplane, and down onto another ramp. Some witnesses knew from the engine noise that the man had made a fatal mistake. As soon as the jump began, he should have closed the throttle to stop driving the rear wheel. Instead, he kept the throttle wide open. When the motorcycle left the ramp and traction disappeared, the rear wheel suddenly spun up to its maximum angular speed ω_{wf} of 160 rad/s and the motorcycle began to rotate around its center of mass. Thus the motorcycle was far from the correct orientation when it reached the second ramp.

Assume the following: The wheel radius was 0.30 m, the rotational inertia I_c of the motorcycle about its center of mass was 20 kg · m², and the rotational inertia I_w of the rear wheel about its center was 0.40 kg · m². Also assume that ω_{wf} was reached immediately after takeoff, and neglect any effect of the man on the motorcycle. Suppose the motorcycle moved from right to left. (a) Did the rear wheel rotate clockwise or counterclockwise? (b) When the motorcycle was in the air, did it rotate clockwise or counterclockwise around its center of mass, and why? (c) Assuming the motorcycle was a particle projectile, how long was it in the air? What were (d) the initial angular speed and (e) initial angular momentum of the rear wheel about its center, just as the motorcycle left the first ramp? (f) The wheel's angular momentum about its center is the same as the wheel's angular momentum about the motorcycle's center of mass. Using that fact and the conservation of angular momentum, find the angular speed of the motorcycle about its center of mass once ω_{wf} was reached. (g) In degrees, through what angle did the motorcycle rotate during the flight?

61. In 1980, over the waters of San Francisco Bay, a large yo-yo was released six or seven times from a crane. The 116 kg yo-yo consisted of two uniform disks of radius 32 cm connected by an axle of radius 3.2 cm. What was the acceleration of the yo-yo during (a) its fall and (b) its rise? (c) What was the tension in the cord on which it rolled? (d) Was that tension near the cord's limit of 52 kN? Suppose you build a scaled-up version of the yo-yo (same shape and materials but larger in size). (e) Is the magnitude of the yo-yo's acceleration as it falls greater than, less than, or the same as the San Francisco yo-yo? (f) How about the tension in the cord?

62. With center and spokes of negligible mass, a certain bicycle wheel has a thin rim of radius 0.350 m and weight 37 N; it can turn on its axle with negligible friction. A man holds the wheel above his head with the axle vertical while he stands on a turntable that is free to rotate without friction; the wheel rotates clockwise, as seen from above, with an angular speed of 57.7 rad/s, and the turntable is initially at rest. The rotational inertia of *wheel + man + turntable* about the common axis of rotation is 2.10 kg · m². The man's free hand suddenly stops the rotation of the wheel (relative to the turntable). Determine the resulting angular velocity (magnitude and direction) of the system.

63. Wheels A and B in Fig. 12-56 are connected by a belt that does not slip. The radius of wheel B is three times the radius of wheel A. What would be the ratio of the rotational inertias I_A/I_B if the two wheels had (a) the same angular momentum about their central axes and (b) the same rotational kinetic energy?

Fig. 12-56 Problem 63.

64. A projectile of mass m is fired from the ground with an initial speed v_0 and an initial angle θ_0 above the horizontal. (a) Find an expression for the magnitude of its angular momentum about the firing point as a function of time. (b) Find the rate at which the angular momentum changes with time. (c) Evaluate the magnitude of $\vec{r} \times \vec{F}$ directly and compare the result with (b). (d) Why should the results of (b) and (c) be identical?

65. A small 0.50 kg block has a horizontal velocity of 3.0 m/s when it slides off a 1.2-m-high frictionless table as shown in Fig. 12-57. Answer the following in unit vectors for a coordinate system in which the origin is at the edge of the table (at point O), the positive x direction is horizontally away from the table, and the positive y

direction is up. What are the angular momenta of the block about point *A* at the foot of the table leg (a) just after the block leaves the table and (b) just before the block strikes the floor? What are the torques on the block about point *A* (c) just after the block leaves the table and (d) just before the block strikes the floor?

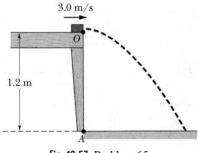

3.0 m/s

1.2 m

O

A

Fig. 12-57 Problem 65.

66. A particle of mass *M* is dropped from a point that is at height *h* above the ground and horizontal distance *s* from an observation point *O*, as shown in Fig. 12-58. What is the magnitude of the angular momentum of the particle with respect to point *O* when the particle has fallen half the distance to the ground? State your answer in terms of *M*, *h*, *s*, and *g*.

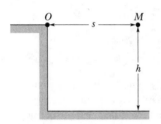

O   *s*   *M*

h

Fig. 12-58 Problem 66.

67. (a) Use the data given in the appendices to compute the total of the magnitudes of the angular momenta of all the planets owing to their revolution about the Sun. (b) What fraction of this total is associated with the planet Jupiter?

68. What is the net torque about the origin on a flea located at coordinates (0, −4.0 m, 5.0 m) when forces $\vec{F}_1 = (3.0 \text{ N})\hat{k}$ and $\vec{F}_2 = (-2.0 \text{ N})\hat{j}$ act on the flea?

69. In Fig. 12-59, a constant horizontal force of 12 N is applied to a uniform solid cylinder by fishing line wrapped around the cylinder. The mass of the cylinder is 10 kg, its radius is 0.10 m, and the cylinder rolls smoothly on the horizontal surface. (a) What is the magnitude of the acceleration of the center of mass of the cylinder? (b) What is the magnitude of the angular acceleration of the cylinder about the center of mass? (c) What are the magnitude and direction of the frictional force acting on the cylinder?

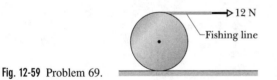

12 N

Fishing line

Fig. 12-59 Problem 69.

70. An automobile has a total mass of 1700 kg. It accelerates from rest to 40 km/h in 10 s. Assume each wheel is a uniform 32 kg disk. Find, for the end of the 10 s interval, (a) the rotational kinetic energy of each wheel about its axle, (b) the total kinetic energy of each wheel, and (c) the total kinetic energy of the automobile.

71. Suppose that the Sun runs out of nuclear fuel and suddenly collapses to form a white dwarf star, with a diameter equal to that of Earth. Assuming no mass loss, what would then be the Sun's new rotation period, which currently is about 25 days? Assume that the Sun and the white dwarf are uniform, solid spheres.

72. A uniform rod rotates in a horizontal plane about a vertical axis through one end. The rod is 6.00 m long, weighs 10.0 N, and rotates at 240 rev/min clockwise when seen from above. Calculate (a) the rotational inertia of the rod about the axis of rotation and (b) the angular momentum of the rod about that axis.

73. A 3.0 kg toy car moves along an *x* axis with a velocity given by $\vec{v} = -2.0t^3\hat{i}$ m/s. For $t > 0$, what are (a) the angular momentum $\vec{L}$ of the car and (b) the torque $\vec{\tau}$ on the car, both calculated about the origin? What are (c) $\vec{L}$ and (d) $\vec{\tau}$ about a point at coordinates (2.0 m, 5.0 m, 0)? What are (e) $\vec{L}$ and (f) $\vec{\tau}$ about a point at coordinates (2.0 m, −5.0 m, 0)?

74. At one instant, a 0.80 kg particle is located at the position $\vec{r} = (2.0 \text{ m})\hat{i} + (3.0 \text{ m})\hat{j}$. The linear momentum of the particle lies in the *xy* plane and has a magnitude of 2.4 kg · m/s and a direction of 115° measured counterclockwise from the positive direction of *x*. What is the angular momentum of the particle about the origin, in unit-vector form?

75. What is the magnitude of the angular momentum, about Earth's center, of an 84 kg person on the equator due to Earth's rotation?

76. A 1200 kg airplane is flying in a straight line at 80 m/s, 1.3 km above the ground. What is the magnitude of its angular momentum with respect to a point on the ground directly under the path of the plane?

77. Given that $\vec{r} = x\hat{i} + y\hat{j} + z\hat{k}$ and $\vec{F} = F_x\hat{i} + F_y\hat{j} + F_z\hat{k}$, show that the torque $\vec{\tau} = \vec{r} \times \vec{F}$ is given by

$$\vec{\tau} = (yF_z - zF_y)\hat{i} + (zF_x - xF_z)\hat{j} + (xF_y - yF_x)\hat{k}.$$

78. A wheel of radius 0.250 m, which is moving initially at 43.0 m/s, rolls to a stop in 225 m. Calculate the magnitudes of (a) its linear acceleration and (b) its angular acceleration. (c) The wheel's rotational inertia is 0.155 kg · m² about its central axis. Calculate the magnitude of the torque about the central axis due to friction on the wheel.

79. A solid sphere of weight 36.0 N rolls up an incline at an angle of 30.0°. At the bottom of the incline the center of mass of the sphere has a translational speed of 4.90 m/s. (a) What is the kinetic energy of the sphere at the bottom of the incline? (b) How far does the sphere travel up along the incline? (c) Does the answer to (b) depend on the sphere's mass?

80. *Left alone, write your own.* For one or more of the following situations, write problems involving physics in this chapter, writing in the style of the sample problems and providing realistic data, graphs of the variables, and explained solutions: (a) two rolling objects race each other along a roller-coaster-like path, (b) multiple forces act on a free-turning merry-go-round, (c) clowns fall onto or off a free-turning merry-go-round, (d) two bodies orbiting each other slowly lose mechanical energy (due to the shifting of materials they cause in each other).

CLUSTERED PROBLEMS

Cluster 1

The problems in this cluster deal with an object rolling with constant linear acceleration due to an applied force and a frictional force.

81. A horizontal force of magnitude 200 N is applied to the axle of a one-wheel cart with a total mass of 50.0 kg, producing an acceleration of 3.00 m/s² along a level path. The cart's wheel, which does not have a uniform distribution of mass, is a long cylinder with a radius of 0.200 m. It rolls smoothly and without sliding, rotating about its axle without friction. What are (a) the magnitude of the frictional force on the wheel from the path, (b) the magnitude of the torque of that force about the rotation axis of the wheel, and (c) the rotational inertia of the wheel about that axis?

82. The cart of Problem 81 is sent rolling down a plane inclined at 30.0° to the horizontal. (a) What is the magnitude of the cart's acceleration? (b) What is the minimum coefficient of static friction needed to keep the cart from sliding? (*Hint:* Use the answer to part (c) of Problem 81; if you cannot get that answer, use the symbol I for the wheel's rotational inertia about the rotation axis.)

83. A disk of rotational inertia I, mass M, and radius R is positioned to roll down a plane inclined at angle θ to the horizontal. The coefficient of static friction between the disk and the inclined plane is μ_s. What is the greatest value of θ at which the disk rolls smoothly down the plane without sliding? (This value is said to be the *critical angle* for rolling objects.)

84. A uniform disk is upright on its edge on the flatbed of a long truck, ready to roll toward the front or rear. The truck then accelerates at magnitude a in the forward direction, and the disk rolls without slipping until it hits the truck's rear wall. Assume that the bed remains horizontal during the acceleration. During the rolling, what are the magnitude and direction of the acceleration of the center of mass of the disk relative to (a) the bed and (b) the road?

Cluster 2

85. Figure 12-60a shows two pulleys, one of radius R_1 and the other of radius R_2, that are rigidly connected and free to turn about a common frictionless axle through their centers. The rotational inertia of this two-pulley device about that axle is I. Fishing line that is wrapped around the circumference of the larger pulley extends to block 2 of mass m_2. Similarly, fishing line that is wrapped around the circumference of the smaller pulley extends to block 1 of mass m_1.

Initially we hold the blocks in place. When we release them, what are the magnitudes of the accelerations of (a) block 1 and (b) block 2? Assume that the direction of positive acceleration is upward for block 2 and downward for block 1, and take the direction of positive angular acceleration of the two-pulley device as counterclockwise.

86. In Fig. 12-60b, a spool of thread with rotational inertia I, radius R, and mass M falls as thread unwraps from the spool's circumference. A hand holds the upper end of the thread. (a) What is the magnitude of the linear acceleration of the spool's center of mass if the hand is stationary? (b) What upward acceleration must the hand give the thread if the spool's center of mass is not to fall?

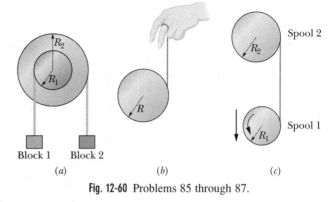

Fig. 12-60 Problems 85 through 87.

87. In Fig. 12-60c, spool 1 falls as thread unwraps from its circumference. At the same time, the upper end of the thread unwraps from the circumference of spool 2, which rotates about a fixed, frictionless axle that coincides with its central axis. Spool 1 has rotational inertia I_1 about its central axis, radius R_1, and mass M. Spool 2 has rotational inertia I_2 about its central axis and radius R_2.

(a) What is the magnitude of the linear acceleration a_1 of the center of mass of spool 1? (*Hint:* Downward acceleration a_1 is equal to the sum of two accelerations: the downward acceleration a_s of the thread that results from its unwrapping from spool 2 and the downward acceleration of spool 1 that results from the thread unwrapping from it. The latter acceleration is like that in Problem 86a.) (b) Show that if $I_2 \gg I_1$, then the expression for a_1 matches that for the acceleration in Problem 86a.

13
Equilibrium and Elasticity

ADDITIONAL SAMPLE PROBLEMS

Sample Problem 13-7

In Fig. 13-50a, a bowler holds a bowling ball of mass $M = 7.2$ kg, with his upper arm vertical and his lower arm horizontal. What are the forces on the lower arm from the biceps muscle and the bony structure of the upper arm? The lower arm (consisting of forearm and hand) has a total mass m of 1.8 kg. The needed dimensions are given in Fig. 13-50a.

SOLUTION: Our system is the lower arm and bowling ball taken together, and the forces on the system are shown in Fig. 13-50b. The

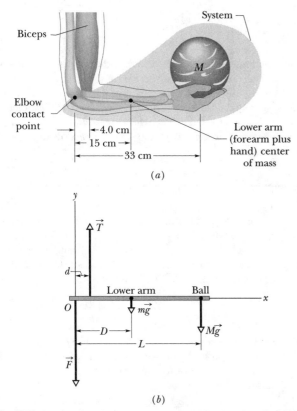

(a)

(b)

Fig. 13-50 Sample Problem 13-7. (a) A hand holds a bowling ball. The system to be analyzed is marked. (b) A free-body diagram of the *lower arm* + *ball* system, showing the forces that act. The vectors are not to scale; the powerful forces exerted by the biceps muscle ($\vec{T}$) and the elbow joint ($\vec{F}$) are many times larger than the gravitational forces on the lower arm (here represented with $m\vec{g}$) and the ball ($M\vec{g}$).

forces on the lower arm are $\vec{T}$ from the biceps muscle, $\vec{F}$ from the bone of the upper arm, and the gravitational force. The last acts at the center of mass of the lower arm as indicated in Fig. 13-50a and is represented by its equivalent $m\vec{g}$ in Fig. 13-50b. The ball is represented by a dot within the boundary of the lower arm, and the gravitational force on the ball is represented by its equivalent $M\vec{g}$.

The **Key Idea** here is that because the system is in static equilibrium, we can apply the balancing equations (Eqs. 13-7 through 13-9) to it. Let us start with the balance of torques ($\tau_{net} = 0$). Note that we have two unknown forces, $\vec{F}$ and $\vec{T}$. To eliminate $\vec{F}$ from the torque balancing equation, we place a rotation axis through O at the elbow, perpendicular to the plane of Fig. 13-50. Then the moment arms about that axis are 0 for $\vec{F}$, d for $\vec{T}$, D for $m\vec{g}$, and L for $M\vec{g}$. Being careful about signs for torques and writing torques in the form $r_\perp F$, we can write the balance of torques ($\tau_{net} = 0$) as

$$(0)(F) + (d)(T) - (D)(mg) - (L)(Mg) = 0.$$

Solving for T yields

$$T = g\frac{mD + ML}{d}$$

$$= (9.8 \text{ m/s}^2)\frac{(1.8 \text{ kg})(15 \text{ cm}) + (7.2 \text{ kg})(33 \text{ cm})}{4.0 \text{ cm}}$$

$$= 648 \text{ N} \approx 650 \text{ N}. \qquad \text{(Answer)}$$

Thus, the biceps muscle must pull up on the forearm with a force that is about nine times larger than the weight of the bowling ball; holding the ball as in Fig. 13-50a is difficult.

Next, to find F, we write the balancing equation $F_{net,y} = 0$ as

$$T - F - mg - Mg = 0.$$

Solving for F and using known values, we find

$$F = T - g(M + m)$$
$$= 648 \text{ N} - (9.8 \text{ m/s}^2)(7.2 \text{ kg} + 1.8 \text{ kg})$$
$$= 560 \text{ N}, \qquad \text{(Answer)}$$

which is about eight times the weight of the bowling ball.

Sample Problem 13-8

In Sample Problem 13-2, let the coefficient of static friction μ_s between the ladder and the pavement be 0.53. How far up the ladder must the firefighter go to put the ladder on the verge of sliding?

SOLUTION: Let q be the fraction of the ladder's length the firefighter climbs to put the ladder on the verge of sliding. She is then a horizontal distance qa from the origin O (distance a is indicated in Fig. 13-6a).

The **Key Idea** here is that the firefighter–ladder system is still in static equilibrium, even when it is on the verge of sliding. Thus, the balancing equations still apply, as does much of our work in Sample Problem 13-2. For the balance of torques in Eq. 13-18, we merely substitute the firefighter's new moment arm qa for her previous moment arm $a/2$, writing

$$-(h)(F_w) + (qa)(Mg) + (a/3)(mg) + (0)(F_{px}) + (0)(F_{py}) = 0,$$

which gives us

$$q = \frac{hF_w}{aMg} - \frac{m}{3M}. \tag{13-31}$$

However, we do not know the force magnitude F_w, so we cannot yet find q.

Let us next examine the force balancing equations to find an expression for F_w. The equation $F_{net,x} = 0$ still gives us Eq. 13-19,

$$F_w = F_{px}, \tag{13-32}$$

but we do not know F_{px}. The equation $F_{net,y} = 0$ still gives us

$$F_{py} = (M + m)g, \tag{13-33}$$

but this does not even contain F_w, so the situation looks hopeless.

However, using physics courage and the following **Key Idea**, we can find an expression for F_w from Eqs. 13-32 and 13-33: When an object is on the verge of sliding, the static frictional force f_s on it has its maximum value $f_{s,max} = \mu_s N$. For the ladder, F_{px} is the static frictional force and F_{py} is the normal force, so here "verge of sliding" means

$$F_{px} = \mu_s F_{py}. \tag{13-34}$$

When we substitute for F_{px} and F_{py} from Eqs. 13-32 and 13-33, this equation becomes

$$F_w = \mu_s(M + m)g. \tag{13-35}$$

Now substituting this into Eq. 13-31, we find

$$q = \frac{h\mu_s(M + m)}{aM} - \frac{m}{3M} \tag{13-36}$$

$$= \frac{(9.3\text{ m})(0.53)(72\text{ kg} + 45\text{ kg})}{(7.6\text{ m})(72\text{ kg})} - \frac{45\text{ kg}}{(3)(72\text{ kg})}$$

$$= 0.85. \tag{Answer}$$

Thus, the firefighter must climb 85% of the way up the ladder to put the ladder on the verge of sliding.

You can show from Eq. 13-36 that the firefighter can climb all the way up the ladder (which corresponds to $q = 1$) without it sliding if $\mu_s > 0.61$. On the other hand, the ladder will slide when the firefighter is on the first rung (which corresponds to $q \approx 0$) if $\mu_s < 0.11$.

Sample Problem 13-9

In an adult male, the minimum cross-sectional area A of the femur (the principal bone of the thigh) is about 6.0×10^{-4} m^2. At what compressive load would the femur shatter?

SOLUTION: The first **Key Idea** here has to do with what is meant by the question. We are to consider a cross-sectional "slice" through the femur, at the narrowest point (which is the most prone to shatter because of its smallest area). Then we assume the slice is being compressed by forces of magnitude F applied to both faces of the slice, perpendicular to the faces. The situation is like that in Fig. 13-11a except that the two forces are directed inward, for compression. The femur is then said to be "under load" or "loaded." The load could be a typical one, as when a person stands, or it could be much larger, as when a person lands feet first from a high jump.

The next **Key Ideas** are that the forces produce a stress on the slice and that the bone shatters when the stress reaches the value of the ultimate strength S_u for human bone in compression. The stress is given by the ratio F/A (the left side of Eq. 13-25), and the value of S_u is given in Table 13-1. We have

$$F = S_u A = (170 \times 10^6\text{ N/m}^2)(6.0 \times 10^{-4}\text{ m}^2)$$

$$= 1.0 \times 10^5\text{ N}. \tag{Answer}$$

Although this is a large force magnitude, it can be encountered, for example, during an unskillful parachute landing on hard ground. The force need not be a sustained force to break the bone; only a few milliseconds are needed.

QUESTIONS

11. *Organizing question:* Figure 13-51 shows a window washer on his rig, which consists of a uniform beam of mass 200 kg and length 8.0 m, suspended by a cable at each end. The 60 kg washer works 2.0 m from the right end. (a) Set up an equation (one equation), complete with known data, to find the tension T_R in the cable at the right end. (b) Suppose, instead, we want the tension T_L in the cable at the left end. Without relying on the answer to (a), set up an equation, complete with known data, to find T_L.

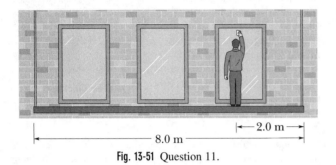

Fig. 13-51 Question 11.

12. Figure 13-52 shows an overhead view of a metal square lying on a frictionless floor. Three forces, which are drawn to scale, act at the corners of the square. (a) Is the first requirement of equilibrium (in Eq. 13-1) satisfied? (b) Is the second requirement of equilibrium satisfied? (c) If the answer to either (a) or (b) is no, could a fourth force acting on the square then satisfy both requirements of equilibrium?

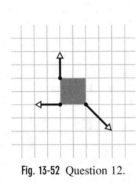

Fig. 13-52 Question 12.

13. *Organizing question:* Figure 13-53 shows a stationary 50 kg rock climber who is on belay via a climbing harness. The line of action of the force on her from the belay line is at perpendicular distances 2.0 cm from her center of mass and 80 cm from her feet. The forces on her feet from the rock face are collectively represented by $\vec{F}_r$. Set up an equation, complete with known data, to find the tension T in the belay line.

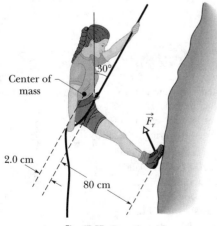

Center of mass

30°

$\vec{F}_r$

2.0 cm

80 cm

Fig. 13-53 Question 13.

14. (a) In Sample Problem 13-3 and Fig. 13-7, if the angle θ were made greater (but the cable still kept horizontal), would the tension

in the cable be greater than, less than, or the same as in the sample problem? (b) Would the magnitude of the net force exerted by the hinge on the beam be greater than, less than, or the same as in the sample problem?

15. A physical therapist gone wild has constructed the (stationary) assembly of massless pulleys and cords seen in Fig. 13-54. One long cord wraps around all the pulleys, and shorter cords suspend pulleys from the ceiling or weights from the pulleys. Except for one, the weights (in newtons) are indicated. (a) What is that last weight? (*Hint:* When a cord loops halfway around a pulley as here, it pulls on the pulley with a net force that is twice the tension in the cord.) (b) What is the tension in the short cord labeled T?

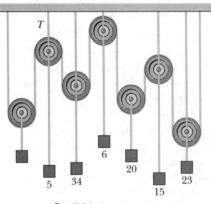

T

6

20

5 34 15 23

Fig. 13-54 Question 15.

16. In Fig. 13-55, a beam of mass M is held horizontal by a massless rod in orientation 1; the beam is attached to a hinge at its other end. Five other orientations of the rod are shown dashed in the figure (the tilted orientations have identical tilts). (a) For which orientations is the rod under tension and for which is it under compression? Rank the six orientations of the rod according to (b) the magnitude of the tension or compression in the rod, (c) the

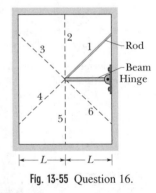

2
3
1 Rod
Beam
Hinge
4
5 6

$\leftarrow L \rightarrow \leftarrow L \rightarrow$

Fig. 13-55 Question 16.

magnitude of the vertical force on the beam from the hinge, and (d) the magnitude of the horizontal force on the beam from the hinge, greatest first.

17. The table gives the initial lengths of three rods and the changes in their lengths when forces are applied to their ends to put them under strain. Rank the rods according to their strain, greatest first.

	Initial Length	Change in Length
Rod A	$2L_0$	ΔL_0
Rod B	$4L_0$	$2\Delta L_0$
Rod C	$10L_0$	$4\Delta L_0$

18. (a) As a quick fix to a bridge that has been partially ruined by floodwater, a concrete slab is laid across a gap in the bridge, supported by remaining vertical structures. As a heavy truck is then driven over the slab and the slab sags, is the slab more likely to rupture on its top surface or its bottom surface?

(b) When an old industrial chimney is taken down, one side of its base is usually knocked out by a bulldozer or blasted out with an explosion. The chimney then rotates around that side of the base toward the ground. During the rotation, the chimney tends to bend backward against the fall until it ruptures at about mid-height. Does the rupture begin on the side facing the ground or on the opposite side?

DISCUSSION QUESTIONS

19. In a simple pendulum, is the bob in equilibrium at any point of its swing?

20. If a certain rigid body is thrown upward without spinning, it does not begin spinning during its flight, provided that air resistance can be neglected. What does this simple result imply about the location of the center of gravity?

21. A wheel rotating at constant angular velocity ω about a fixed axis is in equilibrium because no net external force or torque acts on it. However, the particles that make up the wheel undergo a centripetal acceleration $\vec{a}$ directed toward the axis. Since $\vec{a} \neq 0$, how can the wheel be said to be in equilibrium?

22. Give several examples of bodies that are not in equilibrium, even though the resultant of all the forces acting on them is zero.

23. Which is more likely to break in use, a hammock that is stretched tightly or one that sags quite a bit? Why?

24. A ladder is at rest with its upper end against a wall and the lower end on the ground. Is it more likely to slip when a person stands on it at the bottom or at the top? Explain.

25. A framed rectangular picture hangs by one wire over a frictionless nail. Why is the picture in unstable equilibrium if the wire is "too short" (this is one reason why pictures become skewed), and in stable equilibrium if the wire is "long enough"? Experimentally define "too short" and "long enough" by considering the angle θ that the wire makes at the nail and angle α between the diagonals of the picture, as shown in Fig. 13-56.

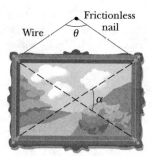

Fig. 13-56 Question 25.

26. Stand facing the edge of an open door, one foot on each side of the door. You will find that you are not able to stand on your toes. Why?

27. Sit in a straight-backed chair and try to stand up without leaning forward. Why can't you do it?

28. How does a long balancing pole help a tightrope walker to maintain balance?

29. A composite block made up of wood and metal rests on a (rough) tabletop. In which of the two orientations of the block in Fig. 13-57 can you tip it over with the least force $\vec{F}$ applied at the same height on the block?

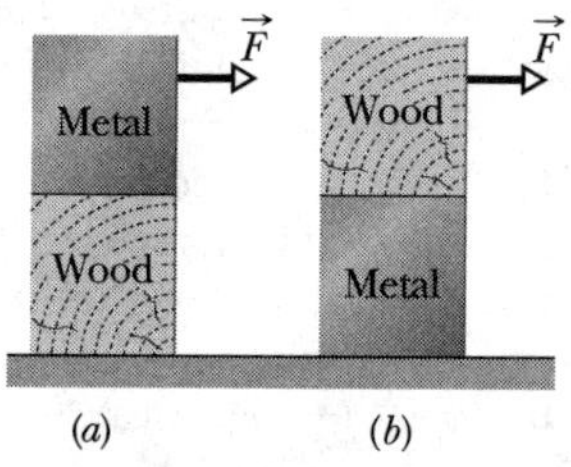

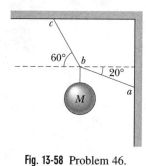

Fig. 13-57 Question 29.

30. If you have a can of soda pop on your tray during a choppy flight, for what levels of fluid content is the can most likely and least likely to tip over?

31. A popular beer drinker's challenge is to balance a can on its edge. How is the stunt accomplished?

32. Why are reinforcing rods used in concrete structures? (Compare the tensile strength of concrete to its compressive strength.)

EXERCISES & PROBLEMS

43. After a fall, a 95 kg rock climber finds himself dangling from the end of a rope that had been 15 m long and 9.6 mm in diameter but has stretched by 2.8 cm. For the rope, calculate (a) the strain, (b) the stress, and (c) the Young's modulus.

44. A mine elevator is supported by a single steel cable 2.5 cm in diameter. The total mass of the elevator cage and occupants is 670 kg. By how much does the cable stretch when the elevator hangs by (a) 12 m of cable, at the surface and (b) 362 m of cable, at the shaft's bottom? (Neglect the mass of the cable.)

45. Suppose the (square) beam in Fig. 13-7a is of Douglas fir. What must be its thickness to keep the compressive stress on it to $\frac{1}{6}$ of its ultimate strength? (See Sample Problem 13-3.)

46. The system shown in Fig. 13-58 is in equilibrium. If $M = 2.0$ kg, what is the tension in (a) string ab and (b) string bc?

47. A uniform ladder is 10 m long and weighs 200 N. The ladder leans against a vertical, frictionless wall at a point 8.0 m above the ground, as shown in Fig. 13-59. A horizontal force $\vec{F}$ is applied to the ladder at a point 2.0 m from the base of the lad-

Fig. 13-58 Problem 46.

der (as measured along the ladder). (a) If $F = 50$ N, what is the force of the ground on the ladder, in terms of the unit vectors shown? (b) If $F = 150$ N, what is the force of the ground on the ladder, in unit-vector form? (c) Suppose the coefficient of static friction between the ladder and the ground is 0.38; for what minimum value of F will the base of the ladder just start to move toward the wall?

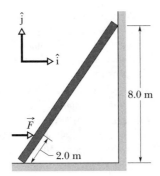

Fig. 13-59 Problem 47.

48. A uniform beam of length 12 m is supported by a horizontal cable (at top) and a hinge as shown in Fig. 13-60. The tension in the cable is 400 N. What are (a) the weight of the beam and (b) the horizontal and vertical components of the force of the hinge on the beam?

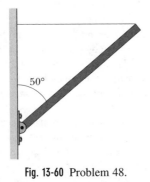

Fig. 13-60 Problem 48.

49. A rectangular slab of slate rests on a 26° incline; see Fig. 13-61. The slab is 43 m long, 2.5 m thick, and 12 m wide. Its density is 3.2 g/cm³. The coefficient of static friction between the slab and the underlying rock is 0.39. (a) Calculate the component of the gravitational force on the slab parallel to the incline. (b) Calculate the magnitude of the static frictional force on the slab. By comparing (a) and (b), you can see that the slab is in danger of sliding and is prevented from doing so only by chance protrusions between the slab and the underlying rock. (c) To stabilize the slab, bolts are to be driven perpendicular to the incline. If each bolt has a cross-sectional area of 6.4 cm² and will snap under a shearing stress of 3.6×10^8 N/m², what is the minimum number of bolts needed? Assume that the bolts do not affect the normal force.

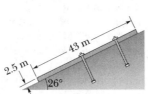

Fig. 13-61 Problem 49.

50. A uniform beam having a weight of 60 N and a length of 3.2 m is hinged at its lower end and acted on by a horizontal force $\vec{F}$ of magnitude 50 N at its upper end (Fig. 13-62). The beam is held vertical by a cable that makes an angle of 25° with the ground. What are (a) the tension in the cable and (b) the horizontal and vertical components of the force of the hinge on the beam?

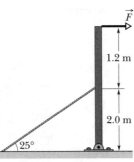

Fig. 13-62 Problem 50.

51. A gymnast with mass 46.0 kg stands on the end of a uniform balance beam as shown in Fig. 13-63. The beam is 5.00 m long and has a mass of 250 kg (excluding the mass of the two supports). Each of the two supports is 0.540 m from its respective end of the beam. What are the magnitude and direction of (a) the force of support 1 on the beam and (b) the force of support 2 on the beam?

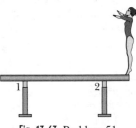

Fig. 13-63 Problem 51.

52. A solid copper cube has an edge length of 85.5 cm. How much pressure must be applied to the cube to reduce the edge length to 85.0 cm? The bulk modulus of copper is 1.4×10^{11} N/m².

53. A uniform cube of side length L rests on a horizontal floor. The coefficient of static friction between cube and floor is μ. A horizontal pull P is applied perpendicular to one of the vertical faces of the cube, at a distance h above the floor on the vertical midline of the cube face. As P is slowly increased, the cube will either (a) begin to slide or (b) begin to tip. What is the condition on μ for (a) to occur? For (b)? (*Hint:* At the onset of tipping, where is the normal force located?)

54. A uniform beam is 5.0 m long and has a mass of 53 kg. The beam is supported in a horizontal position by a hinge and a cable as shown in Fig. 13-64. In unit-vector notation, what is the force on the beam from the hinge?

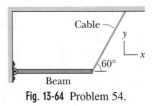

Fig. 13-64 Problem 54.

55. Figure 13-65 is an overhead view of a rigid rod that turns about a vertical axle until the identical rubber stoppers A and B are forced against rigid walls at distances r_A and r_B from the axle. Initially the stoppers touch the walls, without being compressed. Then force $\vec{F}$ is applied perpendicular to the rod at a distance R from the axle. Find expressions for the magnitudes of the forces compressing (a) stopper A and (b) stopper B.

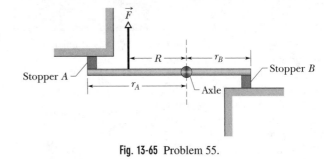

Fig. 13-65 Problem 55.

56. To crack a certain nut in a nutcracker, forces with magnitudes of at least 40 N must act on its shell from both sides. For the nutcracker of Fig. 13-66, what are the force components $F_\perp$ (perpendicular to the handles) corresponding to that 40 N?

57. The system in Fig. 13-67 is in equilibrium, but it begins to slip if any additional mass is added to the 5.0 kg object. What is the coefficient of static friction between the 10 kg block and the plane on which it rests?

58. A trap door in a ceiling is 0.91 m square, has a mass of 11 kg, and is hinged along one side, with a catch at the opposite side. If the center of gravity of the door is 10 cm toward the hinged side from the door's center, what are the magnitudes of the forces on (a) the catch and (b) the hinge from the door?

59. A pan balance is made up of a rigid, massless rod supported at and free to rotate about a point not at the center of the rod. It is balanced by unequal weights that are placed in a pan at each end of the rod. When an unknown mass m is placed in the left-hand pan, it is balanced by a mass m_1 placed in the right-hand pan; and when the mass m is placed in the right-hand pan, it is balanced by a mass m_2 in the left-hand pan. Show that $m = \sqrt{m_1 m_2}$.

60. In Fig. 13-68, two identical, uniform, frictionless spheres, each of mass m, rest in a rigid rectangular container. A line connecting their centers is at 45° to the horizontal. Find the magnitudes of the forces on the spheres from (a) the bottom of the container, (b) the sides of the container, and (c) one another. (*Hint:* The force of one sphere on the other is directed along the line connecting their centers.)

61. A 73 kg man stands on a level bridge of length L. He is $L/4$ from one end. The bridge is uniform and weighs 2.7 kN. What are the vertical forces on the bridge from its supports at (a) the far end and (b) the near end?

62. By means of a turnbuckle G, bar AB of the rigid square frame $ABCD$ in Fig. 13-69 is put in tension, as if its ends A and B were subject to the horizontal, outward

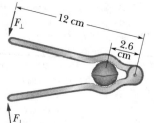

Fig. 13-66 Problem 56.

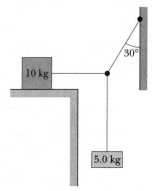

Fig. 13-67 Problem 57.

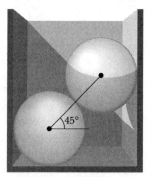

Fig. 13-68 Problem 60.

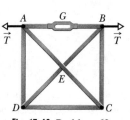

Fig. 13-69 Problem 62.

forces $\vec{T}$ shown. Determine the magnitudes of the forces on the other bars; identify which bars are in tension and which are being compressed. The diagonals AC and BD pass each other freely at E. Symmetry considerations can lead to considerable simplification in this problem.

63. A beam of length L is carried by three men, one man at one end and the other two supporting the beam between them on a crosspiece placed so that the load of the beam is equally divided among the three men. How far from the beam's free end is the crosspiece placed? (Neglect the mass of the crosspiece.)

Tutorial Problems

64. A uniform wood door has mass m, height h, and width w. It is hanging from two hinges attached to one side; the hinges are located $h/3$ and $2h/3$ from the bottom of the door. (a) Sketch this system, showing the positions of the hinges, the center of mass of the door, h, and w. (b) List the forces acting on the door and draw a free-body force diagram for the door. (*Hint:* Show the true lines of action of the forces instead of making them all act at the center of mass.)

(c) State in complete sentences, without the use of mathematical symbols, the net force and net torque conditions for the static equilibrium of the door. (d) Write all the equations that result from the conditions you stated in part (c). If you had chosen the same notation as one of your classmates, would it be possible for the two of you to write correct but different equations (different in a significant way, not just with a different order of terms)? If you think they could have been different, explain why. (e) How many equations did you have in part (d) and how many unknowns? Would it be theoretically possible to determine the numerical values of all the forces if you were given the numerical values of m, h, and w? Why or why not?

(f) It is possible to determine an algebraic expression (in terms of m, h, and w) for at least some, if not all, of the forces. Determine the ones you can. If there are any you cannot determine, provide whatever information you can about the unknown forces, then make up a reasonable extra assumption of your own that will enable you to determine all the forces. (g) Suppose that $m = 20.0$ kg, $h = 2.20$ m, and $w = 1.00$ m. What are the numerical values of the forces whose algebraic expressions you determined in part (f)?

Answers

(a) See Fig. 13-70*a*.

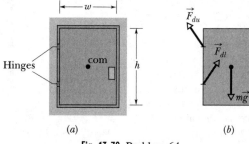

Fig. 13-70 Problem 64.

(b) The forces acting on the door are shown in Fig. 13-70*b*: (1) the gravitational force on the door, which acts at the center of mass of the door (the force is represented with its equivalent $m\vec{g}$); (2) the force $\vec{F}_{du}$ exerted on the door by the upper hinge; and (3) the force $\vec{F}_{dl}$ exerted on the door by the lower hinge.

The forces exerted by the hinges may have both vertical and horizontal components. If we wanted to, we could refer to them as normal and frictional forces, although the term *frictional* may not be appropriate for the vertical component. Let's just denote them as having *x* and *y* components, where the *x* and *y* axes are horizontal and vertical, respectively.

We can guess that at the hinges the vertical components are probably directed upward, so as to support the door. The door is expected to pull away from the top hinge and push onto the bottom hinge, so the horizontal components of the corresponding forces on the door from the hinges should be to the left for the top hinge and to the right for the bottom hinge. Nevertheless, we shall call the horizontal components F_{dux} and F_{dlx} and expect, in the end, to find $F_{dux} < 0$ and $F_{dlx} > 0$.

(c) For the door to be in static equilibrium, two conditions must be met: (1) the net force acting on the door must be zero, since it is not undergoing linear acceleration and (2) the net torque acting on the door must be zero, since it is not undergoing angular acceleration. This must be true for any one point about which the torques are calculated, and then it will be true for any other point.

(d) For the vertical component of the net force to be zero, we must have

$$F_{duy} + F_{dly} - mg = 0. \tag{1}$$

For the horizontal component of the net force to be zero, we must have

$$F_{dux} + F_{dlx} = 0 \quad \text{or} \quad F_{dux} = -F_{dlx}. \tag{2}$$

For the net torque about the center of mass to be zero, we must have

$$-\left(\frac{h}{6}\right)F_{dux} - \left(\frac{w}{2}\right)F_{duy} + \left(\frac{h}{6}\right)F_{dlx} - \left(\frac{w}{2}\right)F_{dly} = 0. \tag{3}$$

Here we take counterclockwise torques as positive and clockwise torques as negative; remember that the force components may be either positive or negative, but only their magnitudes are used to calculate corresponding torques.

Someone else's equations might be different because the net-torque-equals-zero equation might have been determined about a different point. We used the center of mass of the door as our point, so the torque equilibrium condition had no reference to the gravitational force on the door. Had we chosen instead, say, the upper hinge, then the gravitational force on the door, and not the upper hinge forces, would have appeared in the torque equilibrium condition.

(e) There are three equations (two from the net-force-equals-zero equation and one from the net-torque-equals-zero equation), but there are four unknowns (F_{dux}, F_{duy}, F_{dlx}, and F_{dly}). Thus, we do not have enough information to determine the numerical values of the forces, even if we are given the values of *m, h,* and *w*. We need to know something else.

(f) Equation 2 of part (d) shows that $F_{dux} = -F_{dlx}$. Substituting the right side into equation 3 for F_{dux} and rearranging, we have

$$-F_{dlx}\left(\frac{h}{6}\right) - F_{dlx}\left(\frac{h}{6}\right) + F_{duy}\left(\frac{w}{2}\right) + F_{dly}\left(\frac{w}{2}\right) = 0.$$

Collecting terms, and using Equation 1, which tells us that $F_{duy} + F_{dly} = mg$, we have

$$-2F_{dlx}\frac{h}{6} + mg\frac{w}{2} = 0,$$

so

$$F_{dlx} = \frac{3mgw}{2h}.$$

Also, then,

$$F_{dux} = -F_{dlx} = -\frac{3mgw}{2h}.$$

The sign of F_{dlx} (positive) and the sign of F_{dux} (negative) show that the bottom hinge pushes the door away from the frame, while the upper hinge pulls the door toward the frame.

There is no way to determine the vertical components of the hinge forces, although Equation 1 of part (d) shows that they must add up to mg, the weight of the door. A reasonable assumption to make might be that the hinges are identical and identically mounted and that each hinge supports half the mass of the door. In this case we would have

$$F_{duy} = F_{dly} = \frac{mg}{2}.$$

(Another possibility is that the bottom hinge is not screwed into the door frame. That means the door frame could supply a horizontal force component on the door [the same F_{dlx} found previously] but not a vertical component, so then the upper hinge would have to support the whole mass of the door, and we would have $F_{duy} = mg$.)

(g) $F_{dlx} = \dfrac{3mgw}{2h} = \dfrac{3(20.0 \text{ kg})(9.80 \text{ m/s}^2)(1.00 \text{ m})}{2(2.20 \text{ m})} = 134 \text{ N}.$

$$F_{dux} = -F_{dlx} = -134 \text{ N}.$$

$$F_{duy} = F_{dly} = \frac{mg}{2} = \frac{1}{2}(20.0 \text{ kg})(9.80 \text{ m/s}^2) = 98 \text{ N}.$$

In terms of unit-vector notation,

$$\vec{F}_{du} = (-134 \text{ N})\hat{i} + (98 \text{ N})\hat{j}$$

and

$$\vec{F}_{dl} = (+134 \text{ N})\hat{i} + (98 \text{ N})\hat{j}.$$

Note: In problems of this sort, it is often possible to simplify the equations by choosing the point about which the net-torque-is-zero condition is applied. In this problem, if the point about which the torque is measured is the upper hinge, then neither the force at the upper hinge nor the vertical component of the force at the lower hinge will provide a torque. The only torque contributions will be a clockwise torque $mgw/2$ from the gravitational force on the door and a counterclockwise torque $F_{dlx}h/3$ from the horizontal component at the lower hinge. Setting these torques equal gives $F_{dlx} = 3mgw/2h$. Similarly, choosing the torque about the lower hinge immediately leads to $F_{dux} = -3mgw/2h$.

If you find these simpler expressions, great! If you don't, you should still get the right answer if you don't make a mistake. If you wonder why the author of this solution did not do things the easy way, it's to show you that the harder way works and to keep you from asking, "How did he know to take the torque about that point?"

65. A uniform horizontal beam of mass M and length L is supported by two uniform vertical metal rods. Rod 1 has a Young's modulus E_1 and a cross-sectional area A_1 and is located a distance d_1 from the right end of the beam. Rod 2 has a Young's modulus E_2 and a cross-sectional area A_2 and is located a distance d_2 from the right end of the beam. Assume that $0 < d_1 < d_2 < L/2$, so that both rods are on the right half of the beam. Denote the forces on the beam exerted by rods 1 and 2, respectively, by $\vec{F}_1$ and $\vec{F}_2$.

(a) Make a sketch of this physical situation, labeling the distances involved, and draw a force diagram for the beam. (b) Using the torque condition for static equilibrium, derive algebraic expressions for the forces $\vec{F}_1$ and $\vec{F}_2$. (c) Check that these two forces you determined by the torque condition for static equilibrium also satisfy the force condition for static equilibrium. (d) Determine for each rod whether the rod is under compression or tension, and derive algebraic expressions for the magnitudes of the stress and strain in the two rods.

(e) Suppose $M = 5.00$ kg, $L = 80.0$ cm, $A_1 = 4.00$ mm^2, $A_2 = 2.00$ mm^2, $d_1 = 10.0$ cm, $d_2 = 30.0$ cm, and rod 1 is made of steel and rod 2 of aluminum. Determine the numerical values of the forces $\vec{F}_1$ and $\vec{F}_2$ and of the stress and strain in each rod. If the rods are 50.0 cm long, by how much is each rod elongated or compressed?

Answers

(a) See Figs. 13-71a and b. The force diagram for the beam shows that there are three forces acting on the beam: (1) the gravitational force on it, represented with its equivalent $M\vec{g}$, directed down; (2) and (3) the forces $\vec{F}_1$ and $\vec{F}_2$ exerted on the beam by rods 1 and 2, respectively.

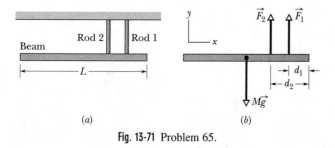

(a)   (b)

Fig. 13-71 Problem 65.

(b) We'll represent counterclockwise torques as positive and clockwise torques as negative. First, we apply the torque condition about the point where rod 1 supports the beam. Letting F_{2y} represent the upward vertical component of $\vec{F}_2$, the torque condition becomes

$$Mg(L/2 - d_1) - F_{2y}(d_2 - d_1) = 0,$$

so

$$F_{2y} = \frac{Mg(L/2 - d_1)}{d_2 - d_1}.$$

Second, we apply the torque condition about the point where rod 2 supports the beam. Letting F_{1y} represent the upward vertical component of $\vec{F}_1$, the torque condition becomes

$$Mg(L/2 - d_2) + F_{1y}(d_2 - d_1) = 0,$$

so

$$F_{1y} = \frac{Mg(L/2 - d_2)}{d_1 - d_2}.$$

(c) The sum of the upward components of the two forces should equal Mg, and it does:

$$\begin{aligned}
F_{2y} + F_{1y} &= \frac{Mg(L/2 - d_1)}{d_2 - d_1} + \frac{Mg(L/2 - d_2)}{d_1 - d_2} \\
&= \frac{Mg(L/2 - d_1)}{d_2 - d_1} - \frac{Mg(L/2 - d_2)}{d_2 - d_1} \\
&= \frac{Mg(L/2 - d_1 - L/2 + d_2)}{d_2 - d_1} \\
&= \frac{Mg(d_2 - d_1)}{d_2 - d_1} = Mg.
\end{aligned}$$

(d) We see that

$$F_{1y} = \frac{Mg(L/2 - d_2)}{d_1 - d_2}$$

must be negative because the numerator is positive while the denominator is negative; so rod 1 is actually pushing down on the beam, and the rod is under compression.

The stress in rod 1 has magnitude

$$\frac{|F_{1y}|}{A_1} = \frac{Mg(L/2 - d_2)}{A_1(d_2 - d_1)},$$

and the strain in it is

$$\frac{\Delta L}{L} = \frac{1}{E_1}\left(\frac{|F_{1y}|}{A_1}\right) = \frac{Mg(L/2 - d_2)}{E_1 A_1(d_2 - d_1)}.$$

Similarly,

$$F_{2y} = \frac{Mg(L/2 - d_1)}{d_2 - d_1}$$

must be positive because both numerator and denominator are positive; so rod 2 is under tension (it is pulling up on the beam).

The stress in rod 2 has magnitude

$$\frac{F_{2y}}{A_2} = \frac{Mg(L/2 - d_1)}{A_2(d_2 - d_1)},$$

and the strain in it is

$$\frac{\Delta L}{L} = \frac{1}{E_2}\left(\frac{F_{2y}}{A_2}\right) = \frac{Mg(L/2 - d_1)}{E_2 A_2(d_2 - d_1)}.$$

(e) The forces are vertical, with y components

$$F_{1y} = \frac{Mg(L/2 - d_2)}{d_1 - d_2}$$

$$= \frac{(5.0 \text{ kg})(9.8 \text{ N/kg})(0.40 \text{ m} - 0.30 \text{ m})}{0.10 \text{ m} - 0.30 \text{ m}}$$

$$= -24.5 \text{ N}$$

and

$$F_{2y} = \frac{Mg(L/2 - d_1)}{d_2 - d_1}$$

$$= \frac{(5.0 \text{ kg})(9.8 \text{ N/kg})(0.40 \text{ m} - 0.10 \text{ m})}{0.30 \text{ m} - 0.10 \text{ m}}$$

$$= +73.5 \text{ N}.$$

Thus, $\vec{F}_1 = -(24.5 \text{ N})\hat{j}$ and $\vec{F}_2 = +(73.5 \text{ N})\hat{j}$, using the usual Cartesian coordinate system in which the $+y$ direction is vertically upward.

The magnitudes of the stress and strain in the two rods, and their elongations ($+$ if under tension, $-$ if under compression) are:

Rod 1   (steel): $E_1 = 200 \times 10^9 \text{ N/m}^2$
Stress $= |F_{1y}|/A_1 = (24.5 \text{ N})/(4.00 \times 10^{-6} \text{ m}^2)$
$= 6.13 \times 10^6 \text{ N/m}^2$
Strain $= (\text{stress})/E_1 = (6.13 \times 10^6 \text{ N/m}^2)$
$\div (200 \times 10^9 \text{ N/m}^2) = 3.1 \times 10^{-5}$
Elongation $= -(\text{strain})(\text{length}) = -(3.1 \times 10^{-5})$
$\times (50.0 \text{ cm}) = -1.6 \times 10^{-3} \text{ cm}$

Rod 2   (aluminum): $E_1 = 70 \times 10^9 \text{ M/m}^2$
Stress $= F_{2y}/A_2 = (73.5 \text{ N})/(2.00 \times 10^{-6} \text{ m}^2)$
$= 36.8 \times 10^6 \text{ N/m}^2$
Strain $= (\text{stress})/E_2 = (36.8 \times 10^6 \text{ N/m}^2)$
$\div (70 \times 10^9 \text{ N/m}^2) = 5.2 \times 10^{-4}$
Elongation $= (\text{strain})(\text{length}) = (5.2 \times 10^{-4})(50.0 \text{ cm})$
$= 2.6 \times 10^{-2} \text{ cm}$

Gravitation

Sample Problem 14-9

In Fig. 14-41, a gray particle of mass $m_1 = 0.67$ kg is a distance $d = 23$ cm from one end of a uniform rod with length $L = 3.0$ m and mass $M = 5.0$ kg. What is the magnitude of the gravitational force $\vec{F}$ on the particle from the rod?

SOLUTION: A **Key Idea** here is that the rod is obviously not a particle, and so we cannot use Eq. 14-1 ($F = Gm_1m_2/r^2$) as written to find $\vec{F}$. However, a second **Key Idea** is that we can split the rod into a great many differential elements. If we treat each element as a particle, then we *can* use Eq. 14-1.

To start, we consider a typical element of the rod, of mass dm, located a distance r from the gray particle and occupying a length dr along the rod (as shown). From Eq. 14-1, the magnitude of the gravitational force $d\vec{F}$ on the gray particle due to this rod element is

$$dF = \frac{Gm_1\, dm}{r^2}. \tag{14-47}$$

The direction of $d\vec{F}$ is to the right in Fig. 14-41. In fact, because the gray particle is located on the longitudinal axis of the rod, each $d\vec{F}$ due to each element of the rod is directed to the right. Thus, we can find the magnitude F of the net force on the gray particle

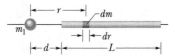

Fig. 14-41 Sample Problem 14-9. A gray particle of mass m_1 is a distance d from one end of a rod of length L. An element of the rod of mass dm is a distance r from the gray particle.

by summing the magnitudes dF of these individual forces. We do so by integrating Eq. 14-47:

$$F = \int dF = \int \frac{Gm_1\, dm}{r^2}. \tag{14-48}$$

However, we want to integrate with respect to coordinate r (not mass m as indicated in the integral), so we must relate an element's mass dm to its length dr along the rod. Because the rod is uniform, the ratio of mass to length is the same for all the elements of the rod and for the rod as a whole. Thus, we can write

$$\frac{\text{element's mass } dm}{\text{element's length } dr} = \frac{\text{rod's mass } M}{\text{rod's length } L},$$

or

$$dm = \frac{M}{L}\, dr.$$

We can use this to remove dm from Eq. 14-48, and then integrate (with respect to r) from the rod's left end (at distance $r = d$ from the gray particle) to its right end (at distance $r = L + d$) to include all the elements of the rod. We find

$$F = \int dF = \int_d^{L+d} \frac{Gm_1}{r^2} \frac{M}{L}\, dr = \frac{Gm_1M}{L} \int_d^{L+d} \frac{dr}{r^2}$$

$$= -\frac{Gm_1M}{L}\left[\frac{1}{r}\right]_d^{L+d} = -\frac{Gm_1M}{L}\left[\frac{1}{L+d} - \frac{1}{d}\right]$$

$$= \frac{Gm_1M}{d(L+d)}$$

$$= \frac{(6.67 \times 10^{-11}\ \text{m}^3/\text{kg} \cdot \text{s}^2)(0.67\ \text{kg})(5.0\ \text{kg})}{(0.23\ \text{m})(3.0\ \text{m} + 0.23\ \text{m})}$$

$$= 3.0 \times 10^{-10}\ \text{N}. \tag{Answer}$$

Sample Problem 14-10

Consider a pulsar, a collapsed star of extremely high density, with a mass M equal to that of the Sun (1.98×10^{30} kg), a radius R of only 12 km, and a rotational period T of 0.041 s. By what percentage does the free-fall acceleration g differ from the gravitational acceleration a_g at the equator of this spherical star?

SOLUTION: The **Key Idea** here is that g differs from a_g on an astronomical body if that body rotates. We want the ratio of the differ-

ence ($a_g - g$) to the gravitational acceleration a_g as a percentage. We can get that ratio if we rearrange Eq. 14-13 and then divide by a_g, to find

$$\frac{a_g - g}{a_g} = \frac{\omega^2 R}{a_g}. \tag{14-49}$$

To evaluate the right side of this equation, we first find ω with Eq. 11-20 ($T = 2\pi/\omega$):

$$\omega = \frac{2\pi}{T} = \frac{2\pi}{0.041 \text{ s}} = 153 \text{ rad/s}.$$

Then we find a_g with Eq. 14-10:

$$a_g = \frac{GM}{R^2} = \frac{(6.67 \times 10^{-11} \text{ m}^3/\text{kg} \cdot \text{s}^2)(1.98 \times 10^{30} \text{ kg})}{(12\,000 \text{ m})^2}$$

$$= 9.17 \times 10^{11} \text{ m/s}^2.$$

With these values for ω and a_g, Eq. 14-49 gives us

$$\frac{a_g - g}{a_g} = \frac{(153 \text{ rad/s})^2(12\,000 \text{ m})}{9.17 \times 10^{11} \text{ m/s}^2}$$

$$= 3.1 \times 10^{-4} = 0.031\%. \qquad \text{(Answer)}$$

Even though a pulsar rotates extremely rapidly, its rotation reduces the free-fall acceleration from the gravitational acceleration only slightly, because its radius is so small.

Sample Problem 14-11

A satellite in circular orbit at an altitude h of 230 km above Earth's surface has a period T of 89 min. Determine Earth's mass from these data.

SOLUTION: The **Key Idea** is that we can relate the mass M of a central body (here Earth) to the orbital radius r and orbital period T of a satellite moving around that body with Kepler's law of periods (Eq. 14-33). From it we have

$$M = \frac{4\pi^2 r^3}{GT^2}. \qquad (14\text{-}50)$$

The orbital radius r is

$$r = R_E + h = 6.37 \times 10^6 \text{ m} + 230 \times 10^3 \text{ m}$$

$$= 6.60 \times 10^6 \text{ m},$$

in which R_E is the radius of Earth. Substituting this value and the period T into Eq. 14-50 yields

$$M = \frac{(4\pi^2)(6.60 \times 10^6 \text{ m})^3}{(6.67 \times 10^{-11} \text{ m}^3/\text{kg} \cdot \text{s}^2)(89 \times 60 \text{ s})^2}$$

$$= 6.0 \times 10^{24} \text{ kg}. \qquad \text{(Answer)}$$

In the same way, we could find the mass of our Sun from the period and radius of Earth's orbit (assumed circular), and the mass of Jupiter from the period and orbital radius of any one of its moons (without knowing the mass of that moon).

Sample Problem 14-12

Two small spaceships, each with mass $m = 2000$ kg, are in the circular Earth orbit of Fig. 14-42, at an altitude h of 400 km. Igor, the commander of one of the ships, arrives at any fixed point in the orbit 90 s ahead of Janeway, the commander of the other ship.

(a) What are the period T_0 and the speed v_0 of the ships?

SOLUTION: A **Key Idea** here is that we can get the period from Kepler's law of periods (Eq. 14-33) if we first find the orbital radius r. Since Earth's radius R_E is 6370 km, that orbital radius must be

$$r = R_E + h = 6370 \text{ km} + 400 \text{ km}$$

$$= 6770 \text{ km} = 6.77 \times 10^6 \text{ m}.$$

Now, using $M = 5.98 \times 10^{24}$ kg for Earth's mass, we obtain from Eq. 14-33

$$T_0 = \sqrt{\frac{4\pi^2 r^3}{GM}}$$

$$= \sqrt{\frac{(4\pi^2)(6.77 \times 10^6 \text{ m})^3}{(6.67 \times 10^{-11} \text{ m}^3/\text{kg} \cdot \text{s}^2)(5.98 \times 10^{24} \text{ kg})}}$$

$$= 5540 \text{ s} = 92.3 \text{ min}. \qquad \text{(Answer)}$$

We get the spaceships' speed v_0 with another **Key Idea**: Because the ships travel in a circular orbit, their speed must be constant and equal to the ratio of the orbit's circumference to the orbit's period (the time to complete one orbit):

$$v_0 = \frac{2\pi r}{T_0} = \frac{(2\pi)(6.77 \times 10^6 \text{ m})}{5540 \text{ s}}$$

$$= 7680 \text{ m/s}. \qquad \text{(Answer)}$$

(b) At point P in Fig. 14-42, Janeway fires an instantaneous burst in the forward direction, *reducing* her ship's speed by 1.00%. After this burst, she follows the elliptical orbit shown dashed in the figure. What are the kinetic energy and potential energy of her ship immediately after the burst?

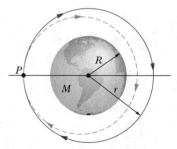

Fig. 14-42 Sample Problem 14-12. Two small spaceships are in a circular Earth orbit of radius r. At point P, the commander of one ship fires an instantaneous burst in the forward direction, slowing the ship. The ship then moves in the dashed elliptical orbit.

SOLUTION: The first **Key Idea** here is that we can get the new kinetic energy from the new speed v, which is $0.99v_0$. Thus, we have

$$K = \tfrac{1}{2}mv^2 = \tfrac{1}{2}m(0.99v_0)^2$$
$$= \tfrac{1}{2}(2000 \text{ kg})(0.99)^2(7680 \text{ m/s})^2$$
$$= 5.78 \times 10^{10} \text{ J.} \qquad \text{(Answer)}$$

We need another **Key Idea** to obtain the new potential energy: Just after the burst and before the ship's distance from Earth has changed appreciably, the potential energy is the same as it was before the burst. However, one more **Key Idea** is that we *cannot* get U by using Eq. 14-43 ($K = -U/2$) and the *new* kinetic energy, because her orbit is now elliptical, not circular. (Equation 14-43 applies only to circular orbits.) Instead, we use Eq. 14-20:

$$U = -\frac{GMm}{r}$$
$$= -\frac{(6.67 \times 10^{-11} \text{ N} \cdot \text{m}^2/\text{kg}^2)(5.98 \times 10^{24} \text{ kg})(2000 \text{ kg})}{6.77 \times 10^6 \text{ m}}$$
$$= -11.8 \times 10^{10} \text{ J.} \qquad \text{(Answer)}$$

(c) In Janeway's new elliptical orbit, what are the total energy E, the semimajor axis a, and the orbital period T?

SOLUTION: One **Key Idea** here is that we *cannot* get E with Eq. 14-45 ($E = -K$) because the orbit is now elliptical, not circular. (Equation 14-45 applies only to circular orbits.) A second **Key Idea** is that E is the same throughout this new orbit (mechanical energy is conserved). Thus, we can evaluate E at point P, just after the burst, where we know both kinetic energy K and potential energy U. We have

$$E = K + U = 5.78 \times 10^{10} \text{ J} + (-11.8 \times 10^{10} \text{ J})$$
$$= -6.02 \times 10^{10} \text{ J.} \qquad \text{(Answer)}$$

A **Key Idea** for finding the semimajor axis is that it is related to the total energy E of the system, as given by Eq. 14-46. Solving that equation for a yields

$$a = -\frac{GMm}{2E}$$
$$= -\frac{(6.67 \times 10^{-11} \text{ m}^3/\text{kg} \cdot \text{s}^2)(5.98 \times 10^{24} \text{ kg})(2000 \text{ kg})}{(2)(-6.02 \times 10^{10} \text{ J})}$$
$$= 6.63 \times 10^6 \text{ m.} \qquad \text{(Answer)}$$

Note that a is smaller, by 2.1%, than the radius r of the original circular orbit.

The **Key Idea** that gets us the period T is the law of periods (Eq. 14-33), with a substituted for r. (Kepler's laws apply to *all* orbits, elliptical and circular.) Solving Eq. 14-33 for T gives us

$$T = \sqrt{\frac{4\pi^2 a^3}{GM}}$$
$$= \sqrt{\frac{(4\pi^2)(6.63 \times 10^6 \text{ m})^3}{(6.67 \times 10^{-11} \text{ m}^3/\text{kg} \cdot \text{s}^2)(5.98 \times 10^{24} \text{ kg})}}$$
$$= 5370 \text{ s} \ (= 89.5 \text{ min}). \qquad \text{(Answer)}$$

This is shorter than the period of Igor's orbit by 170 s. Thus, Janeway will arrive back at point P ahead of Igor by 170 s $-$ 90 s, or 80 s. By slowing down, Janeway managed to get ahead of Igor! To get back into the original circular orbit, still remaining 80 s ahead of Igor, all Janeway has to do is restore her original speed by executing the same burst, but in a backward direction, the next time she is at point P.

Janeway was able to get ahead by slowing down for two reasons: (1) As Fig. 14-42 shows, the length of her new orbit is shorter. (2) By decreasing her speed at point P, she changed her orbit so that, until she returned to P, she dipped closer to the planet, thereby transferring energy from potential energy to kinetic energy and increasing her speed. Table 14-4 compares the properties of the original circular orbit and the new elliptical orbit.

TABLE 14-4   Properties of the Two Orbits in Sample Problem 14-12

Property	Circular Orbit	Elliptical Orbit
Semimajor axis (a, 10^6 m)	6.77	6.63
Closest distance (R_p, 10^6 m)	6.77	6.49
Farthest distance (R_a, 10^6 m)	6.77	6.77
Eccentricity (e)	0	0.021
Period (T, s)	5540	5370
Energy (E, 10^{10} J)	−5.90	−6.02

QUESTIONS

13. *Organizing question:* Figure 14-43 shows four particles fixed in place in a plane. (a) Set up an equation, complete with known data, to find the x component of the net gravitational force on the 1 kg particle at the origin due to the other three particles. (b) Similarly set up an equation to find the corresponding y component.

14. Figure 14-44 shows the orbits of three planets about identical stars; the orbits have identical major axes $2a$. Rank the three orbits according to their (a) eccentricities, (b) perihelion distances,

(c) aphelion distances, and (d) periods of revolution, greatest first. (*Hint:* Roughly locate the star for each orbit.)

15. Rank the orbits in Question 14 and Fig. 14-44 according to (a) the total energy associated with the orbit of each planet, (b) the orbital speed of the planet at perihelion, and (c) the orbital speed of the planet at aphelion, greatest first.

16. In Fig. 14-13, consider three points along the orbit: point 1 at perihelion, point 2 at aphelion, and point 3 midway between perihelion and aphelion. Rank the three points according to (a) the

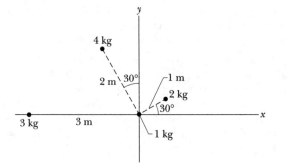

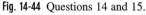

Fig. 14-43 Question 13.

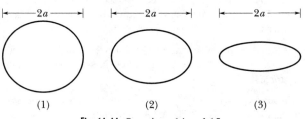

(1)   (2)   (3)

Fig. 14-44 Questions 14 and 15.

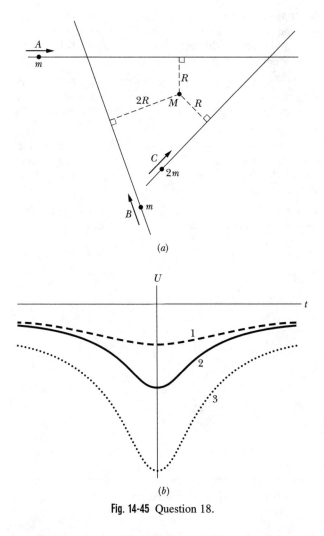

(a)

(b)

Fig. 14-45 Question 18.

angular momentum of the planet there and (b) the speed of the planet there, greatest first.

17. Rank the three points in Question 16 according to (a) the total energy of the planet (associated with the planet's orbiting), (b) the kinetic energy of the planet, and (c) the potential energy of the planet (actually, the planet–star system), greatest first.

18. In Fig. 14-45a, a stationary spacecraft of mass M is passed first by asteroid A of mass m, then by asteroid B of the same mass m, and then by asteroid C of mass $2m$. The asteroids move along the indicated straight paths at the same speed; the perpendicular distances between the spacecraft and the paths are given as multiples of R. Figure 14-45b gives the gravitational potential energy $U(t)$ of the spacecraft–asteroid system during the passage of each asteroid. (a) Where is an asteroid along its path at time $t = 0$? (b) Which asteroid corresponds to which plot of $U(t)$?

19. Which of the orbital paths (for, say, a spy satellite) in Fig. 14-46 do not require continuous adjustments by booster rockets? Orbit 1 is at latitude 60°; orbit 2 is in an equatorial plane; orbit 3 is about the center of Earth, between latitudes 60° N and 60° S.

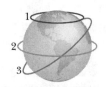

Fig. 14-46 Question 19.

20. Reconsider the situation of Question 11. Would the work done by you be positive, negative, or zero if you moved the particle (a) from A and B, (b) from A to C, and (c) from B to D? (d) Rank those moves according to the absolute value of the work done by your force, greatest first.

21. As your spacecraft travels along an x axis through an asteroid belt, the gravitational potential energy $U(x)$ of the spacecraft–asteroid system is given by the curve in Fig. 14-47. From that curve you can determine the magnitude of the x component F_x of the gravitational force on the spacecraft due to the asteroids. (a) Rank the magnitude of F_x at points A, B, C, D, and E, greatest first. (b) What is the direction of F_x, if any, at those points?

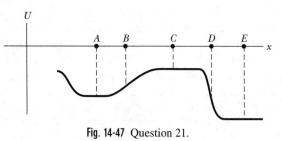

Fig. 14-47 Question 21.

22. On burning out, a (nonrotating) star collapses onto itself from an initial radius R_i. Which curve in Fig. 14-48 best gives the gravitational acceleration a_g on the surface of the star as a function of the radius of the star during the collapse?

23. A satellite, with speed v_1 and mass m, is in a circular orbit about a planet of mass M_1. Another satellite, with speed v_2 and

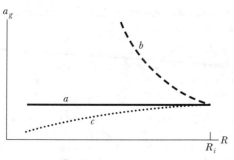

Fig. 14-48 Question 22.

mass $2m$, is in a circular orbit of the same radius about a planet of mass M_2. Is M_2 greater than, less than, or the same as M_1 (a) if the satellites have the same period and (b) if $v_2 > v_1$?

24. The following are the masses and radii of three planets: planet A, $2M$ and R; planet B, $3M$ and $2R$; planet C, $4M$ and $2R$. Rank the planets according to the escape speeds from their surfaces, greatest first.

25. Figure 14-49 shows a particle of mass m that is moved from an infinite distance to the center of a ring of mass M, along the central axis of the ring. For the trip, how does the magnitude of the gravitational force on the particle due to the ring change?

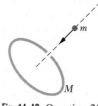

Fig. 14-49 Question 25.

DISCUSSION QUESTIONS

26. If the gravitational force acts on all bodies in proportion to their masses, why doesn't a heavy body fall correspondingly faster than a light body?

27. How does the net gravitational force on a space probe vary as it travels from Earth to the Moon? Would its mass change?

28. Would we have more sugar to the newton at the north pole or at the equator? What about sugar to the kilogram?

29. Because Earth bulges near the equator, the source of the Mississippi River, although high above sea level, is nearer to the center of Earth than is its mouth. How can the river flow "uphill"?

30. One clock uses an oscillating spring; a second clock uses a pendulum. Both are taken to Mars. Will they keep the same time there that they kept on Earth? Will they agree with each other? Explain. (The mass of Mars is one-tenth that of Earth, and its radius is half that of Earth.)

31. Two identical cars travel at the same speed in opposite directions on an east–west highway. Which car presses down harder on the road?

32. As a car speeds around a curve, the passengers tend to be thrown radially outward. Why are astronauts in a space shuttle not similarly affected as their shuttle speeds in orbit around Earth?

33. The gravitational force exerted by the Sun on the Moon is about twice as great as the gravitational force exerted by Earth on the Moon. Why then doesn't the Moon escape from Earth?

34. Explain why the following reasoning is wrong. "The Sun attracts all bodies on Earth. At midnight, when the Sun is directly below, it pulls on an object in the same direction as the pull of Earth on that object; at noon, when the Sun is directly above, it pulls on an object in a direction opposite the pull of Earth. Hence all objects should be heavier at midnight (or at night) than they are at noon (or during the day)."

35. If lunar tides slow down the rotation of Earth (owing to friction), the angular momentum of Earth decreases. What happens to the motion of the Moon as a consequence of the conservation of angular momentum? Does the Sun (and solar tides) play a role here?

36. A satellite in low Earth orbit experiences a small drag force from Earth's atmosphere. What happens to its speed because of this drag force?

37. Would you expect the total mechanical energy of the solar system to be constant? The total angular momentum? Explain your answers.

38. Objects at rest on Earth's surface move in circular paths with a period of 24 h. Are they "in orbit" in the sense that an Earth satellite is in orbit? Why not? What would the period of Earth have to be to put such objects in true orbit?

39. As measured by an observer on Earth, would there be any difference in the periods of two satellites, each in a circular orbit near Earth in an equatorial plane, but one moving eastward and the other westward?

40. After *Sputnik I* was put into orbit, some scientists said that it would not return to Earth but would burn up in its *descent*. Considering the fact that it did not burn up in its *ascent*, how is this possible?

41. What is being plotted in Fig. 14-50? Put numbers with units on each axis and adjust the horizontal axis if needed.

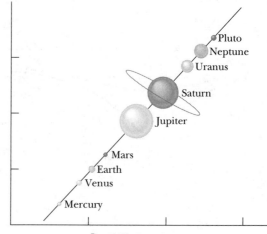

Fig. 14-50 Question 41.

42. An artificial satellite is in a circular orbit about Earth. How will its orbit change if one of its rockets is momentarily fired (a) toward Earth, (b) away from Earth, (c) in a forward direction, (d) in a backward direction, and (e) perpendicular to the plane of the orbit?

43. What advantage does Florida have over California for launching satellites?

44. For a flight to Mars, a rocket is fired in the direction Earth is moving in its orbit. For a flight to Venus, it is fired backward along that orbit. Explain why.

45. Saturn is about six times farther from the Sun than Mars. Which planet has (a) the greater period of revolution, (b) the greater orbital speed, and (c) the greater angular speed?

46. What happens to the angular momentum of an artificial Earth satellite as it descends through the atmosphere?

47. You are a passenger on the SS *Arthur C. Clarke,* the first interstellar spaceship that rotates about a central axis to simulate Earth gravity. If you are in an enclosed cabin, how can you tell that you are not on Earth?

EXERCISES & PROBLEMS

65. The mysterious visitor that appears in the enchanting story, *The Little Prince,* was said to come from a planet that "was scarcely any larger than a house!" Assume that the planet's density is about that of Earth and that the planet does not appreciably spin. Approximate (a) the free-fall acceleration on the planet's surface and (b) the escape speed from the planet.

66. The first known collision between space debris and a functioning satellite occurred in 1996: At an altitude of 700 km, a year-old French spy satellite was hit by a piece of an Ariane rocket that had been in orbit for 10 years. A stabilizing boom on the satellite was demolished, and the satellite was sent spinning out of control. Just before the collision and in kilometers per hour, what was the speed of the rocket piece relative to the satellite if both were in circular orbits and the collision was (a) head on and (b) along perpendicular paths?

67. (a) If the legendary apple of Newton could be released from rest at a height of 2 m from the surface of a neutron star with a mass 1.5 times that of our Sun (see Appendix C) and a radius of 20 km, what would be its speed when it reached the surface of the star? (b) If the apple could rest on the surface of the star, what would be the approximate difference between the gravitational acceleration at the top and at the bottom of the apple? (Choose a reasonable size for an apple; the answer indicates that an apple would never survive near a neutron star.)

68. The most valued orbit for a communications satellite is a *geosynchronous orbit,* in which the satellite remains above a certain point on Earth's equator as it orbits. Such a "geo satellite" would always be in the same position in the sky as seen from the transmitting or receiving equipment of, say, the BBC in London. Unfortunately, room along a geosynchronous orbit is limited because the satellites in such an orbit cannot be too close or the signal quality is degraded. The least separation along the 360° of the circular orbit is 3.00°. What are (a) the maximum number of geo satellites the orbit should contain and (b) their orbital radius? (c) What then is the minimum separation *d* between any two geo satellites along the orbital path?

A magnetic storm can disrupt a satellite as well as move it, either toward or away from Earth. Suppose a magnetic storm turns off geo satellite *Hot Bird 2,* and when ground-based engineers turn it back on, it is east of its assigned orbital spot. (d) Is the period of *Hot Bird 2* now greater than or smaller than that of the other geo satellites? (e) Did the storm move the satellite toward Earth or away from it?

69. We watch two identical astronomical bodies *A* and *B,* each of mass *m,* fall toward each other from rest because of the gravitational force on each from the other. Their initial center-to-center separation is R_i. Assume that we are in an inertial reference frame that is stationary with respect to the center of mass of this two-body system. Use the principle of conservation of mechanical energy ($K_f + U_f = K_i + U_i$) to find the following when the center-to-center separation is $0.5R_i$: (a) the total kinetic energy of the system, (b) the kinetic energy of each body, (c) the speed of each body relative to us, and (d) the speed of body *B* relative to body *A.*

Next assume that we are in a reference frame attached to body *A* (we ride on the body). Now we see body *B* fall from rest toward us. From this reference frame, again use $K_f + U_f = K_i + U_i$ to find the following when the center-to-center separation is $0.5R_i$: (e) the kinetic energy of body *B* and (f) the speed of body *B* relative to body *A.* (g) Why are the answers to (d) and (f) different? Which answer is correct?

70*. The presence of an unseen planet orbiting a distant star can sometimes be inferred from the motion of the star as we see it. As the star and planet orbit the center of mass of the star–planet system, the star moves toward and away from us with what is called the *line of sight velocity,* a motion that can be detected. Figure 14-51 shows a graph of the line of sight velocity versus time for the star 14 Herculis. The star's mass is believed to be 0.90 of the mass of our Sun. Assume that only one planet orbits

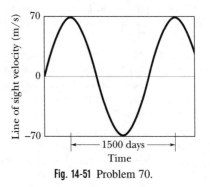

Fig. 14-51 Problem 70.

the star and that our view is along the plane of the orbit. Then approximate (a) the planet's mass in terms of Jupiter's mass m_J and (b) the planet's orbital radius in terms of Earth's orbital radius r_E. (*Hint:* See Sample Problem 14-7 for a similar arrangement of visible star and unseen companion.)

71. In his 1865 science fiction novel *From the Earth to the Moon,* Jules Verne described how three astronauts are shot to the Moon by means of a huge gun constructed below ground in Florida. According to Verne, the aluminum capsule containing the astronauts is accelerated by ignition of nitrocellulose to a speed of 11 km/s along the gun barrel's length of 220 m. (a) In *g* units, what is the average acceleration of the capsule and astronauts in the gun barrel?

A modern version of such gun-launched space craft (although without passengers) has been proposed. In this modern version, called the SHARP (Super High Altitude Research Project) gun, ignition of methane and air shoves a piston down the gun's tube, compressing hydrogen gas that then launches a rocket. During this launch, the rocket moves 3.5 km and reaches a speed of 7.0 km/s. Once launched, the rocket can be fired to gain additional speed. (b) In *g* units, what would be the average acceleration of the rocket within the launcher? (c) How much additional speed is needed (via the rocket engine) if the rocket is to orbit Earth at an altitude of 700 km?

72. With what speed would mail pass through the center of Earth if it were dropped down the tunnel of Sample Problem 14-4?

73. Show that, at the bottom of a vertical mine shaft dug to depth *D*, the gravitational acceleration a_g is

$$a_g = a_{gs}\left(1 - \frac{D}{R}\right),$$

where a_{gs} is the surface value. Assume that Earth is a uniform sphere of radius *R*.

74*. The gravitational force between two particles with masses *m* and *M*, initially at rest at great separation, pulls then together. Show that at any instant the speed of either particle relative to the other is $\sqrt{2G(M + m)/d}$, where *d* is their separation at that instant. (*Hint:* Use the laws of conservation of energy and conservation of linear momentum.)

75. Two 20 kg spheres are fixed in place on a *y* axis, one at *y* = 0.40 m and the other at *y* = −0.40 m. A 10 kg ball is then released from rest at a point on the *x* axis that is at a great distance (effectively infinite) from the spheres. If the only forces acting on the ball are the gravitational forces from the spheres, then when the ball reaches the (*x, y*) point (0.30 m, 0), what are (a) its kinetic energy and (b) the magnitude and direction of the net force on it from the spheres?

76. A projectile is launched directly away from the surface of a planet of mass *M* and radius *R*; the launch speed is $(GM/R)^{1/2}$. Use the principle of conservation of energy to determine the maximum distance from the center of the planet achieved by the projectile. Express your result in terms of *R*.

77. Figure 14-52 shows two uniform rods of the same length *L* and mass *M*, lying on an axis with separation *d*. Using the result of Sample Problem 14-9, set up an integral (with integration limits and ready for integration) to find the gravitational force of one rod on the other.

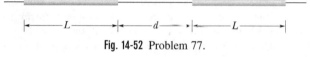

Fig. 14-52 Problem 77.

78. An object lying on Earth's equator is accelerated (a) toward the center of Earth because Earth rotates, (b) toward the Sun because Earth revolves around the Sun in an almost circular orbit, and (c) toward the center of our galaxy because the Sun moves about the galactic center. For the latter, the period is 2.5×10^8 y and the radius is 2.2×10^{20} m. Calculate these three accelerations as multiples of $g = 9.8$ m/s^2.

79. A 6000 kg planetary probe is in a circular polar orbit with a radius of 108×10^6 km (the orbital radius of Venus). The space agency wants to put the probe into the same orbit as Earth, a circular orbit with a radius of 150×10^6 km. The first step is to increase the probe's speed so that the probe is in an elliptical orbit with a perihelion distance equal to the radius of its initial orbit and an aphelion distance equal to the radius of its final orbit. The mass of the Sun is 1.99×10^{30} kg. (a) Calculate the increases in speed and energy that are required. Plot the elliptical orbit. (In polar coordinates, with the origin at the center of the ellipse, the equation for an ellipse is

$$r^2 = \frac{r_p r_a (r_p + r_a)^2}{(r_p + r_a)^2 \sin^2 \theta + 4 r_p r_a \cos^2 \theta},$$

where r_p is the perihelion distance and r_a is the aphelion distance.) (b) When the aphelion position is reached, the speed of the probe is again changed to put it into the desired circular orbit. What changes in the speed and energy are required?

80. In Fig. 14-53, blocks with identical masses *m* hang from strings on a balance at the surface of Earth. The strings have negligible mass and differ in length by *h*. Assume that Earth is spherical, with a density $\rho = 5.5$ g/cm^3. (a) Show that the difference ΔW in the weights of the blocks, due to one block being closer to Earth than the other, is $8\pi G\rho mh/3$. (b) Find the difference in length that will give a ratio $\Delta W/W = 1 \times 10^{-6}$, where *W* is the weight of either block.

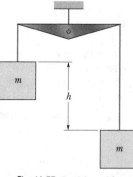

Fig. 14-53 Problem 80.

81. Show how, guided by Kepler's third law (Eq. 14-33), Newton could deduce that the force holding the Moon in its orbit, assumed circular, depends on the inverse square of the Moon's distance from the center of Earth.

82. A planet requires 300 (Earth) days to complete its circular orbit about its sun, which has a mass of 6.0×10^{30} kg. What are (a) its orbital radius and (b) its orbital speed?

83. In a shuttle craft of mass *m* = 3000 kg, Captain Janeway orbits a planet of mass $M = 9.50 \times 10^{25}$ kg, in a circular orbit of radius $r = 4.20 \times 10^7$ m. What are (a) the period of the orbit and (b) the speed of the shuttle craft? Janeway briefly fires a forward-pointing thruster, reducing her speed by 2.00%. Just then, what are (c) the speed, (d) the kinetic energy, (e) the gravitational potential energy, and (f) the mechanical energy of the shuttle craft? (g) What is the semimajor axis of the elliptical orbit now taken by the craft? (h) What is the difference between the period of the original circular orbit and that of the new elliptical orbit? (i) Which orbit has the smaller period?

84. A typical neutron star may have a mass equal to that of the Sun but a radius of only 10 km. (a) What is the gravitational acceleration at the surface of such a star? (b) How fast would an object be moving if it fell from rest through a distance of 1.0 m on such a star? (Assume the star does not rotate.)

85. Two spheres with masses $m_A = 800$ kg and $m_B = 600$ kg are separated by 0.25 m. What is the net gravitational force (in both magnitude and direction) from them on a 2.0 kg sphere located 0.20 m from A and 0.15 m from B?

86. (a) What is the escape speed from the Sun for an object in Earth's orbit (of orbital radius R) but far from Earth? (b) If an object already has a speed equal to Earth's orbital speed, what additional speed must it be given to escape as in (a)? (c) Suppose an object is launched from Earth in the direction of Earth's orbital motion. What speed must it be given during the launch so that when it is far from Earth, it has that additional speed calculated in (b) and thus can escape from the Sun? (This is the speed required for an Earth-launched object to escape from the Sun.)

87. In your calculator, make a list of the periods T for the planets in Table 14-3, and a separate list of the semimajor axes a. Multiply the list of T values by an appropriate factor so that the unit of T is seconds. (a) Store values of T^2 and a^3 in two new lists. Have the calculator do a linear regression fit of T^2 versus a^3. From the parameters of the fit and using the known value of G, determine the mass of the Sun. (b) Store values of $\log T$ in one list and $\log a$ in another list. Have the calculator plot $\log T$ versus $\log a$ and then do a linear regression fit of the plot. From the parameters of this fit and using the known value of G, again determine the mass of the Sun.

88. In a double-star system, two stars of mass 3.0×10^{30} kg each rotate about the system's center of mass at radius 1.0×10^{11} m. (a) What is their common angular speed? (b) If a meteoroid passes through the system's center of mass perpendicular to their orbital plane, what minimum speed must it have at that point if it is to escape to "infinity" from the two-star system?

89. A satellite orbits a planet of unknown mass in a circle of radius 2.0×10^7 m. The magnitude of the gravitational force on the satellite from the planet is 80 N. (a) What is the kinetic energy of the satellite in this orbit? (b) What would be the magnitude of the gravitational force on the satellite from the planet if the radius of the orbit were increased to 3.0×10^7 m?

90. A 50 kg satellite circles planet Cruton every 6.0 h. The magnitude of the gravitational force exerted on the satellite by Cruton is 80 N. (a) What is the radius of the orbit? (b) What is the kinetic energy of the satellite? (c) What is the mass of planet Cruton?

91. The orbit of Earth about the Sun is *almost* circular: The closest and farthest distances are 1.47×10^8 km and 1.52×10^8 km respectively. Determine the corresponding variations in (a) total energy, (b) gravitational potential energy, (c) kinetic energy, and

(d) orbital speed. (*Hint:* Use the laws of conservation of energy and conservation of angular momentum.)

92. Consider a satellite in a circular orbit about Earth. State how the following properties of the satellite depend on the radius r of its orbit: (a) period, (b) kinetic energy, (c) angular momentum, and (d) speed.

93. Show that the escape speed from the Sun at Earth's distance from the Sun is $\sqrt{2}$ times the speed of Earth in its orbit, assumed to be a circle. (This is a specific case of a general result for circular orbits: $v_{\text{esc}} = \sqrt{2}v_{\text{orb}}$.)

94. A sphere of matter, of mass M and radius a, has a concentric cavity of radius b, as shown in cross section in Fig. 14-54. (a) Sketch a curve of the magnitude of the gravitational force F from the sphere on a particle of mass m located a distance r from the center of the sphere, as a function of r in the range $0 \leq r \leq \infty$. Consider $r = 0$, b, a, and ∞ in particular. (b) Sketch the corresponding curve for the potential energy $U(r)$ of the system.

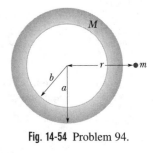

Fig. 14-54 Problem 94.

Tutorial Problems

95. Let's look at the gravitational forces in a system of four particles arranged in a square with sides of length D. Assume they do not interact with any other objects. Let the masses of the particles be $m_1 = M$, $m_2 = 2.00M$, $m_3 = 3.00M$, and $m_4 = 4.00M$. Use $\vec{F}_{41}$ to denote the force on particle 4 due to particle 1; $\vec{F}_{42}$ and $\vec{F}_{43}$ are similarly defined. Let $\vec{F}_4$ denote the total force on particle 4.

(a) Show the forces $\vec{F}_{41}$, $\vec{F}_{42}$, and $\vec{F}_{43}$ on a sketch. (b) Express each of the three gravitational forces on particle 4 in terms of the given quantities (M, D, etc.) using the form $\vec{F} = F\hat{F}$, where F is a magnitude and $\hat{F}$ is a unit vector in the direction of $\vec{F}$. Each $\hat{F}$ should be expressed in unit-vector ($\hat{i}$, $\hat{j}$) notation, using the directions defined by the coordinate axes. (c) Name the principle of physics that can be used to determine the total force $\vec{F}_4$ acting on particle 4 in terms of the forces produced by the other particles. (d) Express $\vec{F}_4$ in $\hat{i}$ and $\hat{j}$ notation. (e) Express $\vec{F}_4$ as a magnitude and direction, giving the direction in terms of an angle. (f) Show the vector $\vec{F}_4$ on the sketch in part (a).

(g) The gravitational field $\vec{g}$ at a point in space is defined as the ratio $\vec{F}_{\text{grav}}/m$, where $\vec{F}_{\text{grav}}$ is the total gravitational force on a mass m at that point. Determine the contribution of each particle to the gravitational field $\vec{g}$ at the center of the square, expressing each as a magnitude and a unit vector. Add them to determine the total value for $\vec{g}$. Show this field on the diagram of part (a).

(h) Consider a point on the positive y axis very far from this system of particles. What is the approximate gravitational field $\vec{g}$ as a function of y? Use unit-vector notation. (i) Suppose the four particles in the square are initially at rest and are allowed to move until they have merged into a single object. Assume that no external forces act on the system. Where will that object be located?

Answers

(a) See Fig. 14-55.

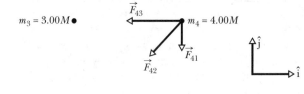

Fig. 14-55 Problem 95.

(b) $\vec{F}_{41}$ is directed from m_4 toward m_1, that is, in the direction $-\hat{j}$, so $\hat{F} = -\hat{j}$. The magnitude is

$$F = \frac{Gm_1m_4}{D^2} = \frac{GM(4.00M)}{D^2} = \frac{4.00GM^2}{D^2}.$$

Thus,

$$\vec{F}_{41} = \frac{4.00GM^2}{D^2}(-\hat{j}).$$

$\vec{F}_{42}$ is directed from m_4 toward m_2, that is, at an angle of $-225°$ from the direction of $+x$; so the vector is in the direction

$$\hat{F} = \cos(225°)\hat{i} + \sin(225°)\hat{j} = -0.707\hat{i} - 0.707\hat{j}.$$

Its magnitude is

$$F = \frac{Gm_2m_4}{(D^2 + D^2)} = \frac{4.00GM^2}{D^2}.$$

Thus,

$$\vec{F}_{42} = F\hat{F} = \frac{4.00GM^2}{D^2}(-0.707\hat{i} - 0.707\hat{j}).$$

$\vec{F}_{43}$ is directed from m_4 toward m_3, that is, in the $-\hat{i}$ direction; so $\hat{F} = -\hat{i}$. The magnitude is

$$F = \frac{Gm_3m_4}{D^2} = \frac{12.0GM^2}{D^2}.$$

Therefore,

$$\vec{F}_{43} = \frac{12.0GM^2}{D^2}(-\hat{i}).$$

(c) To determine the total force on particle 4 we use the principle of superposition: The total force on the particle is the vector sum of all the individual forces acting on it due to the other particles.

(d) In the notation using unit vectors along the axes,

$$\vec{F}_4 = \vec{F}_{41} + \vec{F}_{42} + \vec{F}_{43}$$
$$= -\frac{4.00GM^2}{D^2}\hat{j} + \frac{2.83GM^2}{D^2}(-\hat{i} - \hat{j}) - \frac{12.0GM^2}{D^2}\hat{i}$$
$$= \frac{GM^2}{D^2}(-14.8\hat{i} - 6.83\hat{j}).$$

(e) The magnitude is (using more significant figures)

$$F_4 = \frac{GM^2}{D^2}\sqrt{(-14.828)^2 + (-6.828)^2} = \frac{16.3GM^2}{D^2}.$$

The direction has angle (in the third quadrant)

$$\theta = \tan^{-1}[-6.828/(-14.828)] = \tan^{-1}(0.4605) = 24.7°.$$

Therefore, we can say that the total force on the fourth particle is $16.3(GM^2/D^2)$ in a direction of $24.7°$ below the $-x$ direction.

(g) The center of the square is a distance $D/\sqrt{2}$ from each of the particles, since that distance is half the length $\sqrt{2}D$ of a diagonal of a square with sides of length D. Thus, the factor $1/r^2$ in the gravitational field equals $2/D^2$ in each case. The contribution of m_i to the gravitational field is then $(2Gm_i/D^2)$ times the unit vector that points from the center of the square toward m_i. The contributions are then

$$m_1 = M: \quad (2.00GM/D^2)(+\hat{i} - \hat{j})/\sqrt{2}$$
$$m_2 = 2M: \quad (4.00GM/D^2)(-\hat{i} - \hat{j})/\sqrt{2}$$
$$m_3 = 3M: \quad (6.00GM/D^2)(-\hat{i} + \hat{j})/\sqrt{2}$$
$$m_4 = 4M: \quad (8.00GM/D^2)(+\hat{i} + \hat{j})/\sqrt{2}$$

These add up to $\vec{g} = (8.00GM/\sqrt{2}D^2)\hat{j}$.

(h) Far from the system, the system resembles a single particle of mass

$$m_1 + m_2 + m_3 + m_4 = 10M,$$

so the gravitational field is

$$\vec{g}(0, y, 0) \approx -\frac{10GM}{y^2}\hat{j}.$$

The negative sign here shows that the field points toward the system of particles.

(i) The object will have to be located at the center of mass of the original particles. That point will be located above the center of the square, about $\frac{2}{5}$ of the way to the top edge of the square.

96. Near Earth, the gravitational field (see Problem 95) is mainly due to Earth's mass, with the Sun, Moon, and other astronomical objects making only minor contributions. Earth is approximately spherically symmetric, but its density is not uniform. In this problem we'll use geophysical data to model Earth as a sphere of varying density and determine $g(r)$ both inside Earth ($r < R_E =$ radius of Earth) and outside ($r > R_E$). A key fact to remember is that outside a spherical shell, the gravitational field of the object can be calculated by assuming that the mass is concentrated at the center, while inside the shell, the gravitational field is zero. In other words, to determine the gravitational field at any point inside Earth, you need to take into account only that part of Earth's mass that is closer to the center than that point.

From the behavior of seismic waves and other information, geophysicists have shown that Earth does not have a uniform density but can be modeled as consisting of five spherically symmetric regions, each of which is approximately uniform in density. This model is similar to, but a slight simplification of, the density variation shown in Figure 14-6 in the textbook. The five regions of the model are

the inner core ($r < 1221.5$ km), density = 12.9 g/cm³;
the outer core (1221.5 km $< r <$ 3480 km),
density = 10.9 g/cm³;
the lower mantle (3480 km $< r <$ 5701 km),
density = 4.9 g/cm³;
the upper mantle (5701 km $< r <$ 6346.6 km),
density = 3.6 g/cm³;
the crust/ocean region (6346.6 km $< r <$ 6371 km),
density = 2.4 g/cm³.

(a) First, use Earth's mass (5.98×10^{24} kg) and its mean radius (6371 km) to determine the average density. Which of the five regions are denser than average, and which are less dense than average? (b) Using the information in the preceding list, determine the magnitude $g(r)$ of Earth's gravitational field in the inner core as a function of r. Express your answer as a numerical function of r. What is the functional dependence of $g(r)$ on r (linear, quadratic, inverse square, etc.)?

(c) Determine the value of g at each of the boundaries between the five regions and at Earth's surface. (d) Determine the magnitude $g(r)$ for the region $r > R_E = 6371$ km. (e) Make a plot of $g(r)$ for $0 \le r \le 2R_E = 2 \times 6371$ km, choosing an appropriate scale.

(f) Show that at any value of r inside Earth (or another spherically symmetric object), $g(r)$ will increase with r if and only if the density $\rho(r)$ there exceeds $\frac{2}{3}$ of the average density inside the radius r. Check the data in this problem to show that the maximum of $g(r)$ inside Earth occurs at one of the boundaries between the regions, and determine which one.

Answers

(a) Earth's mass is $M = 5.98 \times 10^{24}$ kg and its volume is

$$V = \tfrac{4}{3}\pi R_E^3 = \tfrac{4}{3}\pi (6.371 \times 10^6 \text{ m})^3 = 1.083 \times 10^{21} \text{ m}^3,$$

so its average density is

$$\rho_{avg} = \frac{M}{V} = \frac{5.98 \times 10^{24} \text{ kg}}{1.083 \times 10^{21} \text{ m}^3}$$
$$= 5500 \text{ kg/m}^3 = 5.5 \text{ g/cm}^3.$$

Only the inner and outer core have a density greater than the average density.

(b) In the inner core, the density has a constant value $\rho = 12.9$ g/cm³. At a distance r from Earth's center, the mass inside r is

$$\rho V = \rho(\tfrac{4}{3}\pi r^3),$$

so the gravitational field there is

$$g(r) = \frac{GM}{r^2} = \frac{G(\tfrac{4}{3}\pi \rho r^3)}{r^2} = \tfrac{4}{3}\pi G \rho r$$
$$= (\tfrac{4}{3}\pi)(6.67 \times 10^{-11} \text{ N} \cdot \text{m}^2/\text{kg}^2)(12\,900 \text{ kg/m}^3)r$$
$$= (3.60 \times 10^{-6} \text{ N} \cdot \text{m})r.$$

This is a linear function of r; in other words, the gravitational field starts at 0 at $r = 0$ and increases linearly up to the boundary of the inner core with the outer core.

(c) We need to determine the total mass M inside each boundary and then use that mass with the equation $g(r) = GM/r^2$. Apart from the inner core, which is a solid sphere, the regions are spherical shells. The volume of a spherical shell that lies between $r = r_1$ and $r = r_2$ is the difference in volume between spheres of radii r_1 and r_2, namely, $\tfrac{4}{3}\pi(r_2^3 - r_1^3)$. Thus, the mass of a region is this quantity multiplied by the density of the region. Results are shown in Table 14-5.

(d) For the region $r > R_E$, the region outside Earth, the gravitational field has the form $g(r) = GM/r^2$, where M is the total mass of Earth. At Earth's surface, $r = R_E$ and $g(R_E) = 9.8$ N/kg. We can thus write, for the region $r > R_E$,

$$g(r) = (9.8 \text{ N/kg})\left(\frac{R_E}{r}\right)^2.$$

(f) To see this, let $M(r)$ be the mass up to radius r, $\rho(r)$ be the density at radius r, and $\rho_{avg}(r)$ be the average density up to radius r. From

$$\rho = \frac{M}{V} = \frac{M}{\tfrac{4}{3}\pi r^3},$$

we have

$$\frac{dM(r)}{dr} = 4\pi r^2 \rho(r)$$

and

$$M(r) = \tfrac{4}{3}\pi r^3 \rho_{avg}(r).$$

TABLE 14-5   Problem 14-96

	Inner Core	Outer Core	Lower Mantle	Upper Mantle	Crust/Ocean
r_1 (km)	0	1221.5	3480	5701	6346.6
r_2 (km)	1221.5	3480	5701	6346.6	6371
Volume (m³)	7.63×10^{18}	1.69×10^{20}	6.00×10^{20}	2.95×10^{20}	1.24×10^{19}
Density (kg/m³)	12 900	10 900	4900	3600	2400
Mass (kg)	9.85×10^{22}	1.84×10^{24}	2.94×10^{24}	1.06×10^{24}	2.98×10^{22}
M (kg)	9.85×10^{22}	1.94×10^{24}	4.88×10^{24}	5.94×10^{24}	5.97×10^{24}
$g(r_2)$ (N/kg)	4.40	10.68	10.01	9.83	9.81

The gravitational acceleration at radius r is $g(r) = GM(r)/r^2$; its derivative with respect to r is

$$\frac{dg}{dr} = -\frac{2GM(r)}{r^3} + \frac{G}{r^2}\frac{dM(r)}{dr} = -\frac{2GM(r)}{r^3} + 4\pi G\rho(r)$$

$$= 4\pi G(\rho(r) - \tfrac{2}{3}\rho_{avg}(r)).$$

This is positive only if $\rho(r) > \tfrac{2}{3}\rho_{avg}(r)$; if $\rho(r) < \tfrac{2}{3}\rho_{avg}(r)$, g will actually decrease with increasing r. The average density of Earth is 5.51 g/cm^3. From the data in this problem we see that the density near the surface is considerably less than $\tfrac{2}{3}$ of this, which would be 3.67 g/cm^3. The density of the lower mantle, 4.9 g/cm^3, is less than $\tfrac{2}{3}$ of the density of either the inner or the outer cores, so g is still increasing with r even in the lower mantle.

15

FLUIDS

Sample Problem 15-10

An enterprising diver reasons that if a typical snorkel tube, which is 20 cm long, works, then a tube 6.0 m long should also work. If he foolishly uses such a tube (Fig. 15-45), what is the pressure difference Δp between the external pressure on him and the air pressure in his lungs? Why is he in danger?

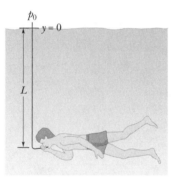

p_0

$y = 0$

L

Fig. 15-45 Sample Problem 15-10. DON'T TRY THIS with a tube that is longer than a standard snorkel tube or the attempt to breathe through the tube could kill you. The reason is that, because the external (water) pressure on your chest can be so much greater than the internal (air) pressure, you might not be able to expand your lungs to inhale.

SOLUTION: First consider the diver at depth $L = 6.0$ m without the snorkel tube. One **Key Idea** here is that the external pressure p on him is greater than normal and is given by Eq. 15-8 as

$$p = p_0 + \rho g L,$$

where p_0 is the atmospheric pressure. His body adjusts to external pressure p by contracting slightly until the internal pressures are in equilibrium with the external pressure. In particular, his average blood pressure increases, and the average air pressure in his lungs increases until it equals the external pressure p.

If he then foolishly uses the 6.0 m tube to breathe, the pressurized air in his lungs will be expelled upward through the tube to the atmosphere, and the air pressure in his lungs will rapidly drop to atmospheric pressure p_0. Assuming he is in fresh water of density 1000 kg/m³, the pressure difference Δp between the lower pressure within his lungs and the higher pressure on his chest will then be

$$\Delta p = p - p_0 = \rho g L$$
$$= (1000 \text{ kg/m}^3)(9.8 \text{ m/s}^2)(6.0 \text{ m})$$
$$= 5.9 \times 10^4 \text{ Pa.} \qquad \text{(Answer)}$$

This pressure difference, about 0.6 atm, is sufficient to collapse the lungs and force the still-pressurized blood into them, a process known to divers as lung squeeze.

Sample Problem 15-11

In one observation, the column in a mercury barometer (as is shown in Fig. 15-5a) has a measured height h of 740.35 mm. The temperature is $-5.0°C$, at which temperature the density of mercury ρ is 1.3608×10^4 kg/m³. The free-fall acceleration g at the site of the barometer is 9.7835 m/s². What is the atmospheric pressure at that site in pascals and in torr (which is the common unit for barometer readings)?

SOLUTION: The **Key Idea** here is that the atmospheric pressure p_0 bearing down on the barometer's mercury pool is equal to the pressure $\rho g h$ at the base of the column of mercury due to that column. That pressure is given by Eq. 15-9:

$$p_0 = \rho g h$$
$$= (1.3608 \times 10^4 \text{ kg/m}^3)(9.7835 \text{ m/s}^2)(0.74035 \text{ m})$$
$$= 9.8566 \times 10^4 \text{ Pa.} \qquad \text{(Answer)}$$

To five significant figures, 1 torr = 133.33 Pa. We have

$$p_0 = (9.8566 \times 10^4 \text{ Pa}) \frac{1 \text{ torr}}{133.33 \text{ Pa}}$$
$$= 739.26 \text{ torr.} \qquad \text{(Answer)}$$

In general, the atmospheric pressure in torr (here, 739.26 torr) is numerically close to the height h of a barometer's mercury column expressed in millimeters (here, 740.35 mm).

QUESTIONS

11. *Organizing question:* Figure 15-46 shows six situations in which milk (density 1030 kg/m^3) flows rightward through a circular pipe that changes radius or elevation. Changes in radius are from 2 cm to either 1 cm or 3 cm, as drawn. Changes in elevation are 0.50 m either up or down, as drawn. In all the situations, the milk's initial speed is 4 m/s and its initial pressure is 2×10^5 Pa. For each situation, set up an equation, complete with known data, to find the speed v_2 of the milk after the change in the pipe. Then, assuming we know that speed, set up an equation, complete with known data, to find the pressure p_2 of milk after the change.

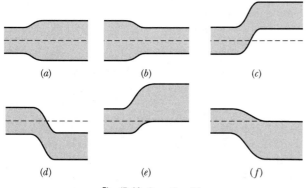

Fig. 15-46 Question 11.

12. The containers of Fig. 15-47 have the same level of water and same base area, and are made of the same material. (a) Rank the containers (and contents) according to their weights on a scale, greatest first. (b) Rank the containers according to the pressure of the water on the base of the container. (c) Does Eq. 15-4 indicate that the answers to (a) and (b) are inconsistent? This apparent inconsistency is known as the *hydrostatic paradox.*

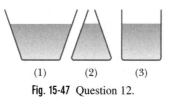

Fig. 15-47 Question 12.

13. Three gas bubbles of identical sizes are submerged in water: Bubble 1 is filled with hydrogen, bubble 2 is filled with helium, and bubble 3 is filled with carbon dioxide. Rank the bubbles according to the magnitude of the buoyant force acting on them, greatest first.

14. Figure 15-48 shows three streamlines in the flow of a fluid. What are the directions of flow of the fluid elements at (a) point B and (b) point C? Rank the fluid elements at points $A, B,$ and C according to (c) the speed of the fluid elements and (d) the pressure on the fluid elements, greatest first.

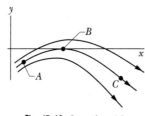

Fig. 15-48 Question 14.

15. Figure 15-49 shows four situations in which an open-tube manometer, like that in Fig. 15-6, is attached to a tank of gas. Rank the situations according to the gauge pressure of the gas within the tank, most positive first, most negative last.

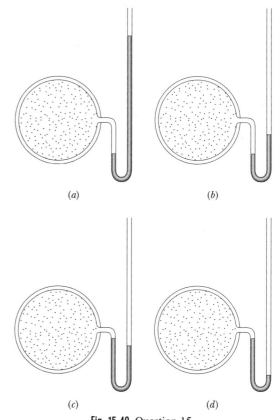

Fig. 15-49 Question 15.

16. A container of water is suspended from a spring balance. Does the weight reading increase, decrease, or remain the same when we, without spilling any water, (a) lower a heavy metal object via string into the water and (b) float a cork in the water?

DISCUSSION QUESTIONS

17. Can you assign a coefficient of static friction between two surfaces, one of which is a fluid surface?

18. Make an estimate of the average density of your body. Explain a way in which you could get an accurate value using ideas in this chapter.

19. Explain the pressure variations in your blood as it circulates through your body.

20. Does Archimedes' principle hold in a vessel in free fall or in a satellite moving in a circular orbit?

21. A spherical bob made of cork floats half submerged in a pot of tea at rest on Earth. Will the cork float or sink aboard a spaceship (a) coasting in free space and (b) on the surface of Jupiter?

22. How does a suction cup work?

23. What, if anything, happens to the buoyant force acting on a helium-filled balloon if you replace the helium with the same volume of hydrogen? (Hydrogen is less dense than helium.)

24. Two identical buckets are filled to the brim with water, but one has a block of wood floating in the water. Which bucket, if either, is heavier?

25. Can you sink an iron ship by siphoning seawater into it?

26. A beaker is exactly full of liquid water at its freezing point and has an ice cube floating in it, also at the freezing point. As the cube melts, what happens to the water level in these three cases: (a) the cube is solid ice; (b) the cube contains some grains of sand; (c) the cube contains some bubbles?

27. A ball floats on the surface of water in a container exposed to the atmosphere. Will the ball remain immersed at its former depth or will it sink or rise somewhat if (a) the container is covered and the air above the water is removed and (b) the container is covered and the air is compressed?

28. Is the buoyant force acting on a submerged submarine the same at all depths? Explain why an inflated balloon will rise only to a certain height once it starts to rise, whereas a submarine will always sink to the very bottom of the ocean once it starts to sink, if no changes are made.

29. Why does a balloon weigh the same when deflated as when inflated with air at atmospheric pressure? Would the weights be the same if measured in a vacuum?

30. During World War II, a damaged freighter that was barely able to float in the salty water of the North Sea sank because it steamed up the Thames estuary toward the London docks. Why?

31. A boat, floating in a swimming pool that is hardly wider than the boat, springs a small leak and gradually sinks until it is completely submerged. Explain what happens to the water level during the sinking.

32. A bucket of water is suspended from a spring balance. Does the balance reading change when a piece of iron suspended from string is immersed in the water? When a piece of cork is put in the water? (No water is spilled in the two situations.)

33. Why does a uniform wooden stick or log float horizontally? If enough iron is added to one end, it will float vertically. Explain this also.

34. Although there are practical difficulties, it is possible in principle to float an ocean liner in a few hundred liters of water. Explain how.

35. Explain why a thin-walled pipe will burst more easily if, when there is a pressure differential between inside and outside, the excess pressure is on the outside.

36. Explain why the height of the liquid in the vertical standpipes in Fig. 15-50 indicates that the pressure drops along the channel, even though the channel has a uniform cross section and even though the flowing liquid is incompressible.

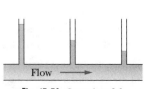

Fig. 15-50 Question 36.

37. In a lecture demonstration a Ping-Pong ball is kept in midair by a vertical jet of air. Is the equilibrium stable, unstable, or neutral? Explain.

38. Two rowboats moving parallel to one another in the same direction tend to be pulled toward one another. Two automobiles moving parallel are also pulled together. Explain such phenomena using Bernoulli's equation.

39. Explain why you cannot remove the filter paper from the funnel of Fig. 15-51 by blowing into the narrow end.

Fig. 15-51 Question 39.

40. Why can a discus be thrown farther *against* a strong wind than with it?

41. On August 2, 1985, during a routine approach to the Dallas–Fort Worth airport, a Delta Airline L-1011 jet suddenly fell and crashed when it was unknowingly flown through a *microburst* (or *wind-shear*), whose flow is shown in Fig. 15-52. The crash killed 136 of the 167 people on board. The speed of the air in the microburst was 75 km/h relative to the ground. As the jet entered the microburst, its airspeed (speed relative to the air) was 300 km/h. The L-1011 jet must have an airspeed exceeding 210 km/h, or it *stalls* (loses lift). Why did the Delta jet crash?

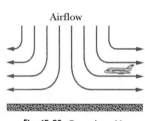

Fig. 15-52 Question 41.

42. Why does the shower curtain in a typical shower arrangement tend to flutter inward during a shower (Fig. 15-53)? (The flutter appears regardless of the water temperature, so circulation due to hot and cold air is not the primary answer.)

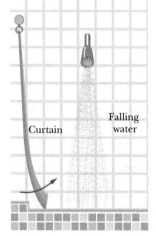

Fig. 15-53 Question 42.

EXERCISES & PROBLEMS

63. Molten chocolate is a viscous fluid consisting of sugar, milk, cocoa butter and, depending on the type of chocolate, perhaps also solid cocoa particles. To produce chocolate products, the molten chocolate (at a temperature of 27–32°C) is normally poured

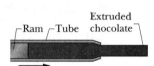

Fig. 15-54 Problem 63.

into a mold. However, long, firm cylinders of chocolate can be formed if cold chocolate (at 24°C) is forced to move through a narrow tube by a ram (Fig. 15-54). The pressure applied by the ram alters the viscosity of the chocolate such that the chocolate *yields* to the pressure and then flows through the tube like molten plastic. The chocolate is said to be *extruded* from the tube.

Assume that in a chocolate extrusion process, the ram applies uniform pressure p across the chocolate on the ram's face, and let ρ be the density of the chocolate. (a) How much work per unit mass is done on the chocolate by the applied pressure during the extrusion? (*Hint:* Bernoulli's equation as presented in the chapter is not helpful here. Consider the definition of the work done in displacing an object, as well as the definitions of pressure, volume, and density.) (b) Find the work done per unit mass for $p = 5.5$ MPa and $\rho = 1200$ kg/m^3.

64. When researchers find a reasonably complete fossil of a dinosaur, they can determine the mass and weight of the living dinosaur with a scale model sculpted from plastic and based on the dimensions of the fossil bones. The scale of the model is 1/20; that is, lengths are 1/20 actual length, areas are $(1/20)^2$ ac-

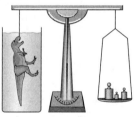

Fig. 15-55 Problem 64.

tual areas, and volumes are $(1/20)^3$ actual volumes. First, the model is suspended from one arm of a balance and weights are added to the other arm until equilibrium is reached. Then the model is fully submerged in water and enough weights are removed from the second arm to reestablish equilibrium (Fig. 15-55). For a model of a particular *T. rex* fossil, 637.76 g had to be removed to reestablish equilibrium. (a) What is the volume of the model? (b) What was the volume of the actual *T. rex?* (c) Assuming that the density of *T. rex* was approximately the density of water, what was its mass? (d) What, approximately, is the ratio of the dinosaur's weight to your weight?

65. During a parachute jump, a folded parachute is released from its package with its mouth downward so that it will become inflated. Air flows into the mouth, through the folded layers of fabric, and then out through an opening at the top. However, this flow does not simply produce an upward force on the fabric layers. More important, it increases the air pressure inside the parachute,

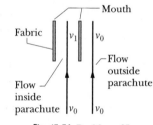

Fig. 15-56 Problem 65.

and it is the pressure difference between the inside and outside of the parachute that inflates the parachute, peeling apart the fabric layers.

Figure 15-56 gives a very simplified idea of the air flow. Two streamlines are shown, one inside the mouth of the parachute and the other outside. Assume that a falling parachutist has a typical speed of 65 m/s relative to the ground when the parachute is released. Then, relative to the parachute, the speed of the air along the two streamlines *below* the mouth is $v_0 = 65$ m/s. At the level of the parachute the air along the outside streamline is still v_0, but the speed v_1 of the air inside the mouth is only a few meters per second. What is the pressure difference between the inside and outside of the parachute mouth?

66. According to one government specification, the "black box" recorders on airplanes must be able to withstand water pressures down to depths of 6.0 km so that they can be recovered in case of a crash over water. Assuming that seawater has a uniform density of 1024 kg/m^3, find the pressure corresponding to that government specification.

67. At a depth of 10.9 km, the Challenger Deep in the Marianas Trench of the Pacific Ocean is the deepest site in any ocean. Yet, in 1960, Donald Walsh and Jacques Piccard reached the Challenger Deep in the bathyscaph *Trieste*. Assuming that seawater has a uniform density of 1024 kg/m^3, approximate the hydrostatic pressure (in atmospheres) that the *Trieste* had to withstand.

68. Figure 15-57 shows a *siphon*, which is a device for removing liquid from a container. Tube *ABC* must initially be filled, but once this has been done, liquid will flow through the tube until the liquid surface in the container is level with the tube opening at *A*. The liquid has density ρ and negligible viscosity. (a) With what speed does the liquid emerge from the tube at *C*? (b) What is the pressure in the liquid at the topmost point *B*? (c) Theoretically, what is the greatest possible height h_1 that a siphon can lift water?

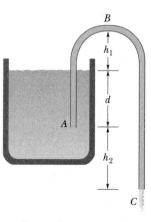

Fig. 15-57 Problem 68.

69. In an experiment, a rectangular block with height h is allowed to float in four separate liquids. In the first liquid, which is water, it floats fully submerged. In liquids *A, B,* and *C,* it floats with heights $h/2$, $2h/3$, and $h/4$ above the liquid surface, respectively. What are the *relative densities* (the densities relative to that of water) of (a) liquid *A,* (b) liquid *B,* and (c) liquid *C?*

70. What gauge pressure must be produced by a machine for it to suck mud of density 1800 kg/m^3 up a tube by a height of 1.5 m?

71. What is the acceleration of a rising hot-air balloon if the ratio of the air density outside the balloon to that inside is 1.39? Neglect the mass of the balloon fabric and the basket.

72. How much work is done by pressure in forcing 1.4 m³ of water through a pipe having an internal diameter of 13 mm if the difference in pressure at the two ends of the pipe is 1.0 atm?

73. A 7.00 kg sphere of radius 5.00 cm is at a depth of 1.20 km in seawater that has an average density of 1025 kg/m³. What are (a) the gauge pressure, (b) the total pressure, and (c) the corresponding total force compressing the sphere's surface? What are (d) the magnitude of the buoyant force on the sphere and (e) the magnitude and direction of the sphere's acceleration if it is free to move?

74. An object hangs from a spring balance. The balance registers 30 N in air, 20 N when this object is immersed in water, and 24 N when the object is immersed in another liquid of unknown density. What is the density of that other liquid?

75. You place a glass beaker, partially filled with water, in a sink (Fig. 15-58). The beaker has a mass of 390 g and an interior volume of 500 cm³. You now start to fill the sink with water and you find, by experiment, that if the beaker is less than half full, it will float; but if it is more than half

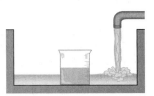

Fig. 15-58 Problem 75.

full, it remains on the bottom of the sink as the water rises to its rim. What is the density of the material of which the beaker is made?

76. A simple open U-tube contains mercury. When 11.2 cm of water is poured into the right arm of the tube, how high above its initial level does the mercury rise in the left arm?

77. Figure 15-59 shows a dam and part of the freshwater reservoir backed up behind it. The dam is made of concrete of density 3.2 g/cm³ and has the dimensions shown on the figure. The water pushes horizontally on the dam face and is resisted by the force of static friction between the dam and the bedrock foundation on which it rests. The coefficient of static friction is 0.47. (a) Calculate the factor of safety against sliding, that is, the ratio

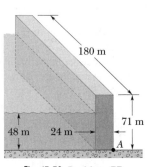

Fig. 15-59 Problem 77.

of the magnitudes of the maximum possible friction force to the force of the water. The water also tries to rotate the dam about a line running along the base of the dam through point A. The torque due to the gravitational force on the dam acts in the opposite sense about that line. (b) Calculate the factor of safety against rotation—that is, the ratio of the magnitudes of the torque due to the gravitational force on the dam to the torque due to the net force from the water and the air.

78. Calculate the height of a column of water that gives a pressure of 1 atm at the bottom.

79. A tin can has a total volume of 1200 cm³ and a mass of 130 g. How many grams of lead shot of density 11.4 g/cm³ could it carry without sinking in water?

80. It has been proposed to move natural gas from the North Sea gas fields in huge dirigibles, using the gas itself to provide lift. Calculate the magnitude of the vertical force required to tether (hold) such an airship to the ground for off-loading when it is fully loaded with 1.0×10^6 m³ of gas at a density of 80 kg/m³. (The gravitational force on the airship is negligible by comparison.)

81. A helium balloon is used to lift a 40 kg payload to an altitude of 27 km, where the air density is 0.035 kg/m³. The balloon has a mass of 15 kg, and the density of the gas in the balloon is 0.0051 kg/m³. What is the volume of the balloon? Neglect the volume of the payload.

82. (a) Consider a container of fluid subject to a *vertical upward* acceleration of magnitude a. Show that the pressure variation with depth in the fluid is given by

$$p = \rho h(g + a),$$

where h is the depth and ρ is the density of the fluid. (b) Show also that if the fluid as a whole undergoes a *vertical downward* acceleration of magnitude a, the pressure at a depth h is given by

$$p = \rho h(g - a).$$

(c) What is the pressure if the fluid is in free fall?

83. Find the absolute pressure, in pascals, at a depth of 150 m in the ocean. The density of seawater is 1.03 g/cm³, and the atmospheric pressure at sea level is 1.01×10^5 Pa.

84. A uniform block of length 5.0 cm, width 4.0 cm, and height 2.0 cm floats in seawater of density 1025 kg/m³. The block has a broad side face down, 1.5 cm below the water surface. What is the mass of the block?

85. The human lungs can operate against a pressure difference (inside and outside the lungs) of up to about 0.050 atm. If a diver uses a snorkel for breathing, about how far below water level can the diver swim?

86. A cube with a surface area of 24 m² floats upright in water. If the density of water is 4.00 times the density of the cube, how far does the cube sink into the water?

87. A lead sinker of volume 0.40 cm³ and density 11.4 g/cm³ is used in fishing. The sinker is suspended from a vertical string whose other end is attached to the bottom of a spherical cork (of density 0.20 g/cm³) that is floating on the surface of a lake. Neglecting the effects of the line, hook, and bait, determine what the radius of the cork must be if it is to float with half its volume submerged.

88. Water flows through a horizontal pipe that widens at two points; initial radius = 0.200 m, intermediate radius = 0.400 m, and final radius = 0.600 m. If the initial speed of the water's flow is 9.00×10^{-2} m/s, what is the final speed of the flow?

89. A liquid of density 900 kg/m³ flows through a horizontal pipe that changes in cross-sectional area from 1.90×10^{-2} m² to 9.50×10^{-2} m². The pressures of the liquids on the two sides of the area change have a difference of 7.20×10^3 Pa. What are (a) the volume flow rate and (b) the mass flow rate through the pipe?

90. A laminar stream of water necks down as it falls, as in Sample Problem 15-7 and Fig. 15-18. At one level in a certain stream, the radius is r_1 and the speed is v_1. At a lower level, the radius is $r_1/2$. (a) What is the speed at that lower level? The *kinetic energy density* of a fluid is measured with the SI unit of joule per cubic meter and is defined as $\frac{1}{2}\rho v^2$, where ρ is the density of the fluid and v is the speed of the fluid. (b) What is the increase in the kinetic energy density of the water as it falls from the higher level to the lower level?

Tutorial Problem

91. A rectangular metal block of height 10.0 cm is suspended by a thin wire of negligible mass. The wire is attached to a spring balance so that the block's weight can be determined; its weight is 353 N in air, and 294 N when it is fully submerged in water. (a) Determine the density, volume, and cross-sectional area of the block.

(b) Suppose the block is suspended in a large container of water, with the top of the block at depth $d = 100$ cm. Draw a free-body force diagram for the block and determine the magnitude and direction of each of the forces acting on it. (c) Now suppose that the block is raised slowly until its top is level with the water's surface. Determine how much work is done on the block by each of the forces you identified in part (b). Check that the sum of the works is correct. (d) Determine what changes, if any, occur in the gravitational potential energy of the block and the water while the block is raised. How are these changes reflected in the works calculated in part (c)?

Answers

(a) Let ρ be the average density of the block and $\rho' = 1000$ kg/m³ be the density of water. Let the block have height $h = 10.0$ cm, width w, and length ℓ, so its volume is $V = hw\ell$. In air, neglecting the density of air (which is only 1.3 kg/m³), the weight of the block is ρgV. In water its weight, as measured by the spring balance, is $(\rho - \rho')gV$.

The ratio of the weights is

$$\frac{\rho}{\rho - \rho'} = \frac{353}{294} \approx 1.20 = \frac{6}{5},$$

which leads to $\rho/\rho' = 6$. Thus,

$$\rho = 6\rho' = 6(1000 \text{ kg/m}^3) = 6000 \text{ kg/m}^3.$$

The volume of the block must then be

$$V = \frac{\rho gV}{\rho g} = \frac{353 \text{ N}}{(6000 \text{ kg/m}^3)(9.8 \text{ N/kg})} = 6.00 \times 10^{-3} \text{ m}^3.$$

Since $h = 0.100$ m, the cross-sectional area is

$$w\ell = \frac{V}{h} = \frac{6.00 \times 10^{-3} \text{ m}^3}{0.10 \text{ m}} = 6.00 \times 10^{-2} \text{ m}^2.$$

The exact values of w and ℓ cannot be determined, but they might be, for example, 20 cm and 30 cm.

(b) There are three forces acting on the block: (1) the downward gravitational force of magnitude $\rho gV = 353$ N; (2) the upward buoyant force $\vec{F}_b$ of magnitude $\rho' gV = 59$ N; and (3) the upward tension force $\vec{T}$ provided by the wire. Since the block is in static equilibrium, the three forces must sum vectorially to zero, so the tension force must have magnitude 353 N − 59 N = 294 N.

(c) The block is raised vertically a distance of 100 cm, either parallel or antiparallel to the directions of the three forces. The work can be computed by multiplying the magnitude of the force by the distance the block is raised, and then including a + or a − sign depending on whether the force and displacement are parallel or antiparallel. The work done by the gravitational force is $(353 \text{ N})(-1.00 \text{ m}) = -353$ J, that by the buoyancy force is $(59 \text{ N})(1.00 \text{ m}) = 59$ J, and that by the tension force of the wire is $(294 \text{ N})(1.00 \text{ m}) = 294$ J. The sum of the works is -353 J $+$ 59 J + 294 J = 0, which is correct, since there is no change in the block's kinetic energy (it is zero initially and finally).

(d) The block is raised 100 cm = 1.00 m, so its gravitational potential energy increases by

$$mg \, \Delta y = \rho gV \, \Delta y = (353 \text{ N})(1.00 \text{ m}) = 353 \text{ J}.$$

This is the negative of the work done by the gravitational force (see part (c)).

The change in the gravitational potential energy of the water can be determined by recognizing that as the block rises 1.00 m, an equal volume of water descends 1.00 m, so its gravitational potential energy decreases. The change in the gravitational potential energy of the water is

$$\rho' gV \, \Delta y = (59 \text{ N})(-1.00 \text{ m}) = -59 \text{ J}.$$

This is the negative of the work done by the buoyancy force.

16
Oscillations

ADDITIONAL SAMPLE PROBLEMS

Sample Problem 16-8

In Fig. 16-40a, a uniform bar with mass m lies symmetrically across two rapidly rotating, fixed rollers, A and B, with distance $L = 2.0$ cm between the bar's center of mass and each roller. The rollers, whose directions of rotation are shown in the figure, slip against the bar with coefficient of kinetic friction $\mu_k = 0.40$. Suppose the bar is displaced horizontally by a distance x, as in Fig. 16-40b, and then released. What is the angular frequency ω of the resulting horizontal simple harmonic (back and forth) motion of the bar?

SOLUTION: This situation seems very different from other SHM situations we have analyzed. However, the **Key Idea** here is that we can find ω if we find an expression for the horizontal acceleration a of the bar as a function of x, and then compare it with Eq. 16-8 ($a = -\omega^2 x$). Since vertical forces (as shown in Fig. 16-40b), horizontal frictional forces, and torques all act on the bar, we shall apply Newton's second law both vertically and horizontally, and then in angular form about one of the contact points between bar and roller.

The vertical forces acting on the bar are the gravitational force $\vec{F}_g$ (with magnitude mg) and supporting forces $\vec{F}_A$ due to roller A and $\vec{F}_B$ due to roller B. Since there is no vertical acceleration of the bar, Newton's second law for components along a vertical y axis ($F_{\text{net},y} = ma_y$) gives us

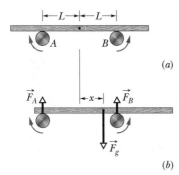

Fig. 16-40 Sample Problem 16-8. (a) A bar is in equilibrium on two rotating rollers, A and B, that slip beneath it. (b) The bar is displaced from equilibrium by a distance x and then released.

$$F_A + F_B - mg = 0. \qquad (16\text{-}45)$$

The horizontal forces acting on the bar are the kinetic frictional forces $f_{kA} = \mu_k F_A$ (toward the right) due to roller A and $f_{kB} = -\mu_k F_B$ (toward the left) due to roller B. Thus, Newton's second law for components along a horizontal x axis ($F_{\text{net},x} = ma_x$) gives us

$$\mu_k F_A - \mu_k F_B = ma$$

or

$$a = \frac{\mu_k F_A - \mu_k F_B}{m}. \qquad (16\text{-}46)$$

We now consider torques about an axis perpendicular to the plane of Fig. 16-40 and through the contact point between the bar and roller A. The bar experiences no angular acceleration about that axis, so Newton's second law for torques about such an axis ($\tau_{\text{net}} = I\alpha$) gives us

$$F_A(0) + F_B 2L - mg(L + x) + f_{kA}(0) + f_{kB}(0) = 0, \qquad (16\text{-}47)$$

where the forces $\vec{F}_A, \vec{F}_B, m\vec{g}, \vec{f}_{kA}$, and $\vec{f}_{kB}$ have moment arms about that axis of 0, $2L$, $L + x$, 0, and 0, respectively.

Solving Eq. 16-47 for F_B, and then solving Eq. 16-45 for F_A, we find

$$F_B = \frac{mg(L + x)}{2L} \quad \text{and} \quad F_A = \frac{mg(L - x)}{2L}.$$

Substituting these results into Eq. 16-46 yields

$$a = -\frac{\mu_k g}{L} x. \qquad (16\text{-}48)$$

Comparison of Eq. 16-48 with Eq. 16-8 reveals that the bar must be undergoing simple harmonic motion with angular frequency ω given by

$$\omega^2 = \frac{\mu_k g}{L}.$$

Thus,

$$\omega = \sqrt{\frac{\mu_k g}{L}} = \sqrt{\frac{(0.40)(9.8 \text{ m/s}^2)}{0.020 \text{ m}}}$$

$$= 14 \text{ rad/s}. \qquad \text{(Answer)}$$

Sample Problem 16-9

A disk whose radius R is 12.5 cm is suspended, as a physical pendulum, from a point at distance h from its center C (Fig. 16-41). Its period T is 0.871 s when $h = R/2$. What is the free-fall acceleration g at the location of the pendulum?

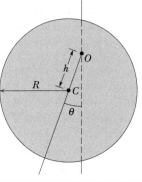

Fig. 16-41 Sample Problem 16-9. A physical pendulum consisting of a uniform disk suspended from point O a distance h from the center C of the disk.

SOLUTION: The **Key Idea** here is that the disk is not a simple pendulum, because its mass is not all concentrated at some distance h from a pivot point. Rather, the disk is a physical pendulum, with a period given by Eq. 16-29 ($T = 2\pi(I/mgh)^{1/2}$). To use that equation, we need the rotational inertia I of the disk about pivot point O. From Table 11-2c, we know that the rotational inertia I_{com} of the disk about an axis through its center of mass and perpendicular to its plane is $\frac{1}{2}mR^2$. From the parallel-axis theorem of Eq. 11-29 ($I = I_{\text{com}} + Mh^2$), we then find that the rotational inertia I about a parallel axis through pivot point O is

$$I = I_{\text{com}} + mh^2 = \tfrac{1}{2}mR^2 + m(\tfrac{1}{2}R)^2 = \tfrac{3}{4}mR^2.$$

If we now put $I = \tfrac{3}{4}mR^2$ and $h = \tfrac{1}{2}R$ in Eq. 16-29, we find

$$T = 2\pi\sqrt{\frac{\frac{3}{4}mR^2}{mg(\frac{1}{2}R)}} = 2\pi\sqrt{\frac{3R}{2g}}.$$

Solving for g then gives us

$$g = \frac{6\pi^2 R}{T^2} = \frac{(6\pi^2)(0.125 \text{ m})}{(0.871 \text{ s})^2} = 9.76 \text{ m/s}^2. \quad \text{(Answer)}$$

QUESTIONS

14. Do the following increase, decrease, or remain the same if the amplitude of a simple harmonic oscillator is doubled: (a) the period, (b) the spring constant, (c) the total energy, (d) the maximum speed, (e) the maximum acceleration?

15. *Organizing question:* Figure 16-42 shows the SHM of a block–spring system in four situations. For each, set up an equation, complete with known data, that gives the position $x(t)$ of the block.

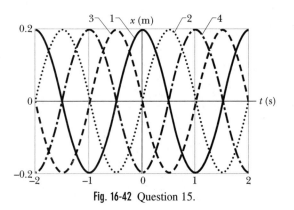

Fig. 16-42 Question 15.

16. Figure 16-43 shows the $x(t)$ curves for three experiments involving a particular spring–box system oscillating in SHM. Rank the curves according to (a) the system's angular frequency, (b) the spring's potential energy at time $t = 0$, (c) the box's kinetic energy at $t = 0$, (d) the box's speed at $t = 0$, and (e) the box's maximum kinetic energy, greatest first.

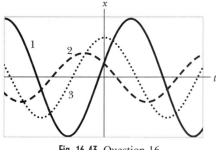

Fig. 16-43 Question 16.

17. When a block with mass m_1 is hung from A and a smaller block with mass m_2 is hung from spring B, the springs are stretched by the same distance. If the two spring–block systems are then put into vertical SHM with the same amplitude, which system has more energy?

18. Figure 16-44 shows three arrangements of a block attached to identical springs that are in the relaxed state when the block is centered. Rank the arrangements according to the frequency of oscillation of the block, greatest first.

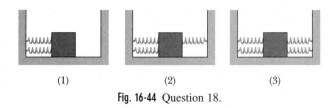

Fig. 16-44 Question 18.

19. Three springs each hang from a ceiling with a stationary block attached to the lower end. The blocks (with $m_1 > m_2 > m_3$) stretch the springs by equal distances. Each spring–block system is then set into vertical SHM. Rank the masses according to the period of oscillation, greatest first.

20. A particle oscillates according to

$$x = x_m \cos(\omega t + \phi).$$

At $t = 0$, is the particle at $-x_m$, at $+x_m$, at 0, between $-x_m$ and 0, or between 0 and $+x_m$ if ϕ is (a) $\pi/2$, (b) $-\pi/3$, (c) $-3\pi/4$, and (d) $3\pi/4$?

21. A new type of ride, the SHM Monster, opens up at your local amusement park with three choices of cars (Fig. 16-45). Each car is attached to a large spring that is pulled from the equilibrium point ($x = 0$) and held fixed at a loading point. Those initial displacements of the identical cars and the spring constants of the springs are given in the figure in terms of basic units d and k. After a passenger is secured in a car at the loading point, the car is released so that its spring can put it and the passenger in SHM along approximately frictionless rails. Rank the three cars according to (a) the time it takes to first reach the equilibrium point, (b) the time it takes to complete 10 cycles of the ride, and (c) the magnitude of the maximum acceleration (and thus fear) that the passenger experiences, greatest first.

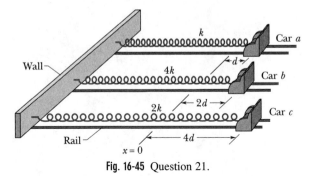

Fig. 16-45 Question 21.

22. Figure 16-46 shows a new variation of a common amusement-park game. A puck is forced to move around a circle of diameter 4 m at a constant angular speed of 2 rad/s. At the right side of the circle (at time $t = 0$), it bumps a pendulum bob from its perch, thus allowing the pendulum to swing to the left. (The pendulum, which is suspended from a point high above the center of the circle, swings through only a small angle.)

You need to choose a pendulum such that the puck next bumps the bob at the left side of the pendulum's swing. Here, for six pendulums, is the coordinate $x(t)$ of the pendulum's bob during a swing (with x in meters and t in seconds). (a) Which pendulum should you choose? (b) Which should you choose if the next bump is to be at the right side?

(1) $x = 4 \cos 2t$ (4) $x = 2 \cos 2t$

(2) $x = 4 \cos 4t$ (5) $x = 2 \cos 4t$

(3) $x = 4 \cos t$ (6) $x = 2 \cos t$

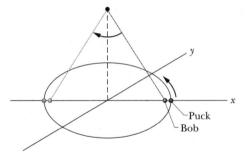

Fig. 16-46 Question 22.

23. A spring and a block are arranged to oscillate (1) on a frictionless horizontal surface as in Fig. 16-5, (2) on a frictionless slope of 45° (with the block at the lower end of the spring), and (3) vertically while hanging from a ceiling. Rank the arrangements according to (a) the rest length of the spring and (b) the frequency of oscillation, greatest first.

24. An object of mass m is suspended from a spring with spring constant k. The spring is then cut in half and the same object is suspended from one of the halves. Which, the original spring or the half-spring, gives a greater frequency of oscillation when the object is put into vertical SHM?

25. A pendulum suspended from the ceiling of an elevator cab has period T when the cab is stationary. Is the period larger, smaller, or the same when the cab moves (a) upward with constant speed, (b) downward with constant speed, (c) downward with constant upward acceleration, (d) upward with constant upward acceleration, (e) upward with constant downward acceleration $a = g$, and (f) downward with constant downward acceleration $a = g$?

26. *Math Tool Time.* (a) Which of the following is the derivative of $x = 4.0 \cos(5t)$ with respect to time t: $-4.0 \sin(5t)$, $4.0 \sin t$, $-20 \sin t$, $20 \sin t$, $-20 \sin(5t)$, or $20 \sin(5t)$? (b) Which of the following is the result of $\ln(e^{-at})$: $\ln(-at)$, $\ln(at)$, at, $-at$, $1/at$, or $-1/at$?

DISCUSSION QUESTIONS

27. Suppose that a system consists of a block of unknown mass and a spring of unknown force constant. Show how we can predict the period of oscillation of this block–spring system simply by measuring the extension of the spring produced by attaching the block to it.

28. Any real spring has mass. If this mass is taken into account, explain qualitatively how this will affect the period of oscillation of a block–spring system.

29. What changes could you make in a block–spring harmonic oscillator that would double the maximum speed of the oscillating block?

30. What would happen to the motion of an oscillating system if the sign of the force term, $-kx$ in Eq. 16-10, were changed?

31. Will the frequency of oscillation of a torsion pendulum change if you take it to the Moon? What about the frequencies of a simple pendulum, a block–spring oscillator, and a physical pendulum, such as a wooden plank swinging from one end?

32. Predict by qualitative arguments whether the period of a pendulum increases or decreases when its amplitude is increased.

33. How can a pendulum be used so that it traces out a sinusoidal curve?

34. Why are damping devices often used on machinery? Give an example.

35. A singer, holding a note of the right frequency, can shatter a glass if the glassware is of high quality. This cannot be done if the glassware quality is low. Explain why in terms of the damping constant of the glass.

EXERCISES & PROBLEMS

65. The physical pendulum in Fig. 16-47 has two possible pivot points A and B; point A has a fixed position, and B is adjustable along the length of the pendulum. The period of the pendulum when suspended from A is found to be T. The pendulum is then reversed and suspended from B, which is moved until the pendulum again has period T. Show that the free-fall acceleration g at the pendulum's location is given by

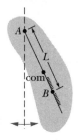

Fig. 16-47 Problem 65.

$$g = \frac{4\pi^2 L}{T^2},$$

in which L is the distance between A and B for equal periods T. (This method can be used to measure g without knowledge of the rotational inertia of the pendulum or any of its dimensions except L.)

66. The center of oscillation of a physical pendulum has this interesting property: If an impulse (assumed horizontal and in the plane of oscillation) acts at the center of oscillation, no oscillations are felt at the point of support. Baseball players (and players of many other sports) know that unless the ball hits the bat at this point (called the "sweet spot" by athletes), the oscillations due to the impact will sting their hands. To prove this property, let the stick in Fig. 16-11a simulate a baseball bat. Suppose that a horizontal force $\vec{F}$ (due to impact with the ball) acts toward the right at P, the center of oscillation. The batter is assumed to hold the bat at O, the pivot point of the stick. (a) What acceleration does the point O undergo as a result of $\vec{F}$? (b) What angular acceleration is produced by $\vec{F}$ about the center of mass of the stick? (c) As a result of the angular acceleration in (b), what linear acceleration does point O undergo? (d) Considering the magnitudes and directions of the accelerations in (a) and (c), convince yourself that P is indeed the "sweet spot."

67. Although California is known for earthquakes, it also includes large regions that are dotted with *precariously balanced rocks* (Fig. 16-48) that would be easily toppled by even a mild earthquake. Evidence shows that the rocks have stood this way for thousands of years, suggesting that major earthquakes have not occurred in those regions for that long. If an earthquake were to put such a rock into sinusoidal oscillations (parallel to the ground) with a frequency of, say, 2.2 Hz, an oscillation amplitude of 1.0 cm would cause the rock to topple. What would be the magnitude of the corresponding maximum acceleration of the oscillation, in terms of g?

Fig. 16-48 Problem 67.

68. A particle undergoes simple harmonic motion along an x axis about $x = 0$ with a period of 0.40 s and an amplitude of 0.10 m. At $t = 0$, the particle is at its position of maximum negative displacement—that is, at $x = -0.10$ m. Write expressions, as functions of time, for (a) the particle's position and (b) its velocity.

69. A damped harmonic oscillator consists of a block ($m = 2.00$ kg), a spring ($k = 10.0$ N/m), and a damping force $F = -bv$. Initially, it oscillates with an amplitude of 25.0 cm; because of the damping, the amplitude falls to three-fourths of this initial value at the completion of four oscillations. (a) What is the value of b? (b) How much energy has been "lost" during these four oscillations?

70. (a) Find the ratio of the maximum damping force ($b \, dx/dt$) to the maximum spring force (kx) during the first oscillation for the data of Sample Problem 16-7. (b) Does this ratio change appreciably during later oscillations?

71. The fact that g varies from place to place over Earth's surface drew attention when Jean Richer in 1672 took a pendulum clock from Paris to Cayenne, French Guiana, and found that it lost 2.5 min/day. If $g = 9.81 \text{ m/s}^2$ in Paris, what is g in Cayenne?

72. An amateur scientist is making a precise measurement of g at a certain point in the Indian Ocean (on the equator) by timing the swings of a pendulum of accurately known construction. To provide a stable base, the measurements are conducted in a submerged submarine. It is observed that a slightly different result for g is obtained when the submarine is moving eastward than when it is moving westward, the speed in each case being 16 km/h. Account for this difference, and calculate the fractional error $\Delta g/g$ in g for either travel direction.

73. A simple harmonic oscillator consists of a block attached to a spring of spring constant $k = 200$ N/m. The block slides back and forth along a straight line on a frictionless surface, with equilibrium point $x = 0$ and amplitude 0.20 m. A graph of the velocity v of the block as a function of time t is shown in Fig. 16-49. What are (a) the period of the simple harmonic motion, (b) the mass of the block, (c) the displacement of the block at $t = 0$, (d) the acceleration of the block at $t = 0.10$ s, and (e) the maximum kinetic energy attained by the block?

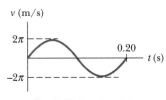

Fig. 16-49 Problem 73.

74. A block weighing 20 N oscillates at one end of a vertical spring for which $k = 100$ N/m; the other end of the spring is attached to a ceiling. At a certain instant the spring is stretched 0.30 m beyond its relaxed length (the length when no object is attached) and the block has zero velocity. (a) What is the net force on the block at this instant? What are (b) the amplitude and (c) the period of the resulting simple harmonic motion? (d) What is the maximum kinetic energy of the block as it oscillates?

75. A 0.10 kg block oscillates back and forth along a straight line on a frictionless horizontal surface. Its displacement from the origin is given by

$$x = (10 \text{ cm}) \cos[(10 \text{ rad/s})t + \pi/2 \text{ rad}].$$

(a) What is the oscillation frequency? (b) What is the maximum speed acquired by the block? At what value of x does this occur? (c) What is the magnitude of the maximum acceleration of the block? At what value of x does this occur? (d) What force, applied to the block by the spring, results in the given oscillation?

76. To alleviate the traffic congestion between two cities such as Boston and Washington, DC, engineers have proposed building a rail tunnel along a chord line connecting the cities (Fig. 16-50). A train, unpropelled by any engine and starting from rest, would fall through the first half of the tunnel and then move up the second half. Assuming Earth to be a uniform sphere and ignoring air drag and friction, (a) show that the travel between cities is half a cycle of simple harmonic motion and (b) find the travel time between cities.

Fig. 16-50 Problem 76.

77. The end point of a spring oscillates with a period of 2.0 s when a block with mass m is attached to it. When this mass is increased by 2.0 kg, the period is found to be 3.0 s. Find m.

78. A 2.00 kg block hangs from a spring. A 300 g body hung below the block stretches the spring 2.00 cm farther. (a) What is the spring constant? (b) If the 300 g body is removed and the block is set into oscillation, find the period of the motion.

79. A 50.0 g stone is attached to the bottom of a vertical spring and set vibrating. If the maximum speed of the stone is 15.0 cm/s and the period is 0.500 s, find (a) the spring constant of the spring, (b) the amplitude of the motion, and (c) the frequency of oscillation.

80. When a 20 N can is hung from the bottom of a vertical spring, it causes the spring to stretch 20 cm. (a) What is the spring constant? (b) This spring is now placed horizontally on a frictionless table. One end of it is held fixed, and the other end is attached to a 5.0 N can. The can is then moved (stretching the spring) and released from rest. What is the period of the resulting oscillation?

81. The vibration frequencies of atoms in solids at normal temperatures are of the order of 10^{13} Hz. Imagine the atoms to be connected to one another by springs. Suppose that a single silver atom in a solid vibrates with this frequency and that all the other atoms are at rest. Compute the effective spring constant. One mole of silver (6.02×10^{23} atoms) has a mass of 108 g.

82. A 4.00 kg block hangs from a spring, extending it 16.0 cm from its unstretched position. (a) What is the spring constant? (b) The block is removed and a 0.500 kg body is hung from the same spring. If the spring is then stretched and released, what is its period of oscillation?

83. A block sliding on a horizontal frictionless surface is attached to a horizontal spring with a spring constant of 600 N/m. The block executes SHM about its equilibrium position with a period of 0.40 s and an amplitude of 0.20 m. As the block slides through its equilibrium position, a 0.50 kg putty wad is dropped vertically onto the block. If the putty wad sticks to the block, determine (a) the new period of the motion and (b) the new amplitude of the motion.

84. If a simple pendulum with length 1.50 m makes 72.0 oscillations in 180 s, what is the free-fall acceleration g at its location?

85. A 2.0 kg tuna can executes SHM while attached to a spring of spring constant 200 N/m. The maximum speed of the can as it slides on a horizontal frictionless surface is 3.0 m/s. What are (a) the amplitude of the block's motion, (b) the magnitude of the block's maximum acceleration, and (c) the magnitude of its minimum acceleration? (d) How long does the can take to complete 7.0 cycles of its motion?

86. In Fig. 16-51, a block hangs in equilibrium from a spring of spring constant k and stretches the spring by a distance h from its relaxed length. Show that if the block is pulled down and released, its resulting simple harmonic motion has the same frequency as a simple pendulum of length h.

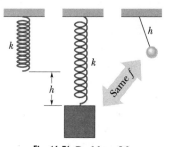

Fig. 16-51 Problem 86.

87. What is the length of a simple pendulum whose period is 1.50 s at a point at which $g = 9.75$ m/s^2?

88. A uniform rod with length L swings from a pivot as a physical pendulum. How far, in terms of L, should the pivot be from the center of mass to minimize the period?

89. A physical pendulum consists of two meter-long sticks joined together as shown in Fig. 16-52. What is the pendulum's period of oscillation about a pin inserted through point A?

Fig. 16-52 Problem 89.

90. A grandfather clock has a pendulum that consists of a thin brass disk of radius 15 cm and mass 1.0 kg that is attached to a long thin rod that may be considered massless. The pendulum swings freely about an axis perpendicular to the rod and through the end of the rod opposite the disk, as shown in Fig. 16-53. If the pendulum is to have a period of 2.0 s for small oscillations at a place where $g = 9.80$ m/s^2, what must be the length L of the rod? Express your answer to the nearest tenth of a millimeter.

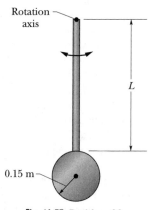

Fig. 16-53 Problem 90.

91. A block weighing 10 N is attached to the lower end of a vertical spring ($k = 200$ N/m), the other end of which is attached to a ceiling. The block oscillates vertically and has a kinetic energy of 2.0 J as it passes through the point at which the spring is unstretched. (a) What is the period of the oscillation? (b) Use the law

of conservation of energy to determine the maximum distance the block moves both above and below the point at which the spring is unstretched. (These are not necessarily the same.) (c) What is the amplitude of the oscillation? (d) What is the maximum kinetic energy of the block as it oscillates?

92. A simple harmonic oscillator consists of a block (mass = 0.80 kg) attached to a spring ($k = 200$ N/m). The block slides on a horizontal frictionless surface about the equilibrium point $x = 0$ with a total mechanical energy of 4.0 J. (a) What is the amplitude of the oscillation? (b) How many oscillations does the block complete during a 10 s interval? (c) What is the maximum kinetic energy attained by the block as it oscillates? (d) What is the speed of the block at the instant it is displaced 0.15 m from its equilibrium position?

93. A thin uniform rod (mass = 0.50 kg) swings about an axis that passes through one end of the rod and is perpendicular to the plane of the swing. The rod swings with a period of 1.5 s and an angular amplitude of 10°. (a) What is the length of the rod? (b) Use the law of conservation of energy to determine the maximum kinetic energy of the rod as it swings.

94. A 95 kg solid sphere with a 15 cm radius is suspended by a vertical wire. A torque of 0.20 N · m is required to rotate the sphere through an angle of 0.85 rad and then maintain that orientation. What is the period of the oscillations that result when the sphere is then released?

95. An engineer has an odd-shaped object of mass 10 kg and needs to find its rotational inertia about an axis through its center of mass. The object is supported on a wire that is stretched along the desired axis. The wire has a torsion constant $\kappa = 0.50$ N · m. If this torsion pendulum oscillates through 20 complete cycles in 50 s, what is the rotational inertia of the odd-shaped object?

96. A massless spring with spring constant 19 N/m hangs vertically. A body of mass 0.20 kg is attached to its free end and then released. Assume that the spring was unstretched before the body was released. Find (a) how far below the initial position the body descends, and (b) the frequency and (c) the amplitude of the resulting motion, assumed to be simple harmonic.

97. A meter stick swinging from one end oscillates with frequency f_0. What would be its oscillation frequency, in terms of f_0, if the bottom half of the stick were cut off?

98. A 1.2 kg block sliding on a horizontal frictionless surface is attached to a horizontal spring with a spring constant of 480 N/m. Let x measure the displacement of the block from the position at which the spring is unstretched. At $t = 0$ the block passes through $x = 0$ with a speed of 5.2 m/s in the positive x direction. What are (a) the frequency and (b) the amplitude of the block's motion? (c) Write an expression for the block's displacement x as a function of time.

99. In Fig. 16-54, a solid cylinder is attached to a horizontal spring so that it can roll without slipping along a horizontal surface. The spring constant k is 3.0 N/m. If the system is released from rest at a position in which the spring is stretched by 0.25 m, find (a) the translational kinetic energy and (b) the rotational kinetic energy of the cylinder as it passes through the equilibrium position. (c) Show that under these conditions the center of mass of the cylinder executes simple harmonic motion with period

$$T = 2\pi \sqrt{\frac{3M}{2k}},$$

where M is the mass of the cylinder. (*Hint:* Find the time derivative of the total mechanical energy.)

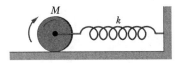

Fig. 16-54 Problem 99.

100. A simple pendulum with length L is swinging freely with small angular amplitude. As the pendulum passes its central (or equilibrium) position, its cord is suddenly and rigidly clamped at its midpoint. In terms of the original period T of the pendulum, what is the new period?

101. A 3.0 kg particle is in simple harmonic motion in one dimension and moves according to the equation

$$x = (5.0 \text{ m}) \cos[(\pi/3 \text{ rad/s})t - \pi/4 \text{ rad}],$$

with t in seconds. (a) At what value of x is the potential energy of the particle equal to half the total energy? (b) How long does the particle take to move to this position x from the equilibrium position?

102. A simple harmonic oscillator consists of a block of mass 0.50 kg attached to a spring. The block slides back and forth along a straight line on a frictionless surface with equilibrium point $x = 0$. At $t = 0$ the block is at its equilibrium point and is moving in the direction of increasing x. A graph of the magnitude of the net force $\vec{F}$ on the block as a function of its position is shown in Fig. 16-55. What are (a) the amplitude and (b) the period of the simple harmonic motion, (c) the magnitude of the maximum acceleration experienced by the block, and (d) the maximum kinetic energy attained by the block?

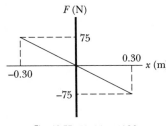

Fig. 16-55 Problem 102.

103. A particle executes linear SHM with frequency 0.25 Hz about the point $x = 0$. At $t = 0$, it has displacement $x = 0.37$ cm and zero velocity. For the motion, determine (a) the period, (b) the angular frequency, (c) the amplitude, (d) the displacement $x(t)$, (e) the velocity $v(t)$, (f) the maximum speed, (g) the magnitude of the maximum acceleration, (h) the displacement at $t = 3.0$ s, and (i) the speed at $t = 3.0$ s.

104. In Edgar Allen Poe's masterpiece of terror, "The Pit and the Pendulum," a prisoner who is strapped flat on a floor spies a seemingly motionless pendulum 12 m above him. Then, to his horror, he realizes that the pendulum consists of " a crescent of glittering steel, . . . the under edge as keen as that of a razor" and that it is gradually descending. As hours go by, the pendulum's motion becomes mesmerizing, with the left–right sweep and the speed at the lowest point of each swing both increasing. The pendulum's intent becomes clear: to sweep directly across the prisoner's heart. "Down—steadily down it crept. I took a frenzied pleasure in contrasting its downward with its lateral velocity. To the right to the left—far and wide—with the shriek of a damned spirit! . . . Down—certainly, relentlessly down!"

Assume that the pendulum is ideal, consisting of a particle of mass m on the end of a massless cord of length r, and that it descends via small increases in r. Take the initial cord length r_0 to be 0.80 m and the initial maximum angular speed $\omega_{0,\text{max}}$ (when the pendulum passes through $\theta = 0$) to be 1.30 rad/s. Assume also that the pendulum descends only in small steps and only as it passes through $\theta = 0$. (a) Show that this last assumption means that the angular momentum of the pendulum does not change during each step of the descent.

In terms of the cord length r, find (b) the maximum angular speed ω_{max} during a swing and (c) the maximum kinetic energy K_{max} during a swing. (d) As r increases, does K_{max} increase, decrease, or stay the same?

As the pendulum swings upward for any given value of r, assume that its total mechanical energy is conserved. (e) Find the maximum gravitational potential energy U_{max} attained by the pendulum during the upward swing, first in terms of the maximum kinetic energy K_{max} and then in terms of r and the maximum angle θ_{max} reached during a swing.

(f) Using the results of (d) and (e), find θ_{max} as a function of r and given data. (g) Graph θ_{max} versus r. Does θ_{max} increase or decrease with the increases in r? (h) For what value of r has θ_{max} changed by a factor of 2 from its initial value (is it twice or half as much)?

(i) Find the horizontal sweep Δx of the pendulum in terms of r and θ_{max}. (j) Graph Δx versus r. Does Δx increase or decrease with the increases in r? (k) For what value of r has Δx changed by a factor of 2 from its initial value?

(l) Find the maximum speed v_{max} during a swing in terms of r. (m) Graph v_{max} versus r. Does v_{max} increase or decrease with the increases in r? (n) For what value of r has v_{max} changed by a factor of 2 from its initial value? (o) Compare these results with Poe's description.

Tutorial Problem

105. An object of mass 0.25 kg oscillates in SHM along an x axis with its displacement given by

$$x(t) = 0.25 \text{ m} + (0.50 \text{ m}) \cos[(2.0 \text{ rad/s})t + \pi/6].$$

(a) Make a rough sketch of the displacement of this object as a function of time for several periods. (b) What is the equilibrium position of the object? (c) Determine the amplitude of the motion and the maximum and minimum values of $x(t)$. (d) What is the period of the motion?

(e) Write expressions for the velocity and acceleration of the object as a function of time. Remember to use notation consistent with the vector nature of these quantities. (f) Write two expressions for the force acting on the object, one as a function of time and the other as a function of x. (g) What is the total mechanical energy of the object? (h) Determine the position, velocity, and acceleration of the object at time $t = 0.0$ s.

Answers

(a) Sketch a sine curve between displacements of 0.75 m and -0.25 m, with a displacement of 0.68 m at $t = 0$.

(b) The equilibrium position of the object is given by the non-oscillatory term in $x(t)$ and thus is $x_{eq} = 0.25$ m.

(c) The amplitude of the motion is the factor in front of the trigonometric function: 0.50 m. The maximum value of $x(t)$ is 0.25 m $+ 0.50$ m $= 0.75$ m, and the minimum value is 0.25 m $- 0.50$ m $= -0.25$ m.

(d) The expression for the oscillatory motion should be a trigonometric function of ωt, and from the given expression we see that $\omega = 2.0$ rad/s. The period is then $T = 2\pi/\omega = (2\pi$ rad$)/(2.0$ rad/s$) = 3.1$ s.

(e) The linear velocity of the object has the x component

$$v_x(t) = \frac{dx(t)}{dt} = -(0.50 \text{ m})(2.0 \text{ rad/s}) \sin[(2.0 \text{ rad/s})t + \pi/6]$$

$$= -(1.00 \text{ m/s}) \sin[(2.0 \text{ rad/s})t + \pi/6].$$

The linear acceleration of the object then has the x component

$$a_x(t) = \frac{dv_x(t)}{dt} = -(1.00 \text{ m/s})(2.0 \text{ rad/s}) \cos[(2.0 \text{ rad/s})t + \pi/6]$$

$$= -(2.00 \text{ m/s}^2) \cos[(2.0 \text{ rad/s})t + \pi/6].$$

(f) The force as a function of time is most easily determined by

$$F_x(t) = ma_x(t) = (0.25 \text{ kg})(-(2.00 \text{ m/s}^2) \cos[(2.0 \text{ rad/s})t + \pi/6])$$

$$= -(0.50 \text{ N}) \cos[(2.0 \text{ rad/s})t + \pi/6].$$

To determine the force as a function of x we can use $F_x(x) = -k(x - x_{eq})$, but we would need first to calculate the value of k corresponding to this problem. Another approach is to note that the statement of this problem gives x as a function of t, which can be rearranged to yield

$$\cos[(2.0 \text{ rad/s})t + \pi/6] = \frac{x - 0.25 \text{ m}}{0.50 \text{ m}}.$$

Thus,

$$F_x(x) = -(0.50 \text{ N})\left(\frac{x - 0.25 \text{ m}}{0.50 \text{ m}}\right)$$

$$= -(1.00 \text{ N/m})(x - 0.25 \text{ m}).$$

We see that $k = 1.00$ N/m in this situation.

(g) The total mechanical energy of the object can most easily be determined as the value of the maximum kinetic energy of the object. From part (e) we see that $v_{max} = 1.00$ m/s, so

$$E = K_{max} = \tfrac{1}{2}mv_{max}^2 = (0.5)(0.25 \text{ kg})(1.00 \text{ m/s})^2 = 0.125 \text{ J}.$$

Note that the potential energy of this system will be of the form $U = \tfrac{1}{2}k(x - x_{eq})^2 = \tfrac{1}{2}k(x - 0.25 \text{ m})^2$, not $\tfrac{1}{2}kx^2$.

(h) We can use the general expressions given here for $x(t)$, $v_x(t)$, and $a_x(t)$. At $t = 0.0$ s, we have

$$x = 0.25 \text{ m} + (0.50 \text{ m}) \cos(\pi/6)$$

$$= 0.25 \text{ m} + (0.50 \text{ m})(0.866) = 0.68 \text{ m}.$$

$$v_x = -(1.00 \text{ m/s}) \sin(\pi/6)$$

$$= -(1.00 \text{ m/s})(0.50) = -0.50 \text{ m/s}.$$

$$a_x = -(2.00 \text{ m/s}^2) \cos(\pi/6)$$

$$= -(2.00 \text{ m/s}^2)(0.866) = -1.73 \text{ m/s}^2.$$

Waves—1

Sample Problem 17-8

A sinusoidal wave moving along a string in the positive direction of an x axis was recorded on video tape. The curves in Fig. 17-37 represent the wave in two freeze-frames of the tape, first as the solid curve and then, 1.0 ms later, as the dotted curve. The grid lines along the x axis are 1.0 cm apart. Each string element oscillated vertically (perpendicular to the x axis) a total distance of 4.0 mm as the wave passed through it. The wave moved a distance $d = 3.16$ cm to the right in the 1.0 ms time interval. Write an equation for this wave in the form

$$y(x, t) = y_m \sin(kx \pm \omega t), \quad (17\text{-}57)$$

where $\pm$ indicates that the proper sign must be determined.

SOLUTION: We need to find values for y_m, k, and ω in Eq. 17-57. To

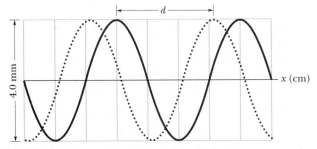

Fig. 17-37 Sample Problem 17-8. A sinusoidal string wave traveling in the positive direction of an x axis is shown at two instants, between which the wave travels distance d.

find the amplitude y_m, we use the **Key Idea** that y_m is half the total oscillation distance of 4.0 mm, so we have

$$y_m = 2.0 \text{ mm}.$$

To find the angular wave number k, we use two **Key Ideas**: First, k is related to the wavelength λ by Eq. 17-5 ($k = 2\pi/\lambda$). Second, we can determine λ from Fig. 17-37 by measuring the distance (parallel to the x axis) between repetitions of the wave shape. Let's use, say, the solid curve and pick the point at which it crosses the x axis at the third vertical grid line from the left. That curve makes an identical crossing 4 grid lines to the right of the first point. Thus, $\lambda = 4 \times 1.0$ cm $= 4.0$ cm, and so

$$k = \frac{2\pi}{\lambda} = \frac{2\pi}{0.040 \text{ m}} = 157 \text{ m}^{-1} \approx 160 \text{ m}^{-1}.$$

To find angular frequency ω, we again need two **Key Ideas**: First, ω is related to k and the wave speed v by Eq. 17-12 ($v = \omega/k$). Second, v is the ratio of the distance d traveled by the wave to the time interval Δt required for that travel. From Eq. 17-12 and the given data, we get

$$\omega = kv = k\frac{d}{\Delta t} = (157 \text{ m}^{-1})\frac{0.0316 \text{ m}}{0.0010 \text{ s}} = 4961 \text{ s}^{-1} \approx 5000 \text{ s}^{-1}.$$

An additional **Key Idea** is that, because the wave is moving in the positive direction of the x axis, we must use the minus sign in Eq. 17-57. We can now rewrite that equation as

$$y(x, t) = (2.0 \text{ mm}) \sin[(160 \text{ m}^{-1})x - (5000 \text{ s}^{-1})t], \quad \text{(Answer)}$$

with y in millimeters, x in meters, and t in seconds.

Sample Problem 17-9

Two sinusoidal waves with the same amplitude of 4.0 mm and the same wavelength traveled together along a string that is stretched along an x axis; the *resultant* wave due to their interference was recorded on video tape. The curves in Fig. 17-38 represent the resultant wave in two freeze-frames, first as the solid curve and then, 1.0 ms later, as the dotted curve. The grid lines along the x axis are 1.0 cm apart, and the string elements oscillated vertically (perpendicular to the x axis) by 6.0 mm as the resultant wave passed through them. That wave moved a distance $d = 4.20$ cm to the right in the 1.0 ms time interval. Write equations for the two interfering waves and for their resultant wave.

SOLUTION: From Fig. 17-38, we see that the resultant wave is a sinusoidal wave that moves in the positive direction of the x axis. A first **Key Idea** here is that the interfering waves must both move in that direction and, from Eqs. 17-34 and 17-35, we can write them as

$$y_1(x, t) = y_m \sin(kx - \omega t) \quad (17\text{-}58)$$

and

$$y_2(x, t) = y_m \sin(kx - \omega t + \phi), \quad (17\text{-}59)$$

where $y_m = 4.0$ mm. From Eq. 17-38, we can then write the resultant wave as

$$y'(x, t) = [2y_m \cos\tfrac{1}{2}\phi] \sin(kx - \omega t + \tfrac{1}{2}\phi). \quad (17\text{-}60)$$

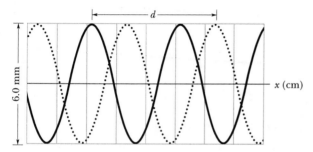

Fig. 17-38 Sample Problem 17-9. The resultant wave of two sinusoidal string waves traveling along an x axis is shown at two instants, between which the resultant wave travels distance d.

The amplitude $2y_m \cos \frac{1}{2}\phi$ of the resultant wave is half the total oscillation distance of 6.0 mm, so we have

$$2y_m \cos \tfrac{1}{2}\phi = 3.0 \text{ mm}.$$

Substituting $y_m = 4.0$ mm and solving for ϕ, we get

$$\phi = 2 \cos^{-1} \frac{3.0 \text{ mm}}{2(4.0 \text{ mm})} = 136° \approx 2.4 \text{ rad}.$$

Thus, the phase constant in Eq. 17-59 is $\phi = 2.4$ rad and the phase constant in Eq. 17-60 is $\frac{1}{2}\phi = \frac{1}{2}(2.4 \text{ rad}) \approx 1.2$ rad.

To find the angular wave number k, we use two **Key Ideas**. First, the common value of k in Eqs. 17-58, 17-59, and 17-60 is related to the common wavelength λ by Eq. 17-5 ($k = 2\pi/\lambda$). Second, the wavelength λ can be measured in Fig. 17-38 as the distance (par-

allel to the x axis) between repetitions of the wave shape. Let's use the solid curve and pick any point at which it crosses the x axis. That curve makes an identical crossing 3.0 cm to the right (or left) from the first point. Thus, $\lambda = 3.0$ cm, and

$$k = \frac{2\pi}{\lambda} = \frac{2\pi}{0.030 \text{ m}} = 209 \text{ m}^{-1} \approx 210 \text{ m}^{-1}.$$

To find the common angular frequency ω, we use two more **Key Ideas**: The common value of ω in Eqs. 17-58, 17-59, and 17-60 is related to k and the common wave speed v by Eq. 17-12 ($v = \omega/k$). That wave speed v is the ratio of the distance d traveled by the resultant wave to the time interval Δt required for that travel. Thus, we have

$$\omega = kv = k\frac{d}{\Delta t} = (209 \text{ m}^{-1})\frac{0.0420 \text{ m}}{0.0010 \text{ s}} = 8778 \text{ s}^{-1} \approx 8800 \text{ s}^{-1}.$$

We can now write Eqs. 17-58 and 17-59 for the interfering waves as

$$y_1(x, t) = (4.0 \text{ mm}) \sin(210x - 8800t)$$

and

$$y_2(x, t) = (4.0 \text{ mm}) \sin(210x - 8800t + 2.4 \text{ rad}), \quad \text{(Answer)}$$

with x in meters and t in seconds. We can also write Eq. 17-60 for the resultant wave as

$$y'(x, t) = (3.0 \text{ mm}) \sin(210x - 8800t + 1.2 \text{ rad}), \quad \text{(Answer)}$$

again with x in meters and t in seconds.

Sample Problem 17-10

Two sinusoidal waves with the same amplitude and wavelength travel along a string that is stretched along an x axis; the *resultant wave* due to their interference was recorded on video tape. The curves in Fig. 17-39 represent the resultant wave in four freeze-frames in the sequence of a, b, c, and d, with 1.0 ms elapsing between curves a and d. The grid lines along the x axis are 1.0 cm apart, and the string elements oscillated perpendicular to the x axis by 16.0 mm (between the extreme displacements shown by curves a and d) as the resultant wave passed through them. Write equations for the two interfering waves and for their resultant wave.

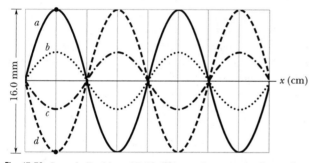

Fig. 17-39 Sample Problem 17-10. The resultant wave of two sinusoidal string waves traveling along an x axis is shown at four instants (a, b, c, and d).

SOLUTION: A first **Key Idea** here is that Fig. 17-39 shows a standing wave and so the two waves must be traveling in opposite directions and have the forms given by Eqs. 17-45 and 17-46:

$$y_1(x, t) = y_m \sin(kx - \omega t) \qquad (17\text{-}61)$$

and

$$y_2(x, t) = y_m \sin(kx + \omega t). \qquad (17\text{-}62)$$

From Eq. 17-47, we can then write the resultant wave as

$$y'(x, t) = (2y_m \sin kx) \cos \omega t. \qquad (17\text{-}63)$$

We need to find the common values of y_m, k, and ω in these three equations.

To find y_m, we use the **Key Idea** that the amplitude of a standing wave is a maximum at an antinode, and there the amplitude is $2y_m$. Here the maximum amplitude of the standing wave is 8.0 mm, so

$$y_m = 4.0 \text{ mm}.$$

To find the angular wave number k, we use two more **Key Ideas**: First, the common angular wave number k in Eqs. 17-61, 17-62, and 17-63 is related to the common wavelength λ by Eq. 17-5 ($k = 2\pi/\lambda$). Second, the nodes in a standing wave are separated by 0.50λ. Because the antinodes in Fig. 17-39 are separated by 2.0 cm, we have $\lambda = 2 \times 2.0$ cm $= 4.0$ cm. Then

$$k = \frac{2\pi}{\lambda} = \frac{2\pi}{0.040 \text{ m}} = 157 \text{ m}^{-1} \approx 160 \text{ m}^{-1}.$$

To find the angular frequency ω, we again use two **Key Ideas:** We can relate the common value of ω in Eqs. 17-61, 17-62, and 17-63 to the period T of the standing wave with Eq. 17-8 ($\omega = 2\pi/T$). A string element at an antinode takes time T to move through a full oscillation from one extreme displacement to the opposite extreme and back to the first extreme. According to the given data, the string element at the dot on curve a in Fig. 17-39 takes 1.0 ms to move to the position of the dot on curve d, which is one-half a full oscillation. Thus, $T = 2.0$ ms, and

$$\omega = \frac{2\pi}{T} = \frac{2\pi}{0.0020 \text{ s}} = 3142 \text{ s}^{-1} \approx 3100 \text{ s}^{-1}.$$

Now we can write Eqs. 17-61 and 17-62 for the interfering waves as

$$y_1(x, t) = (4.0 \text{ mm}) \sin(160x - 3100t)$$

and

$$y_2(x, t) = (4.0 \text{ mm}) \sin(160x + 3100t), \quad \text{(Answer)}$$

with x in meters and t in seconds. We can also write the standing wave of Eq. 17-63 as

$$y'(x, t) = (8.0 \text{ mm})(\sin 160x)(\cos 3100t), \quad \text{(Answer)}$$

again with x in meters and t in seconds.

QUESTIONS

12. The first harmonic is set up on the string in Fig. 17-21, and the oscillations of the string have period T. If time $t = 0$ when the string is in its extreme upward configuration (the solid line in Fig. 17-21a), at what time, in terms of T, does the string (a) first become horizontal, (b) first reach its extreme downward configuration (the dashed line in Fig. 17-21a), and (c) again become horizontal?

13. *Organizing question:* Two series of snapshots taken of a wave traveling along a string under tension are shown in Figs. 17-40a and b. (The snapshots span less than a period of the wave.) For each series, set up an equation, complete with known data, for the traveling wave. (This question continues with Question 14.)

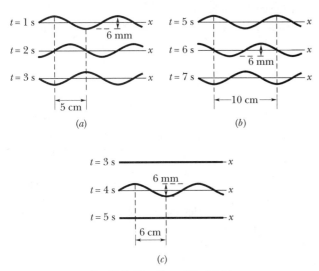

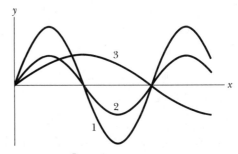

Fig. 17-40 Questions 13 and 14.

14. Question 13 continued: Figure 17-40c shows three snapshots taken of a string as traveling waves pass along it. Set up equations, complete with known data, for the traveling waves.

15. (a) Three waves, all having the same amplitude and wavelength, travel along a string in the same direction. Take one to have a phase constant of $0°$. What are the phase constants of the other waves if all three together produce fully destructive interference? (*Hint:* Construct a phasor diagram.) (b) What are the phase constants if there are four such waves? Now there are many (different) answers.

16. Here are some more phase differences between the two waves of Sample Problem 17-5, expressed in radians: $\pi/4$, $7\pi/4$, $-\pi/4$, and $-7\pi/4$. (a) Without written calculation, rank them in terms of the amplitude of the resultant wave, greatest first. (b) What type of interference occurs with each phase difference?

17. Figure 17-41 shows three waves that are *separately* sent along a string that is stretched under a certain tension along an x axis. Rank the waves according to (a) their wavelengths, (b) their speeds, and (c) their angular frequencies, greatest first.

Fig. 17-41 Question 17.

18. Do the wavelength and frequency, respectively, of the second harmonic on a string stretched between two supports increase, decrease, or remain the same if we (a) increase the distance between the supports without increasing the tension, (b) increase the tension in the string, and (c) switch to a string of greater linear density?

19. The following table gives the wave speeds (in terms of v_0) and the frequencies (in terms of f_0) for three waves traveling along different strings. The waves all have the same amplitude, and the strings have the same linear density. Rank the waves according to the average rate at which they transfer energy along the strings, greatest first.

Wave	1	2	3
Wave speed	$2v_0$	$4v_0$	$6v_0$
Frequency	$4f_0$	f_0	$2f_0$

20. The nodes of a standing wave on a 10 m string stretched along an x axis happen to be at $x = 0.5, 1.5, 2.5$ m, where $x = 0$ is somewhere near the middle of the string. Is there a node, an antinode, or some intermediate state at (a) $x = -0.5$ m and (b) $x = 3.0$ m?

21. Four waves are separately sent along a string that is stretched under a certain tension. The displacement $y(t)$ of a particular string element during the passage of each wave is given in Fig. 17-42. Rank the waves according to the average rate at which they transfer energy along the strings, greatest first.

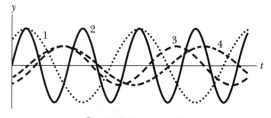

Fig. 17-42 Question 21.

22. Two strings of equal lengths but unequal linear densities are tied together with a knot and stretched between two supports. A particular frequency happens to produce a standing wave on each length, with a node at the knot, as shown in Fig. 17-43. Which string has the greater linear density?

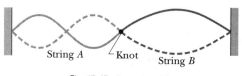

Fig. 17-43 Question 22.

23. A string is stretched between two fixed supports separated by distance L. (a) For which harmonic numbers is there a node at the point located $L/3$ from one of the supports? Is there a node, antinode, or some intermediate state at a point located $2L/5$ from one support if (b) the fifth harmonic is set up and (c) the tenth harmonic is set up?

24. Waves of the same amplitude and frequency are sent along three strings. The following table gives the tensions and linear densities of the strings. Rank the strings according to the average rate at which energy is transferred along them, greatest first.

String	1	2	3
Tension	$2\tau_0$	$4\tau_0$	$3\tau_0$
Linear density	$4\mu_0$	$2\mu_0$	$6\mu_0$

25. Guitar players know that, prior to a show, a guitar must be played and the strings then tightened, because during the first few minutes of playing the strings warm and loosen slightly. Does that loosening increase or decrease the resonant frequencies of the strings?

26. In Fig. 17-44, four strings are placed under tension by one or two suspended blocks, all of the same mass. Strings A, B, and C have the same linear density; string D has a greater linear density. Rank the strings according to the speed that waves will have when sent along them, greatest first.

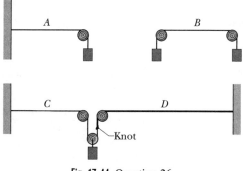

Fig. 17-44 Question 26.

DISCUSSION QUESTIONS

27. How can you prove experimentally that energy can be transported by a wave?

28. Energy can be transferred by particles as well as by waves. How can we experimentally distinguish between these methods of energy transfer?

29. Can a wave motion be generated in which the particles of the medium oscillate with angular simple harmonic motion? If so, explain how and describe the wave.

30. The following functions, in which A is a constant, are of the form $h(x \pm vt)$:

$$y = A(x - vt), \qquad y = A(x + vt)^2,$$
$$y = A\sqrt{x - vt}, \qquad y = A \ln(x + vt).$$

Explain why these functions are not useful in wave motion.

31. Can one produce on a string a wave form that has a discontinuity in slope at a point—that is, one having a sharp corner? Explain.

32. Compare and contrast the behavior of (a) the block of a block–spring system, oscillating in simple harmonic motion, and (b) an element of a stretched string through which a traveling sinusoidal wave is passing. Discuss from the point of view of displacement, velocity, acceleration, and energy transfers.

33. A passing motorboat creates a wake that causes waves to wash ashore. As time goes on, the period of the arriving waves grows shorter and shorter. Why?

34. When two waves interfere, does one alter the progress of the other? Explain.

35. When waves interfere, is there a loss of energy? Explain your answer.

36. If two waves differ only in amplitude and travel in opposite directions through a medium, will they produce standing waves? Is energy transported? Are there any nodes?

37. In the discussion of transverse waves in a string, we have dealt only with displacements in a single plane, the *xy* plane. If all displacements lie in one plane, the wave is said to be *plane polarized*. Can there be displacements in a plane other than the single plane with which we dealt? If so, can two differently plane-polarized waves be combined? What appearance would such a combined wave have?

38. A wave transmits energy. Does it transfer linear momentum? Can it transfer angular momentum?

EXERCISES & PROBLEMS

53. In the 1990s, the bridge Le Pont de Normandie was built across the Seine River (France) with the longest cable-stayed span in the world (Fig. 17-45*a*). However, engineers noticed that standing waves were produced on some of the cables by even mild winds. Such oscillations are of concern because they can fatigue the metal strands in the cables, eventually leading to premature failure. To fix the problem, mountain climbers scaled the bridge to attach ropes between adjacent cables.

Consider the simple arrangement of two vertical cables in Fig. 17-45*b*. Cable *A* of length L and cable *B* of length $2L$ have the same linear density, are under the same tension, and are connected by a horizontal rope at the midpoint of the shorter cable. Initially the rope is neither slack nor under tension. Suppose that at time $t = 0$ a wind gust starts the cables oscillating in their fundamental modes by moving the cables rightward. Let T_A be the oscillation period of cable *A*, and assume that the cables would oscillate with the same amplitude if not connected. For which of the following times is the rope (a) under tension, (b) back in its initial state, and (c) slack: $0.25T_A$, $0.5T_A$, $0.75T_A$, T_A, $1.25T_A$, $1.5T_A$, $1.75T_A$, $2T_A$? Fig. 17-45*c* shows a second choice of rope placement. (d) Do the two designs damp out the cable oscillations if only the fundamental oscillations can occur?

54. A certain transverse sinusoidal wave of wavelength 20 cm is moving in the positive *x* direction. The transverse velocity of the particle at $x = 0$ as a function of the time is shown in Fig. 17-46. What are (a) the wave speed, (b) the amplitude, and (c) the frequency of the wave? (d) Sketch the wave between $x = 0$ and $x = 20$ cm at $t = 2.0$ s.

(a)

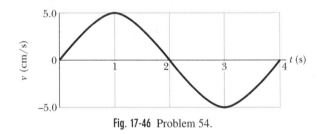

(b)   (c)

Fig. 17-45 Problem 53.

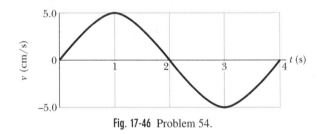

Fig. 17-46 Problem 54.

55. Two waves are described by

$$y_1 = 0.30 \sin[\pi(5x - 200)t]$$

and $$y_2 = 0.30 \sin[\pi(5x - 200t) + \pi/3],$$

where y_1, y_2, and x are in meters and t is in seconds. When these two waves are combined, a traveling wave is produced. What are (a) the amplitude, (b) the wave speed, and (c) the wavelength of that traveling wave?

56. Two sinusoidal waves of the same wavelength travel in the same direction along a stretched string with amplitudes of 4.0 and 7.0 mm and phase constants of 0 and 0.8π rad, respectively.

What are (a) the amplitude and (b) the phase constant of the resultant wave?

57. A traveling wave on a string is described by

$$y = 2.0 \sin\left[2\pi\left(\frac{t}{0.40} + \frac{x}{80}\right)\right],$$

where x and y are in centimeters and t is in seconds. (a) For $t = 0$, plot y as a function of x for $0 \le x \le 160$ cm. (b) Repeat (a) for $t = 0.05$ s and $t = 0.10$ s (c) From your graphs, what is the wave speed, and in which direction is the wave traveling?

58. The equation of a transverse wave traveling along a string is given by

$$y = (2.0 \text{ mm}) \sin[(20 \text{ m}^{-1})x - (600 \text{ s}^{-1})t].$$

Find the (a) amplitude, (b) frequency, (c) velocity, and (d) wavelength of the wave. (e) Find the maximum transverse speed of a particle in the string.

59. A single pulse, whose wave shape is given by the function $h(x - 5t)$, is shown in Fig. 17-47 for $t = 0$. Here x is in centimeters and t is in seconds. What are the (a) speed and (b) direction of travel of the pulse? (c) Plot $h(x - 5t)$ as a function of x for $t = 2$ s. (d) Plot $h(x - 5t)$ as a function of t for $x = 10$ cm.

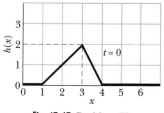

Fig. 17-47 Problem 59.

60. Energy is transmitted at rate P_1 by a wave of frequency f_1 on a string with tension τ_1. What is the new energy transmission rate P_2 in terms of P_1 (a) if the tension of the string is increased to $\tau_2 = 4\tau_1$ and (b) if, instead, the frequency of the wave is decreased to $f_2 = f_1/2$?

61. The speed of a transverse wave on a string is 170 m/s when the string tension is 120 N. To what value must the tension be changed to raise the wave speed to 180 m/s?

62. A 1.50 m wire has a mass of 8.70 g and is under a tension of 120 N. The wire is held rigidly at both ends and set into oscillation. Calculate (a) the speed of waves on the wire, (b) the wavelengths of the waves that produce one- and two-loop standing waves on the string, and (c) the frequencies of the waves that produce one- and two-loop standing waves.

63. The equation of a transverse wave traveling along a string is

$$y = 0.15 \sin(0.79x - 13t),$$

in which x and y are in meters and t is in seconds. (a) What is the displacement y at $x = 2.3$ m, $t = 0.16$ s? (b) Write the equation of a wave that, when added to the given one, produces standing waves on the string. (c) What is the displacement of the resultant standing wave at $x = 2.3$ m, $t = 0.16$ s?

64. (a) Write an equation describing a sinusoidal transverse wave traveling on a cord in the positive y direction with an angular wave number of 60 cm^{-1}, a period of 0.20 s, and an amplitude of 3.0 mm. Take the transverse direction to be the z direction. (b) What is the maximum transverse speed of a point on the cord?

65. A 120 cm length of string is stretched between fixed supports. What are the three longest possible wavelengths for traveling waves on the string that can produce standing waves? Sketch the corresponding standing waves.

66. Two waves on a string are described by these equations:

$$y_1 = (0.10 \text{ m}) \sin 2\pi [(0.50 \text{ m}^{-1})x + (20 \text{ s}^{-1})t],$$
$$y_2 = (0.20 \text{ m}) \sin 2\pi [(0.50 \text{ m}^{-1})x - (20 \text{ s}^{-1})t].$$

Graph y versus t for the point on the string at $x = 3.0$ m.

67. When played in a certain manner, the lowest resonant frequency of a certain violin string is concert A (440 Hz). What are the frequencies of the second and third harmonics of the string?

68. A wave has a speed of 240 m/s and a wavelength of 3.2 m. What are the (a) frequency and (b) period of the wave?

69. A wave on a string is described by

$$y(x, t) = 15 \sin(\pi x/8 - 4\pi t),$$

where x and y are in centimeters and t is in seconds. (a) What is the transverse speed for a point on the string at $x = 6.0$ cm when $t = 0.25$ s? (b) What is the maximum transverse speed of any point on the string? (c) What is the magnitude of the transverse acceleration for a point on the string at $x = 6.0$ cm when $t = 0.25$ s? (d) What is the magnitude of the maximum transverse acceleration for any point on the string?

70. A transverse sinusoidal wave is moving along a string in the positive x direction with a speed of propagation of 80 m/s. At $t = 0$, the string particle at $x = 0$ has a transverse displacement of 4.0 cm from its equilibrium position and is not moving. The maximum transverse speed of the string particle at $x = 0$ is 16 m/s. (a) What is the frequency of the wave? (b) What is the wavelength of the wave? (c) Write an equation describing the wave.

71. The amplitudes of two sinusoidal waves traveling in the same direction along a stretched string are 3.0 and 5.0 mm; their phase constants are 0° and 70°, respectively. The waves have the same wavelength. What are (a) the amplitude and (b) the phase constant of the resultant wave?

72. Two waves,

$$y_1 = (2.50 \text{ mm}) \sin[(25.1 \text{ rad/m})x - (440 \text{ rad/s})t]$$

and $y_2 = (1.50 \text{ mm}) \sin[(25.1 \text{ rad/m})x + (440 \text{ rad/s})t],$

travel along a stretched string. (a) Plot the resultant wave as a function of t for $x = 0$, $\lambda/8$, $\lambda/4$, $3\lambda/8$, and $\lambda/2$, where λ is the wavelength. The graphs should extend from $t = 0$ to a little over one period. (b) The resultant wave is the superposition of a standing wave and a traveling wave. In which direction does the traveling wave travel? (c) How can you change the original waves so the resultant wave is the superposition of standing and traveling waves with the same amplitudes as before but with the traveling wave moving in the opposite direction? (d) Use your graphs to find the place at which the amplitude of oscillation is the largest and the smallest. (e) How are the maximum and minimum amplitudes related to the amplitudes 2.50 mm and 1.50 mm of the original two traveling waves?

Tutorial Problem

73. Figure 17-48 is a graph of a sinusoidal mechanical wave at a particular time, giving displacement y versus position x. (Note the difference in scales on the axes.) (a) The graphed curve gives (at the particular instant) the wave function $y(x)$—that is, y as a function of x. What does this wave function represent physically? (Make no assumption as to whether the wave is a transverse wave or a longitudinal wave.)

(b) Suppose the wave is a transverse wave. Sketch the displacements of the particles (through which the wave moves) at the positions $x = 0, 1, 2, 3, 4, 5,$ and 6 cm. (c) Repeat part (b), but now suppose the wave is a longitudinal wave. (d) What are the numerical values of the amplitude and wavelength of the wave? Show these quantities in Fig. 17-48. (e) Graph the wave function for a time that is half a period after the time of Fig. 17-48. Does your graph's appearance depend on whether the wave of Fig. 17-48 is traveling in the positive direction or the negative direction of x?

(f) Now suppose that the wave is traveling toward increasing x with a speed of 4.0 cm/s. On the graph of part (e) include a dashed curve to show the wave function 0.25 s later than the time assumed in drawing that graph. (g) Define in words what is meant by the period of a wave. What is the period of the wave of Fig. 17-48? (h) Define in words what is meant by the frequency of a wave. What is the frequency of the wave of Fig. 17-48? (i) Draw a graph of the wave as a function of time t for a particular position, say, $x = 1.0$ cm. On the graph, indicate the period T.

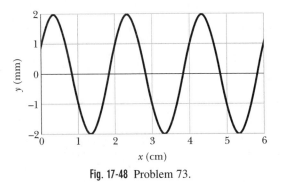

Fig. 17-48 Problem 73.

Answers

(a) The wave function $y(x)$ represents the displacements (from equilibrium) of the particles along which the wave travels. The displacements can be perpendicular to the wave's direction of travel (transverse wave) or parallel to that direction (longitudinal wave).

(b) See Fig. 17-49a.

(c) See Fig. 17-49b.

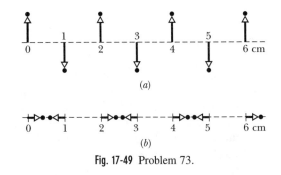

Fig. 17-49 Problem 73.

(d) The amplitude of the wave is 2.0 mm because the peaks and valleys are 2.0 mm from the equilibrium position. In Fig. 17-48, the wavelength is the horizontal distance between successive crests, or twice the horizontal distance between a peak and an adjacent valley. We see that the latter is 1.0 cm, so the wavelength is 2.0 cm.

(e) Peaks have become valleys, and vice versa, regardless of whether the wave is traveling in the positive direction or the negative direction of x.

(f) In the 0.25 s the wave moves (4.0 cm/s)(0.25 s) = 1.0 cm toward increasing x, so shift the curve of part (e) 1.0 cm to the right.

(g) The period of a wave is the time interval between successive peaks at any given position. Here, the wavelength is 2.0 cm while the wave is traveling with a speed of 4.0 cm/s, so this time interval is (2.0 cm)/(4.0 cm/s) = 0.50 s.

(h) The frequency of a wave is the rate at which peaks occur at a particular point along the wave. Here, the time between successive peaks is 0.50 s, and so the frequency is 1/(0.50 s) = 2.0 s^{-1} = 2.0 Hz.

(i) Sketch an inverted sine function with an amplitude of 2 mm and successive peaks separated by 0.50 s.

Waves—II

Sample Problem 18-9

A sound source emits two sinusoidal sound waves, both of wavelength λ, along paths A and B in Fig. 18-38. The sound traveling along path B is reflected from five surfaces as shown and then merges at point Q with the sound that traveled along path A. For what values of d (in wavelengths) are the waves exactly out of phase when they merge at point Q? For what values of d are the waves exactly in phase at point Q?

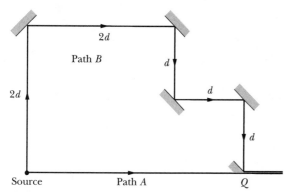

Fig. 18-38 Sample Problem 18-9. A sound source sends sound waves along paths A and B. The wave moving along path B is reflected five times and then travels with the wave moving along path A.

SOLUTION: Because the two waves emerge from the same source, they must emerge in phase with each other. The **Key Idea** here is that when they merge at point Q their phase difference can depend only on the path length difference ΔL between the two paths they followed. To reach Q, the wave following path A travels distance $3d$, and the wave following path B travels distance $7d$. Thus, their path length difference ΔL is $4d$.

Equation 18-25 tells us that if the waves are to be exactly out of phase at Q, we must have

$$\frac{\Delta L}{\lambda} = 0.5, \ 1.5, \ 2.5, \ \ldots \ .$$

Substituting $\Delta L = 4d$ and solving for d then give us

$$d = \tfrac{1}{8}\lambda, \ \tfrac{3}{8}\lambda, \ \tfrac{5}{8}\lambda, \ \ldots \ . \qquad \text{(Answer)}$$

Similarly, Eq. 18-23 tells us that if the waves are to be exactly in phase at point Q, we must have

$$\frac{\Delta L}{\lambda} = 0, \ 1, \ 2, \ 3, \ \ldots \ .$$

We can eliminate 0 from the list (which implies that there is no path difference in Fig. 18-38). Then solving for d, we find

$$d = \tfrac{1}{4}\lambda, \ \tfrac{1}{2}\lambda, \ \tfrac{3}{4}\lambda, \ \ldots \ . \qquad \text{(Answer)}$$

Sample Problem 18-10

The two pipes in Fig. 18-39 are submerged in seawater at 20°C. Pipe A, with length $L_A = 1.50$ m and one open end, contains a small sound source that is to set up the standing wave with the second lowest resonant frequency of that pipe.

(a) What is that resonant frequency?

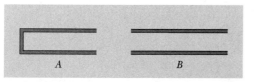

Fig. 18-39 Sample Problem 18-10. Two pipes submerged in seawater. A sound source in pipe A sets up acoustic resonance in that pipe. Sound then leaks from pipe A into pipe B.

SOLUTION: The **Key Idea** here is that only odd harmonics can exist in a pipe with one open end. The resonant frequencies are given by Eq. 18-41 ($f = nv/4L$) for harmonic numbers $n = 1, 3, 5, \ldots$. Because resonant frequency increases with increasing n, the second lowest resonant frequency corresponds to $n = 3$, the second lowest choice of n. *Caution:* The frequency is the *second* lowest but, because $n = 3$, it is for the *third* harmonic of pipe A. Using the speed of sound in seawater from Table 18-1, we then have

$$f = \frac{nv}{4L_A} = \frac{(3)(1522 \text{ m/s})}{(4)(1.50 \text{ m})} = 761 \text{ Hz}. \qquad \text{(Answer)}$$

(b) Sound leaking from pipe A is to set up resonance in pipe B, which has two open ends. That resonance is to be at the second lowest resonant frequency of pipe B. What is the required length L_B of pipe B?

SOLUTION: The **Key Idea** here is that any harmonic can exist in a pipe with two open ends. The resonant frequencies are given by Eq. 18-39 ($f = nv/2L$) for harmonic numbers $n = 1, 2, 3, \ldots$. Because resonant frequency increases with increasing n, the second lowest resonant frequency corresponds to $n = 2$, the second lowest choice of n. Therefore, we can write Eq. 18-39 as

$$f = \frac{2v}{2L_B}.$$

Solving for L_B and substituting known data yield

$$L_B = \frac{v}{f} = \frac{1522 \text{ m/s}}{761 \text{ Hz}} = 2.00 \text{ m.} \qquad \text{(Answer)}$$

Sample Problem 18-11

Suppose a horseshoe bat flies toward a moth at speed $v_b = 9.0$ m/s relative to the air, while the moth flies toward the bat with speed $v_m = 8.0$ m/s relative to the air. From its nostrils, the bat emits ultrasonic waves of frequency f_{be} that reflect from the moth back to the bat with frequency f_{bd}. The bat adjusts the emitted frequency f_{be} until the returned frequency f_{bd} is 83 kHz, at which the bat's hearing is best.

(a) What is the frequency f_m of the waves reflected by the moth when f_{bd} is 83 kHz?

SOLUTION: Our **Key Ideas** here are these:

1. The moth is the source of this sound (because it sends the sound back to the bat), and the bat is the detector.

2. The motions of the bat and the moth relative to the air change the frequency of the sound via the Doppler effect, as given by Eq. 18-47.

3. The motion of the detector (the bat) toward the source (the moth) tends to increase the frequency. The motion of the source toward the detector also tends to increase the frequency. The *detected* sound has the increased frequency of 83 kHz.

This third idea tells us to use the plus sign in the numerator of Eq. 18-47 to get a frequency increase due to the bat's motion toward the moth, and to use the minus sign in the denominator to get a frequency increase due to the moth's motion toward the bat. Thus, we have

$$f_{bd} = f_m \frac{v + v_b}{v - v_m},$$

or

$$83 \text{ kHz} = f_m \frac{343 \text{ m/s} + 9.0 \text{ m/s}}{343 \text{ m/s} - 8.0 \text{ m/s}},$$

from which

$$f_m = 78.99 \text{ kHz} \approx 79 \text{ kHz.} \qquad \text{(Answer)}$$

(b) What is the frequency f_{be} emitted by the bat when f_{bd} is 83 kHz?

SOLUTION: Our **Key Ideas** are the same as in (a) except that the bat is now the source (emitting sound at frequency f_{be}) and the moth is now the detector (detecting sound at frequency $f_m = 78.99$ Hz, the frequency of the sound it reflects). In Eq. 18-47, we again use the plus sign in the numerator to get a frequency increase, but now the increase is due to the moth's motion toward the bat. Also, we again use the minus sign in the numerator to get a frequency increase, but now the increase is due to the bat's motion toward the moth. We then have

$$f_m = f_{be} \frac{v + v_m}{v - v_b}$$

or

$$78.99 \text{ kHz} = f_{be} \frac{343 \text{ m/s} + 8.0 \text{ m/s}}{343 \text{ m/s} - 9.0 \text{ m/s}},$$

from which

$$f_{be} = 75 \text{ kHz.} \qquad \text{(Answer)}$$

The bat determines the relative speed of the moth (17 m/s) from the 8 kHz (= 83 kHz − 75 kHz) it must lower its emitted frequency to hear an echo with a frequency of 83 kHz (at which it hears best).

QUESTIONS

15. In Fig. 18-40, three long tubes (A, B, and C) are filled with different gases under different pressures. The ratio of the bulk modulus to the density is indicated for each gas in terms of a basic value B_0/ρ_0. Each tube has a piston at its left end that can send a sound pulse through the tube (as in Fig. 17-2). The three pulses are sent simultaneously. Rank the tubes according to the time of arrival of the pulses at the open right ends of the tubes, earliest first.

16. Figure 18-41 is a graph, at a certain time t, of the displacement functions $s(x, t)$ of three sound waves that travel along an x axis through air. Rank the waves according to their pressure amplitudes, greatest first.

17. The following three pairs of sound waves each produce a resultant wave due to their interference. (Amplitudes are in microm-

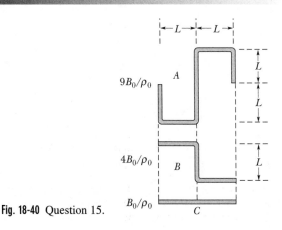

Fig. 18-40 Question 15.

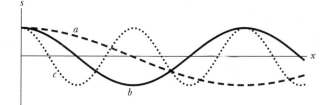

Fig. 18-41 Questions 16 and 21.

eters, x is in meters, and t is in centiseconds.) Without computation, rank the pairs according to the amplitudes of the resultant waves, greatest first.
(a) $s_1 = 2 \sin(3x - 4t)$ and $s_2 = 2 \sin(3x - 4t + \pi/2)$
(b) $s_1 = 3 \sin(5x - 6t)$ and $s_2 = 3 \sin(5x - 6t + \pi/2)$
(c) $s_1 = 7 \sin(3x - 5t)$ and $s_2 = 7 \sin(3x - 5t + \pi)$

18. Figure 18-42 indicates the Mach cones that form on an airplane flying at a certain speed at three different altitudes. Rank the altitudes according to (a) the speed of sound there and (b) the Mach number of the airplane there, greatest first. (c) Assuming that the speed of sound in air varies as the square root of the air temperature (in kelvins), rank the altitudes according to the air temperature, greatest first.

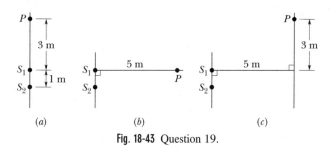

Fig. 18-42 Question 18.

19. *Organizing question:* In Figs. 18-43a through c, two point sources S_1 and S_2 emit sound of wavelength λ uniformly in all directions. Three points P are indicated. For each point, set up an equation, complete with given data, to calculate the phase difference (in wavelengths) between the waves arriving at the point from S_1 and S_2.

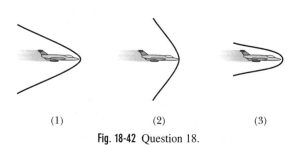

Fig. 18-43 Question 19.

20. (a) In Sample Problem 18-3, suppose that the sources are not in phase, but that instead S_1 emits earlier than S_2 by $0.2T$, where T is the period of the sound waves. Will the two points of 0λ in Fig. 18-9b be positioned as they are now, below the horizontal

dashed line, or above that dashed line? Now suppose that S_1 emits earlier than S_2 by $0.5T$. (b) What type of interference then occurs at point P_1 on the perpendicular bisector, and (c) what is the phase difference there in terms of wavelength λ? (d) What type of interference occurs at point P_2 on the line through the sources, and (e) what is the phase difference there?

21. Figure 18-41 shows graphs, at a certain time t, of the displacement function $s(x, t)$ of three sound waves traveling along an x axis through air. Rank the waves according to their intensity on a surface perpendicular to that axis, greatest first.

22. The third harmonic is set up inside a pipe of length L by a small internal sound source; the pipe has both ends closed and is filled with air (like the external air). Is there a displacement node or an antinode (a) across the ends and (b) across the middle? (c) Is the frequency of that third harmonic greater than, less than, or the same as the frequency of the third harmonic of a similarly filled pipe of the same length L with both ends open?

23. If you first set up the third harmonic in a pipe and then switch to the fourth harmonic, does the spacing between adjacent antinodes increase, decrease, or remain the same?

24. In Fig. 18-44, pipe A is made to oscillate in its third harmonic by a small internal sound source. Sound emitted at the right end happens to resonate four nearby pipes, each with only one open end (they are not drawn to scale). Pipe B oscillates in its lowest harmonic, pipe C in its second lowest harmonic, pipe D in its third lowest harmonic, and pipe E in its fourth lowest harmonic. Without computation, rank all five pipes according to their length, greatest first. (*Hint:* Draw the standing waves to scale and then draw the pipes to scale.)

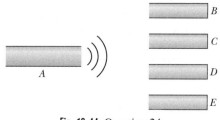

Fig. 18-44 Question 24.

25. Without using a calculator, determine the increase in a sound level when the intensity of the sound source becomes 10^7 times what it was previously.

26. You are given four tuning forks. The fork with the lowest frequency oscillates at 500 Hz. By striking two tuning forks at a time, you can produce the following beat frequencies: 1, 2, 3, 5, 7, and 8 Hz. What are the possible frequencies of the other three forks? (There are two sets of answers.)

27. *Organizing question:* A bat flies at a speed of 9.0 m/s directly toward a moth while emitting sound with a certain frequency f. The moth flies directly away from the bat at a speed of 7.0 m/s. If you use Eq. 18-47 to determine the frequency f' heard by the moth, (a) what speed should you substitute for v_D, (b) what sign should you use in front of that substitution, (c) what speed should you substitute for v_S, and (d) what sign should you use in front of that substitution?

Then, to determine the frequency of the echo heard by the bat from the moth, again with Eq. 18-47, (e) what speed should you substitute for v_D, (f) what sign should you place in front of that substitution, (g) what speed should you substitute for v_S, and (h) what sign should you place in front of that substitution?

28. A source emitting a sound wave at a certain frequency moves along an x axis (Fig. 18-45a). The source moves directly toward detector A and directly away from detector B. The superimposed three plots of Fig. 18-45b indicate the displacement function $s(x)$ at some time t of the sound wave as measured by detector A, by detector B, and by someone (C) in the rest frame of the source. Which plot corresponds to which measurement?

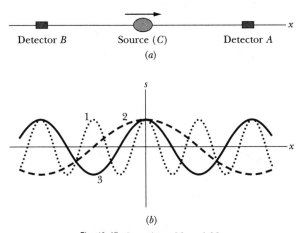

(a)

(b)

Fig. 18-45 Questions 28 and 29.

29. A sound wave of a certain frequency is emitted by a source moving along an x axis (Fig. 18-45a). The source moves directly toward detector A and directly away from detector B. Which of the following wave functions best correspond to (a) the wave detected by A, (b) the wave detected by B, and (c) the wave detected by someone (C) in the rest frame of the source and in front of the source? (Amplitudes are in micrometers, x is in meters, and t is in centiseconds.)

$$s_1 = 2 \sin(2x + 6t) \qquad s_2 = 2 \sin(x - 3t)$$
$$s_3 = 2 \sin(\tfrac{2}{3}x - 2t) \qquad s_4 = 2 \sin(2x - 4t)$$
$$s_5 = 2 \sin(2x - 6t) \qquad s_6 = 2 \sin(\tfrac{2}{3}x + 4t)$$
$$s_7 = 3 \sin(\tfrac{2}{3}x + 2t) \qquad s_8 = 2 \sin(\tfrac{2}{3}x + 2t)$$

DISCUSSION QUESTIONS

30. In some science fiction movies, the explosion of a starship can apparently be heard in another starship, while both are in the vacuum of space. Is this possible? Is there any way the explosion can produce sound inside the second ship?

31. Ultrasonic waves can be used to reveal internal structures of the body. They can, for example, distinguish between liquid and soft human tissues far better than can x rays. Why?

32. What experimental evidence is there for assuming that the speed of sound in air is the same for all wavelengths?

33. In practice, the inverse square law does not apply exactly to the decrease in intensity of sounds with distance. Why not?

34. Could the reference intensity for audible sound be set so as to permit negative sound levels in decibels? If so, how?

35. What is the common purpose of the valves of a cornet and the slide of a trombone?

36. The bugle has no valves. How then can we sound different notes on it? To what notes is the bugler limited? Why?

37. When you strike one prong of a tuning fork, the other prong also oscillates, even if the bottom end of the fork is clamped firmly in a vise. How can this happen? How does the second prong "get the word" that somebody has struck the first prong?

38. How can a sound wave travel down an organ pipe and be reflected at its open end? It seems that there is nothing there to reflect it.

39. How can we experimentally locate the positions of nodes and antinodes on a string, in an air column, and on an oscillating surface?

40. What physical properties of a sound wave correspond to the perceptions of pitch, loudness, and tone quality?

41. Does your singing really sound better in a shower? If so, what are the physical reasons?

42. Explain the audible tone produced by drawing a wet finger around the rim of a wine glass.

43. A lightning flash dissipates an enormous amount of energy and is essentially instantaneous. How is that energy transformed into the sound waves of thunder?

44. Sound waves can be used to measure the speed at which blood flows in arteries and veins. Explain how.

45. Suppose that George blows a whistle and Gloria hears it. She hears an increased frequency whether she is running toward George or George is running toward her. Are the increases in frequency the same in each case? Assume the same running speeds.

46. Suppose that, in the Doppler effect for sound, the source and receiver are at rest in some reference frame, but the air is moving with respect to this frame. Will there be a change in the wavelength (or frequency) received?

47. Jenny, sitting on a bench, sees Lew, also sitting on a bench, across the campus. She blows a whistle to attract his attention. A steady wind is blowing from Jenny to Lew. How does the presence of the wind affect the frequency of the sound that Lew hears? How does the wind affect the sound travel time?

48. How might the Doppler effect be used in an instrument to detect the fetal heartbeat? (Such measurements are routinely made.)

49. A satellite emits radio waves of constant frequency. These waves are picked up on the ground and made to beat against some standard frequency. The beat frequency is then sent through a loudspeaker and one "hears" the satellite signals. Describe how the sound changes as the satellite approaches, passes overhead, and recedes from the detector on the ground.

EXERCISES & PROBLEMS

61. When you "crack" a knuckle, you suddenly widen the knuckle cavity, allowing more volume for the synovial fluid inside it and causing a gas bubble suddenly to appear in the fluid. The sudden production of the bubble, called "cavitation," produces a sound pulse—the cracking sound. Assume the sound is transmitted uniformly in all directions and that it fully passes from the knuckle interior to the outside. If the pulse has a sound level of 62 dB at your ear, estimate the rate at which energy is produced by the cavitation.

62. The crest of a *Parasaurolophus* dinosaur skull contains a nasal passage in the shape of a long, bent tube, open at both ends. The dinosaur may have used the passage to produce sound by setting up the fundamental mode in it. (a) If the nasal passage in a certain *Parasaurolophus* fossil is 2.0 m long, what frequency would have been produced? (b) If that dinosaur could be recreated (as in *Jurassic Park*), would a person with a normal hearing range of 60 Hz to 20 kHz be able to hear that fundamental mode and, if so, would the sound be high or low frequency? Fossil skulls that contain shorter nasal passages are thought to be those of the female *Parasaurolophus*. (c) Would that make the female's fundamental frequency higher or lower than the male's?

63. Approximately a third of people with normal hearing have ears that continuously emit a low-intensity sound outward through the ear canal. A person with such *spontaneous otoacoustic emission* is rarely aware of the sound, except perhaps in a noise-free environment, but occasionally the emission is loud enough to be heard by someone else nearby. In one observation, the sound wave had a frequency of 1665 Hz and a pressure amplitude of 1.13×10^{-3} Pa. What were (a) the displacement amplitude and (b) the intensity of the wave emitted by the ear? (See Table 15-1 for the density of air.)

64. The males of the bullfrog *Rana catesbeiana* are known for their loud mating calls. These calls are emitted not by their mouths but by their eardrums, which lie on the surface of the head. If the emitted sound has a frequency of 260 Hz and a sound level of 85 dB (near the eardrum), what is the amplitude of the eardrum's oscillation? (See Table 15-1 for the density of air.)

65. *Waterfall acoustics.* The turbulent impact of the water in a waterfall causes the surrounding ground to oscillate in a wide range of low frequencies. If the water falls freely (instead of hitting rock on the way down), the oscillations are greatest in amplitude at a particular frequency f_m. This fact suggests that acoustic resonance is involved and f_m is the fundamental frequency. The following table gives, for nine U.S. and Canadian waterfalls, measured values for f_m and for the length L of the water's free fall. Is the proposed resonance closer to that of Fig. 18-14 (thus Eq. 18-39) or to that of Fig. 18-15b (thus Eq. 18-41)? (*Hint:* Plot the data. The speed of sound in still water is about 1400 m/s; in turbulent water filled with air bubbles, it can be 25% less.)

WATERFALL	1	2	3	4	5	6	7	8	9
f_m (Hz)	5.6	3.8	8.0	6.1	8.8	6.0	19	21	40
L (m)	97	71	53	49	35	24	13	11	8

66. A sperm whale vocalizes by producing a series of clicks. Actually, the whale makes only a single sound near the front of its head to start the series (Fig. 18-46a). Part of that sound then emerges from the head into the water to become the first click of the series. The rest of the sound travels backward through the spermaceti sac (a body of fat), reflects from the frontal sac (an air layer), and then travels forward through the spermaceti sac. When it reaches the distal sac (another air layer) at the front of the head, some of the sound escapes into the water to form the second click, and the rest is sent back through the spermaceti sac (and ends up forming later clicks).

Fig. 18-46b shows a strip-chart recording of a series of clicks detected by an underwater microphone. A unit time interval of 1.0 ms is indicated on the chart. Assuming that the speed of sound in the spermaceti sac is 1372 m/s, find the length of the spermaceti sac. From such a calculation, marine scientists can estimate the length of a whale from its click series.

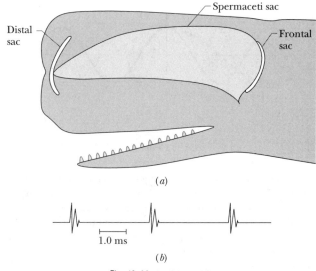

(a)

(b)

Fig. 18-46 Problem 66.

67. To search for a fossilized dinosaur embedded in rock, paleontologists can use sound waves to produce a computer image of the dinosaur. The image then guides the paleontologists as they dig the dinosaur out of the rock. (The technique is shown in the opening scenes of the movie *Jurassic Park*.) The basic idea of the detection technique is that a strong pulse of sound is emitted by a source (a seismic gun) at

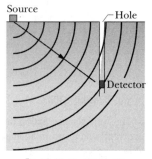

Fig. 18-47 Problem 67.

ground level and then detected by hydrophones that lie at evenly spaced depths in a bore hole drilled into the ground. The source and one hydrophone are shown in Fig. 18-47. If the sound wave travels from the source to the hydrophone through only rock as in Fig. 18-47, it travels at a known speed V and takes a certain time T. If, instead, it travels through a fossilized bone along the way, it takes slightly more time because it travels slower in the bone than in the rock. By measuring the difference Δt between the expected and measured travel times, the distance d traveled in the bone can be determined. After this procedure is repeated for many locations of the source and hydrophones, a computer can transform the many computed distances d into an image of the fossil.

(a) Let the speed of sound through fossilized bone be $V - \Delta V$, where ΔV is small relative to V. Show that the distance d is given by

$$d \approx \frac{V^2 \, \Delta t}{\Delta V}.$$

(b) For $V = 5000$ m/s and $\Delta V = 200$ m/s, what typical value of Δt can be expected if the sound passes along the diameter of a leg bone of an adult *T. rex*? (Estimate the bone's diameter.)

68. On July 10, 1996, a granite block broke away from a wall in Yosemite Valley and, as it began to slide down the wall, was launched into projectile motion. Seismic waves produced by its impact with the ground triggered seismographs as far away as 200 km. Later measurements indicated that the block had a mass of 7.3×10^7 kg to 1.7×10^8 kg and that it landed 500 m vertically below the launch point and 30 m horizontally from it. (The launch angle for the projectile motion is not known.) (a) Estimate the block's kinetic energy just before it landed.

Consider two types of seismic waves that spread from the impact point—a hemispherical *body wave* traveled through the ground in an expanding hemisphere and a cylindrical *surface wave* traveled along the ground in an expanding shallow vertical cylinder (Fig. 18-48). Assume that the impact lasted 0.50 s, the vertical cylinder had a depth d of 5.0 m, and each wave type received 20% of the energy the block had just before impact. Neglecting the loss of mechanical energy of the waves as they traveled, determine the intensities of (b) the body wave and (c) the surface wave when they reached a seismograph 200 km from the impact point. (d) On the basis of only these results, which type of wave is more easily detected on that distant seismograph?

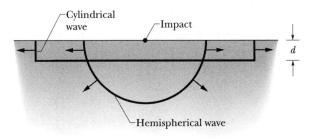

Fig. 18-48 Problem 68.

69. An avalanche of sand along some rare desert sand dunes can produce a booming that is loud enough to be heard 10 km away. The booming apparently results from a periodic oscillation of the sliding layer of sand—the layer's thickness expands and contracts. If the emitted frequency is 90 Hz, what are (a) the period of the thickness oscillation and (b) the wavelength of the sound?

70. Ultrasound, which consists of sound waves with frequencies above the human audible range, can be used to produce an image of the interior of a human body. Moreover, ultrasound can be used to measure the speed of the blood in the body; it does so by comparing the frequency of the ultrasound sent into the body with the frequency of the ultrasound reflected back to the body's surface by the blood. As the blood pulses, this detected frequency varies.

Suppose that an ultrasound image of the arm of a patient shows an artery that is angled at $\theta = 20°$ to the ultrasound's line of travel (Fig. 18-49). Suppose also that the frequency of the ultrasound reflected by the blood in the artery is increased by a maximum of 5495 Hz from the original ultrasound frequency of 5.000 000 MHz. (a) In Fig. 18-49, is the direction of the blood flow rightward or leftward? (b) The speed of sound in the human arm is 1540 m/s. What is the maximum speed of the blood? (*Hint:* The Doppler effect is caused by the component of the blood's velocity along the ultrasound's direction of travel.) (c) If angle θ were greater, would the reflected frequency be greater or less?

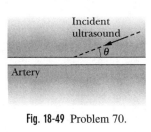

Fig. 18-49 Problem 70.

71. When the door of the Chapel of the Mausoleum in Hamilton, Scotland, is slammed shut, the last echo heard by someone standing just inside the door reportedly comes 15 s later. If that echo were due to a single reflection off a wall opposite the door, how far from the door would that wall be? If, instead, the wall is 25.7 m away, how many reflections (back and forth) correspond to the last echo?

72. In Fig. 18-50, sound waves A and B, both of wavelength λ, are initially in phase and traveling rightward, as indicated by the two rays. One of the waves is reflected from four surfaces but ends up traveling in its original direction. Expressed in terms of wavelength λ, what is the least value of the distance L in the figure that puts waves A and B exactly out of phase with each other after the reflections?

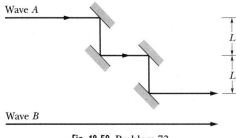

Fig. 18-50 Problem 72.

73. In a discussion of Doppler shifts of ultrasonic waves used in medical diagnosis, the authors remark: "For every millimeter per second that a structure in the body moves, the frequency of the incident ultrasonic wave is shifted approximately 1.30 Hz per MHz." What speed of ultrasonic waves in tissue do you deduce from this statement?

74. A sound wave of frequency 1000 Hz propagating through air has a pressure amplitude of 10.0 Pa. What are the (a) wavelength, (b) particle displacement amplitude, and (c) maximum particle speed? (d) An organ pipe open at both ends has this frequency as a fundamental. How long is the pipe?

75. A whistle used to call a dog has a frequency of 30 kHz. The dog, however, ignores it. The owner of the dog, who cannot hear sounds above 20 kHz, wants to use the Doppler effect to make certain that the whistle is working. She asks a friend to blow the whistle from a moving car while the owner remains stationary and listens. (a) How fast would the car have to move and in what direction for the owner to hear the whistle at 20 kHz? (b) Repeat for a whistle frequency of 22 kHz instead of 30 kHz.

76. Two identical tuning forks can oscillate at 440 Hz. A person is located somewhere on the line between them. Calculate the beat frequency as measured by this individual if (a) she is standing still and the tuning forks move in the same direction along the line at 3.00 m/s, and (b) the tuning forks are stationary and the listener moves along the line at 3.00 m/s.

77. A person on a railroad car blows a trumpet note at 440 Hz. The car is moving toward a wall at 20.0 m/s. Calculate (a) the frequency of the sound as received at the wall and (b) the frequency of the reflected sound arriving back at the trumpeter.

78. A submarine, near the water surface, moves north at speed 75.0 km/h in a northbound current of speed 30.0 km/h, where both speeds are relative to the ocean floor. The sub emits a sonar signal (sound wave), of frequency $f = 1000$ Hz and speed 5470 km/h, that is detected by a destroyer north of the sub. What is the difference between the detected frequency and f if the destroyer (a) drifts with the current at 30.0 km/h and (b) is stationary relative to the ocean floor?

79. The sound intensity is 0.0080 W/m² at a distance of 10 m from an isotropic point source of sound. (a) What is the power of the source? (b) What is the sound intensity at a distance of 5.0 m from the source? (c) What is the sound level in decibels at a distance of 10 m from the source?

80. A guitar player tunes the fundamental frequency of a guitar string to 440 Hz. (a) What will be the fundamental frequency if she then increases the tension in the string by 20%? (b) What is it if, instead, she decreases the length along which the string oscillates by sliding her finger from the tuning key one-third of the way down the string toward the bridge at the lower end?

81. A listener at rest (with respect to the air and the ground) hears a signal of frequency f_1 from a source moving toward him with a velocity of 15 m/s, due east. If the listener then moves toward the approaching source with a velocity of 25 m/s, due west, he hears a frequency f_2 that differs from f_1 by 37 Hz. What is the frequency of the source? (The speed of sound in air is 340 m/s.)

82. Figure 18-51 shows two point sources S_1 and S_2 that emit sound of wavelength $\lambda = 2.0$ m. The emissions are isotropic and in phase, and the separation between the sources is $d = 16$ m. At any point P on the x axis, the wave from S_1 and the wave from S_2 interfere. When P is very far away ($x \approx \infty$), what are (a) the phase difference between the arriving waves from S_1 and S_2 and (b) the type of interference they produce (approximately fully constructive or fully destructive)? (c) As we then move P along the x axis toward S_1, does the phase difference between the waves from S_1 and S_2 in-

crease or decrease? (d) Produce a table that gives the positions x at which the phase differences are 0, 0.50λ, 1.00λ, . . . , 2.50λ, and for each indicate the corresponding type of interference, either as fully destructive (fd) or fully constructive (fc).

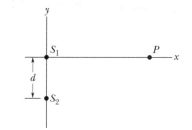

Fig. 18-51 Problem 82.

83. The average density of Earth's crust 10 km beneath the continents is 2.7 g/cm³. The speed of longitudinal seismic waves at that depth, found by timing their arrival from distant earthquakes, is 5.4 km/s. Use this information to find the bulk modulus of Earth's crust at that depth. For comparison, the bulk modulus of steel is about 16×10^{10} Pa.

84. A man strikes one end of a thin rod with a hammer. The speed of sound in the rod is 15 times the speed of sound in air. A woman, at the other end with her ear close to the rod, hears the sound of the blow twice with a 0.12 s interval between; one sound comes through the rod and the other comes through the air alongside the rod. If the speed of sound in air is 343 m/s, what is the length of the rod?

85. (a) A conical loudspeaker has a diameter of 15.0 cm. At what frequency will the wavelength of the sound it emits in air be equal to its diameter? Ten times its diameter? One-tenth its diameter? (b) Make the same calculations for a speaker of diameter 30.0 cm.

86. At a certain point, two waves produce pressure variations given by

$$\Delta p_1 = \Delta p_m \sin \omega t,$$
$$\Delta p_2 = \Delta p_m \sin(\omega t - \phi).$$

What is the pressure amplitude of the resultant wave at this point when $\phi = 0$, $\phi = \pi/2$, $\phi = \pi/3$, and $\phi = \pi/4$?

87. A certain loudspeaker system emits sound isotropically with a frequency of 2000 Hz and an intensity of 0.960 mW/m² at a distance of 6.10 m. Assume that there are no reflections. (a) What is the intensity at 30.0 m? At 6.10 m, what are (b) the displacement amplitude and (c) the pressure amplitude of the sound?

88. At a distance of 10 km, a 100 Hz horn, assumed to be an isotropic point source, is barely audible. At what distance would it begin to cause pain?

89. What is the bulk modulus of oxygen if 32.0 g of oxygen occupies 22.4 L and the speed of sound in the oxygen is 317 m/s?

90. Remarkably detailed images of transistors can be formed by an acoustic microscope. If the sound waves in such a microscope have a frequency of 4.2 GHz and a speed (in the liquid helium in which the specimen is immersed) of 240 m/s, what is their wavelength?

91. Two point sources, separated by a distance of 5.00 m, emit sound waves at the same amplitude and frequency (300 Hz), but the sources are exactly out of phase. At what points along the line between the sources do the interfering sound waves result in maximum oscillations of the air molecules? (*Hint:* One such point is midway between the sources. Try sketching how the molecules there tend to oscillate due to each wave.)

92. A hand-clap on stage in an amphitheater (Fig. 18-52) sends out sound waves that scatter from terraces of width $w = 0.75$ m. The sound returns to the stage as a periodic series of pulses, one from each terrace; the parade of pulses sounds like a played note. (a) Assuming that all the rays in Fig. 18-52 are horizontal, find the frequency at which the pulses return (that is, the frequency of the perceived note). (b) If the width w of the terraces were smaller, would the frequency be higher or lower?

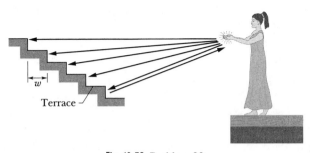

Fig. 18-52 Problem 92.

93. An audio engineer has designed a loudspeaker that is spherical and emits sound isotropically. The speaker emits sound at 10 W into a room with completely absorbent walls, floor, and ceiling (an *anechoic chamber*). (a) What is the intensity of the sound 3.0 m from the center of the source? (b) How does the wave amplitude A_4 at 4.0 m from the center of the source compare with the wave amplitude A_3 at 3.0 m?

94. You are standing at a distance D from an isotropic point source of sound. You walk 50.0 m toward the source and observe that the intensity of the sound has doubled. Calculate the distance D.

95. A violin string, oscillating in its fundamental mode, generates a sound wave with wavelength λ. By what multiple must the tension be increased if the string, still oscillating in its fundamental mode, is to generate a sound wave with wavelength $\lambda/2$?

96. In 1845, Buys Ballot first tested the Doppler effect for sound. He put a trumpet player on a flatcar drawn by a locomotive and another player ahead of the train near the tracks. If each player blew a 440 Hz note and there were 4.0 beats/s as they approached each other, what was the speed of the flatcar?

97. A siren emitting a sound of frequency 1000 Hz moves away from you toward the face of a cliff at a speed of 10 m/s. Take the speed of sound in air as 330 m/s. (a) What is the frequency of the sound you hear coming directly from the siren? (b) What is the frequency of the sound you hear reflected off the cliff? (c) What is the beat frequency between the two sounds? Is it perceptible (it must be less than 20 Hz to be perceived)?

98. A pipe 0.60 m long and closed at one end is filled with an unknown gas. The third lowest harmonic frequency for the pipe is 750 Hz. (a) What is the speed of sound in the unknown gas? (b) What is the fundamental frequency for this pipe when it is filled with the unknown gas?

99. Passengers in an auto traveling at 16 m/s toward the east hear a siren frequency of 950 Hz from an emergency vehicle approaching them from behind at a speed (relative to the air and ground) of 40 m/s. The speed of sound in air is 340 m/s. (a) What siren frequency does a passenger riding in the emergency vehicle hear? (b) What frequency do the passengers in the auto hear after the emergency vehicle passes them?

100. *Coffee-cup acoustics.* If you stir water in a ceramic or glass mug with a metal spoon, striking the inner wall with the spoon, you hear a certain resonant frequency f corresponding to a standing sound wave in the water. If you then stir in a powder, such as cocoa or coffee, the resonant frequency quickly and noticeably shifts to a different value f_{shift} before it gradually returns to its former value. The shift in resonant frequency is due to the change in the speed of sound in the water owing to the formation of air bubbles as the powder enters the water. The shift decreases as the air bubbles rise and pop open, until all the bubbles have disappeared and the frequency has returned to f.

(a) Does the frequency increase or decrease when you stir in the powder? (b) Show that the speed of sound v in the water–bubble mixture is related to the density ρ and volume V of the mixture by

$$\frac{1}{v^2} = \frac{\rho}{V}\frac{dV}{dp},$$

where dV/dp is the change in volume due to the change in pressure of a sound wave.

This volume change consists of a volume change dV_w in the water and a volume change dV_a in the air bubbles: $dV = dV_w + dV_a$. The volume V of the mixture is approximately the volume V_w of the water alone because the volume V_a of the air bubbles is small; that is, the ratio $r = V_a/V_w$ is small. Similarly, the density ρ of the mixture is approximately the density ρ_w of the water because the density ρ_a of air is so small.

(c) Show that

$$v = 1/[1/v_w^2 + r(\rho_w/\rho_a)/v_a^2]^{0.5},$$

where v_a is the speed of sound in air and v_w is the speed of sound in water. (d) Substituting appropriate values for the constants in this expression, show that

$$f_{\text{shift}}/f = 1/(1 + 15400r)^{0.5}.$$

In normal circumstances, r can range from 4.0×10^{-3} (many bubbles have just formed) to 0 (the bubbles have dissipated). (e) Graph f_{shift}/f versus r for this range. (f) From the graph, find the value of r that gives $f_{\text{shift}}/f = 1/3$. (Adapted from "The Hot Chocolate Effect," by Frank S. Crawford, *American Journal of Physics*, 1982, Vol. 50, pp. 398–404.)

Tutorial Problem

101. In this problem you will look at the Doppler effect and how motion by the source or detector affects the wavelength of sound in air and the apparent frequency heard at the detector. The Doppler effect is simpler when the relative motion between the source and the detector is along the line between them, and that is the only case that will be discussed quantitatively in this problem.

Consider a police car with a siren emitting sound at a frequency of 1200 Hz. (a) What does that statement mean? After all, the apparent frequency depends on the motion of the police car. (b) Draw a qualitative sketch showing the police car moving to the right and how the wavefronts are spaced in front of and behind the car. (c) If the police car were at rest, what would be the wavelength (in air) of the siren sound? Use reasonable values for any physical quantities needed to calculate this wavelength. (d) If the car were moving at a speed of 28 m/s, what would be the wavelength (in air) of the siren sound in front of the car and behind the car? (e) What would be the apparent frequency heard by a person standing behind the police car? Use the results of part (d) to make this calculation.

The sound of the siren reflects off a building directly in front of the police car and travels back to the left. The building then is the source of the reflected sound. (f) What are the wavelength of the reflected sound and the frequency of the sound "emitted" by the source (the building)? Explain your reasoning thoroughly, using complete sentences. (g) What is the apparent frequency of the reflected sound as heard by the person of part (e)? (h) What is the apparent frequency of the reflected sound as heard by the police in their speeding car? (i) What are (1) the apparent frequency of the direct siren sound and (2) the apparent frequency of the reflected siren sound as heard by someone in a car that is trailing the police car at a speed of 16 m/s?

Answers

(a) The statement means that the siren is oscillating 1200 times a second. The apparent frequency heard by an observer may be equal to, more than, or less than that oscillation frequency. The police in the police car hear a frequency equal to the oscillation frequency.

(b) The wavefronts are closer together in front of the car and farther apart behind the car.

(c) If the police car were at rest, the wavelength of the 1200 Hz sound would be

$$\lambda_0 = \frac{v}{f} = \frac{340 \text{ m/s}}{1200 \text{ Hz}} = 0.283 \text{ m},$$

where $v = 340$ m/s is the speed of sound in air.

(d) If the police car were moving, the wavelength in front of it would be reduced and would be

$$\lambda = \lambda_0 \left(\frac{v - v_S}{v} \right)$$

$$= (0.283 \text{ m}) \left(\frac{340 \text{ m/s} - 28 \text{ m/s}}{340 \text{ m/s}} \right) = 0.260 \text{ m}.$$

The wavelength behind the police car would be increased and would be

$$\lambda = \lambda_0 \left(\frac{v + v_S}{v} \right)$$

$$= (0.283 \text{ m}) \left(\frac{340 \text{ m/s} + 28 \text{ m/s}}{340 \text{ m/s}} \right) = 0.306 \text{ m}.$$

(e) The apparent frequency f' at the receiver may be found from the wavelength λ by

$$f' = \frac{v}{\lambda} = \frac{340 \text{ m/s}}{0.306 \text{ m}} = 1111 \text{ Hz}.$$

(f) The wavelength in air of the reflected sound is just the wavelength of the sound in front of the police car: 0.260 m, as calculated in part (d). This means that the frequency emitted by the (stationary) building is

$$f = \frac{v}{\lambda} = \frac{340 \text{ m/s}}{0.260 \text{ m}} = 1308 \text{ Hz}.$$

(g) The person hears the sound at its reflected frequency, 1308 Hz, since neither the source (the building) nor the detector (the person) is moving.

(h) The police car acts as a detector that moves toward a stationary source. The apparent frequency heard by the police is then

$$f' = \frac{v + v_{car}}{\lambda_{reflected}} = \frac{340 \text{ m/s} + 28 \text{ m/s}}{0.260 \text{ m}} = 1415 \text{ Hz}.$$

(i) (1) The motion of the police car tends to decrease the frequency of the siren's direct sound, but the motion of the trailing car tends to increase it. We can calculate the frequency that is actually heard by using the wavelength we calculated in part (d)—that is, the wavelength of the sound behind the police car. The frequency of the direct sound is

$$f_c = \frac{v + v_c}{\lambda} = \frac{340 \text{ m/s} + 16 \text{ m/s}}{0.306 \text{ m}} = 1163 \text{ Hz},$$

where the subscript refers to the trailing car. (2) The sound of the siren as reflected by the building is increased in apparent frequency by the motion of the car; its apparent frequency is then

$$f_{cr} = \frac{v + v_c}{\lambda} = \frac{340 \text{ m/s} + 16 \text{ m/s}}{0.260 \text{ m}} = 1369 \text{ Hz},$$

where we are now using the wavelength (0.260 m) of the reflected sound wave (hence, the extra subscript r).

19

Temperature, Heat, and the First Law of Thermodynamics

ADDITIONAL SAMPLE PROBLEMS

Sample Problem 19-8

The bulb of a gas thermometer is filled with nitrogen to a pressure of 120 kPa. What provisional value (see Fig. 19-6) would this thermometer yield for the boiling point of water, and what is the error in this value?

SOLUTION: From Fig. 19-6, the curve for nitrogen shows that, at 120 kPa, the provisional temperature for the boiling point of water would be about 373.44 K. The actual boiling point of water (found by extrapolation on Fig. 19-6) is 373.125 K. Thus, using the provisional temperature leads to an error of 0.315 K, or 315 mK.

Sample Problem 19-9

You are putting up an external wall measuring 2.0 m by 3.0 m and consisting of a layer of white pine of thickness $L_p = 1.9$ cm and a layer of rock wool. You want to put in enough rock wool so that the thermal conduction rate P_{cond} through the wall does not exceed 120 W during a winter night when the external temperature is 36 K below the indoor temperature.

(a) What thickness L_w of rock wool is needed?

SOLUTION: The **Key Idea** is that, if we assume the conduction through the wall is steady, the conduction rate through the rock wool must equal that through the pine. We then have the situation of Eq. 19-34, so the conduction rate is given by Eq. 19-36. Here we can write Eq. 19-36 as

$$P_{cond} = \frac{A(T_H - T_C)}{\dfrac{L_w}{k_w} + \dfrac{L_p}{k_p}}, \qquad (19\text{-}42)$$

where face area A is 6.0 m^2, the temperature difference $T_H - T_C$ is 36 K, and k_w and k_p are the thermal conductivities of the rock wool and the pine, respectively. Solving for L_w, using values for k_w and k_p from Table 19-6, and substituting known data, we find

$$L_w = k_w\left(\frac{A\,\Delta T}{P_{cond}} - \frac{L_p}{k_p}\right)$$

$$= (0.043 \text{ W/m} \cdot \text{K})\left[\frac{(6.0 \text{ m}^2)(36 \text{ K})}{120 \text{ W}} - \frac{0.019 \text{ m}}{0.11 \text{ W/m} \cdot \text{K}}\right]$$

$$= 0.070 \text{ m} = 7.0 \text{ cm}. \qquad \text{(Answer)}$$

(b) If rock wool is sold only in layers of 2.0 cm, what R-value (in metric units) would you expect to be listed on the packages?

SOLUTION: The **Key Idea** here is that the R-value depends on the thickness of the rock wool. We relate R to the thickness $L = 2.00$ cm with Eq. 19-33, finding

$$R = \frac{L}{k_w} = \frac{0.020 \text{ m}}{0.043 \text{ W/m} \cdot \text{K}} = 0.47 \text{ m}^2 \cdot \text{K/W}. \qquad \text{(Answer)}$$

Sample Problem 19-10

During an extended wilderness hike, you have a terrific craving for ice. Unfortunately, the air temperature drops to only 6.0°C each night—too high to freeze water. However, because a clear, moonless night sky acts like a blackbody radiator at a temperature of $T_s = -23°$C, perhaps you can make ice by letting a shallow layer of water radiate energy to such a sky. To start, you thermally insulate a container from the ground by placing a poorly conducting layer of, say, foam rubber or straw beneath it. Then you pour water into the container, forming a thin, uniform layer with mass $m = 4.5$ g, top surface area $A = 9.0$ cm^2, depth $d = 5.0$ mm, emissivity $\varepsilon = 0.90$, and initial temperature 6.0°C. Find the time required for the water to freeze via radiation. Can the freezing be accomplished during one night?

SOLUTION: One **Key Idea** is that the water cannot freeze at a temperature above the freezing point. Therefore, the radiation must first reduce the water temperature from 6.0°C to the freezing point of 0°C. Using Eq. 19-14 and Table 19-3, we find the associated energy loss to be

$$Q_1 = cm(T_f - T_i)$$
$$= (4190 \text{ J/kg} \cdot \text{K})(4.5 \times 10^{-3} \text{ kg})(0°C - 6.0°C)$$
$$= -113 \text{ J}.$$

Thus, 113 J must be radiated away by the water to drop its temperature to the freezing point.

A second **Key Idea** is that the water must next have an additional energy loss Q_2 for all of it to freeze. Using Eq. 19-16 ($Q = mL_F$), taking the value of L_F from Eq. 19-18 or Table 19-4, and inserting a minus sign to indicate an energy loss, we find

$$Q_2 = -mL_F = -(4.5 \times 10^{-3} \text{ kg})(3.33 \times 10^5 \text{ J/kg})$$
$$= -1499 \text{ J}.$$

The total required energy loss is thus

$$Q_{tot} = Q_1 + Q_2 = -113 \text{ J} - 1499 \text{ J} = -1612 \text{ J}.$$

A third **Key Idea** is that, while the water loses energy by radiating to the sky, it also absorbs energy radiated to it from the sky. In a total time t, we want the net energy of this exchange to be the energy loss Q_{tot}, so we want the power of this exchange to be

$$\text{power} = \frac{\text{net energy}}{\text{time}} = \frac{Q_{tot}}{t}. \quad (19\text{-}43)$$

The power of such an energy exchange is also the net rate P_{net} of thermal radiation, as given by Eq. 19-40, so the time t required for the energy loss to be Q_{tot} is

$$t = \frac{Q}{P_{net}} = \frac{Q}{\sigma \varepsilon A(T_s^4 - T^4)}. \quad (19\text{-}44)$$

Although the temperature T of the water decreases slightly while the water is cooling, we can approximate T as being the freezing point, 273 K. With $T_s = 250$ K, the denominator of Eq. 19-44 is

$$(5.67 \times 10^{-8} \text{ W/m}^2 \cdot \text{K}^4)(0.90)(9.0 \times 10^{-4} \text{ m}^2)$$
$$\times [(250 \text{ K})^4 - (273 \text{ K})^4] = -7.57 \times 10^{-2} \text{ J/s},$$

and Eq. 19-44 gives us

$$t = \frac{-1612 \text{ J}}{-7.57 \times 10^{-2} \text{ J/s}} = 2.13 \times 10^4 \text{ s} = 5.9 \text{ h}. \quad (\text{Answer})$$

Because t is less than a night, freezing water by having it radiate to the dark sky is feasible. In fact, in some parts of the world people used this technique long before the introduction of electric freezers.

QUESTIONS

13. Figure 19-44 shows three different arrangements of materials 1, 2, and 3 to form a wall. The thermal conductivities are $k_1 > k_2 > k_3$. The left side of the wall is 20 C° higher than the right side. Rank the arrangements according to (a) the (steady state) rate of energy conduction through the wall and (b) the temperature difference across material 1, greatest first.

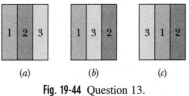

(a)   (b)   (c)

Fig. 19-44 Question 13.

14. The p-V diagram of Fig. 19-45 indicates four processes that take a gas between state A and state B. Rank the four processes according to (a) the work W done by the gas and (b) the energy Q transferred as heat between the gas and a thermal reservoir, most positive first, most negative last. (c) Rank the four processes according to the magnitude (or absolute value) of $Q - W$.

Fig. 19-45 Question 14.

15. Figure 19-46a gives the temperature along an x axis that extends directly through a wall consisting of three layers. The air temperature on one side of the wall differs from that on the other side; thermal conduction through the wall is constant (steady). (a) Rank the three layers according to their thermal conductivities, greatest first. (b) Figures 19-46b, c, and d are similar in intent but may be impossible. Indicate which are impossible and why. For any that are possible, rank the three layers of the wall according to their thermal conductivities, greatest first.

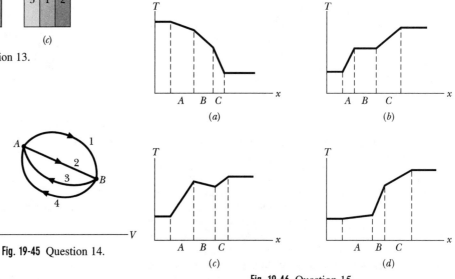

Fig. 19-46 Question 15.

16. A ball of surface temperature T is in thermal equilibrium with its environment. (a) Which of the curves in Fig. 19-47 gives the energy E radiated by the sphere versus time t? (b) Which curve gives the energy E absorbed by the sphere versus time t?

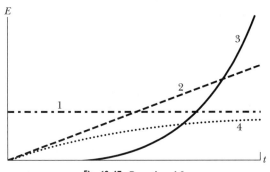

Fig. 19-47 Question 16.

17. *Organizing question:* A ball at temperature $+20°C$ is in an environment at temperature $-20°C$. The radius of the ball is r and its emissivity is 0.8. Which of the following expressions give the ball's net rate P_{net} of energy exchange with the environment due to thermal radiation? For each wrong expression, explain why it is wrong.
(a) $\sigma(0.8)\pi r^2[(-20°C)^4 - (20°C)^4]$
(b) $\sigma(0.8)4\pi r^2[(-20°C)^4 - (20°C)^4]$
(c) $\sigma(0.8)\pi r^2(-20°C - 20°C)^4$
(d) $\sigma(0.8)4\pi r^2(-20°C - 20°C)^4$
(e) $\sigma(0.8)\pi r^2[(273°C - 20°C)^4 - (273°C + 20°C)^4]$
(f) $\sigma(0.8)4\pi r^2[(273°C - 20°C)^4 - (273°C + 20°C)^4]$
(g) $\sigma(0.8)\pi r^2[(253\ K)^4 - (293\ K)^4]$
(h) $\sigma(0.8)4\pi r^2[(253\ K)^4 - (293\ K)^4]$
(i) $\sigma(0.8)\pi r^2(253\ K - 293\ K)^4$
(j) $\sigma(0.8)4\pi r^2(253\ K - 293\ K)^4$

18. Figure 19-48 shows a horizontal cross section (top view) of a square room surrounded on four sides by thick walls. The walls are all made of the same material and all have the same face area. They have thicknesses of either L, $2L$, or $3L$ as shown, and they are well insulated along their lengths. The faces forming the room are maintained at 5°C, and the conduction of energy outward through the walls is steady. Rank the walls according to the rate of conduction through them, greatest first.

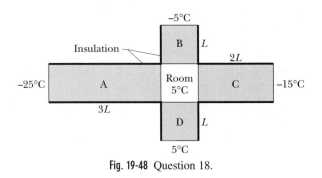

Fig. 19-48 Question 18.

19. The following pairs give the temperatures of an object and its environment, respectively, for three situations: (1) 300 K and

350 K; (2) 350 K and 400 K; and (3) 400 K and 450 K. Without computation, rank the situations according to the object's net rate of energy exchange P_{net}, greatest first.

20. A rod, initially at room temperature, is to be heated and cooled in six steps. The changes ΔL in its length, expressed in terms of some basic unit, will be $+7$, $+5$, $+3$, -4, -6, and -4. (a) Is the final temperature of the rod equal to room temperature, above it, or below it? (b) Is there a sequence of the steps such that the rod is at room temperature after one of the intermediate steps?

21. Rank the Celsius, Kelvin, and Fahrenheit temperature scales according to the heat required to raise the temperature of one gram of water by one degree on each scale, greatest first.

22. Suppose a block of wood and a block of metal are at the *same* temperature. When the blocks feel cold to your fingers, the metal feels colder than the wood; when the blocks feel hot to your fingers, the metal feels hotter than the wood. At what temperature will the blocks feel the same?

23. Three different materials of identical masses are placed, in turn, in a special oven where a material absorbs energy at a certain constant rate. During the heating process, each material begins in the liquid state and ends in the gaseous state; Fig. 19-49 gives the temperature T versus time t for the three materials. (a) For material 1, is the specific heat for the liquid state greater than or less than that for the gaseous state? Rank the three materials according to (b) their boiling-point temperatures, (c) their specific heats in the liquid state, and (d) their heats of vaporization, all greatest first.

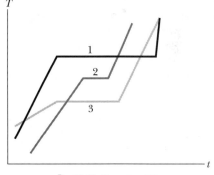

Fig. 19-49 Question 23.

24. In four experiments, a material A at a particular low temperature T_C and a material B at a particular high temperature T_H are placed in an isolated and insulated container. When they reach thermal equilibrium with each other (no phase change occurs), their final common temperature T is measured. The masses m_A and m_B and specific heats c_A and c_B of the materials are given in the table. Without written calculation, rank the four experiments according to the final temperature T, greatest first.

Experiment	m_A	c_A	m_B	c_B
1	m	c	m	c
2	m	c	$2m$	c
3	m	$2c$	m	c
4	$2m$	c	m	$2c$

25. Three spheres of the same radius and emissivity have different (constant) surface temperatures $T_1 > T_2 > T_3$. Three of the curves in Fig. 19-50 pertain to the energy radiated by the surfaces versus time t. Which curve corresponds to which surface temperature?

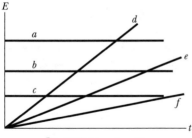

Fig. 19-50 Question 25.

DISCUSSION QUESTIONS

26. Is temperature a microscopic or macroscopic concept?

27. Are there physical quantities other than temperature that tend to equalize if two different systems are joined?

28. A piece of ice and a warmer thermometer are suspended in an insulated evacuated enclosure so that they are not in contact. Why does the thermometer reading decrease for a time?

29. Let p_3 be the pressure in the bulb of a constant-volume gas thermometer when the bulb is at the triple-point temperature of 273.16 K. Also let p be the pressure when the bulb is at room temperature. You are given three constant-volume gas thermometers: For A the gas is oxygen and $p_3 = 20$ cm Hg; for B the gas is also oxygen but $p_3 = 40$ cm Hg; for C the gas is hydrogen and $p_3 = 30$ cm Hg. The measured values of p for the three thermometers are p_A, p_B, and p_C. (a) An approximate value of the room temperature T can be obtained with each of the thermometers using

$$T_A = 273.16 \text{ K} \frac{p_A}{20 \text{ cm Hg}}; \quad T_B = 273.16 \text{ K} \frac{p_B}{40 \text{ cm Hg}};$$

$$T_C = 273.16 \text{ K} \frac{p_C}{30 \text{ cm Hg}}.$$

Mark each of the following statements true or false. (1) With the method described, all three thermometers will give the same value of T. (2) The two oxygen thermometers will agree with each other but not with the hydrogen thermometer. (3) Each of the three will give a different value of T. (b) In the event that there is disagreement among the three thermometers, explain how you would change the method of using them to cause all three to give the same value of T.

30. A student, when told that the temperature at the center of the Sun was thought to be about 1.5×10^7 degrees, asked whether that was on the Celsius or the Kelvin scale. How would you answer? How would you reply if he had asked whether it was on the Celsius or the Fahrenheit scale?

31. The editor-in-chief of a well-known business magazine, discussing possible warming effects associated with the increasing concentration of carbon dioxide in Earth's atmosphere, wrote: "The polar regions might be three times warmer than now, ..." What

do you suppose he meant? (From "Warmth and Temperature: A Comedy of Errors," by Albert A. Bartlett, *The Physics Teacher*, November 1984.)

32. Although the absolute zero of temperature seems to be experimentally unattainable, temperatures as low as 0.000 000 002 K have been achieved in the laboratory. Isn't this low enough for all practical purposes? Why would physicists (as indeed they do) strive to obtain still lower temperatures?

33. Can a temperature be assigned to a vacuum?

34. Does our "temperature sense" have a built-in sense of direction; that is, does hotter necessarily mean higher temperature, or is this just an arbitrary convention? Anders Celsius (for whom the scale is named) originally chose the boiling point of water as 0°C and the freezing point as 100°C.

35. Many medicine labels inform the user to store below 86°F. Why 86? (*Hint:* Change to Celsius.)

36. How would you suggest measuring the temperature of (a) the Sun, (b) Earth's upper atmosphere, (c) an insect, (d) the Moon, (e) the ocean floor, and (f) liquid helium?

37. Is one gas any better than another for purposes of a standard constant-volume gas thermometer? What properties are desirable in a gas for such purposes?

38. State some objections to using water-in-glass as a thermometer. Is mercury-in-glass an improvement?

39. Explain why the column of mercury first descends and then rises when a mercury-in-glass thermometer is put in a flame.

40. What do the Celsius and Fahrenheit temperature scales have in common?

41. What are the dimensions of α, the coefficient of linear expansion? Does the value of α depend on the unit of length used? When degrees Fahrenheit are used instead of degrees Celsius as the unit of temperature change, how does the numerical value of α change?

42. A metal ball can pass through a metal ring. When the ball is heated, however, it gets stuck in the ring. What would happen if the ring, rather than the ball, were heated?

43. Two strips, one of iron and one of zinc, are riveted together side by side to form a straight bar that curves when heated. Why is the iron on the inside of the curve?

44. Explain how the period of a pendulum clock can be kept constant with temperature by attaching vertical tubes of mercury to the bottom of the pendulum.

45. Why should a chimney be freestanding—that is, not part of the structural support of the house?

46. Water expands when it freezes. Can we define a coefficient of volume expansion for the freezing process?

47. Explain why the expansion of a liquid in a glass bulb during heating is not the true expansion of the liquid.

48. Temperature and heat are often confused, as in, "bake in an oven at moderate heat." By example, distinguish between these two concepts as carefully as you can.

49. Give an example of a process in which no energy is transferred to or from a system as heat but the temperature of the system changes.

50. Can thermal energy be considered a form of stored (or potential) energy? Would such an interpretation contradict the concept of heat as energy in the process of transfer because of a temperature difference?

51. Energy can be added as heat to a substance without causing the temperature of the substance to rise. Does this contradict the concept of heat as energy in the process of transfer because of a temperature difference?

52. Explain the fact that the presence of a large body of water nearby, such as a sea or ocean, tends to moderate the temperature extremes of the climate on adjacent land.

53. An electric fan not only does not cool the air it circulates but heats it slightly. How then can it cool you?

54. Both heat conduction and wave propagation involve the transfer of energy. Is there any difference in principle between these two phenomena? Explain.

55. When a hot object warms a cool one, are their temperature changes equal in magnitude? Give examples.

56. How can you best use a spoon to cool a cup of coffee? Stirring—which involves doing work—would seem to heat the coffee rather than to cool it.

57. How does a layer of snow protect plants during cold weather? During freezing spells, citrus growers in Florida often spray their fruit with water, hoping that it will freeze. How does that help?

58. Explain the wind-chill effect that radio and TV meteorologists mention during cold weather.

59. You put your hand in a hot oven to remove a casserole and burn your fingers on the hot dish. However, the air in the oven is at the same temperature as the casserole dish but it does not burn your fingers. Why not?

60. Why is thicker insulation used in an attic than in the walls of a house?

61. Is ice always at 0°C? Can it be colder? Can it be warmer? What about an ice–water mixture?

62. On a winter day the temperature of the inside surface of a house wall is much lower than room temperature, and that of the wall's outside surface is much higher than the outdoor temperature. Explain.

63. What requirements for thermal conductivity, specific heat, and coefficient of expansion should be satisfied by a material to be used in cooking utensils?

64. Can you tell whether the internal energy of a body is due to an energy transfer as heat or as work?

65. If only the pressure and volume of a system are given, is the temperature always uniquely determined?

66. Discuss the process by which water freezes, from the point of view of the first law of thermodynamics. Remember that ice occupies a greater volume than an equal mass of water.

67. A thermos bottle contains coffee. The thermos bottle is vigorously shaken. Consider the coffee as the system. (a) Does its temperature rise? (b) Has energy been added to it as heat? (c) Has work been done on it? (d) Has its internal energy changed?

68. Can energy be transferred through matter by radiation? If so, give an example. If not, explain why.

69. Why does stainless steel cookware often have a layer of copper or aluminum on the bottom?

70. Consider that energy can be transferred by convection and radiation, as well as by conduction, and explain why a thermos bottle is double-walled, evacuated, and silvered (like a mirror).

71. A lake freezes first at its upper surface. Is convection involved? What about conduction and radiation?

72. You put two uncovered pails of water, one containing hot water and one containing an equal amount of warm water, outside in below-freezing weather. The pail with the hot water may actually develop ice first. Why? What would happen if you covered the pails?

EXERCISES & PROBLEMS

69. Three metal rods, one copper, one aluminum, and one brass, are each 6.00 cm long and 1.00 cm in diameter. These rods are placed end to end, with the aluminum between the other two. The free ends of the copper and brass rods are maintained at the boiling point and the freezing point of water, respectively. Find the steady-state temperatures of the copper–aluminum junction and the aluminum–brass junction. The thermal conductivity of brass is 109 W/m · K.

70. A large cylindrical water tank with a bottom 1.7 m in diameter is made of iron boilerplate 5.2 mm thick. Water in the tank is heated from below by a gas burner that is able to maintain a temperature difference of 2.3 C° between the top and bottom surfaces of the bottom plate. How much energy is conducted through that plate in 5.0 min? (Iron has a thermal conductivity of 67 W/m · K.)

71. Suppose that you intercept 5.0×10^{-3} of the energy radiated by a hot sphere that has a radius of 0.020 m, an emissivity of 0.80, and a surface temperature of 500 K. How much energy do you intercept in 2.0 min?

72. In the extrusion of cold chocolate from a tube (see Problem 63 in Chapter 15 in this supplement), work is done on the chocolate by the applied pressure that is due to a ram forcing the chocolate through the tube. The work per unit mass of extruded chocolate is equal to p/ρ, where p is the difference between the applied pressure and the pressure where the chocolate emerges from the tube, and ρ is the density of the chocolate. This work appears to result in the melting of cocoa fats in the chocolate rather than in increasing the temperature of the chocolate. These fats have a heat of fusion of 150 kJ/kg. Assume that all of the work goes into that melting and that these fats make up 30% of the chocolate's mass. What percentage of the fats melt during the extrusion if $p = 5.5$ MPa and $\rho = 1200$ kg/m³?

73. A sample of gas expands from 1.0 m³ to 4.0 m³ along path B in the p-V diagram in Fig. 19-51. It is then compressed back to 1.0 m³ along either path A or path C. Compute the net work done by the gas for the complete cycle in each case.

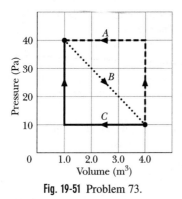

Fig. 19-51 Problem 73.

74. A sample of gas undergoes a transition from an initial state a to a final state b by three different paths (processes), as shown in the p-V diagram in Fig. 19-52. The energy transferred to the gas as heat in process 1 is $10p_iV_i$. In terms of p_iV_i, what are (a) the energy transferred to the gas as heat in process 2 and (b) the change in internal energy that the gas undergoes in process 3?

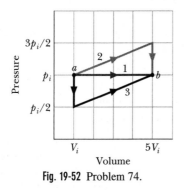

Fig. 19-52 Problem 74.

75. Figure 19-53a shows a cylinder containing gas and closed by a movable piston. The cylinder is kept submerged in an ice–water mixture. The piston is *quickly* pushed down from position 1 to position 2 and then held at position 2 until the gas is again at the temperature of the ice–water mixture; it then is *slowly* raised back to position 1. Figure 19-53b is a p-V diagram for the process. If 100 g of ice is melted during the cycle, how much work has been done *on* the gas?

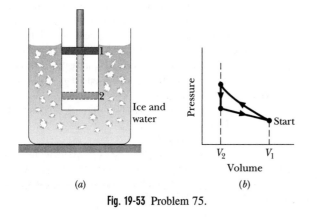

Fig. 19-53 Problem 75.

76. An object of mass 6.00 kg falls through a height of 50.0 m and, by means of a mechanical linkage, rotates a paddle wheel that stirs 0.600 kg of water. Assume that the initial gravitational potential energy of the object is fully transferred to thermal energy of the water, which is initially at 15.0°C. What is the temperature rise of the water?

77. A steel rod has a length of exactly 20 cm at 30°C. How much longer is it at 50°C?

78. The Stanford linear accelerator contains hundreds of brass disks tightly fitted into a steel tube. The system was assembled by cooling the disks in dry ice (at -57.00°C) to enable them to slide into the close-fitting tube. If the diameter of a disk is 80.00 mm at 43.00°C, what is its diameter in the dry ice?

79. At 20°C, a brass cube has an edge length of 30 cm. What is the increase in the cube's surface area when it is heated from 20°C to 75°C?

80. If your doctor tells you that your temperature is 310 kelvins, should you worry? Explain your answer.

81. (a) The temperature of the surface of the Sun is about 6000 K. Express this on the Fahrenheit scale. (b) Express normal human body temperature, 98.6°F, on the Celsius scale. (c) In the continental United States, the lowest officially recorded temperature is -70°F at Rogers Pass, Montana. Express this on the Celsius scale. (d) Express the normal boiling point of oxygen, -183°C, on the Fahrenheit scale. (e) At what Celsius temperature would you find a room to be uncomfortably warm?

82. Icebergs in the North Atlantic present hazards to shipping, causing the lengths of shipping routes to be increased by about 30% during the iceberg season. Attempts to destroy icebergs include planting explosives, bombing, torpedoing, shelling, ramming, and coating with black soot. Suppose that direct melting of the iceberg, by placing heat sources in the ice, is tried. How much heat is required to melt 10% of an iceberg with a mass of 200 000 metric tons? (Use 1 metric ton = 1000 kg.)

83. Soon after Earth was formed, heat released by the decay of radioactive elements raised the average internal temperature from 300 to 3000 K, at about which value it remains today. Assuming an average coefficient of volume expansion of 3.0×10^{-5} K^{-1}, by how much has the radius of Earth increased since the planet was formed?

84. A glass window pane is exactly 20 cm by 30 cm at 10°C. By how much has its area increased when its temperature is 40°C, assuming that it can expand freely?

85. The thermal conductivity of Pyrex glass at temperature 0°C is 2.9×10^{-3} cal/cm · C° · s. Express this quantity in (a) W/m · K and (b) Btu/ft · F° · h. (c) Using your result in (a), find the R-value for a Pyrex sheet of thickness 6.4 mm.

86. By how much does the volume of an aluminum cube 5.00 cm on an edge increase when the cube is heated from 10.0°C to 60.0°C?

87. A cube of edge length 6.0×10^{-6} m, emissivity 0.75, and temperature -100°C floats in an environment of temperature -150°C. What is the cube's net thermal radiation transfer rate?

88. A steel rod at 25.0°C is bolted securely at both ends and then cooled. At what temperature will it rupture? Use Table 13-1.

89. The timing of a certain kind of electric watch is governed by the oscillations of a small quartz tuning fork. The frequency of the fork is inversely proportional to the square root of the length of the fork. What is the fractional gain or loss in time for a watch with a quartz tuning fork 8.00 mm long at (a) $-40.0°F$ and (b) $+120°F$ if it keeps perfect time at $25.0°F$?

90. How many 20 g ice cubes, whose initial temperature is $-10°C$, must be added to 1.0 L of hot tea, whose initial temperature is 90°C, for the final mixture to have a temperature of 10°C? Assume that all the ice is melted in the final mixture, the specific heat of tea is the same as that of water, and the system is isolated.

91. A sample of gas expands from an initial pressure and volume of 10 Pa and 1.0 m³ to a final volume of 2.0 m³. During the expansion, the pressure and volume are related by the equation $p = aV^2$, where $a = 10$ N/m⁸. Determine the work done by the gas during this expansion.

92. A composite bar of length $L = L_1 + L_2$ is made from a bar of material 1 and length L_1 attached to a bar of material 2 and length L_2, as shown in Fig. 19-54. (a) Show that the coefficient of linear expansion α for this composite bar is given by $\alpha = (\alpha_1 L_1 + \alpha_2 L_2)/L$. (b) Using steel and brass, design such a composite bar whose length is 52.4 cm and whose coefficient of linear expansion is $13.0 \times 10^{-6}/C°$.

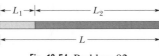

Fig. 19-54 Problem 92.

93. A solid cylinder of radius $r_1 = 2.5$ cm, length $h_1 = 5.0$ cm, emissivity 0.85, and temperature 30°C is suspended in an environment of temperature 50°C. (a) What is the cylinder's net thermal radiation transfer rate P_1? (b) If the cylinder is stretched until its radius is $r_2 = 0.50$ cm, its net thermal radiation transfer rate becomes P_2. What is the ratio P_2/P_1?

94. A *flow calorimeter* is a device used to measure the specific heat of a liquid. Energy is added as heat at a known rate to a stream of the liquid as it passes through the calorimeter at a known rate. Measurement of the resulting temperature difference between the inflow and the outflow points of the liquid stream enables us to compute the specific heat of the liquid. Suppose a liquid of density 0.85 g/cm³ flows through a calorimeter at the rate of 8.0 cm³/s. When energy is added at the rate of 250 W by means of an electric heating coil, a temperature difference of 15 C° is established in steady-state conditions between the inflow and the outflow points. What is the specific heat of the liquid?

95. Show that the temperature T_X at the interface of a compound slab (Fig. 19-19) is given by

$$T_X = \frac{R_1 T_H + R_2 T_C}{R_1 + R_2}.$$

96. Figure 19-55a shows a closed cycle for a gas. From c to b, 40 J is transferred from the gas as heat. From b to a, 130 J is transferred from the gas as heat, and the magnitude of the work done by the gas is 80 J. From a to c, 400 J is transferred to the gas

as heat. What is the work done by the gas from a to c? (*Hint:* You need to supply the plus and minus signs for the given data.)

97. Figure 19-55b shows a closed cycle for a gas. The change in internal energy along the path from c to a is -160 J. The energy transferred to the gas as heat is 200 J along the path from a to b, and 40 J along the path from b to c. How much work is done by the gas along (a) path *abc* and (b) path *ab*?

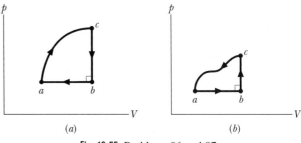

Fig. 19-55 Problems 96 and 97.

98. A thermometer of mass 0.0550 kg and of specific heat 0.837 kJ/kg · K reads 15.0°C. It is then completely immersed in 0.300 kg of water, and it comes to the same final temperature as the water. If the thermometer then reads 44.4°C, what was the temperature of the water before insertion of the thermometer?

99. (a) What is the coefficient of linear expansion of aluminum per Fahrenheit degree? (b) Use your answer in (a) to calculate the change in length of a 6.0 m aluminum rod if the rod is heated from 40°F to 95°F.

100. A 1.28-m-long vertical glass tube is half filled with a liquid at 20°C. How much will the height of the liquid column change when the tube is heated to 30°C? Take $\alpha_{glass} = 1.0 \times 10^{-5}/K$ and $\beta_{liquid} = 4.0 \times 10^{-5}/K$.

101. A 15.0 kg sample of ice is initially at a temperature of $-20.0°C$. Then 7.0×10^6 J is added as heat to the sample, which is otherwise isolated. What then is the sample's temperature?

102. A cube with edge length 2.0×10^{-5} m is at a temperature of 50°C and has an emissivity of 0.80. It is suspended in an environment that has a temperature of 20°C. What is the net rate at which its energy changes by thermal radiation?

103. Calculate the specific heat of a metal from the following data. A container made of the metal has a mass of 3.6 kg and contains 14 kg of water. A 1.8 kg piece of the metal initially at a temperature of 180°C is dropped into the water. The container and water initially have a temperature of 16.0°C, and the final temperature of the entire system is 18.0°C.

104. A block of ice, at its melting point and of initial mass 50.0 kg, slides along a horizontal surface of thermally insulated material, starting at a speed of 5.38 m/s and finally coming to rest after traveling 28.3 m. Compute the mass of ice melted as a result of friction between the block and the surface. (Assume that all the energy lost by mechanical energy due to the friction is transferred to thermal energy of the block of ice.)

105. A 3.00 kg piece of copper at temperature 70.0°C is placed in 4.00 kg of water at temperature 10.0°C. The copper–water system is isolated. What is the final (equilibrium) temperature of the system?

106. In a certain solar house, energy from the Sun is stored in barrels filled with water. In a particular winter stretch of five cloudy days, 1.00×10^6 kcal is needed to maintain the inside of the house at 22.0°C. Assuming that the water in the barrels is at 50.0°C and that the water has a density of 1.00×10^3 kg/m³, what volume of water is required?

107. An athlete needs to lose weight and decides to do it by "pumping iron." (a) How many times must an 80.0 kg weight be lifted a distance of 1.00 m in order to burn off 1 lb of fat, assuming that that much fat is equivalent to 3500 Cal? (b) If the weight is lifted once every 2.00 s, how long does the task take?

108. On an X temperature scale, water freezes at −125.0°X and boils at 375.0°X. On a Y temperature scale, water freezes at −70.00°Y and boils at −30.00°Y. A temperature of 50.00°Y corresponds to what temperature on the X scale?

109. An idealized representation of the air temperature as a function of distance from a single-pane window on a calm winter day is shown in Fig. 19-56. The window dimensions are 60 cm × 60 cm × 0.50 cm. Assume that energy is conducted along a path that is perpendicular to the window, from points 8.0 cm from the window on one side to points 8.0 cm from it on the other side. (a) At what rate is energy conducted through the window? (*Hint:* The temperature drop across the window glass is very small.) (b) Estimate the difference in temperature between the inner and outer glass surfaces.

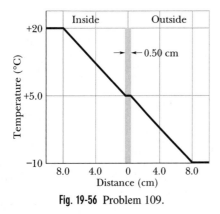

Fig. 19-56 Problem 109.

110. A rectangular plate of glass initially has the dimensions 0.200 m by 0.300 m. The coefficient of linear expansion for the glass is 9.00×10^{-6}/K. What is the change in the plate's area if its temperature is increased by 20 K?

Tutorial Problem

111. A system consists of 200 g of water that is initially ice at −20°C. Energy is added to the system until the water is converted to steam at 100°C. (The system is otherwise isolated from its environment.) (a) Describe in words the four processes in which energy must be added to the system during this heating. For each, state the equation that gives the required energy, and find that energy. (b) What is the total energy required for the heating, and what

fraction of the total is added in each process? (c) Graph the energy added as heat versus temperature. Identify which portions of the graph correspond to which process. (d) Suppose that the energy is provided by an electric heater at the rate of 240 W. Graph the temperature of the system versus time.

Answers

(a) *Process 1:* warming the ice to the melting temperature (0°C) by adding energy in the amount

$$mc_{ice}\,\Delta T = (0.200 \text{ kg})(2220 \text{ J/kg} \cdot \text{C°})[0°\text{C} - (-20°\text{C})]$$
$$= 8.88 \text{ kJ}.$$

Process 2: melting the ice into liquid water at the melting temperature by adding energy in the amount

$$mL_F = (0.200 \text{ kg})(333 \text{ kJ/kg}) = 66.6 \text{ kJ}.$$

Process 3: warming the liquid water from its freezing temperature to the vaporization temperature (100°C) by adding energy in the amount

$$mc_{liq}\,\Delta T = (0.200 \text{ kg})(4190 \text{ J/kg} \cdot \text{C°})(100°\text{C} - 0°\text{C})$$
$$= 83.8 \text{ kJ}.$$

Process 4: vaporizing the water to form steam at the vaporization temperature by adding energy in the amount

$$mL_V = (0.200 \text{ kg})(2256 \text{ kJ/kg}) = 451 \text{ kJ}.$$

(b) The total energy added as heat is $(8.88 + 66.6 + 83.8 + 451)$ kJ = 610 kJ.

Process 1: (8.88 kJ)/(610 kJ) = 0.015 or 1.5%;

Process 2: (66.6 kJ)/(610 kJ) = 0.11 or 11%;

Process 3: (83.8 kJ)/(610 kJ) = 0.14 or 14%;

Process 4: (451 kJ)/(610 kJ) = 0.74 or 74%.

(c) From −20°C to 0°C, plot the function given in (a) for process 1. Then plot a vertical line to account for process 2. Then, from 0°C to 100°C, plot the function given in (a) for process 3. Then plot a vertical line to account for process 4.

(d) The time taken by each process is as follows:

Process 1: (8.88 kJ)/(0.240 kW) = 37 s;

Process 2: (66.6 kJ)/(0.240 kW) = 278 s;

Process 3: (83.8 kJ)/(0.240 kW) = 349 s;

Process 4: (451 kJ)/(0.240 kW) = 1880 s.

(The total time is 2540 s, which equals the sum of the individual times and is also found by dividing the total heat 610 kJ by the power 0.240 kW.) Now for the graph of temperature T versus time t:

   Process 1 causes a linear increase in temperature for 37 s;
   Process 2 causes no change in temperature for 278 s;
   Process 3 causes a linear increase in temperature for 349 s;
   Process 4 causes no change in temperature for 1880 s.

20
The Kinetic Theory of Gases

ADDITIONAL SAMPLE PROBLEMS

Sample Problem 20-10

(a) What are the average translational kinetic energies K_{avg} of the oxygen molecules and nitrogen molecules in the air that you breathe? Assume that the air is an ideal gas at a temperature of 300 K, which is the traditional value taken for room temperature (although that is a bit warm).

SOLUTION: The **Key Idea** here is that, for an ideal gas, K_{avg} depends only on the temperature and not on the type of molecule. From Eq. 20-24, the value of K_{avg} for both oxygen and nitrogen molecules is

$$K_{avg} = \tfrac{3}{2}kT = (\tfrac{3}{2})(1.38 \times 10^{-23} \text{ J/K})(300 \text{ K})$$
$$= 6.21 \times 10^{-21} \text{ J}. \qquad \text{(Answer)}$$

(b) The mass of an oxygen molecule differs from that of a nitrogen molecule. Do the oxygen molecules and the nitrogen molecules in air have the same rms speed v_{rms}? Use the value of K_{avg} to find v_{rms}.

SOLUTION: The **Key Idea** here is that v_{rms} is related to K_{avg} by Eq. 20-23:

$$K_{avg} = \tfrac{1}{2}mv_{rms}^2. \qquad (20\text{-}65)$$

Although the two types of molecules have the same K_{avg}, they have different masses m and thus their rms speeds must differ.

To find their rms speeds, we solve Eq. 20-65 for v_{rms} and then substitute for molecular mass m with Eq. 20-4 ($M = mN_A$), because molar masses M can be found in Appendix F. We obtain

$$v_{rms} = \sqrt{\frac{2K_{avg}N_A}{M}}. \qquad (20\text{-}66)$$

Caution: Any molar mass listed in Appendix F is the mass of a mole of monatomic molecules (that is, single atoms). In the air that you breathe, oxygen and nitrogen are diatomic molecules (O_2 and N_2)—each molecule consists of two atoms. Thus, we must double the listed molar masses, finding that the molar mass M_{O_2} of diatomic oxygen is 2(15.9994 g/mol) $\approx$ 0.0320 kg/mol and the molar mass M_{N_2} of diatomic nitrogen is 2(14.0067 g/mol) $\approx$ 0.0280 kg/mol.

Now we can substitute known data into Eq. 20-66. For oxygen we find

$$v_{rms} = \sqrt{\frac{2(6.21 \times 10^{-21} \text{ J})(6.02 \times 10^{23} \text{ mol}^{-1})}{0.0320 \text{ kg/mol}}}$$
$$= 483 \text{ m/s}. \qquad \text{(Answer)}$$

Similarly, for nitrogen we find

$$v_{rms} = 517 \text{ m/s}. \qquad \text{(Answer)}$$

The nitrogen molecules, with their smaller mass (and the same average kinetic energy), have the greater rms speed.

Sample Problem 20-11

When 10.61 J is added as heat Q to a particular ideal gas at a constant pressure of 1.01×10^5 Pa, the volume of the gas increases by 3.00×10^{-5} m^3.

(a) What is the change ΔE_{int} in the internal energy of the gas?

SOLUTION: First note that we cannot simply use Eq. 20-45 ($\Delta E_{int} = nC_V \Delta T$) because we do not know the number of moles n or the change ΔT in the temperature. Also, because we do not know whether the gas is monatomic, diatomic, or polyatomic, we do not know C_V.

However, we can use the **Key Idea** that ΔE_{int} is related to the given Q by the first law of thermodynamics ($\Delta E_{int} = Q - W$). Because the pressure is constant, we can use Eq. 20-16 to find the work W done by the gas as it expands:

$$W = p \, \Delta V = (1.01 \times 10^5 \text{ Pa})(3.00 \times 10^{-5} \text{ m}^3)$$
$$= 3.03 \text{ J}.$$

The first law of thermodynamics then gives us

$$\Delta E_{int} = Q - W = 10.61 \text{ J} - 3.03 \text{ J}$$
$$= 7.58 \text{ J}. \qquad \text{(Answer)}$$

(b) Does the gas consist of monatomic, diatomic, or polyatomic molecules?

SOLUTION: The **Key Idea** here is that we can determine the type of molecule if we know the molar specific heats C_V and C_p of the gas. Unfortunately, we lack enough information to use either Eq. 20-46

$(Q = nC_p \, \Delta T)$ or Eq. 20-45 $(\Delta E_{int} = nC_V \, \Delta T)$ to evaluate C_V and C_p. However, we can also determine the type of molecule if we know the ratio C_p/C_V. To get this ratio, we divide Eq. 20-46 by Eq. 20-45,

$$\frac{Q}{\Delta E_{int}} = \frac{nC_p \, \Delta T}{nC_V \, \Delta T}.$$

Solving for C_p/C_V, we then find

$$\frac{C_p}{C_V} = \frac{Q}{\Delta E_{int}} = \frac{10.61 \text{ J}}{7.58 \text{ J}} = 1.4.$$

From the two rightmost columns of Table 20-3, we see that diatomic molecules have such a ratio. Thus, the gas here consists of diatomic molecules.

Sample Problem 20-12

An ideal monatomic gas with an initial pressure p_0 is allowed to expand freely until its volume is 2.00 times its initial volume. Next, the gas is slowly and adiabatically compressed back to its initial volume. What then is its pressure?

SOLUTION: To find the pressure p_2 of the gas at the end of the compression, we first must find its pressure p_1 at the beginning of the compression, which is the pressure at the end of the free expansion. Thus, we must first consider the free expansion.

One **Key Idea** here is that when a gas undergoes a free expansion, its temperature cannot change, so its pressure and volume are related by Eq. 20-63 $(p_iV_i = p_fV_f)$. Here we have

$$p_0V_0 = p_1V_1,$$

which gives us

$$p_1 = p_0 \frac{V_0}{V_1} = p_0 \frac{V_0}{2.00V_0} = \tfrac{1}{2}p_0.$$

A second **Key Idea** is that when a gas undergoes an adiabatic

process, we know from Eq. 20-53 that the product pV^γ does not change. Thus, for the adiabatic compression from V_1 to volume $V_2 \, (= V_0)$, we have

$$p_1V_1^\gamma = p_2V_2^\gamma,$$

or

$$p_2 = p_1\left(\frac{V_1}{V_2}\right)^\gamma. \qquad (20\text{-}67)$$

To evaluate γ, we use C_p and C_V listed in Table 20-3 for a monatomic gas, finding

$$\gamma = \frac{C_p}{C_V} = \frac{\tfrac{5}{2}R}{\tfrac{3}{2}R} = 1.667.$$

Substituting this value, $p_1 = \tfrac{1}{2}p_0$, $V_1 = 2.00V_0$, and $V_2 = V_0$ into Eq. 20-67 gives us

$$p_2 = \tfrac{1}{2}p_0\left(\frac{2.00V_0}{V_0}\right)^{1.667} = 1.59p_0. \qquad \text{(Answer)}$$

QUESTIONS

13. *Organizing question:* The dot in Fig. 20-25a represents the initial state of a gas, and the vertical line through the dot divides the p-V diagram into regions 1 and 2. For the following processes, determine whether the work W done by the gas is positive, negative, or zero: (a) the gas moves up along the vertical line, (b) it moves down along the vertical line, (c) it moves to anywhere in region 1, and (d) it moves to anywhere in region 2.

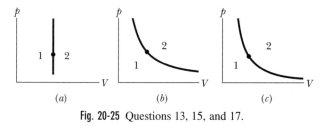

Fig. 20-25 Questions 13, 15, and 17.

14. An instrument that can monitor molecular speeds in a narrow range dv is placed in a gas of molecules. Initially no molecules have speeds in the monitored range; then the number of molecules in the range gradually increases as the temperature of the gas is changed. (a) Is the monitored range initially above or below the root-mean-square speed of the molecules? (b) Is the gas being cooled or warmed during the observations?

15. *Organizing question:* The dot in Fig. 20-25b represents the initial state of a gas, and the isotherm through the dot divides the p-V diagram into regions 1 and 2. For the following processes, determine whether the change ΔE_{int} in the internal energy of the gas is positive, negative, or zero: (a) the gas moves up along the isotherm, (b) it moves down along the isotherm, (c) it moves to anywhere in region 1, and (d) it moves to anywhere in region 2.

Now determine whether the heat Q involved is greater than, less than, or equal to the work W done by the gas for the same processes: (e) the gas moves up along the isotherm, (f) it moves down along the isotherm, (g) it moves to anywhere in region 1, and (h) it moves to anywhere in region 2.

16. A certain gas is taken to the five states represented by dots in Fig. 20-26; the plotted lines are isotherms. Rank the states according to (a) the most probable speed v_P of the molecules and (b) their average speed v_{avg}, greatest first.

17. *Organizing question:* The dot in Fig. 20-25c represents the initial state of a gas, and the adiabat

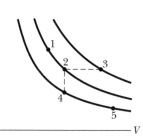

Fig. 20-26 Question 16.

through the dot divides the p-V diagram into regions 1 and 2. For the following processes, determine whether the corresponding heat Q is positive, negative, or zero: (a) the gas moves up along the adiabat, (b) it moves down along the adiabat, (c) it moves to anywhere in region 1, and (d) it moves to anywhere in region 2.

18. Two rooms of equal volume are connected by an open passageway and are maintained at different temperatures. Which room contains more air molecules?

19. The p-V diagram of Fig. 20-27 (not to scale) shows eight separate transformations (lettered a to h) that an ideal gas undergoes in proceeding from an initial state to a final state. The table shows eight sets of values for Q, W, and ΔE_{int} (in joules), numbered 1 to 8. Which set of energy changes goes with which transformation? (Hint: Compare the paths in Fig. 20-27 to those in Fig. 20-14.)

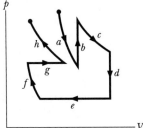

Fig. 20-27 Question 19.

	1	2	3	4	5	6	7	8
Q	10	5	−20		−10	−12		10
W	10		−10	10		−12	−10	
ΔE_{int}		2		−10	−10		10	10

20. In Figs. 20-28a through d, ideal gases with identical numbers of moles and identical volumes are confined to identical cylinders. Each cylinder has a movable piston with identical shot on top but in different numbers. Rank the gases according to their temperature, greatest first.

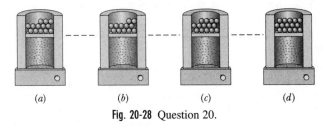

(a)   (b)   (c)   (d)

Fig. 20-28 Question 20.

21. Figure 20-29 shows the initial state of an ideal gas and an isotherm through that state. Which of the paths shown result in (a) an increase in the root-mean-square speed of the gas molecules and (b) a decrease in their average translational kinetic energy?

22. A one-dimensional gas is a hypothetical gas with molecules that can move along only a single axis. The following table gives, for four situations, the velocities in meters per second of such a gas having four molecules. (The plus and minus signs refer to the direction of the velocity along the axis, as usual.) Without written

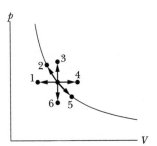

Fig. 20-29 Question 21.

calculation or a calculator, rank the four situations according to the root-mean-square speed of the molecules, greatest first.

Situation	Velocities			
a	−2	+3	−4	+5
b	+2	−3	+4	−6
c	+2	+3	+4	+5
d	+3	+3	−4	−5

23. Suppose that the volume of a certain gas is to be doubled by one of the following processes: (1) isothermal expansion, (2) adiabatic expansion, (3) free expansion, or (4) expansion at constant pressure. Rank those processes according to the change in the average kinetic energy of the gas molecules, most positive change first, most negative last.

24. In each of six situations, 30 J is transferred as heat to an ideal gas under the listed circumstances. The gases have the same number of moles. Rank the situations according to (a) the increase in the temperature of the gas, (b) the increase in the average kinetic energy of the gas molecules, (c) the work done by the gas, and (d) the increase in the internal energy of the gas, greatest first.

Situation	Gas Type	Circumstances
1	monatomic	constant volume
2	monatomic	constant pressure
3	diatomic	constant volume, no rotation
4	diatomic	constant pressure, no rotation
5	diatomic	constant volume, with rotation
6	diatomic	constant pressure, with rotation

DISCUSSION QUESTIONS

25. A sealed rubber balloon contains a very light gas. The balloon is released and it rises high into the atmosphere. Describe and explain the temperature of the gas and the size of the balloon.

26. In kinetic theory we assume that there are a large number of molecules in a gas. Real gases behave like an ideal gas at low densities. Are these statements contradictory? If not, what conclusion can you draw from them?

27. We have assumed that the collisions of gas molecules with the walls of their container are elastic. Actually, the collisions may be inelastic. Why does this make no difference as long as the walls are at the same temperature as the gas?

28. On a humid day, some say that the air is "heavy." How does the density of humid air compare with that of dry air at the same temperature and pressure?

29. Where does the average speed of air molecules in still air at room temperature fit into this sequence: 0; 2 m/s (walking speed); 30 m/s (fast car); 500 m/s (supersonic airplane); 1.1×10^4 m/s (escape speed from Earth); 3×10^8 m/s (speed of light)?

30. Molecular motions are maintained by no outside force, yet continue indefinitely with no sign of diminishing speed. Why doesn't friction bring these tiny particles to rest, as it does other moving particles?

31. What justification is there for neglecting the changes in gravitational potential energy of molecules in a gas?

32. We have assumed that the force exerted by gas molecules on the wall of a container is steady in time. How is this justified?

33. The average velocity of the molecules in a gas must be zero if the gas as a whole and its container are not in translational motion. Explain how it can be that the *average speed* is not zero.

34. Consider a hot, stationary golf ball sitting on a tee and a cold golf ball just moving off the tee after being hit. The total kinetic energy of the molecules in the balls, relative to the tee, can be the same in the two cases. Explain how. What is the difference between the two cases?

35. Justify the fact that the pressure of a gas depends on the *square* of the speed of its particles by explaining the dependence of pressure on the collision frequency and the momentum transfer of the particles.

36. Why does the boiling temperature of a liquid increase with pressure?

37. How is the speed of sound in a gas related to the pressure and temperature of the gas?

38. Far above Earth's surface the gas temperature is reported to be on the order of 1000 K. However, a person placed in such an environment would freeze to death rather than vaporize. Explain.

39. At the top of the atmosphere molecules will occasionally be headed out with a speed exceeding the escape speed. Why doesn't Earth's atmosphere leak away? Isn't it just a matter of time?

40. Titan, one of Saturn's many moons, has an atmosphere, but our own Moon does not. Why?

41. As energy is added as heat to ice, the ice melts to become liquid, and then the liquid eventually boils to become a vapor (or gas). However, as solid carbon dioxide is heated it goes directly to the vapor state—we say it *sublimes*—without passing through a liquid state. How could liquid carbon dioxide be produced?

42. What direct evidence do we have for the existence of atoms? What indirect evidence is there?

43. How, if at all, would you expect the composition of the atmosphere to change with altitude?

44. We often say that we see the steam emerging from the spout of a kettle in which water is boiling. However, steam itself is a colorless gas. What is it that we really see?

45. Why does smoke rise, rather than fall, from a lighted candle? Explain in terms of molecular collisions.

46. Would a gas whose molecules were true geometric points obey the ideal gas law?

47. If you fill a saucer with water at room temperature, the water, under normal conditions, will evaporate completely. It is easy to believe that some of the more energetic molecules can escape from the water surface, but how can *all* of them eventually escape? Many of them—in fact the vast majority—do not have enough energy to do so.

48. Give a qualitative explanation of the connection between the mean free path of ammonia molecules in air and the time it takes to smell the ammonia when a bottle is opened across the room.

49. List effective ways of increasing the number of molecular collisions per unit time in a gas.

50. If the molecules of a gas are not spherical, what meaning can we give to d in Eq. 20-25 for the mean free path? In which gases would the molecules act the most like rigid spheres?

51. The two opposite walls of a container of gas are kept at different temperatures. The air between the panes of glass in a storm window is a familiar example. Describe in terms of kinetic theory the mechanism of heat conduction through the gas.

52. A gas can transmit only those sound waves whose wavelength is long compared with the mean free path. Can you explain this? Describe a situation for which this limitation would be important.

53. Justify qualitatively the statement that, in a mixture of molecules of different kinds, in complete equilibrium, each kind of molecule has the same Maxwellian distribution in speed that it would have if the other kinds were not present.

54. Is it possible for a gas to consist of molecules that all have the same speed?

55. What observation is good evidence that not all molecules in a body at a certain temperature are moving with the same speed?

56. The fraction of molecules within a given range Δv of the rms speed decreases as the temperature of a gas rises. Explain.

57. (a) Do half the molecules in a gas in thermal equilibrium have speeds greater than v_P? Than v_{avg}? Than v_{rms}? (b) Which speed, v_P, v_{avg}, or v_{rms}, corresponds to a molecule having average kinetic energy?

58. Keeping in mind that the internal energy of a body consists of kinetic energy and potential energy of its particles, how would you distinguish between the internal energy of a body and its temperature?

59. The gases in two identical containers are at 1 atm pressure and room temperature. One contains helium gas (monatomic, molar mass 4 g/mol) and the other contains an equal number of moles of argon gas (monatomic, molar mass 40 g/mol). If 1 cal of heat added to the helium gas increases its temperature by a certain amount, what amount of heat must be added to the argon gas to increase its temperature by the same amount?

60. Explain how we might keep a gas at a constant temperature during a thermodynamic process.

61. Why is it more common to excite radiation from gaseous atoms by use of electrical discharge than by thermal methods?

62. Explain why the temperature of a gas drops in an adiabatic expansion.

63. If hot air rises, why is it cooler at the top of a mountain than near sea level? (*Hint:* Air is a poor conductor of heat.)

EXERCISES & PROBLEMS

62. For a certain ideal gas, C_V is 6.00 cal/mol · K. The temperature of 3.0 mol of the gas is raised 50 K by each of three different processes: at constant volume, at constant pressure, and by an adiabatic compression. Complete the table, showing for each process the heat Q, the work W done by the gas, the change ΔE_{int} in internal energy of the gas, and the change ΔK in total translational kinetic energy of the gas.

Process	Q	W	ΔE_{int}	ΔK
Constant volume	———	———	———	———
Constant pressure	———	———	———	———
Adiabatic	———	———	———	———

63. An ideal gas undergoes an adiabatic compression from $p = 1.0$ atm, $V = 1.0 \times 10^6$ L, $T = 0.0°C$ to $p = 1.0 \times 10^5$ atm, $V = 1.0 \times 10^3$ L. (a) Is the gas monatomic, diatomic, or polyatomic? (b) What is its final temperature? (c) How many moles of gas are present? (d) What is the total translational kinetic energy per mole before and after the compression? (e) What is the ratio of the squares of the rms speeds before and after the compression?

64. A sample of ideal gas expands from an initial pressure and volume of 32 atm and 1.0 L to a final volume of 4.0 L. The initial temperature of the gas is 300 K. What are the final pressure and temperature of the gas and how much work is done by the gas during the expansion, if the expansion is (a) isothermal, (b) adiabatic and the gas is monatomic, and (c) adiabatic and the gas is diatomic?

65. An ideal gas, at initial temperature T_1 and initial volume 2.0 m³, is expanded adiabatically to a volume of 4.0 m³, then expanded isothermally to a volume of 10 m³, and then compressed adiabatically back to T_1. What is its final volume?

66. The speed of sound in different gases at the same temperature depends only on the molar masses of the gases. Show that $v_1/v_2 = \sqrt{M_2/M_1}$ (at constant T), where v_1 is the speed of sound in a gas of molar mass M_1 and v_2 is the speed of sound in a gas of molar mass M_2. (*Hint:* See Exercise 58.)

67. The molar mass of iodine is 127 g/mol. A standing wave in a tube filled with iodine gas at 400 K has nodes that are 6.77 cm apart when the frequency is 1400 Hz. (a) What is γ for iodine gas? (b) Is iodine gas monatomic or diatomic? (*Hint:* See Exercise 58.)

68. Knowing that a gas has $C_V = 5.0R$, calculate the ratio of the speed of sound in that gas to the rms speed of its molecules at temperature T. (*Hint:* See Exercise 58.)

69. A weather balloon is loosely inflated with helium at a pressure of 1.0 atm (= 760 torr) and a temperature of 20°C, to a volume of 2.2 m³. When the balloon reaches an elevation of 6.1 km, the atmospheric pressure is down to 380 torr and the helium has expanded, but the balloon is still loosely inflated. At this elevation the gas temperature is −48°C. What is the gas volume now?

70. A steel tank contains 300 g of ammonia gas (NH_3) at a pressure of 1.35×10^6 Pa and a temperature of 77°C. (a) What is the volume of the tank in liters? (b) When the tank is checked later, the temperature is 22°C and the pressure is 8.7×10^5 Pa. How many grams of gas have leaked out of the tank?

71. An ideal gas is suddenly allowed to expand freely so that the ratio of its new volume V_1 to its initial volume V_0 is $V_1/V_0 = 5.00$. The gas is then adiabatically compressed back to its initial volume V_0, leaving it with a pressure p_2 that is $(5.00)^{0.40}$ times its initial pressure p_0. (a) Is the gas monatomic, diatomic with no rotation of the molecules, diatomic with rotating molecules, or polyatomic? In terms of the initial average kinetic energy K_0 of the molecules, what are the average kinetic energies after (b) the free expansion and (c) the adiabatic compression?

72. Derive an expression, in terms of N/V, v_{avg}, and d, for the collision frequency of a gas atom or molecule.

73. The mass of a helium atom is 6.66×10^{-27} kg. Compute the specific heat at constant volume for (monatomic) helium gas (in J/kg · K) from the molar specific heat at constant volume.

74. (a) What is the molar volume (volume per mole) of an ideal gas at standard conditions (0.00°C, 1.00 atm)? (b) Calculate the ratio of the root-mean-square speed of helium atoms to that of neon atoms under these conditions. (c) What is the mean free path of helium atoms under these conditions? Assume the atomic diameter d of helium to be 1.00×10^{-8} cm. (d) What is the mean free path of neon atoms under these conditions? Assume they have the same atomic diameter as helium. (e) Comment on the results of (c) and (d) in view of the fact that the helium atoms are traveling faster than the neon atoms.

75. Figure 20-30 shows two paths that may be taken by a gas from an initial point i to a final point f. Path 1 consists of an isothermal expansion (work is 50 J in magnitude), an adiabatic expansion (work is 40 J in magnitude), an isothermal compression (work is 30 J in magnitude), and then an adiabatic compression (work is 25 J in magnitude). What is the change in the internal energy of the gas if the gas goes from point i to point f along path 2?

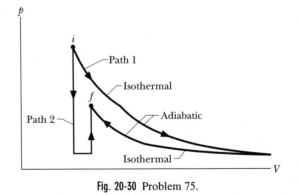

Fig. 20-30 Problem 75.

76. Figure 20-31 shows a cycle consisting of five paths: AB is isothermal at 300 K, BC is adiabatic with work = 5.0 J, CD is at a constant pressure of 5 atm, DE is isothermal, and EA is adiabatic with a change in internal energy of 8.0 J. What is the change in internal energy of the gas along path CD?

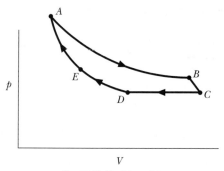

Fig. 20-31 Problem 76.

77. At what temperature do the atoms of helium gas have the same rms speed as molecules of hydrogen gas have at 20.0°C? (The molar masses are given in Table 20-1.)

78. A quantity of an ideal monatomic gas consists of n moles initially at temperature T_1. The pressure and volume are then slowly doubled in such a manner as to trace out a straight line on a p-V diagram. In terms of n, R, and T_1, what are (a) W, (b) ΔE_{int}, and (c) Q for this process? (d) If one were to define a molar specific heat for the gas as it undergoes this process, what would be its value?

79. (a) What is the volume occupied by 1.00 mol of an ideal gas at standard conditions—that is, at 1.00 atm $(= 1.01 \times 10^5$ Pa) and 0°C $(= 273$ K)? (b) Show that the number of molecules per cubic centimeter (the *Loschmidt number*) at standard conditions is 2.69×10^{19}.

80. An ideal gas initially has a volume of 4.00 m³, a pressure of 5.67 Pa, and a temperature of -56°C. The gas is then expanded to 7.00 m³, leaving it with a temperature of 40°C. What then is the pressure?

81. Figure 20-32 shows a hypothetical speed distribution for particles of a certain gas: $P(v) = Cv^2$ for $0 < v \le v_0$ and $P(v) = 0$ for $v > v_0$. Find (a) an expression for C in terms of v_0, (b) the average speed of the particles, and (c) their rms speed.

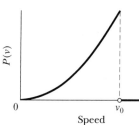

Fig. 20-32 Problem 81.

82. The temperature of 3.00 mol of an ideal diatomic gas is increased by 40.0 C° without the pressure of the gas changing. The molecules in the gas rotate but do not oscillate. (a) How much energy is transferred to the gas as heat? (b) What is the change in the internal energy of the gas? (c) How much work is done by the gas? (d) By how much does the translational kinetic energy of the gas increase?

83. At what temperature is the average translational kinetic energy of a molecule in an ideal gas equal to 4.0×10^{-19} J?

84. In an interstellar gas cloud of temperature 50 K, the pressure is 1.00×10^{-8} Pa. Assuming that the molecular diameters of the gases in the cloud are all 20.0 nm, what is their mean free path?

85. In a hot-air balloon, the fully inflated envelope (which contains the hot air) has a volume of 2.18×10^3 m³ and the basket and envelope have a combined mass of 249 kg. Find the temperature

required of the hot air so that the buoyancy it provides exceeds the gravitational force on the balloon (basket, envelope, and hot air) by 2.70 kN. Assume that the surrounding air has a density of 1.21 kg/m³, the air is an ideal gas of molar mass 28.0 g/mole (corresponding to nitrogen), and the air pressure inside and outside the envelope is 1.00 atm.

86. An ideal gas undergoes isothermal compression from an initial volume of 4.00 m³ to a final volume of 3.00 m³. There are 3.5 moles of the gas, and the gas temperature is 10°C. (a) How much work is done by the gas? (b) How much energy is transferred as heat between the gas and its environment?

87. (a) What is the number of molecules per cubic meter in air at 20°C and at a pressure of 1.0 atm $(= 1.01 \times 10^5$ Pa)? (b) What is the mass of 1 m³ of this air? Assume that 75% of the molecules are nitrogen (N_2) and 25% are oxygen (O_2).

88. Show that about 10^{13} air molecules (diameter ≈ 0.2 nm) are needed to cover the period that closes this sentence. Show that about 10^{21} air molecules collide with that period each second.

89. An ideal gas initially at 300 K is compressed at a constant pressure of 25 N/m² from a volume of 3.0 m³ to a volume of 1.8 m³. In the process, 75 J is lost by the gas as heat. What are (a) the change in internal energy of the gas and (b) the final temperature of the gas?

90. An ideal gas with 3.00 mol is initially at 425 K when it is then compressed until it reaches 350 K, without its pressure changing. The gas loses 4670 J via heat. What is the change in the internal energy of the gas?

91. Figure 20-33 represents an adiabatic compression of an ideal gas from 15 m³ to 12 m³, followed by an isothermal compression at 300 K to a final volume of 3.0 m³. There are 2.0 moles of the gas. What is the total energy transferred as heat?

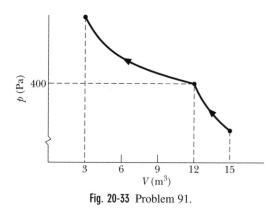

Fig. 20-33 Problem 91.

92. *The hots.* Figure 20-34 shows a temperature scale that is laid out as blocks on a game board. A sample of material begins at a temperature in one of the blocks, and then, in the steps given below, energy is added to or removed from the sample as heat. With each step, you are to determine where the sample is on the game board by giving a block's label. (*Caution:* One wrong move means all further moves are probably wrong.)

The material is fictitious, and its properties, listed below, were invented just for this game. In the gaseous state, the material is diatomic and all changes occur at constant volume unless otherwise stated. The initial temperature of the sample is 80.00°C. Remove, in this order, (a) 300.0 kJ, (b) 400.0 kJ, (c) 820.0 kJ, (d) 250.0 kJ, and finally (e) 670.0 kJ. Now add, in this order, (f) 1240.0 kJ, (g) 1280 kJ, (h) 820.0 kJ, (i) 1000 kJ, (j) 583.1 kJ, (k) 166.2 kJ, (l) 277.0 kJ, (m) 581.7 kJ at constant pressure, and finally (n) 249.3 kJ at constant pressure. (*Hint:* For the last step, see Fig. 20-12 and use Eq. 20-49.)

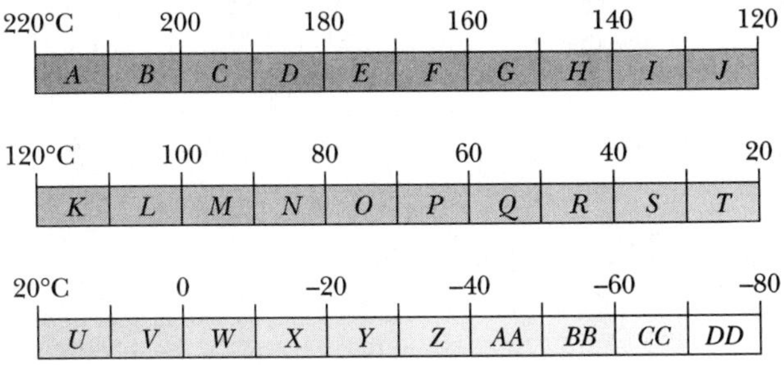

220°C	200		180		160		140		120
A	B	C	D	E	F	G	H	I	J

120°C	100		80		60		40		20
K	L	M	N	O	P	Q	R	S	T

20°C	0		−20		−40		−60		−80
U	V	W	X	Y	Z	AA	BB	CC	DD

Fig. 20-34 Problem 92.

Sample's mass = 4.000 kg

Molar mass = 3.000 g/mol

Freezing point = −15.00°C

Boiling point = 105.0°C

Heat of fusion = 150 kJ/kg

Heat of vaporization = 500 kJ/kg

Molecular rotation: $T \geq 135°C$

Molecular oscillation: $T \geq 185°C$

Specific heat: liquid = 4.000 kJ/kg · K

solid = 2.000 kJ/kg · K

Tutorial Problems

93. An ideal gas of 1.00 mol is initially at a pressure of 1.00×10^5 Pa and a temperature of 47°C. (a) What are the gas's initial temperature in kelvins and its initial volume? Plot graphs of pressure p versus volume V, p versus temperature T, and V versus T for the following processes that the gas could undergo from its initial state: (b) an *isobaric process* (a constant-pressure process) to a temperature of 480 K, (c) a constant-volume process to a temperature of 480 K, and (d) an isothermal process to twice the initial volume. On each graph, mark the initial and final states and give their values on the axes, draw the path accurately, and add an arrowhead to the path to indicate the path's direction.

Answers

(a) The temperature is (47°C + 273 K) = 320 K. From the ideal gas law, the volume is

$$V = \frac{nRT}{p} = \frac{(1.00 \text{ mol})(8.31 \text{ J/K} \cdot \text{mol})(320 \text{ K})}{1.00 \times 10^5 \text{ Pa}}$$

$$= 0.0266 \text{ m}^3.$$

(b) From the ideal gas law, the final volume is 0.0399 m³. For p versus V, plot a horizontal line at the constant pressure, from the initial volume to the final volume. For p versus T, plot a horizontal

line at the constant pressure, from the initial temperature to the final temperature. For V versus T, plot a straight line from the initial volume and temperature to the final volume and temperature. This line is straight because here, from the ideal gas law, $V = nRT/p =$ (a constant)T.

(c) From the ideal gas law, the final pressure is 1.50×10^5 Pa. For p versus V, plot a vertical line at the constant volume, from the initial pressure to the final pressure. For p versus T, plot a straight line from the initial pressure and temperature to the final pressure and temperature. This line is straight because here, from the ideal gas law, $p = nRT/V =$ (a constant)T. For V versus T, plot a horizontal line at the constant volume, from the initial temperature to the final temperature.

(d) From the ideal gas law, the final pressure is 0.50×10^5 Pa. For p versus V, plot the equation

$$p = \frac{nRT}{V} = \frac{(1.0 \text{ mol})(8.31 \text{ J/K} \cdot \text{mol})(320 \text{ K})}{V}$$

$$= \frac{(2659 \text{ Pa} \cdot \text{m}^3)}{V}$$

from the initial pressure and volume to the final pressure and volume. For p versus T, plot a vertical line at the constant temperature, from the initial pressure to the final pressure. For V versus T, plot a vertical line at the constant temperature, from the initial volume to the final volume.

94. Consider a system containing 1.50 moles of an ideal gas. Initially this system has a pressure of 1.00×10^5 N/m² and a temperature of 27°C = 300 K. Let's suppose that this gas has an internal energy E_{int} that depends on the absolute temperature T as $E_{int} = 3.5nRT$. For each of the processes listed below, which all start from the given initial state: (1) Determine the final pressure, volume, and temperature of the system. Explain your procedure in complete sentences, making any equations part of a sentence. Don't simply list equations and plug into them. Also, indicate whether you are using a law or definition to determine these quantities. (2) Next, determine the change ΔE_{int} in internal energy of the system, the heat Q added to the system, and the work W done by the system. Explain logically and completely how you determined these quantities, again using complete sentences. Determine ΔE_{int}, Q, and W in a logical order, which might differ from one process to another. (3) Last, determine the path of the process on a p-V diagram. Remember that all graphs have the same initial state. Of course, the graph should be done after you have determined the appropriate numbers.

(a) The gas undergoes a constant-pressure process to a final temperature of 400 K. (b) It undergoes a constant-volume process to a final temperature of 400 K. (c) It undergoes an isothermal process to a final pressure of 1.20×10^5 N/m².

Answers

(a) First, let's calculate the initial volume using the ideal gas law:

$$V_i = \frac{nRT_i}{p_i} = \frac{(1.50 \text{ mol})(8.31 \text{ J/mol} \cdot \text{K})(300 \text{ K})}{1.00 \times 10^5 \text{ Pa}}$$

$$= 0.0374 \text{ m}^3.$$

We know that $p_f = p_i = 1.00 \times 10^5$ Pa and $T_f = 400$ K, so

$$V_f = \frac{400 \text{ K}}{300 \text{ K}} V_i = 0.0499 \text{ m}^3,$$

which tells us that the gas expands. During the expansion, the change in the internal energy is

$$\Delta E_{int} = 3.5 nR \, \Delta T$$
$$= (3.5)(1.5 \text{ mol})(8.31 \text{ J/mol} \cdot \text{K})(100 \text{ K})$$
$$= 4.36 \text{ kJ}$$

and the work done by the gas is

$$W = p \, \Delta V$$
$$= (1.00 \times 10^5 \text{ Pa})(0.0499 \text{ m}^3 - 0.0374 \text{ m}^3)$$
$$= 1.25 \text{ kJ}.$$

Then, by the first law of thermodynamics, we know that the heat is

$$Q = \Delta E_{int} + W = 4.36 \text{ kJ} + 1.25 \text{ kJ} = 5.61 \text{ kJ}.$$

(b) We know that $V_f = V_i = 0.0374$ m^3 and $T_f = 400$ K. From the ideal gas law with constant volume, the pressure is proportional to the absolute temperature, so $p_f = \frac{4}{3}p_i = 1.33 \times 10^5$ N/m^2. The change in the internal energy is 4.36 kJ, the same as in part (a) because the final temperature is the same. And $W = 0$ because the volume does not change. Thus, by the first law of thermodynamics,

$$Q = \Delta E_{int} = 4.36 \text{ kJ}.$$

(c) We know that $T_f = T_i = 300$ K, which means that $\Delta E_{int} = 0$. We also know that $p_f = 1.20 \times 10^5$ N/m^2. Then, using the ideal gas law,

$$V_f = \frac{nRT}{p_f} = \frac{V_i}{1.20} = \frac{0.0374 \text{ m}^3}{1.20} = 0.0312 \text{ m}^3.$$

From Eq. 20-14, the work done is

$$W = nRT \ln(V_f/V_i)$$
$$= (1.5 \text{ mol})(8.31 \text{ J/mol} \cdot \text{K})(300 \text{ K}) \ln(1/1.20)$$
$$= -0.682 \text{ kJ}.$$

Then, by the first law of thermodynamics, since $\Delta E_{int} = 0$,

$$Q = W = -0.682 \text{ kJ}.$$

21

Entropy and the Second Law of Thermodynamics

Sample Problem 21-7

In Sample Problem 21-4, assume for the moment that the inventor's engine is approximately a Carnot engine. Then determine the entropy change ΔS that would occur within the closed system consisting of the engine and its reservoirs during one cycle. Compute ΔS in terms of $|Q_H|$, the magnitude of the energy transferred during a cycle as heat from the high-temperature reservoir to the working substance.

SOLUTION: The **Key Idea** here is that net entropy change is equal to the sum of the entropy changes of the two reservoirs (ΔS_H and ΔS_L) and that of the working substance (ΔS_{ws}):

$$\Delta S = \Delta S_H + \Delta S_L + \Delta S_{ws}. \quad (21\text{-}23)$$

Because the working substance returns to its initial state at the end of each cycle, its entropy change ΔS_{ws} during one cycle is zero. The high-temperature reservoir loses energy $|Q_H|$ as heat at temperature T_H, and the low-temperature reservoir gains energy $|Q_L|$ as heat at temperature T_L. Thus, from Eq. 21-2, their entropy changes are

$$\Delta S_H = \frac{-|Q_H|}{T_H} \quad \text{and} \quad \Delta S_L = \frac{+|Q_L|}{T_L}.$$

Substituting these changes into Eq. 21-23 yields

$$\Delta S = -\frac{|Q_H|}{T_H} + \frac{|Q_L|}{T_L} + 0. \quad (21\text{-}24)$$

We can substitute for $|Q_L|$ with Eq. 21-10 and then substitute the claimed efficiency ε of the engine. First, we find

$$|Q_L| = |Q_H| (1 - \varepsilon).$$

Now, combining this with Eq. 21-24, we have, after some algebra,

$$\Delta S = |Q_H| \left(\frac{1 - \varepsilon}{T_L} - \frac{1}{T_H} \right).$$

Substituting the data from Sample Problem 21-4 then gives us

$$\Delta S = |Q_H| \left(\frac{1 - 0.75}{273 \text{ K}} - \frac{1}{373 \text{ K}} \right)$$
$$= -0.0018 \, |Q_H|.$$

Thus, if the inventor's claim were correct, the entropy of the system would decrease with each cycle. This is impossible. If every process that occurs in the engine were reversible, ΔS would be zero. If some were irreversible, ΔS would be greater than zero. No set of processes in a closed system can lead to ΔS less than zero.

Sample Problem 21-8

A Carnot refrigerator, with coefficient of performance $K = 4.7$, extracts energy as heat from the cold chamber at the rate of 250 J/cycle.

(a) How much work per cycle is required to operate the refrigerator?

SOLUTION: The cold chamber is the low-temperature reservoir, and the magnitude of the energy extracted as heat from it in one cycle is $|Q_L| = 250$ J. The **Key Idea** here is that we can relate the magnitude $|W|$ of the work done per cycle to the known energy $|Q_L|$ extracted per cycle and the coefficient K with the definition of K (Eq. 21-12). Solving that equation for $|W|$ yields

$$|W| = \frac{|Q_L|}{K} = \frac{250 \text{ J}}{4.7} = 53 \text{ J.} \quad \text{(Answer)}$$

(b) How much energy per cycle is discharged as heat to the room?

SOLUTION: The room is the high-temperature reservoir. Let $|Q_H|$ represent the magnitude of the energy discharged to it in a cycle. Then we need two **Key Ideas**. First, because this is a Carnot refrigerator, the processes that make up its operating cycle are reversible, with no energy transfers other than $|Q_H|$, $|Q_L|$, and $|W|$. Second, we can relate $|Q_H|$ to the known values of $|W|$ and $|Q_L|$ with the first law of thermodynamics ($\Delta E_{int} = Q - W$); that is, we can write

$$\Delta E_{int} = (|Q_H| - |Q_L|) - |W|.$$

Here $\Delta E_{int} = 0$ because the working substance operates in a cycle. Solving for $|Q_H|$ and inserting known data then yield

$$|Q_H| = |W| + |Q_L| = 53 \text{ J} + 250 \text{ J}$$
$$= 303 \text{ J} \approx 300 \text{ J.} \quad \text{(Answer)}$$

QUESTIONS

12. In a Carnot refrigerator, is the magnitude $|Q_H|$ of the energy transferred per cycle to the high-temperature reservoir always less than, always greater than, or always equal to the magnitude of the work W done per cycle?

13. *Organizing question:* Block A and Block B are placed together in an insulated box and allowed to come to thermal equilibrium. Their initial temperatures are $T_A = -10°C$ and $T_B = 15°C$. The materials making up the blocks have melting temperatures of $T_{A,F} = 0°C$ and $T_{B,F} = 20°C$. Block A has a mass of 0.20 kg and its material has a specific heat of 4190 J/kg · K and a heat of fusion of 333 kJ/kg. Set up equations, complete with known data, for finding the change in the entropy of block A when the block reaches thermal equilibrium, if the equilibrium temperature is (a) $-5.0°C$ and (b) $5.0°C$.

14. (a) Which of the curves in Fig. 21-31 gives the limiting efficiency of an engine as a function of the temperature ratio T_H/T_L? (b) Which curve gives the limiting coefficient of performance for a refrigerator?

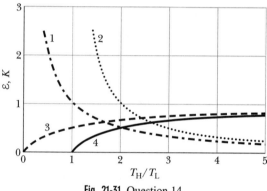

Fig. 21-31 Question 14.

15. *Organizing question:* Figure 21-32 shows the elements of refrigerators (heat pumps) or engines in six general arrangements, each drawn in the style of Chapter 21. Which arrangement corresponds to (a) a Carnot engine, (b) a real engine, (c) a perfect engine, (d) a Carnot refrigerator, (e) a real refrigerator, (f) a perfect refrigerator, (g) an engine that produces only waste heat, and (h) an arrangement in which the work done is transformed entirely to thermal energy by friction? (i) For each of these devices [(a) through (h)], determine whether the net entropy change for the entire closed system during one cycle is positive, negative, or zero. (j) Which of these devices can be built; which are the theoretical limit of what can be built; and which cannot be built?

16. A basic assumption of statistical mechanics is that all microstates are equally probable. How would the air in your bedroom behave if all configurations were equally probable (they aren't)?

17. A sample of ice at $-10°C$ is placed on a controllable thermal reservoir. The water is then taken through three steps: (1) the ice is warmed to the melting point, (2) the ice is melted, and (3) the liquid is warmed to 10°C. Without written computation or a calculator, rank the three steps according to the resulting change in

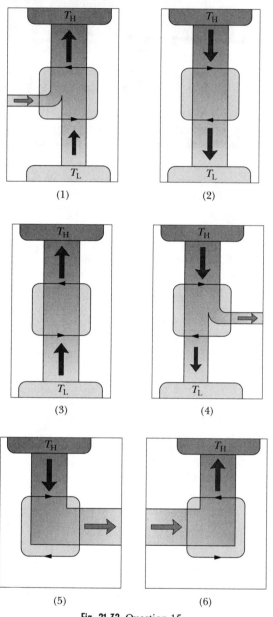

Fig. 21-32 Question 15.

entropy of the sample, greatest first. (The specific heat of ice is 2220 J/kg · K, the specific heat of liquid water is 4190 J/kg · K, and the heat of fusion of water is 333 000 J/kg.)

18. Figure 21-33 shows four reversible paths along which an ideal gas, in contact with a controllable thermal reservoir, can be taken from the common initial state i (lower left corner) to a final state of either f_1 (upper left corner) or f_2 (upper right corner). Rank the paths according to the magnitudes of the resulting entropy changes of (a) the gas, (b) the reservoir, and (c) the gas–reservoir system, greatest first.

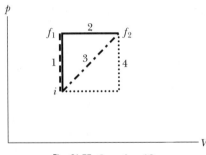

Fig. 21-33 Question 18.

19. The following table gives the masses, the heats of fusion, and the melting points (in kelvins) for four samples. The samples are initially frozen and at their melting points; energy is then transferred to them as heat until they are completely unfrozen. Rank the samples according to the entropy change during this melting process, greatest first.

Sample	Mass	Heat of Fusion	Melting Point
1	M	$2L$	T
2	$2M$	$L/2$	$2T$
3	$M/2$	$2L$	$T/2$
4	$2M$	L	$T/2$

20. An ideal gas, which is in contact with a controllable thermal reservoir, is taken reversibly from initial state i to final state f via the process plotted in Fig. 21-34. Does the entropy of (a) the gas, (b) the reservoir, and (c) the gas–reservoir system increase, decrease, or remain the same?

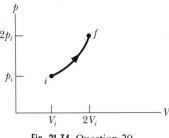

Fig. 21-34 Question 20.

21. *Math Tool Time.* Which of the following is equivalent to $\ln(2a/b^3)$: (a) $2 \ln a - 3 \ln b$, (b) $2 \ln a + 3 \ln b$, (c) $\ln 2a - 3 \ln b$, (d) $2 \ln a - \frac{1}{3} \ln b$?

DISCUSSION QUESTIONS

22. Is a human being a heat engine? Explain.

23. Could we not define the efficiency of any engine as $\varepsilon = |W|/|Q_L|$ rather than as $\varepsilon = |W|/|Q_H|$? Why don't we?

24. The efficiencies of nuclear power plants are lower than those of fossil-fuel plants. Why?

25. Can a given amount of mechanical energy be converted completely into heat? If so, give an example.

26. An inventor suggested a house might be heated in the following manner: A system resembling a refrigerator draws energy as heat from the ground and rejects it to the house. He claimed that the energy supplied as heat to the house can exceed the work done by the engine of the system. What is your comment?

27. Give a qualitative explanation of how frictional forces between moving surfaces increase the temperature of those surfaces. Why does the reverse process not occur?

28. Are any of the following processes reversible: (a) breaking an empty soda bottle, (b) mixing a cocktail, (c) melting an ice cube in a glass of iced tea, (d) burning a log of firewood, (e) puncturing an automobile tire, (f) finishing the "Unfinished Symphony," (g) writing this book?

29. Give some examples of natural processes that are nearly reversible.

30. Can we calculate the work done during an irreversible process in terms of an area on a p-V diagram? Is any work done?

31. Suggest a reversible process whereby energy can be added as heat to a system. Why would adding energy by means of a Bunsen burner not be a reversible process?

32. In a Carnot cycle, we need not start at point a in Fig. 21-8 but may equally well start at point b, c, or d, or any intermediate point. Explain.

33. If a Carnot engine is independent of the working substance, then perhaps real engines should be similarly independent, to a certain extent. Why then, for real engines, are we so concerned to find suitable fuels such as coal, gasoline, or fissionable material? Why not use stones as a fuel?

34. Under what conditions would a Carnot engine be 100% efficient?

35. What factors reduce the efficiency of a heat engine from its ideal value?

36. You wish to increase the efficiency of a Carnot engine as much as possible. You can do this by increasing T_H a certain amount while keeping T_L constant, or by decreasing T_L the same amount while keeping T_H constant. Which would you do?

37. Why do you get poorer gasoline mileage from your car in winter than in summer?

38. From time to time inventors claim to have perfected a device that does useful work but consumes no (or very little) fuel. What do you think is most likely true in such cases: (a) the claimants are right, (b) the claimants are mistaken in their measurements, or (c) the claimants are swindlers? (d) Do you think that such a claim should be examined closely by a panel of scientists and engineers? In your opinion, would the time and effort be justified?

39. We have seen that real engines always discard substantial amounts of heat to their low-temperature reservoirs. It seems a shame to throw this energy away. Why not use it to run a second engine, with the low-temperature reservoir of the first engine serving as the high-temperature reservoir of the second?

40. Give examples in which the entropy of a system decreases, and explain why the second law of thermodynamics is not violated.

41. Do living things violate the second law of thermodynamics? As a chicken grows from an egg, for example, it becomes more and more ordered and organized. Increasing entropy, however, calls

for disorder and decay. Is the entropy of a chicken actually decreasing as it grows?

42. Does a change in entropy occur in purely mechanical motions?

43. Show that total entropy increases when kinetic energy is dissipated by friction between sliding surfaces. Describe the increase in disorder.

44. Two pieces of molding clay of equal mass are moving directly toward each other with equal speeds. They strike and stick together. Treat the two pieces as a single system and state whether each of the following quantities is positive, negative, or zero for this process: ΔE_{int}, W, Q, and ΔS. Justify your answers.

45. Is it true that the energy of the universe is steadily growing less available? If so, why?

46. Discuss the following comment of the physicists Panofsky and Phillips: "From the standpoint of formal physics there is only one concept which is asymmetric in time, namely, entropy. But this makes it reasonable to assume that the second law of thermodynamics can be used to ascertain the sense of time independently in any frame of reference; that is, we shall take the positive direction of time to be that of statistically increasing disorder, or increasing entropy." (See, in this connection, "The Arrow of Time," by David Layzer, *Scientific American*, December 1975.)

EXERCISES & PROBLEMS

46. An inventor claims to have invented four engines, each of which operates between constant-temperature reservoirs at 400 and 300 K. Data on each engine, per cycle of operation, are: engine A, $Q_H = 200$ J, $Q_L = -175$ J, and $W = 40$ J; engine B, $Q_H = 500$ J, $Q_L = -200$ J, and $W = 400$ J; engine C, $Q_H = 600$ J, $Q_L = -200$ J, and $W = 400$ J; engine D, $Q_H = 100$ J, $Q_L = -90$ J, and $W = 10$ J. Of the first and second laws of thermodynamics, which (if either) does each engine violate?

47. A car engine delivers 8.2 kJ of work per cycle. Before a tuneup, its efficiency is 25%. Assume the engine is reversible. Calculate, per cycle, (a) the heat the engine takes from the combustion of fuel and (b) the energy lost as heat by the engine. (c) After a tune-up, the engine's efficiency is 31%. What are the new values of the quantities calculated in (a) and (b) when 8.2 kJ of work is delivered per cycle?

48. Check your calculator against Stirling's approximation, with N equal to (a) 50, (b) 100, and (c) 250. (d) Is there a trend?

49. Suppose you put 100 pennies in a cup, shake the cup, and then toss the pennies onto the floor. (a) How many different head–tail arrangements (microstates) are possible for the 100 pennies? (*Hint:* See Exercise 43.) What is the probability of finding exactly (b) 50 heads, (c) 48 heads, (d) 48 tails, (e) 40 heads, and (f) 30 heads?

50. A box contains N molecules. Consider two configurations: configuration A with an equal division of the molecules between the two halves of the box, and configuration B with 60.0% of the molecules in the left half of the box and 40.0% in the right half. For $N = 50$, what are the multiplicities of (a) configuration A and (b) configuration B? (c) What is the ratio f of the time the system spends in configuration B to the time it spends in configuration A? (d) Repeat (a) through (c) for $N = 100$. (e) Repeat (a) through (c) for $N = 200$. (f) With increasing N, does f increase, decrease, or remain the same?

51. Construct plots of T versus S, S versus E_{int}, and S versus V for the isothermal expansion process of Fig. 21-4.

52. An ideal gas in contact with a constant-temperature reservoir undergoes a reversible isothermal expansion to twice its initial volume. Show that the reservoir's change in entropy is independent of its temperature.

53. A 600 g lump of copper at 80.0°C is placed in 70.0 g of water at 10.0°C in an insulated container. (See Table 19-3 for specific heats.) (a) What is the equilibrium temperature of the copper–water

system? What entropy changes do (b) the copper, (c) the water, and (d) the copper–water system undergo in reaching the equilibrium temperature?

54. A 45.0 g block of tungsten at 30.0°C and a 25.0 g block of silver at −120°C are placed together in an insulated container. (See Table 19-3 for specific heats.) (a) What is the equilibrium temperature? What entropy changes do (b) the tungsten, (c) the silver, and (d) the tungsten–silver system undergo in reaching the equilibrium temperature?

55. A Carnot engine whose high-temperature reservoir is at 400 K has an efficiency of 0.300. By how much should the temperature of the low-temperature reservoir be changed to increase the efficiency to 0.400?

56. A diatomic gas of 2.00 mol is initially at 300 K. Its molecules do not rotate or oscillate both initially and throughout the following complete cycle of three processes: (1) the gas is heated at constant volume to a temperature of 800 K, (2) it is then allowed to expand isothermally to its initial pressure, and (3) it is then compressed at constant pressure to its initial state. During the cycle, what are (a) the net energy transferred as heat to the gas and (b) the net work done by the gas? (c) What is the efficiency of the cycle?

57. Suppose that 260 J is conducted from a constant-temperature reservoir at 400 K to another reservoir at (a) 100 K, (b) 200 K, (c) 300 K, and (d) 360 K. What are the net changes in entropy of the reservoirs for those four cases? (e) What is the trend in those changes?

58. The temperature of 1.0 mol of a monatomic ideal gas is raised reversibly from 300 K to 400 K, with its volume kept constant. What is the entropy change of the gas?

59. Repeat Problem 58, with the pressure now kept constant.

60. The temperature of 25 mol of an ideal diatomic gas, originally 300 K, is increased by placing the gas in contact with a thermal reservoir at 800 K. The pressure remains constant throughout the increase. (a) Calculate the change in the entropy of the gas, the change in the entropy of the reservoir, and the change in the total entropy of the gas–reservoir system.

Suppose the same change in temperature at constant pressure is carried out by placing the gas in contact successively with n intermediate reservoirs, equally spaced in temperature in the range 300 K to 800 K. Calculate the total change in the entropy of the reservoirs and the change in the entropy of the closed system consisting of the gas and the reservoirs for (b) $n = 1$, (c) $n = 10$, (d) $n = 50$, and (e) $n = 100$. With each increase in n, the process more closely approximates a reversible process.

61. An ideal gas of 4.00 mol is taken reversibly through a cycle of three processes: (1) an adiabatic expansion that gives the gas 2.00 times its initial volume, (2) a constant-volume process, (3) an isothermal compression back to the initial state of the gas. We do not know whether the gas is monatomic or diatomic; if it is diatomic, we do not know whether the molecules are rotating or oscillating. What are the entropy changes for (a) the complete cycle, (b) process 1, (c) process 3, and (d) process 2?

62. Calculate the efficiency of a fossil-fuel power plant that consumes 380 metric tons of coal each hour to produce useful work at the rate of 750 MW. The heat of combustion of coal (the heat due to the burning of coal) is 28 MJ/kg.

63. System A of three particles and system B of five particles are in insulated boxes like that in Fig. 21-16. What is the least multiplicity W of (a) system A and (b) system B? What is the greatest multiplicity W of (c) system A and (d) system B? Finally, what is the greatest entropy of (e) system A and (f) system B?

64. One mole of a monatomic ideal gas initially at a volume of 10 L and a temperature of 300 K is heated at constant volume to a temperature of 600 K, allowed to expand isothermally to its initial pressure, and finally compressed at constant pressure to its original volume, pressure, and temperature. During the cycle, what are (a) the net heat to the system (the gas) and (b) the net work done by the gas? (c) What is the efficiency of the cycle?

65. The entropy change ΔS between two states of a system, which is $\int (dQ/T)$, depends only on the initial and final states. In other words, this integral is path independent, as is shown for an ideal gas in Section 21-2. Show similarly that the integrals $\int dQ$, $\int (T\, dQ)$, and $\int (dQ/T^2)$ are not path independent and thus do not define new state properties for an ideal gas.

66. Energy can actually be removed from water as heat at and then below the normal freezing point (0.0°C at atmospheric pressure) without causing the water to freeze; the water is then said to be *supercooled*. Suppose a 1.00 g water drop is supercooled until its temperature is that of the surrounding air, which is at −5.00°C. The drop then suddenly and irreversibly freezes, transferring energy to the air until the drop is again at −5.00°C. What is the entropy change for the drop? (*Hint:* Use a three-step reversible process as if the water were taken through the normal freezing point.) The specific heat of ice is 2220 J/kg · K.

67. An apparatus that liquefies helium is in a room maintained at 300 K. If the helium in the apparatus is at 4.0 K, what is the minimum ratio of energy delivered as heat to the room to energy removed as heat from the helium?

68. To make ice, a freezer that is a reversed Carnot engine extracts 42 kJ as heat at −15°C during each cycle. The freezer has a coefficient of performance of 5.7. The room temperature is 30.3°C. (a) How much energy per cycle is delivered as heat to the room? (b) How much work per cycle is required to run the freezer?

69. (a) A Carnot engine operates between a hot reservoir at 320 K and a cold reservoir at 260 K. If it absorbs 500 J as heat per cycle at the hot reservoir, how much work per cycle does it deliver? (b) If the same engine, working in reverse, functions as a refrigerator between the same two reservoirs, how much work per cycle must be supplied to remove 1000 J as heat from the cold reservoir?

Tutorial Problem

70. An ideal gas (the system) is taken around a cycle back to its initial state through four processes (steps). Initially, the gas has a pressure of 1.00×10^5 Pa, a temperature of 300 K, and a volume of 0.900 m³. The internal energy is given by $E_{\text{int}} = 2.5nRT$, where T is in kelvins. (a) Determine how many moles of gas are in the system.

For each of the following four steps, find the final pressure, volume, and temperature of the system. Also find the change ΔE_{int} in the system, the energy Q absorbed or lost as heat by the system, and the work W done by or on the system. Finally, graph the path taken by the system on a p-V diagram, labeling the path with the step number and indicating its direction with an arrowhead. For notation, use i as a subscript for initial values and the step number as a subscript for the values at the end of each step. (b) Step 1 of the cyclic process is a constant-volume process with a final pressure of 2.00×10^5 Pa. (c) Step 2 is a constant-pressure process with a final temperature of 800 K. (d) Step 3 is a constant-volume process back to the initial pressure of 1.00×10^5 Pa. (e) Step 4 is a constant-pressure process back to the initial state.

(f) For the entire cycle, what are the net quantities ΔE_{int}, Q, and W? Explain how you can determine W directly from a p-V diagram of the cycle.

The thermal efficiency of a cyclic process is defined as the ratio of the net work done during the cycle to the total positive heat absorbed (counting Q only when $Q > 0$). (g) What is the thermal efficiency of the cycle? Suppose that the cycle were run in reverse; that is, do step 4, then 3, then 2, and then 1, each in reverse. (h) What would be the effect (if any) on the net quantities of part (f)? Imagine a different cyclic process with the same steps 1 and 2 but with steps 3 and 4 replaced by a shortcut directly back to the initial state along a straight line on a p-V diagram. (i) Describe in words how the net quantities determined in part (f) would be affected.

Answers

(a) Use the ideal gas law to write

$$n = \frac{p_i V_i}{RT_i}$$
$$= \frac{(1.00 \times 10^5 \text{ Pa})(0.900 \text{ m}^3)}{(8.31 \text{ J/K} \cdot \text{mol})(300 \text{ K})}$$
$$= 36.1 \text{ mol.}$$

(b) The volume remains 0.900 m³. At the pressure of 2.00×10^5 Pa, the new temperature is

$$T_1 = \frac{p_1 V_1}{nR}$$
$$= \frac{(2.00 \times 10^5 \text{ Pa})(0.900 \text{ m}^3)}{(36.1 \text{ mol})(8.31 \text{ J/K} \cdot \text{mol})}$$
$$= 600 \text{ K.}$$

The work done by the system is zero because there is no change in volume: $W_1 = 0$. The change in internal energy is

$$\Delta E_{int,1} = 2.5nR\,\Delta T$$
$$= (2.5)(36.1\text{ mol})(8.31\text{ J/K}\cdot\text{mol})(600\text{ K} - 300\text{ K})$$
$$= 225\text{ kJ}.$$

The heat is

$$Q_1 = \Delta E_{int,1} + W_1$$
$$= 225\text{ kJ} + 0 = 225\text{ kJ}.$$

Because the result is positive, the heat is absorbed by the system. On a p-V diagram this step forms a vertical line at the constant volume, from pressure 1.00×10^5 Pa to pressure 2.00×10^5 Pa.

(c) Since the pressure remains 2.00×10^5 Pa and the final temperature T_2 is 800 K, the final volume must be

$$V_2 = \frac{nRT_2}{p_2}$$
$$= \frac{(36.1\text{ mol})(8.31\text{ J/K}\cdot\text{mol})(800\text{ K})}{2.00 \times 10^5\text{ Pa}}$$
$$= 1.20\text{ m}^3.$$

Since the pressure remains constant, the work done by the system is

$$W_2 = p_2\,\Delta V$$
$$= (2.00 \times 10^5\text{ Pa})(1.20\text{ m}^3 - 0.900\text{ m}^3) = 60\text{ kJ}.$$

The change in internal energy is

$$\Delta E_{int,2} = 2.5nR\,\Delta T$$
$$= (2.5)(36.1\text{ mol})(8.31\text{ J/K}\cdot\text{mol})(800\text{ K} - 600\text{ K})$$
$$= 150\text{ kJ}.$$

Thus, the heat is

$$Q_2 = \Delta E_{int,2} + W_2$$
$$= 150\text{ kJ} + 60\text{ kJ} = 210\text{ kJ}.$$

Because the result is positive, the heat is absorbed by the system. On a p-V diagram this step forms a horizontal line at the constant pressure, from volume 0.900 m^3 to volume 1.20 m^3.

(d) The volume remains 1.20 m^3. At a pressure of 2.00×10^5 Pa, the temperature at the end of step 3 is

$$T_3 = \frac{p_3 V_3}{nR}$$
$$= \frac{(1.00 \times 10^5\text{ Pa})(1.20\text{ m}^3)}{(36.1\text{ mol})(8.31\text{ J/K}\cdot\text{mol})}$$
$$= 400\text{ K}.$$

The work done by the system is zero since there is no change in volume: $W_3 = 0$. The change in internal energy is

$$\Delta E_{int,3} = 2.5nR\,\Delta T$$
$$= (2.5)(36.1\text{ mol})(8.31\text{ J/K}\cdot\text{mol})(400\text{ K} - 800\text{ K})$$
$$= -300\text{ kJ}.$$

The heat is

$$Q_3 = \Delta E_{int,3} + W_3 = -300\text{ kJ} + 0 = -300\text{ kJ},$$

where the minus sign means that 300 kJ is removed from the system. On a p-V diagram this step forms a vertical line at the constant volume, from pressure 2.00×10^5 Pa to pressure 1.00×10^5 Pa.

(e) The final pressure is $p_f = p_i = 1.00 \times 10^5$ Pa, the final temperature is $T_f = T_i = 300$ K, and the final volume is $V_f = V_i = 0.900\text{ m}^3$. Since the pressure does not change during this step, the work done by the system is

$$W_4 = p_f\,\Delta V$$
$$= (1.00 \times 10^5\text{ Pa})(0.900\text{ m}^3 - 1.20\text{ m}^3) = -30\text{ kJ}.$$

The change in internal energy is

$$\Delta E_{int,4} = 2.5nR\,\Delta T$$
$$= (2.5)(36.1\text{ mol})(8.31\text{ J/K}\cdot\text{mol})(300\text{ K} - 400\text{ K})$$
$$= -75\text{ kJ}.$$

The heat is then

$$Q_4 = \Delta E_{int,4} + W_4$$
$$= -75\text{ kJ} - 30\text{ kJ} = -105\text{ kJ}.$$

On a p-V diagram this step forms a horizontal line at the constant pressure 1.00×10^5 Pa, from volume 1.20 m^3 to volume 0.900 m^3.

(f) The net change in internal energy should be zero. Let's check:

$$\Delta E_{int} = \Delta E_{int,1} + \Delta E_{int,2} + \Delta E_{int,3} + \Delta E_{int,4}$$
$$= 225\text{ kJ} + 150\text{ kJ} - 300\text{ kJ} - 75\text{ kJ} = 0.$$

The net heat to the system is

$$Q = Q_1 + Q_2 + Q_3 + Q_4$$
$$= 225\text{ kJ} + 210\text{ kJ} - 300\text{ kJ} - 105\text{ kJ} = 30\text{ kJ}.$$

The net work done by the system is

$$W = W_1 + W_2 + W_3 + W_4$$
$$= 0 + 60\text{ kJ} + 0 - 30\text{ kJ} = 30\text{ kJ}.$$

W is also just the area inside the path of the cyclic process on a p-V diagram:

$$W = (1.00 \times 10^5\text{ Pa})(0.30\text{ m}^3) = 30\text{ kJ}.$$

(g) The net work done is 30 kJ. The energy absorbed as heat during steps 1 and 2, which are the only steps in which $Q > 0$, is $225\text{ kJ} + 210\text{ kJ} = 435\text{ kJ}$. The thermal efficiency of this cycle is then $(30\text{ kJ})/(435\text{ kJ}) = 0.069$, or 6.9%.

(h) Running the cyclic process in reverse would reverse all the quantities ΔE_{int}, Q, and W from their values in the corresponding processes. Thus, the net quantities would also be reversed, so we would have $\Delta E_{int} = 0$, $Q = -30\text{ kJ}$, and $W = -30\text{ kJ}$.

(i) The shortcut would not affect the fact that $\Delta E_{int} = 0$, but it would halve the area inside the curve on a p-V diagram. Consequently, the energy absorbed as heat, which equals W for a closed path, would also be halved.

22
Electric Charge

Sample Problem 22-5

Simplifying with symmetry. Figure 22-22a shows an arrangement of six fixed charged particles, where $a = 2.0$ cm and $\theta = 30°$. All six particles have the same magnitude of charge, $q = 3.0 \times 10^{-6}$ C; their electrical signs are as indicated. What is the net electrostatic force $\vec{F}_{1,\text{net}}$ acting on q_1 due to the other charges?

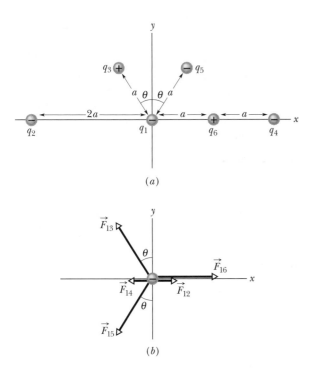

(a)

(b)

Fig. 22-22 Sample Problem 22-5. (a) An arrangement of six charged particles. (b) The electrostatic forces acting on particle 1 due to the other five particles.

SOLUTION: The **Key Idea** here is that $\vec{F}_{1,\text{net}}$ is the vector sum of forces $\vec{F}_{12}, \vec{F}_{13}, \vec{F}_{14}, \vec{F}_{15}$, and $\vec{F}_{16}$, which are the electrostatic forces acting on q_1 due to the other charges. We need to find and sum those forces. Because q_2 and q_4 are equal in magnitude and are both a distance $r = 2a$ from q_1, we have from Eq. 22-4

$$F_{12} = F_{14} = \frac{1}{4\pi\varepsilon_0}\frac{|q_1||q_2|}{(2a)^2}. \tag{22-16}$$

Similarly, since q_3, q_5, and q_6 are equal in magnitude and are each a distance $r = a$ from q_1, we have

$$F_{13} = F_{15} = F_{16} = \frac{1}{4\pi\varepsilon_0}\frac{|q_1||q_3|}{a^2}. \tag{22-17}$$

Another **Key Idea** is that we can use symmetry to simplify the summation of forces. Figure 22-22b is a free-body diagram for q_1. It and Eq. 22-16 show that $\vec{F}_{12}$ and $\vec{F}_{14}$ are equal in magnitude but opposite in direction; thus, those forces cancel and we need not evaluate them. Inspection of Fig. 22-22b and Eq. 22-17 reveals that the y components of $\vec{F}_{13}$ and $\vec{F}_{15}$ also cancel (and need not be evaluated), and that their x components are identical in magnitude and both are in the negative direction of the x axis. Figure 22-22b also shows us that $\vec{F}_{16}$ is in the positive direction of the x axis. Thus, $\vec{F}_{1,\text{net}}$ must be parallel to the x axis; its magnitude is the difference between F_{16} and twice the x component of $\vec{F}_{13}$:

$$F_{1,\text{net}} = F_{16} - 2F_{13}\sin\theta$$
$$= \frac{1}{4\pi\varepsilon_0}\frac{|q_1||q_6|}{a^2} - \frac{2}{4\pi\varepsilon_0}\frac{|q_1||q_3|}{a^2}\sin\theta.$$

Setting $q_3 = q_6$ and $\theta = 30°$, we find

$$F_{1,\text{net}} = \frac{1}{4\pi\varepsilon_0}\frac{|q_1||q_6|}{a^2} - \frac{2}{4\pi\varepsilon_0}\frac{|q_1||q_6|}{a^2}\sin 30° = 0. \quad \text{(Answer)}$$

Note that the presence of q_6 along the line between q_1 and q_4 does not alter the electrostatic force exerted by q_4 on q_1.

Sample Problem 22-6

If a current of 0.0010 A entered your body through, say, a fingertip, you would feel a mild tingling in that finger. How many conduction electrons would pass through the fingertip in 10 s?

SOLUTION: The **Key Idea** here is that the current through the fingertip is the amount of charge transferred by the electrons that pass through that point per unit time. From the relation of charge and current in Eq. 22-3 ($dq = i\,dt$), we can write

$$Q = it,$$

where Q is the charge transferred in time t. The number n of electrons transferring that charge is Q/e, where e is the magnitude of one electron's charge. Thus, we find

$$n = \frac{Q}{e} = \frac{it}{e} = \frac{(0.0010 \text{ A})(10 \text{ s})}{1.60 \times 10^{-19} \text{ C}} = 6.3 \times 10^{16}. \quad \text{(Answer)}$$

Sample Problem 22-7

An electrically neutral penny, of mass $m = 3.11$ g, contains equal amounts of positive and negative charge.

(a) Assuming that the penny is made entirely of copper, what is the magnitude q of the total positive (or negative) charge in the coin?

SOLUTION: One **Key Idea** here is that a neutral atom has a negative charge of magnitude Ze associated with its electrons and a positive charge of the same magnitude associated with the protons in its nucleus, where Z is the *atomic number* of the element in question. For copper, Appendix F tells us that Z is 29, which means that an atom of copper has 29 protons and, when electrically neutral, 29 electrons.

A second **Key Idea** is that the charge magnitude q we seek is equal to NZe, in which N is the number of atoms in the penny. To find N, we multiply the number of moles of copper in the penny by the number of atoms in a mole (Avogadro's number, $N_A = 6.02 \times 10^{23}$ atoms/mol). The number of moles of copper in the penny is m/M, where M is the molar mass of copper, 63.5 g/mol (from Appendix F). Thus, we have

$$N = N_A \frac{m}{M} = 6.02 \times 10^{23} \text{ atoms/mol} \frac{3.11 \text{ g}}{63.5 \text{ g/mol}}$$

$$= 2.95 \times 10^{22} \text{ atoms.}$$

We then find the magnitude of the total positive or negative charge in the penny to be

$$q = NZe$$

$$= (2.95 \times 10^{22})(29)(1.60 \times 10^{-19} \text{ C})$$

$$= 1.37 \times 10^5 \text{ C.} \qquad \text{(Answer)}$$

This is an enormous charge. (For comparison, if you rub a plastic rod with fur, you will be lucky to deposit any more than 10^{-9} C on the rod.)

(b) Suppose that the positive charge and the negative charge in a penny could be concentrated into two separate bundles, 100 m apart. What attractive force would act on each bundle?

SOLUTION: The **Key Idea** here is that the bundles would be small relative to their separation, so we can treat them as particles. Then from Eq. 22-4 we have

$$F = \frac{1}{4\pi\varepsilon_0} \frac{q^2}{r^2}$$

$$= \frac{(8.99 \times 10^9 \text{ N} \cdot \text{m}^2/\text{C}^2)(1.37 \times 10^5 \text{ C})^2}{(100 \text{ m})^2}$$

$$= 1.69 \times 10^{16} \text{ N.} \qquad \text{(Answer)}$$

This would be a huge force—from just the charges contained in a penny. However, it is actually impossible to disturb the electrical neutrality of ordinary matter very much. If we try to remove any sizable fraction of the charge of one electrical sign from a body, a large electrostatic force appears automatically, tending to pull it back.

QUESTIONS

12. Figure 22-23 shows three charged particles. Which of the 12 vectors in the figure best gives the force on the particle of charge $+q_2$ due to the presence of (a) the particle of charge $-q_1$ and (b) the particle of charge $+q_3$?

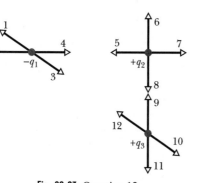

Fig. 22-23 Question 12.

13. *Organizing question:* Figure 22-24 shows four particles fixed in place in a plane. (a) Set up an equation, complete with known data, to find the x component of the net electrostatic force on the particle at the origin due to the other three particles. Here, $q_1 =$ 1.6×10^{-19} C, $q_2 = 3.2 \times 10^{-19}$ C, $q_3 = 1.6 \times 10^{-19}$ C, $q_4 = -3.2 \times 10^{-19}$ C. (b) Similarly set up an equation to find the corresponding y component.

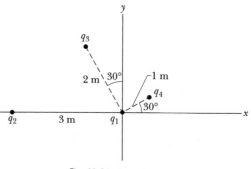

Fig. 22-24 Question 13.

14. Figure 22-25 shows three small spheres that have charges of equal magnitudes and rest on a frictionless surface. Spheres y and z are fixed in place and are equally distant from sphere x. If sphere x is released from rest, which of the five paths shown will it take?

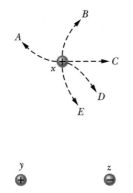

Fig. 22-25 Question 14.

15. In a simple model of the helium atom, two electrons orbit a nucleus consisting of two protons. Is the magnitude of the force exerted on the nucleus by one of the electrons greater than, less than, or the same as the magnitude of the force exerted on that electron by the nucleus?

16. You are given two identical neutral metal spheres A and B mounted on portable insulating supports, as well as a thin conducting wire and a glass rod that you can rub with silk. You can attach the wire between the spheres or between a sphere and the ground. You cannot touch the rod to a sphere. How can you give the spheres charges of (a) equal magnitudes and the same signs and (b) equal magnitudes and opposite signs?

17. Four identical conducting spheres A, B, C, and D have charges of $-8.0Q$, $-6.0Q$, $-4.0Q$, and $8.0Q$, respectively. Which should be connected together (by thin wire) to produce two or more spheres with charges of (a) $-2.0Q$ and (b) $-2.5Q$? (c) What sequence of connections will produce two spheres with charges of $-3.0Q$?

18. Figure 22-26 shows a pair of particles of charge Q and another pair of particles of charge q. The particle at the origin is free to move; the others are fixed in place. Should q be positive or negative if the net force on the free particle is to be zero and Q is (a) positive and (b) negative?

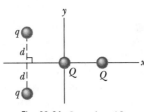

Fig. 22-26 Question 18.

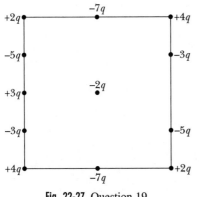

Fig. 22-27 Question 19.

19. In Fig. 22-27, a central particle of charge $-2q$ is surrounded by a square array of charged particles, separated by either distance d or $d/2$ along the perimeter of the square. What are the magnitude and direction of the net electrostatic force on the central particle due to the other particles?

20. In Fig. 22-28a, a particle of charge Q is fixed in place on an x axis. As an electron (not shown) moves slowly along the axis, the electrostatic force $\vec{F}$ on it due to Q is measured and graphed. When the force is in the positive direction of the axis, a positive value is graphed; when it is in the negative direction, a negative value is graphed. (a) Does Fig. 22-28b correspond to the situation in which Q is positive or the one in which Q is negative, or is the graph physically impossible? (b) Repeat (a) for Fig. 22-28c.

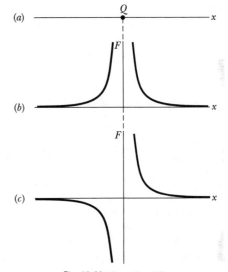

Fig. 22-28 Question 20.

21. Figure 22-29 shows three situations involving a charged particle and a uniformly charged spherical shell. The charges are given, and the radii of the shells are indicated. Rank the situations according to the magnitude of the force on the particle due to the presence of the shell, greatest first.

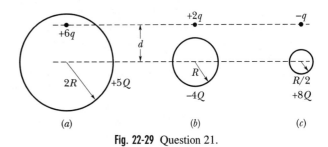

Fig. 22-29 Question 21.

DISCUSSION QUESTIONS

22. An electrically charged rod attracts bits of dry cork dust which, after touching the rod, often jump violently away from it. Explain.

23. The experiments described in Section 22-2 could be explained by postulating four kinds of charge—that is, one each for the glass, silk, plastic, and fur. What is the argument against this?

24. A charged insulator can be discharged by passing it just above a flame. Explain why.

25. If you rub a coin briskly between your fingers, it will not become charged. Why not?

26. Why do electrostatic experiments not work well on humid days?

27. How could you determine the sign of the charge on a charged, isolated rod?

28. The quantum of charge is 1.60×10^{-19} C. Is there a corresponding quantum of mass?

29. In Sample Problem 22-4 we show that the electrical force is about 10^{39} times stronger than the gravitational force. Can you conclude from this that a galaxy, a star, or a planet must be essentially neutral electrically?

30. How do we know that electrostatic forces are not the cause of the attraction between Earth and the Moon?

31. If the electrons in a metal such as copper are free to move about, they must often find themselves headed toward the metal surface. Why don't they keep on going and leave the metal?

32. Would it have made any important difference if Benjamin Franklin had interchanged his positive and negative labels for electric charge? Why, or why not?

33. Coulomb's law predicts that the force exerted by one point charge on another is proportional to the product of the two charges. How might you go about testing this aspect of the law in the laboratory?

34. "The charge of a particle is a constant characteristic of the particle, independent of its state of motion." Explain how you could test this statement by making a rigorous experimental check of whether the hydrogen atom is truly electrically neutral.

35. *Earnshaw's theorem* says that no particle can be in *stable* equilibrium under the action of electrostatic forces alone. Consider, however, point P at the center of a square of four particles, with equal positive charges, that are fixed in place, as in Fig. 22-30. If you put a positively charged particle at P, is it not in stable equilibrium? Explain.

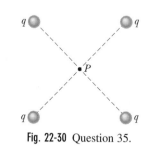

Fig. 22-30 Question 35.

EXERCISES & PROBLEMS

31. What would be the electrostatic force between two 1.00 C point charges separated by a distance of (a) 1.00 m and (b) 1.00 km if such a configuration could be set up?

32. If a cat repeatedly rubs against your cotton pants or skirt on a dry day, the charge transfer between the cat hair and the cotton can leave you with an excess charge of -2.00 μC. (a) How many charged particles are transferred between you and the cat?

You will gradually discharge via the floor, but if instead of waiting, you immediately reach toward a faucet, a painful spark can suddenly appear as your fingers near the faucet. (b) In that spark, do electrons flow from you to the faucet or vice versa? (c) Just before the spark appears, do you induce positive or negative charge in the faucet? (d) If, instead, the cat reaches a paw toward the faucet, which way do electrons flow in the resulting spark? (e) If you stroke a cat with a bare hand on a dry day, you should take care not to bring your fingers near the cat's nose or you will hurt it with a spark. Considering that cat hair is an insulator, explain how the spark can appear.

33. In the return stroke of a typical lightning bolt, a current of 2.5×10^4 A exists for 20 μs. How much charge is transferred in this event?

34. Two small, positively charged spheres have a combined charge of 5.0×10^{-5} C. If each sphere is repelled from the other by an electrostatic force of 1.0 N when the spheres are 2.0 m apart, what is the charge on each sphere?

35. Figure 22-31a shows two point charges, q_1 and q_2, held a fixed distance d apart. (a) What is the magnitude of the electrostatic force that acts on q_1? Assume that $q_1 = q_2 = 20.0$ μC and $d = 1.50$ m. (b) A third point charge $q_3 = 20.0$ μC is brought in and placed as shown in Fig. 22-31b. What now is the magnitude of the electrostatic force on q_1?

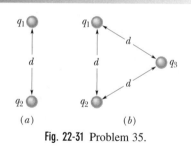

Fig. 22-31 Problem 35.

36. In Figure 22-32, three identical conducting spheres A, B, and C form an equilateral triangle of side length d and have initial charges of $-2Q$, $-4Q$, and $8Q$, respectively. (a) What is the magnitude of the electrostatic force between spheres A and C? The following steps are then taken: A and B are connected by a thin wire and then disconnected; B is grounded by the wire and the wire is then removed; B and C are connected by the wire and then disconnected. What now are the magnitudes of the electrostatic force (b) between spheres A and C and (c) between spheres B and C?

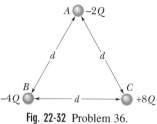

Fig. 22-32 Problem 36.

37. Two engineering students, John with a mass of 90 kg and Mary with a mass of 45 kg, are 30 m apart. Suppose each has a 0.01% imbalance in the amount of positive and negative charge, one student being positive and the other negative. Estimate *roughly* the electrostatic force of attraction between them by replacing each student with a sphere of water having the same mass as the student.

38. How far apart must two protons be if the magnitude of the electrostatic force acting on either one due to the other is equal to the magnitude of the gravitational force on a proton at Earth's surface?

39. A neutron consists of one "up" quark of charge $+2e/3$ and two "down" quarks each having charge $-e/3$. If the down quarks are 2.6×10^{-15} m apart inside the neutron, what is the magnitude of the electrostatic force between them?

40. In the radioactive decay of ^{238}U (see Eq. 22-13), the center of the emerging ^{4}He particle is, at a certain instant, 9.0×10^{-15} m from the center of the daughter nucleus ^{234}Th. At this instant, (a) what is the magnitude of the electrostatic force on the ^{4}He particle, and (b) what is that particle's acceleration?

41. Two point charges of -80 μC and 40 μC are held fixed on an x axis, at the origin and at $x = 20$ cm, respectively. What are the magnitude and direction of the total electrostatic force on a third charge of 20 μC placed at (a) $x = 40$ cm and (b) $x = 80$ cm? (c) Where on the x axis can the third charge be placed such that the total electrostatic force on it is zero?

42. Three charged particles form a triangle: particle 1 with charge $Q_1 = 80.0$ nC is at xy coordinates (0, 3.00 mm), particle 2 with charge Q_2 is at (0, -3.00 mm), and particle 3 with charge $q = 18.0$ nC is at (4.00 mm, 0). What are the magnitude and direction of the electrostatic force on particle 3 due to the other two particles if Q_2 is equal to (a) 80.0 nC and (b) -80.0 nC?

43. Two point charges, 40 μC and Q, are fixed on an x axis at the points $x = -2.0$ cm and $x = 3.0$ cm, respectively. A third point charge of magnitude 20 μC is released from rest at a point on the y axis at $y = 2.0$ cm. (a) What is the value of Q for which the total electrostatic force on this third charge is initially in the positive x direction? (b) What is the value of Q for which the total electrostatic force on this third charge is initially in the positive y direction?

44. Point charges of $+6.0$ μC and -4.0 μC are placed on an x axis, at $x = 8.0$ m and $x = 16$ m, respectively. What charge must be placed at $x = 24$ m so that any charge placed at the origin would experience no electrostatic force?

45. Two point charges of 30 nC and -40 nC are held fixed on an x axis, at the origin and at $x = 72$ cm, respectively. A particle with a charge of 42 μC is released from rest at $x = 28$ cm. If the initial acceleration of the particle has a magnitude of 100 km/s^2, what is its mass?

46. A nonconducting spherical shell, with an inner radius of 4.0 cm and an outer radius of 6.0 cm, has charge spread nonuniformly through its volume (between its inner and outer surfaces). The *volume charge density* ρ is the charge per unit volume, with the unit coulomb per cubic meter. For this shell it is given by $\rho = b/r$, where r is the distance in meters from the center of the shell and $b = 3.0$ μC/m^2. What is the net charge within the shell? (*Hint:* Because ρ is not uniform over the shell, you must integrate.)

47. A charged nonconducting rod, with a length of 2.00 m and a cross-sectional area of 4.00 cm^2, lies along the positive side of an x axis with one end at the origin. The *volume charge density* ρ is the charge per unit volume, with the unit coulomb per cubic meter. How many excess electrons are on the rod if the rod's volume charge density is (a) uniform, with a value of -4.00 μC/m^3,

and (b) nonuniform, with a value given by $\rho = bx^2$, where $b = -2.00$ μC/m^5?

48. The initial charges on the three identical metal spheres in Fig. 22-33 are the following: sphere A, Q; sphere B, $-Q/4$; and sphere C, $Q/2$, where $Q = 2.00 \times 10^{-14}$ C. Spheres A and B are fixed in place, with a center-to-center separation of $d = 1.20$ m, which is much larger than the spheres. Sphere C is touched first to sphere A and then to sphere B and is then removed. What then is the magnitude of the electrostatic force between spheres A and B?

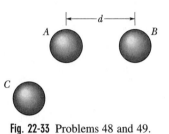

Fig. 22-33 Problems 48 and 49.

49. In Fig. 22-33, three identical conducting spheres initially have the following charges: sphere A, $4Q$; sphere B, $-6Q$; and sphere C, 0. Spheres A and B are fixed in place, with a center-to-center separation that is much larger than the spheres. Two experiments are conducted. In experiment 1, sphere C is touched to sphere A and then (separately) to sphere B, and then it is removed. In experiment 2, starting with the same initial states, the procedure is reversed: Sphere C is touched to sphere B and then (separately) to sphere A, and then it is removed. What is the ratio of the electrostatic force between A and B at the end of experiment 2 to that at the end of experiment 1?

50. In Fig. 22-34 the particles with charges q_1 and q_2 are fixed in place but the third particle is free to move. If the net electrostatic force on that third particle due to the other particles is zero, what is q_1 in terms of q_2?

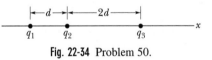

Fig. 22-34 Problem 50.

51. In Fig. 22-35, if $Q = +3.20 \times 10^{-19}$ C, $q = 1.60 \times 10^{-19}$ C, and $a = 2.00$ cm, what is the net electrostatic force on the particle at the origin due to the other charged particles?

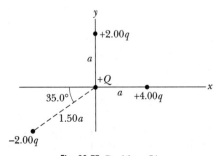

Fig. 22-35 Problem 51.

Tutorial Problem

52. Let's look at the electrostatic forces in a system of four charged particles that form a square with sides of length L (Fig. 22-36). The variable Q is a positive charge of unspecified magnitude. Let $\vec{F}_{12}$, $\vec{F}_{13}$, and $\vec{F}_{14}$ be the forces on particle 1 due to particles 2, 3, and 4, respectively.

(a) Show $\vec{F}_{12}$, $\vec{F}_{13}$, and $\vec{F}_{14}$ on a figure similar to Fig. 22-36. (b) Express each of these forces in the form $\vec{F} = F\hat{F}$, where F is a magnitude and $\hat{F}$ is a unit vector in the direction of $\vec{F}$. Each $\hat{F}$ should be expressed in $\hat{i}$-$\hat{j}$ notation. (c) Name the principle of physics that allows us to find the total force $\vec{F}_{1,net}$ from the individual forces $\vec{F}_{12}$, $\vec{F}_{13}$, and $\vec{F}_{14}$. Express $\vec{F}_{1,net}$ in (d) $\hat{i}$-$\hat{j}$ notation and (e) as a magnitude and direction. (f) Show $\vec{F}_{1,net}$ on a figure similar to Fig. 22-36.

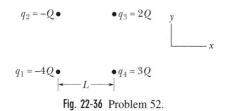

$q_2 = -Q$ $\bullet$ $\bullet$ $q_3 = 2Q$

$q_1 = -4Q$ $\bullet$ $\bullet$ $q_4 = 3Q$

$\longmapsto L \longrightarrow$

Fig. 22-36 Problem 52.

(g) Now assume that only charges q_3 and q_4 are present ($q_1 = q_2 = 0$) and that a positive charge q is placed on the line between charges q_3 and q_4. At what point on the line could the charge be located so that the net electrostatic force on q is zero? (h) Is the point you found one of stable, unstable, or neutral equilibrium? If the charge q were negative instead of positive, would the answers to this question and part (g) be the same or different? Explain.

Answers

(b) $\vec{F}_{14}$ is an attractive force because the two charges have opposite signs, so the force is directed from q_1 toward q_4—that is, in the direction $+\hat{i}$. The magnitude of the force is

$$F_{14} = \frac{k|q_4||q_1|}{L^2} = \frac{k(3Q)(4Q)}{L^2} = \frac{12kQ^2}{L^2},$$

where we have used k from Eq. 22-5. Thus,

$$\vec{F}_{14} = \frac{12kQ^2}{L^2}\,\hat{i}.$$

$\vec{F}_{13}$ is an attractive force because the two charges have opposite signs, so the force is directed from q_1 toward q_3—that is, at an angle of $45°$ from the x axis. Thus, the force is in the direction

$$\hat{F} = (\cos 45°)\hat{i} + (\sin 45°)\hat{j} = 0.707\hat{i} + 0.707\hat{j}.$$

The magnitude of the force is

$$F_{13} = \frac{k|q_3||q_1|}{L^2 + L^2} = \frac{4kQ^2}{L^2}.$$

Thus, $\vec{F}_{13} = \frac{2.828kQ^2}{L^2}\,(\hat{i} + \hat{j}).$

This can be written in the form $F\hat{F}$ as

$$\vec{F}_{13} = \frac{4kQ^2}{L^2}\,\frac{(\hat{i} + \hat{j})}{\sqrt{2}}.$$

$\vec{F}_{12}$ is a repulsive force, directed away from q_2, or in the $-\hat{j}$ direction, so $\hat{F} = -\hat{j}$. The magnitude of the force is

$$F_{12} = \frac{k|q_2||q_1|}{L^2} = \frac{4kQ^2}{L^2},$$

so

$$\vec{F}_{12} = -\frac{4kQ^2}{L^2}\,\hat{j}.$$

(c) To determine the total force on q_1 we use the principle of superposition: The total force on q_1 is the vector sum of all the individual forces acting on q_1 due to the other particles.

(d) In the notation using the unit vectors along the axes,

$$\vec{F}_{1,net} = \vec{F}_{12} + \vec{F}_{13} + \vec{F}_{14}$$

$$= -\frac{4kQ^2}{L^2}\,\hat{j} + \frac{2.828kQ^2}{L^2}\,(\hat{i} + \hat{j}) + \frac{12kQ^2}{L^2}\,\hat{i}$$

$$= \frac{kQ^2}{L^2}\,(14.828\hat{i} - 1.172\hat{j}).$$

(e) The magnitude of $\vec{F}_{1,net}$ is

$$F_{1,net} = \left(\frac{kQ^2}{L^2}\right)\sqrt{(14.828)^2 + (1.172)^2} = 14.9\left(\frac{kQ^2}{L^2}\right).$$

Its direction has angle (in the fourth quadrant)

$$\theta = \tan^{-1}\left(\frac{-1.172}{14.828}\right) = -4.5° \text{ or } 355.5°.$$

Therefore, we can say the total force is $14.9(kQ^2/L^2)$ in a direction at an angle $355.5°$ from the positive direction of the x axis (corresponding to $4.5°$ below the positive direction of the x axis).

(g) Suppose the charge q is a distance y from q_4, and thus a distance $L - y$ from q_3. For a positive charge q the force on q due to q_4 would be directed up and the force on q due to q_3 would be directed down. The magnitudes of these forces would be proportional to $|q_4|/y^2$ or $3Q/y^2$ and to $|q_3|/(L - y)^2$ or $2Q/(L - y)^2$, respectively. Thus, the condition for the net force to be zero is

$$\frac{3Q}{y^2} = \frac{2Q}{(L - y)^2},$$

which leads to

$$y^2 - 6Ly + 3L^2 = 0.$$

This is a quadratic equation with two roots: $y = 3L \pm \sqrt{6}L$. One of these is $y = 0.55L$, which is the root of interest, and the other is $y = 5.45L$, which is outside the range of interest. Thus, the point at which the net force on q is zero is about 55% of the way from q_4 to q_3. Reasonably, the equilibrium point is closer to q_3, which has the smaller magnitude of charge.

(h) The point is a point of unstable equilibrium, because if the charge q were displaced from the equilibrium point, the net force on it would not be back toward the equilibrium point. This may be seen by imagining a displacement along the line between q_3 and q_4; a displacement toward either of these charges would result in a net force toward that charge and not back to the equilibrium point.

If the charge q were negative instead of positive, the net force would be zero at the same point. The equilibrium point would be unstable in this case, too, as may be seen by considering a displacement perpendicular to the line between q_3 and q_4; the charge would be repelled farther away from that line.

23

Electric Fields

Sample Problem 23-6

Figure 23-41a shows a particle with charge $+9q$ at the origin of an x axis and a particle with charge $-q$ at $x = L$. At which points is the net electric field due to these two charges zero?

SOLUTION: The **Key Idea** here is that, if $\vec{E}_1$ is the electric field due to the charge $+9q$, and $\vec{E}_2$ is the electric field due to the charge $-q$, then the points we seek are where $\vec{E}_1 + \vec{E}_2 = 0$. This condition requires that

$$\vec{E}_1 = -\vec{E}_2. \qquad (23-41)$$

This tells us that at the points we seek, the electric field vectors due to the two charges must be of equal magnitude,

$$E_1 = E_2, \qquad (23-42)$$

and the vectors must have opposite directions.

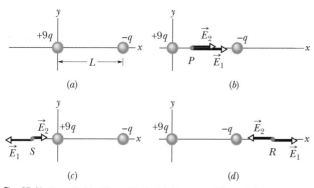

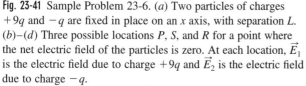

Fig. 23-41 Sample Problem 23-6. (a) Two particles of charges $+9q$ and $-q$ are fixed in place on an x axis, with separation L. (b)–(d) Three possible locations P, S, and R for a point where the net electric field of the particles is zero. At each location, $\vec{E}_1$ is the electric field due to charge $+9q$ and $\vec{E}_2$ is the electric field due to charge $-q$.

Recall that an electric field vector due to a positive charge is directed away from the positive charge, and that due to a negative charge is directed toward the negative charge. Thus, $\vec{E}_1$ and $\vec{E}_2$ could have opposite directions only for a location (or locations) on the x axis. At any location on the x axis between the two charges, such as P in Fig. 23-41b, $\vec{E}_1$ and $\vec{E}_2$ are in the same direction and thus cannot meet the requirement of Eq. 23-41.

At any location on the x axis to the left of the charge $+9q$, such as point S in Fig. 23-41c, vectors $\vec{E}_1$ and $\vec{E}_2$ point in opposite directions. However, Eq. 23-3 tells us that $\vec{E}_1$ and $\vec{E}_2$ cannot have equal magnitudes there: E_1 must be larger than E_2, because E_1 is produced by a closer charge (smaller r) of larger magnitude ($9q$ versus q).

Finally, at any point on the x axis to the right of the charge $-q$, vectors $\vec{E}_1$ and $\vec{E}_2$ are again in opposite directions. However, because now the charge of larger magnitude is *farther* away than the charge of smaller magnitude, there is a point at which E_1 is equal to E_2. Let x be the coordinate of this point, labeled R in Fig. 23-41d. Then with the aid of Eq. 23-3, we can rewrite Eq. 23-42 as

$$\frac{1}{4\pi\varepsilon_0}\frac{9q}{x^2} = \frac{1}{4\pi\varepsilon_0}\frac{q}{(x-L)^2}. \qquad (23-43)$$

(Note that only the magnitudes of the charges are used in Eq. 23-43.) Rearranging Eq. 23-43 gives us

$$\left(\frac{x-L}{x}\right)^2 = \frac{1}{9}.$$

After taking the square root of both sides, we have

$$\frac{x-L}{x} = \frac{1}{3},$$

which gives us

$$x = 1.5L. \qquad \text{(Answer)}$$

Sample Problem 23-7

The nucleus of a uranium atom has a radius R of 6.8 fm. Assuming that the positive charge of the nucleus is distributed uniformly within the nucleus, determine the electric field at a point on the surface of the nucleus due to that charge.

SOLUTION: The nucleus has a positive charge of Ze, where the atomic number Z ($= 92$) is the number of protons within the nucleus, and e ($= 1.60 \times 10^{-19}$ C) is the charge of a proton. One **Key Idea** here is that, because this charge is distributed uniformly, the electrostatic force on a positive test charge placed near the surface of the nucleus is the same as if the nuclear charge were concentrated at the nuclear center.

A second **Key Idea** is that, from Eq. 23-1, the electric field produced by the nucleus is also the same as if the nuclear charge were concentrated at the nuclear center. From Eq. 23-3, which applies to such a pointlike concentration of charge, we can write, for the magnitude of the field,

$$E = \frac{1}{4\pi\varepsilon_0} \frac{Ze}{R^2}$$

$$= \frac{(8.99 \times 10^9 \text{ N} \cdot \text{m}^2/\text{C}^2)(92)(1.60 \times 10^{-19} \text{ C})}{(6.8 \times 10^{-15} \text{ m})^2}$$

$$= 2.9 \times 10^{21} \text{ N/C}. \qquad \text{(Answer)}$$

Since the charge of the nucleus is positive, the electric field vector $\vec{E}$ points outward, away from the center of the nucleus.

Sample Problem 23-8

A molecule of water vapor causes an electric field in the surrounding space as if it were an electric dipole like that of Fig. 23-8. Its electric dipole moment has a magnitude $p = 6.2 \times 10^{-30}$ C · m. What is the magnitude of the electric field due to this electric dipole at a distance $z = 1.1$ nm from the molecule's dipole center, on its dipole axis?

SOLUTION: We shall assume (correctly) that the given distance $z = 1.1$ nm is large relative to the separation of positive and negative charge centers in the molecule. Then the **Key Idea** here is that the

magnitude of the electric field due to the electric dipole, at a point on the dipole axis, is related to distance z and the magnitude p of the dipole moment by Eq. 23-9. From that equation we obtain

$$E = \frac{1}{2\pi\varepsilon_0} \frac{p}{z^3}$$

$$= \frac{6.2 \times 10^{-30} \text{ C} \cdot \text{m}}{(2\pi)(8.85 \times 10^{-12} \text{ C}^2/\text{N} \cdot \text{m}^2)(1.1 \times 10^{-9} \text{ m})^3}$$

$$= 8.4 \times 10^7 \text{ N/C}. \qquad \text{(Answer)}$$

Sample Problem 23-9

The disk of Fig. 23-12 has a surface charge density σ of $+5.3$ μC/m^2 on its upper face. (This, incidentally, is a reasonable value for the surface charge density on the photosensitive cylinder of a photocopying machine.)

(a) What is the electric field at the surface of the disk?

SOLUTION: The **Key Idea** here is that at points near the surface of a uniformly charged disk, the magnitude E of the electric field is like that of an infinite sheet of charge. Thus, we can use Eq. 23-27 to find E, without the complications of Eq. 23-26. Equation 23-27 gives us

$$E = \frac{\sigma}{2\varepsilon_0} = \frac{5.3 \times 10^{-6} \text{ C/m}^2}{(2)(8.85 \times 10^{-12} \text{ C}^2/\text{N} \cdot \text{m}^2)}$$

$$= 3.0 \times 10^5 \text{ N/C}. \qquad \text{(Answer)}$$

This value holds for all points that are close to the surface of the disk but not near its edge.

(b) Using the binomial theorem, find an expression for the electric field at a point far from the disk and on its central axis.

SOLUTION: The phrase *far from the disk* means that the distance z is much greater than the size of the disk, as measured by, say, its radius. As you will see, this allows us to use the binomial theorem to approximate the square-root term in Eq. 23-26 for E.

From Appendix E, we write the general form of the binomial theorem as

$$(1 + x)^n = 1 + \frac{n}{1!}x + \frac{n(n-1)}{2!}x^2 + \cdots, \qquad (23\text{-}44)$$

where $|x| \ll 1$. To get ready for the approximation, we rewrite the

square-root term as

$$\frac{z}{\sqrt{z^2 + R^2}} = \frac{z}{z\sqrt{1 + \frac{R^2}{z^2}}} = \left(1 + \frac{R^2}{z^2}\right)^{-1/2},$$

which is in the proper form for use of the binomial theorem with $x = R^2/z^2$ and $n = -\frac{1}{2}$. Because z is much greater than R, the condition $|x| \ll 1$ is satisfied.

Applying Eq. 23-44, we now write

$$\left(1 + \frac{R^2}{z^2}\right)^{-1/2} = 1 + \frac{-\frac{1}{2}}{1!}\frac{R^2}{z^2} + \frac{-\frac{1}{2}(-\frac{1}{2} - 1)}{2!}\frac{R^4}{z^4} + \cdots.$$

Successive terms on the right side are progressively less because the ratio R/z, which is less than 1, is being raised to progressively greater powers. In fact, we can approximate the required result closely enough by discarding terms smaller than R^2/z^2, which leaves us

$$\frac{z}{\sqrt{z^2 + R^2}} = 1 - \frac{R^2}{2z^2}.$$

Substituting this expression in Eq. 23-26 gives us

$$E = \frac{\sigma}{2\varepsilon_0}\left[1 - \left(1 - \frac{R^2}{2z^2}\right)\right]$$

$$= \frac{\sigma}{4\varepsilon_0}\frac{R^2}{z^2}. \qquad \text{(Answer)}$$

We can rewrite this in terms of the charge q on the upper face of the disk by noting that $\sigma = q/A$ and, for the disk, $A = \pi R^2$. Then

$$E = \frac{\sigma}{4\varepsilon_0}\frac{R^2}{z^2} = \frac{q}{4\varepsilon_0 \pi R^2}\frac{R^2}{z^2}$$

$$= \frac{1}{4\pi\varepsilon_0}\frac{q}{z^2}. \qquad \text{(Answer)} \quad (23\text{-}45)$$

Equation 23-45 tells us that at points on the central axis where $z \gg R$, the electric field produced by the charge q spread over the face of the disk is the same as that produced by a particle of the same charge q. The disk "looks" like a particle of charge q from far away.

Sample Problem 23-10

In the Millikan oil-drop apparatus of Fig. 23-13, a drop of radius $R = 2.76\ \mu\text{m}$ has an excess charge of three electrons. What are the magnitude and direction of the electric field that is required to balance the drop so it remains stationary in the apparatus? The density ρ of the oil is 920 kg/m^3.

SOLUTION: The **Key Idea** here is that, to balance the drop, the electrostatic force acting on it must be upward and have a magnitude equal to the magnitude mg of the gravitational force on the drop. From Eqs. 23-28 and 23-29, we can write the *magnitude* of the electrostatic force as $F = (3e)E$. We can also write the mass m of the drop as the product of its volume $\frac{4}{3}\pi R^3$ and its density. Thus, the balance of forces gives us

$$(\tfrac{4}{3}\pi R^3 \rho)g = (3e)E.$$

Solving for E yields

$$E = \frac{4\pi R^3 \rho g}{9e}$$

$$= \frac{(4\pi)(2.76 \times 10^{-6}\ \text{m})^3(920\ \text{kg/m}^3)(9.80\ \text{m/s}^2)}{(9)(1.60 \times 10^{-19}\ \text{C})}$$

$$= 1.65 \times 10^6\ \text{N/C}. \qquad \text{(Answer)}$$

Because the drop is negatively charged, Eq. 23-28 tells us that the electric field $\vec{E}$ and the electrostatic force $\vec{F}$ are in opposite directions: $\vec{F} = -3e\vec{E}$. Since $\vec{F}$ is directed upward, the electric field must be directed downward.

QUESTIONS

13. Figure 23-42 shows two charged particles. Which of the figure's 12 vectors best depicts the electric field at point P due to (a) the particle of charge $+q$ and (b) the particle of charge $-q$?

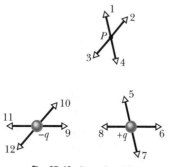

Fig. 23-42 Question 13.

14. Dust devils, vortexes of swirling air and dust that can extend upward a considerable distance from the ground, are electrified by the transfer of charge between dust grains and between the grains and the ground. The center of negative charge and the center of positive charge in a dust devil can then separate vertically to form an electric dipole.

Figure 23-43a depicts an experiment that revealed the dipole in a dust devil. The vertical component E_z of the atmosphere's electric field was measured as the dust devil moved past the measuring equipment at approximately constant velocity. (In clear weather, E_z is normally downward.) Figure 23-43b gives E_z versus time t during the experiment. (a) Which was higher in the dust devil, the center of negative charge or the center of positive charge?

(*Hint:* See Fig. 23-8.) (b) Was the electric dipole moment directed upward or downward? (c) Was the lower center of charge near ground level or well above ground level?

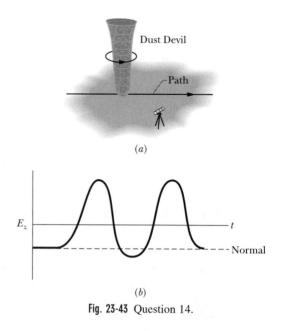

Fig. 23-43 Question 14.

15. *Organizing question:* Figure 23-44 shows a rod of uniform positive charge density λ lying along an x axis; it also shows a differential charge element dq located at distance x from one end of the rod. The net electric field $\vec{E}$ at point P due to the rod is to be evaluated. (a) What is the direction of the differential electric field $d\vec{E}$ at P due to element dq? In the expression

$$dE = \frac{1}{4\pi\varepsilon_0}\frac{dq}{r^2},$$

what should be substituted for (b) dq and (c) r^2 to prepare for the required integration? (d) What are the limits of the integration? (e) If $\vec{E}$ is to be expressed in terms of the charge Q on the rod, what should be substituted for the charge density?

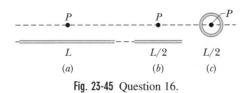

Fig. 23-44 Question 15.

16. Figure 23-45 shows three rods, each with the same charge Q spread uniformly along its length. Rods a (of length L) and b (of length $L/2$) are straight, and points P are aligned with their midpoints. Rod c (of length $L/2$) forms a complete circle about point P. Rank the rods according to the magnitude of the electric field they create at points P, greatest first.

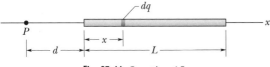

Fig. 23-45 Question 16.

17. Figure 23-46 shows two disks and a flat ring, each with the same uniform charge Q. Rank the objects according to the magnitude of the electric field they create at points P (which are at the same vertical heights), greatest first.

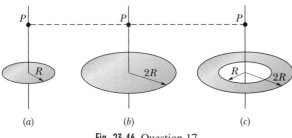

Fig. 23-46 Question 17.

18. An electron is released in a uniform electric field given by $\vec{E} = 4\hat{k}$, where $\hat{k}$ is the usual unit vector and $\vec{E}$ is in newtons per coulomb. What is the direction of (a) the force on the electron, (b) the acceleration of the electron, and (c) the rate of change $d\vec{p}/dt$ of the electron's momentum?

19. Figure 23-47 shows five protons that are launched in a uniform electric field $\vec{E}$; the magnitude and direction of the launch velocities are indicated. Rank the protons according to the magnitude of their accelerations due to the field, greatest magnitude first.

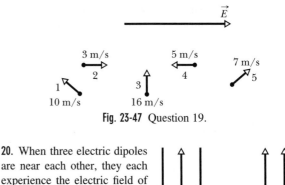

Fig. 23-47 Question 19.

20. When three electric dipoles are near each other, they each experience the electric field of the other two, and the three-dipole system has a certain potential energy. Figure 23-48 shows two arrangements in which three electric dipoles are side by side. All three dipoles have the same electric dipole moment magnitude, and the spacings between adjacent dipoles are identical. In which arrangement is the potential energy of the three-dipole system greater?

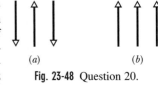

Fig. 23-48 Question 20.

DISCUSSION QUESTIONS

21. Name several scalar fields and vector fields.

22. We used a *positive* test charge to explore electric fields. Could we have used a negative test charge? Explain.

23. Electric field lines never cross. Why not?

24. In Fig. 23-4, why do the field lines around the edge of the figure appear to radiate uniformly from the center of the figure?

25. A point charge q of mass m is released from rest in a nonuniform field. Will it necessarily follow the line of force that passes through its release point?

26. A point charge is moving in an electric field perpendicular to the lines of force. Does any electrostatic force act on it?

27. Two point charges of unknown magnitude and sign are a distance d apart. The electric field is zero at one point on the line joining them. What can you conclude about the charges?

28. In Sample Problem 23-6, a charge placed at point R in Fig. 23-41d is in equilibrium because no force acts on it. Is the equilibrium stable (a) for displacements along the x axis and (b) for displacements perpendicular to this axis?

29. Two point charges of unknown sign and magnitude are fixed on an axis a distance L apart. Can we have $\vec{E} = 0$ at any off-axis points (excluding ∞)? Explain.

30. In Fig. 23-5, the force on the lower charge is directed up and is finite. The crowding of the lines of force, however, suggests that E is infinitely great at the site of this (point) charge. A charge immersed in an infinitely great field should have an infinitely great force acting on it. What is the solution to this dilemma?

31. A positive charge and a negative charge of the same magnitude lie on a long straight line. What is the direction of the electric field $\vec{E}$ due to these charges at points on this line that lie (a) between the charges, (b) outside the charges, on the side with the positive charge, and (c) outside the charges, on the side with the negative charge? (d) What is the direction of $\vec{E}$ for points off the line and in the median plane of the charges?

32. At points in the median plane of an electric dipole, is the electric field parallel or antiparallel to the electric dipole moment $\vec{p}$?

33. (a) Two identical electric dipoles are placed on a straight line, as shown in Fig. 23-49a. What is the direction of the electrostatic force on each dipole owing to the presence of the other? (b) Suppose that the dipoles are rearranged as in Fig. 23-49b. What now is the direction of the force on each dipole?

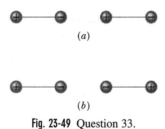

(a)

(b)

Fig. 23-49 Question 33.

34. What mathematical difficulties would you encounter if you were to calculate the electric field of a charged ring (or disk) at points *not* on the axis?

35. The field lines of Fig. 23-3 imply that $\vec{E}$ has the same value for all points in front of an infinite uniformly charged sheet, no matter how far they are from the sheet. Is this reasonable? One might think that the field should be stronger nearer the sheet, because of the proximity of the charges. Explain.

36. You turn an electric dipole end for end in a uniform electric field. How does the work you do depend on the initial orientation of the dipole with respect to the field?

37. For what orientations of an electric dipole in a uniform electric field is the potential energy of the dipole (a) greatest and (b) least?

38. An electric dipole is placed in a nonuniform electric field. Is there a net force on it?

39. An electric dipole is placed at rest in a uniform external electric field, as in Fig. 23-18a, and released. Discuss its motion.

EXERCISES & PROBLEMS

49. A clock face has negative point charges $-q, -2q, -3q, \ldots,$ $-12q$ fixed at the positions of the corresponding numerals. The clock hands do not perturb the net field due to the point charges. At what time does the hour hand point in the same direction as the electric field vector at the center of the dial? (*Hint:* Use symmetry.)

50. An electric dipole with dipole moment

$$\vec{p} = (3.00\hat{i} + 4.00\hat{j})(1.24 \times 10^{-30} \text{ C} \cdot \text{m})$$

is in an electric field $\vec{E} = (4000 \text{ N/C})\hat{i}$. (a) What is the potential energy of the electric dipole? (b) What is the torque acting on it? (c) If an external agent turns the dipole until its electric dipole moment is

$$\vec{p} = (-4.00\hat{i} + 3.00\hat{j})(1.24 \times 10^{-30} \text{ C} \cdot \text{m}),$$

how much work is done by the agent?

51. An electron is placed at each corner of an equilateral triangle having sides 20 cm long. What is the magnitude of the electric field at the midpoint of one of the sides?

52. Two particles, each of positive charge q, are fixed in place on a y axis, one at $y = d$ and the other at $y = -d$. (a) Write an expression that gives the magnitude E of the net electric field at points on the x axis given by $x = \alpha d$. (b) Graph E versus α for the range $0 < \alpha < 4$. From the graph, determine the values of α that give (c) the maximum value of E and (d) half the maximum value of E.

53. (a) What is the magnitude of the acceleration of an electron in a uniform electric field of 1.40×10^6 N/C? (b) How long would it take for the electron, starting from rest, to attain one-tenth the speed of light? (c) How far would it travel in that time?

54. A charge of 20 nC is uniformly distributed along a straight rod of length 4.0 m that is bent into a circular arc with a radius of 2.0 m. What is the magnitude of the electric field at the center of curvature of the arc?

55. A charge (uniform linear density = 9.0 nC/m) lies on string that is stretched along an x axis from $x = 0$ to $x = 3.0$ m. Determine the magnitude of the electric field at $x = 4.0$ m on the x axis.

56. Two charged particles are fixed in place in Fig. 23-50. What is the x coordinate of the point at which the net electric field is zero?

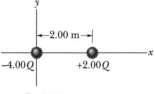

Fig. 23-50 Problem 56.

57. What are the magnitude and direction of the net electric field at point P in Fig. 23-51?

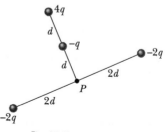

Fig. 23-51 Problem 57.

58. Figure 23-52 shows an uneven arrangement of electrons (e) and protons (p) on a circular arc of radius $r = 2.00$ cm. What are the magnitude and direction of the net electric field produced by the particles at the center of the arc?

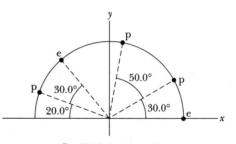

Fig. 23-52 Problem 58.

59. In Fig. 23-53, an electric dipole swings from an initial orientation i to a final orientation f in a uniform external electric field $\vec{E}$. The electric dipole moment is 1.60×10^{-27} C · m; the field magnitude is 3.00×10^6 N/C. What is the change in the dipole's potential energy?

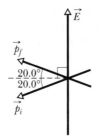

Fig. 23-53 Problem 59.

60. In Fig. 23-54, a pendulum is hung from the higher of two large horizontal plates. The pendulum consists of a small nonconducting sphere of mass m and charge $+q$ and an insulating thread of length l. What is the period of the pendulum if a uniform electric field $\vec{E}$ is set up between the plates by (a) charging the top plate negatively and the lower plate positively and (b) vice versa? In both cases, the field is directly away from one plate and toward the other plate.

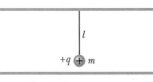

Fig. 23-54 Problem 60.

61. A spherical water drop 1.20 μm in diameter is suspended in calm air owing to a downward-directed atmospheric electric field of magnitude $E = 462$ N/C. (a) What is the magnitude of the gravitational force on the drop? (b) How many excess electrons does it have?

62. (a) What total (excess) charge q must the disk in Sample Problem 23-9 (Fig. 23-12) have for the electric field on the surface of the disk at its center to have a magnitude of 3×10^6 N/C, at which air breaks down electrically, producing sparks? Take the disk radius as 2.5 cm, and use the listing for air in Table 23-1. (b) Suppose that each atom at the surface has an effective cross-sectional area of 0.015 nm². How many atoms are needed to make up the disk's surface? (c) The charge in (a) results from some of the surface atoms having one excess electron. What fraction of the surface atoms must be so charged?

63. In Fig. 23-55, charged particles are placed at the vertices of an equilateral triangle. For what value of Q (both sign and magnitude) does the total electric field vanish at C, the center of the triangle?

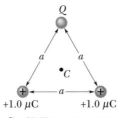

Fig. 23-55 Problem 63.

64. In Fig. 23-56, four charged particles form the corners of a square and four more charged particles lie at the midpoints of the sides of the square. The distance between adjacent particles on the perimeter of the

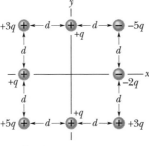

Fig. 23-56 Problem 64.

square is d. What are the magnitude and direction of the electric field at the center of the square?

65. A proton and an electron form two corners of an equilateral triangle of side length 2.0×10^{-6} m. What is the magnitude of their net electric field at the third corner?

66. A particle of charge $-q_1$ is located at the origin of an x axis. (a) At what location should a second particle of charge $-4q_1$ be placed so that the net electric field of the two particles is zero at $x = 2.0$ mm? (b) If, instead, a particle of charge $+4q_1$ is placed at that location, what is the direction of the net electric field at $x = 2.0$ mm?

67. A particle of charge Q lies at the origin of an xy coordinate system. A particle of charge q lies at coordinates $(4a, 0)$. We are concerned about the electric field these particles produce at point P at coordinates $(4a, 3a)$. What is Q in terms of q if, at P, (a) the x and y components of the electric field are equal and (b) the electric field has no y component?

68. The electric field in the xy plane produced by a positively charged particle is $7.2(4\hat{i} + 3\hat{j})$ N/C at the point $(3.0, 3.0)$ cm and $100\hat{i}$ N/C at the point $(2.0, 0)$ cm. What are (a) the x and y coordinates and (b) the charge of the particle?

Tutorial Problem

69. Let's look at a system of four point charges arranged in a square with sides of length L, with charges $q_1 = Q$, $q_2 = -2Q$, $q_3 = 3Q$, and $q_4 = -4Q$, as shown in Fig. 23-57. We are interested in determining the electric field at the center of the square. Then we will determine how this electric field would affect a charged particle placed at the center.

(a) On a diagram similar to Fig. 23-57, sketch the contribution to the electric field vector from each of the four charges by drawing vectors in the correct directions and approximately to scale. Label these vectors, using a fitting notation, such as $\vec{E}_1$ for the contribution to the electric field from charge q_1, $\vec{E}_2$ for charge q_2, and so on, reserving $\vec{E}$ for the total (net) electric field.

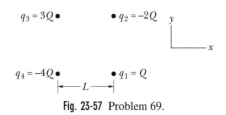

Fig. 23-57 Problem 69.

(b) Determine the contribution of each charge to the electric field algebraically (with symbols, such as Q and L), using the standard unit-vector notation. (c) Name the principle of physics that is used to determine the net electric field from the individual contributions found in part (b). Determine the total electric field at the center of the square. Show this vector on your diagram. (d) Now assume that $Q = 0.150$ μC and $L = 60.0$ cm. What is the numerical value of the electric field at the center of the square?

(e) Suppose an electron were placed at the center of the square. What would be the force on the electron? In which direction would the electron accelerate? What would be its initial acceleration?

(f) Suppose, instead, a proton were placed at the center of the square. What would be the force on it? In which direction would it accelerate? What would be its initial acceleration? Compare the force and the acceleration of the proton with those of the electron. (g) At a great distance from this system of charges, what does the system appear to be?

Answers

(b) The individual charges q_i contribute $\vec{E}_i = -(kq_i/r_i^2)\hat{r}_i$ to the electric field at the center, which we can take as the origin, $\hat{r}_i$ being the unit vector directed toward the charge; the minus sign indicates that the electric field points away from a positive charge q_i. To determine these contributions, we can make a table listing all the terms that show up in $\vec{E}_i$:

i	q_i	r_i	$\hat{r}_i$	$\vec{E}_i = -(kq_i/r_i^2)\hat{r}_i$
1	Q	$L/\sqrt{2}$	$(+\hat{i} - \hat{j})/\sqrt{2}$	$kQ(\sqrt{2}/L^2)(-\hat{i} + \hat{j})$
2	$-2Q$	$L/\sqrt{2}$	$(+\hat{i} + \hat{j})/\sqrt{2}$	$2kQ(\sqrt{2}/L^2)(+\hat{i} + \hat{j})$
3	$3Q$	$L/\sqrt{2}$	$(-\hat{i} + \hat{j})/\sqrt{2}$	$3kQ(\sqrt{2}/L^2)(+\hat{i} - \hat{j})$
4	$-4Q$	$L/\sqrt{2}$	$(-\hat{i} - \hat{j})/\sqrt{2}$	$4kQ(\sqrt{2}/L^2)(-\hat{i} - \hat{j})$

(c) The principle of superposition can be used to find the total electric field at the center of the square from the individual contributions found in part (b). The sum of the four electric fields is

$$\vec{E} = \vec{E}_1 + \vec{E}_2 + \vec{E}_3 + \vec{E}_4$$
$$= \frac{\sqrt{2}kQ}{L^2}[(-1 + 2 + 3 - 4)\hat{i} + (1 + 2 - 3 - 4)\hat{j}]$$
$$= -\frac{4\sqrt{2}kQ}{L^2}\hat{j}.$$

Note that the x components of the electric field cancel one another.

(d) Substituting the given values into the result of part (c), we find

$$\vec{E} = -\frac{4\sqrt{2}kQ}{L^2}\hat{j}$$
$$= -\frac{4\sqrt{2}(8.99 \times 10^9 \text{ N} \cdot \text{m}^2/\text{C}^2)(0.150 \times 10^{-6} \text{ C})}{(0.600 \text{ m})^2}\hat{j}$$
$$= -(2.12 \times 10^4 \text{ N/C})\hat{j}.$$

(e) The force on a charge q at a point where the electric field is $\vec{E}$ is $\vec{F} = q\vec{E}$. For an electron, $q = -e = -1.602 \times 10^{-19}$ C, so the force on the electron at the center of the square would be

$$\vec{F} = -e\vec{E} = -(1.602 \times 10^{-19} \text{ C})(-2.12 \times 10^4 \text{ N/C})\hat{j}$$
$$= +(3.40 \times 10^{-15} \text{ N})\hat{j}.$$

This force is in the $+\hat{j}$ direction, so the electron, which is attracted to the positive charges and repelled by the negative charges, is accelerating toward the midpoint of the top side of the square. The initial acceleration of the electron is found from Newton's second law:

$$\vec{a} = \frac{\vec{F}}{m} = \frac{(3.40 \times 10^{-15} \text{ N})\hat{j}}{9.11 \times 10^{-31} \text{ kg}} = (3.73 \times 10^{15} \text{ m/s}^2)\hat{j}.$$

(f) The force on a charge q at a point where the electric field is $\vec{E}$ is $\vec{F} = q\vec{E}$. For a proton, $q = +e = +1.602 \times 10^{-19}$ C, so the force on the proton at the center of the square would be

$$\vec{F} = +e\vec{E} = +(1.602 \times 10^{-19} \text{ C})(-2.12 \times 10^4 \text{ N/C})\hat{j}$$
$$= -(3.40 \times 10^{-15} \text{ N})\hat{j}.$$

It would accelerate toward the midpoint of the bottom side of the square. The initial acceleration of the proton is found from Newton's second law:

$$\vec{a} = \frac{\vec{F}}{m} = \frac{-(3.40 \times 10^{-15} \text{ N})\hat{j}}{1.67 \times 10^{-27} \text{ kg}}$$
$$= -(2.04 \times 10^{12} \text{ m/s}^2)\hat{j}.$$

The force on the proton is equal in magnitude but opposite in direction to the force experienced by the electron in part (e). The initial acceleration of the proton is smaller in magnitude and opposite in direction to the acceleration of the electron in part (e).

(g) At a great distance from this system of charges, the system will appear to be a single charge equal to the net charge: $(+Q - 2Q + 3Q - 4Q) = -2Q$. In other words, at a great distance from this system of charges, the electric field will be almost exactly the same as the electric field of a single charge $-2Q$; the difference would be difficult to detect.

24

Gauss' Law

Sample Problem 24-7

Figure 24-37 shows a long plastic cylinder of radius $R_c = 5.0$ mm and linear charge density $\lambda_c = -7.0$ μC/m that is coaxial with a long wire of radius $R_w = 0.10$ mm and linear charge density $\lambda_w = 5.0$ μC/m.

(a) What are the magnitude and direction of the electric field $\vec{E}_1$ between the wire and the cylinder?

SOLUTION: We want $\vec{E}_1$ for radial distances $R_w < r < R_c$. One **Key Idea** here is that, because of the cylindrical symmetry, we can apply Gauss' law with a Gaussian surface in the form of a cylinder. We place this Gaussian cylinder so it is coaxial with the wire and give it a radius r such that $R_w < r < R_c$. A second **Key Idea** is that the *only* charge enclosed by the Gaussian cylinder is on the wire; thus, the charge on the plastic cylinder does not affect $\vec{E}_1$, and the situation is like that of Fig. 24-12. The cylindrical symmetry then leads us to Eq. 24-12 ($E = \lambda/2\pi\varepsilon_0 r$), which in this situation becomes

$$E_1 = \frac{\lambda_w}{2\pi\varepsilon_0 r} = \frac{5.0 \times 10^{-6} \text{ C/m}}{2\pi(8.85 \times 10^{-12} \text{ C}^2/\text{N} \cdot \text{m}^2)r}$$
$$= \frac{9.0 \times 10^4 \text{ N} \cdot \text{m/C}}{r}, \qquad \text{(Answer)}$$

with r in meters and $R_w < r < R_c$. Because the charge on the wire is positive, $\vec{E}_1$ is directed radially outward from the wire.

(b) What are the magnitude and direction of the electric field $\vec{E}_2$ outside the plastic cylinder?

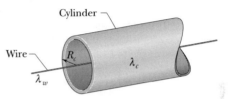

Fig. 24-37 Sample Problem 24-7. A long plastic cylinder of radius R_c and linear charge density λ_c is coaxial with a long wire of radius R_w and linear charge density λ_w.

SOLUTION: Now we want $\vec{E}_2$ for radial distances $r > R_c$. The **Key Ideas** of part (a) apply here also, but this time we place a coaxial Gaussian cylinder around both wire and plastic cylinder. Then the situation is again like that of Fig. 24-12 except now the charge enclosed by the Gaussian cylinder includes both the charge on the wire and the charge on the cylinder. Thus, we now write Eq. 24-12 as

$$E_2 = \frac{\lambda_w + \lambda_c}{2\pi\varepsilon_0 r}$$
$$= \frac{5.0 \times 10^{-6} \text{ C/m} + (-7.0 \times 10^{-6} \text{ C/m})}{2\pi(8.85 \times 10^{-12} \text{ C}^2/\text{N} \cdot \text{m}^2)r}$$
$$= \frac{-3.6 \times 10^4 \text{ N} \cdot \text{m/C}}{r}, \qquad \text{(Answer)}$$

with r in meters and $r > R_c$. The minus sign tells us that $\vec{E}_2$ is directed radially inward toward the wire and plastic cylinder. (Note that the net charge enclosed by the Gaussian cylinder is negative.)

Sample Problem 24-8

The nucleus of an atom of gold has a radius $R = 6.2 \times 10^{-15}$ m and a positive charge $q = Ze$, where the atomic number Z of gold is 79. Plot the magnitude of the electric field due to the nucleus, from the center of the nucleus outward to a distance of about twice its radius. Assume that the nucleus is spherical with a uniform charge distribution.

SOLUTION: One **Key Idea** here is to separate the problem into two parts, one dealing with the electric field outside the nucleus and one dealing with the electric field inside the nucleus. A second **Key Idea** is that, because of the spherical symmetry, we can apply the results for Gauss' law that we found in Section 24-9.

Outside the nucleus: The **Key Idea** here is that this situation is like that of Fig. 24-19a, for which we can use either Eq. 23-3 or Eq. 24-15 for the electric field due to a point charge:

$$E = \frac{1}{4\pi\varepsilon_0} \frac{q}{r^2}. \qquad (24\text{-}21)$$

A plot of this equation would give the field magnitude at points outside the nucleus and at distances r from the nuclear center. We would, of course, need to substitute values of r and the total charge q of the nucleus, which is

$$q = Ze = (79)(1.60 \times 10^{-19} \text{ C}) = 1.264 \times 10^{-17} \text{ C}.$$

For a point on the surface of the nucleus, for example, Eq. 24-21 yields

$$E = \frac{1.264 \times 10^{-17} \text{ C}}{(4\pi)(8.85 \times 10^{-12} \text{ C}^2/\text{N} \cdot \text{m}^2)(6.2 \times 10^{-15} \text{ m})^2}$$

$$= 3.0 \times 10^{21} \text{ N/C}.$$

Inside the nucleus: The **Key Idea** now is that this situation is like that of Fig. 24-19b, for which we would use Eq. 24-20:

$$E = \left(\frac{q}{4\pi\varepsilon_0 R^3}\right) r. \qquad (24\text{-}22)$$

A plot of this equation would give us the field magnitude at points within the nucleus.

The quantity in parentheses in Eq. 24-22 is a constant, so, within the nucleus, E is directly proportional to r and is zero at the nuclear center. (Comparison of Eqs. 24-21 and 24-22 shows that they give the same result, 3.0×10^{21} N/C, at $r = R$. This simply

tells us that the "inside equation" and the "outside equation" are compatible where they both apply.) Figure 24-38 shows these results graphically. To obtain the figure, we plot Eq. 24-22 for $0 \le r \le 6.2 \times 10^{-15}$ m, and Eq. 24-21 for $r > 6.2 \times 10^{-15}$ m.

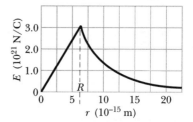

Fig. 24-38 Sample Problem 24-8. The variation of electric field magnitude with distance from the center of the nucleus of a gold atom. The positive charge is assumed to be distributed uniformly throughout the volume of the nucleus, which has radius R.

QUESTIONS

11. *Organizing question:* Figure 24-39 shows a section of a very long cylindrical rod of nonconducting material in which the volume charge density ρ is uniform. It also shows two cylindrical Gaussian surfaces A and B that are coaxial with the rod. Surface A is inside the rod; surface B is outside. The radius of the rod is R, that of surface A is r_A, and that of surface B is r_B. Let L be a length along the axis of the rod, and let E_A and E_B be the magnitudes of the electric field on surfaces A and B, respectively.

In terms of these symbols and ε_0, set up the left and right sides of Gauss' law (Eq. 24-7) for (a) surface A and (b) surface B. (For the left side of Eq. 24-7 you need to replace the integral with its equivalent in terms of the given symbols; also, for the right side, you need to replace q_{enc} with its equivalent in terms of the given symbols.)

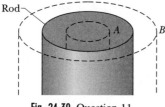
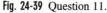

Fig. 24-39 Question 11.

12. Figure 24-40 shows four Gaussian surfaces consisting of identical cylindrical midsections but different end caps. The surfaces are in a uniform electric field $\vec{E}$ that is directed parallel to the central axis of the cylindrical midsections. The end caps of surface S_1 are convex hemispheres; those of surface S_2 are concave hemispheres; those of surface S_3 are cones; and those of surface S_4 are flat disks. Rank the surfaces according to (a) the net electric flux through them and (b) the electric flux through the top end caps, greatest first.

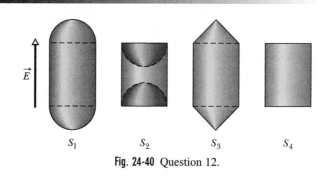

Fig. 24-40 Question 12.

13. You are given a collection of eight particles with charges $+2q$, $+3q$, $+4q$, $+5q$, $-2q$, $-3q$, $-4q$, and $-5q$. You are also given the goal of enclosing one or more of them with various Gaussian surfaces in turn, so that the net fluxes through the surfaces are 0, $+q/\varepsilon_0$, $+2q/\varepsilon_0$, . . . , $+14q/\varepsilon_0$. Which is impossible to produce?

14. Figure 24-41 shows, in section, three long charged cylinders centered on the same axis. Central cylinder A has a uniform charge $q_A = +3q_0$. What uniform charges q_B and q_C should be on cylinders B and C so that (if possible) the net electric field is zero (a) at point 1, (b) at point 2, and (c) at point 3?

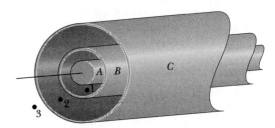

Fig. 24-41 Question 14.

15. *Organizing question:* Figure 24-42 shows a ball of nonconducting material in which the volume charge density ρ is uniform. It also shows, in cross section, two spherical Gaussian surfaces A and B centered on the ball. The radius of the ball is R, that of surface A is r_A, and that of surface B is r_B. Let E_A and E_B be the magnitudes of the electric fields on surfaces A and B, respectively.

Fig. 24-42 Question 15.

In terms of these symbols and ε_0, set up the left and right sides of Gauss' law (Eq. 24-7) for (a) surface A and (b) surface B. (For the left side of Eq. 24-7 you need to replace the integral with its equivalent in terms of the given symbols; also, for the right side, you need to replace q_{enc} with its equivalent in terms of the given symbols.)

16. In Checkpoint 4, what are the magnitude and direction of the electric field at a point that is a distance r from the center of the ball and spherical shell if the point is (a) between the ball and the shell, (b) within the metal of the shell, and (c) outside the shell?

17. A spherical nonconducting balloon has a uniform positive charge on its surface. If the balloon is expanded, does the magnitude of the electric field due to the charge increase, decrease, or remain the same at points that (a) are inside the balloon, (b) are on the balloon's surface, (c) were outside and are now inside, and (d) were and still are outside?

18. Figure 24-43 shows, in cross section, two Gaussian spheres and two Gaussian cubes that are centered on a positively charged particle. (a) Rank the net flux through the four Gaussian surfaces, greatest first. (b) Rank the magnitudes of the electric fields on the surfaces, greatest first, and indicate whether the magnitudes are uniform or variable along each surface.

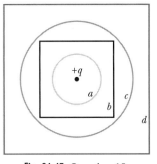

Fig. 24-43 Question 18.

19. Figure 24-44 shows three hollow conducting spheres of the same size; the net charge of each sphere is given. Rank the spheres according to the magnitudes of the electric fields they produce, greatest first, at (a) points P_1, which

are at the same radial distance within the hollows; (b) points P_2, which are at the same radial distance within the material of the spheres; and (c) points P_3, which are at the same radial distance outside the spheres.

20. In Fig. 24-45, an electron is released between two infinite nonconducting sheets that are horizontal and have uniform surface charge densities $\sigma_{(+)}$ and $\sigma_{(-)}$, as indicated. The electron is subjected to the following three situations involving surface charge densities and sheet separations. Rank the magnitudes of the electron's acceleration in these situations, greatest first:

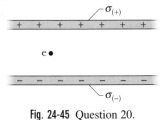

Fig. 24-45 Question 20.

Situation	$\sigma_{(+)}$	$\sigma_{(-)}$	Separation
1	$+4\sigma$	-4σ	d
2	$+7\sigma$	$-\sigma$	$4d$
3	$+3\sigma$	-5σ	$9d$

DISCUSSION QUESTIONS

21. What is the basis for the statement that electric field lines begin and end only on electric charges?

22. Positive charges are sometimes called "sources" and negative charges "sinks" of electric field. How would you justify this terminology? Are there sources and/or sinks of gravitational field?

23. Can Gauss' law be rewritten so that it pertains to the flow of water? Consider various Gaussian surfaces intersecting or enclosing a fountain or a waterfall in different ways. What would correspond to positive and negative charges in this case?

24. A surface encloses an electric dipole. What can you say about the net electric flux through this surface?

25. Suppose that a Gaussian surface encloses no net charge. Does Gauss' law require that $\vec{E}$ equal zero for all points on the surface? Is the converse of this statement true; that is, if $\vec{E}$ equals zero everywhere on the surface, does Gauss' law require that there be no net charge inside the surface?

26. Would Gauss' law be useful in calculating the field due to three equal charges located at the corners of an equilaeral triangle? Explain why or why not.

27. A total charge Q is distributed uniformly throughout a cube of edge length a. Is the resulting electric field at an external point P, a distance r from the center of the cube, given by $E = Q/4\pi\varepsilon_0 r^2$? See Fig. 24-46. If not, can E be found by constructing a "concentric" cubical Gaussian surface? If not, explain why not. Can you say anything about E if $r \gg a$?

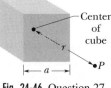

Fig. 24-46 Question 27.

28. Is $\vec{E}$ necessarily zero inside a charged rubber (insulating) balloon if the balloon is (a) spherical or (b) sausage shaped? For each

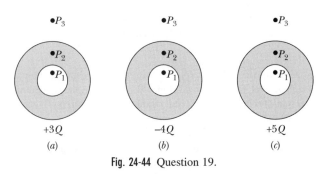

$\bullet P_3$　　　　$\bullet P_3$　　　　$\bullet P_3$

$+3Q$　　　　$-4Q$　　　　$+5Q$

(a)　　　　(b)　　　　(c)

Fig. 24-44 Question 19.

shape, assume the charge to be distributed uniformly over the surface. How would the situation change, if at all, if the balloon had a thin layer of conducting paint on its outside surface?

29. In Section 24-5 you saw that Coulomb's law can be derived from Gauss' law. Does this necessarily mean that Gauss' law can be derived from Coulomb's law?

30. A large, insulated, hollow conductor has a positive charge. A small metal ball carrying a negative charge of the same magnitude is lowered by a thread through a small opening in the top of the conductor, allowed to touch the inner surface, and then withdrawn. What is then the charge on (a) the conductor's inside surface, (b) its outside surface, and (c) the ball?

31. Can we deduce from the argument of Section 24-6 that the electrons in the wires of a house electrical wiring system move along the surfaces of those wires when current flows? If not, why not?

32. Does Gauss' law, as applied in Section 24-6, require that all the conduction electrons in an insulated conductor reside on the surface?

33. Suppose that you have a Gaussian surface in the shape of a

donut and that it encloses a single point charge. Does Gauss' law hold? If not, why not? If so, is there enough symmetry in the situation to apply Gauss' law usefully?

34. A positive point charge q is located at the center of a hollow metal sphere. What charges appear on (a) the inner surface and (b) the outer surface of the sphere? (c) If you bring an (uncharged) metal object near the sphere, will it change your answer in (a) or (b)? Will it change the way charge is distributed over the sphere?

35. Explain why the symmetry of Fig. 24-12 restricts us to consideration of only radial components of $\vec{E}$.

36. In Section 24-7, the *total* charge on the infinite rod is infinite. Why is not E also infinite? After all, according to Coulomb's law, if q is infinite, so is E.

37. Explain why the symmetry of Fig. 24-15 restricts us to consideration of only components of $\vec{E}$ directed away from the sheet. Why, for example, could $\vec{E}$ not have components parallel to the sheet?

38. Explain why the spherical symmetry of Fig. 24-8 restricts us to consideration of only radial components of field $\vec{E}$.

EXERCISES & PROBLEMS

49. A spherical conducting shell with a net charge of $-3.00Q$ contains a particle of charge $+5.00Q$ in its hollow. The shell has an inner radius of 0.80 m and an outer radius of 1.4 m. What are the magnitudes and directions of the electric fields at (a) point A at radius 0.500 m, (b) point B at radius 1.00 m, and (c) point C at radius 2.00 m?

50. Point P is between two infinite, parallel, nonconducting sheets of charge, at distance d from one sheet with surface charge density $-3.00\sigma_1$ and at distance $3.00d$ from the other sheet with surface charge density $+\sigma_1$. What are the magnitude and direction of the electric field at P due to the two sheets?

51. A conducting sphere with positive charge Q is surrounded by a spherical conducting shell. (a) What is the net charge on the inner surface of the shell? (b) Another positive charge q is placed outside the shell. Now what is the net charge on the inner surface of the shell? (c) If q is moved to a position between the shell and the sphere, what then is the net charge on the inner surface of the shell? (d) Are your answers valid if the sphere and the shell are not concentric?

52. Find the electric flux Φ through (a) the flat base and (b) the curved surface of a hemisphere of radius R. The field $\vec{E}$ is uniform and perpendicular to the flat base of the hemisphere, and the field lines enter through the flat base.

53. The net electric flux through each face of a die (singular of dice) has a magnitude in units of 10^3 N · m²/C that is exactly equal to the number of spots N on the face (1 through 6). The flux is inward for N odd and outward for N even. What is the net charge inside the die?

54. An electric field given by $\vec{E} = 4\hat{i} - 3(y^2 + 2)\hat{j}$ pierces the Gaussian cube of Fig. 24-5. (E is in newtons per coulomb and x is in meters.) What is the electric flux through (a) the top face,

(b) the bottom face, (c) the left face, and (d) the back face? (e) What is the net electric flux through the cube?

55. What net charge is enclosed by the Gaussian cube of Problem 54?

56. Figure 24-47 shows, in cross section, three infinitely large nonconducting sheets on which charge is uniformly spread. The surface charge densities are $\sigma_1 = +2.0$ μC/m², $\sigma_2 = +4.0$ μC/m², and $\sigma_3 = -5.0$ μC/m², and distance $L = 1.50$ cm. What are the magnitude and direction of the net electric field at point P?

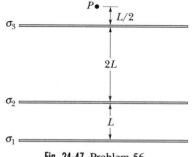

Fig. 24-47 Problem 56.

57. A charge of 6.00 pC is spread uniformly throughout the volume of a sphere of radius $r = 4.00$ cm. What are the magnitude and direction of the electric field at a radial distance of (a) 6.00 cm and (b) 3.00 cm?

58. The electric field in a particular space is $\vec{E} = (x + 2)\hat{i}$ N/C, with x measured in meters. Consider a cylinder of radius = 20 cm that is coaxial with the x axis. One end of the cylinder is positioned at $x = 0$, and the other end at $x = 2.0$ m. (a) What is the magnitude of the electric flux through the end of the cylinder at $x = 2.0$ m? (b) What net charge is enclosed within the cylinder?

59. A uniform charge density of 500 nC/m³ is distributed throughout a spherical volume of radius 6.0 cm. Consider a cubical surface with its center at the center of the sphere. What is the electric flux through this cubical surface if its edge length is (a) 4.0 cm and (b) 14 cm?

60. In Fig. 24-48, the charge on a neutral isolated conductor is separated by a nearby positively charged rod. What is the net flux through each of the five Gaussian surfaces shown in cross section? Assume that the charges enclosed by S_1, S_2, and S_3 are equal in magnitude.

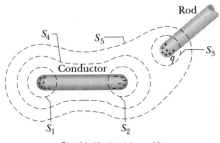

Fig. 24-48 Problem 60.

61. A thin-walled metal sphere has a radius of 25 cm and a charge of 2.0×10^{-7} C. Find E for a point (a) inside the sphere, (b) just outside the sphere, and (c) 3.0 m from the center.

62. Charge of uniform density 8.0 nC/m² is distributed over the entire xy plane; charge of uniform density 3.0 nC/m² is distributed over the parallel plane defined by $z = 2.0$ m. Determine the magnitude of the electric field at any point with the coordinate (a) $z = 1.0$ m and (b) $z = 3.0$ m.

63. In Fig. 24-49, a point charge $+q$ is a distance $d/2$ directly above the center of a square of side d. What is the magnitude of the electric flux through the square? (*Hint:* Think of the square as one face of a cube with edge d.)

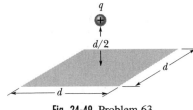

Fig. 24-49 Problem 63.

64. Charge of uniform volume density $\rho = 1.2$ nC/m³ fills an infinite slab between $x = -5.0$ cm and $x = +5.0$ cm. What is the magnitude of the electric field at any point with the coordinate (a) $x = 4.0$ cm and (b) $x = 6.0$ cm?

65. A point charge $q = 1.0 \times 10^{-7}$ C is at the center of a spherical cavity of radius 3.0 cm in a chunk of metal. Use Gauss' law to find the electric field (a) at point P_1, halfway from the center to the surface of the cavity, and (b) at point P_2, within the metal wall.

66. A thin, metallic, spherical shell of radius a has a charge

q_a. Concentric with it is another thin, metallic, spherical shell of radius b (where $b > a$) and charge q_b. Find the electric field at points a distance r from the common center, where (a) $r < a$, (b) $a < r < b$, and (c) $r > b$. (d) Discuss the criterion you would use to determine how the charges are distributed on the inner and outer surfaces of the shells.

67. In a laboratory experiment, the gravitational force on an electron is just balanced by the force on the electron from an electric field. The electric field is due to charges on two large, parallel, nonconducting plates, oppositely charged but equal in charge magnitudes and separated by 2.3 cm. (a) What is the magnitude of the surface charge density, assumed to be uniform, on the plates? (b) What is the field's direction?

68. A spherical conducting shell has a charge of $-14 \, \mu$C on its outer surface and a charged particle in its hollow. If the net charge on the shell is $-10 \, \mu$C, what is the charge (a) on the inner surface of the shell and (b) of the particle?

Tutorial Problem

69. A solid metal sphere of radius a is concentric with, and inside, a metal shell of inner radius b_1 and outer radius b_2. The sphere has a net positive charge Q, and the shell has a net negative charge $-2Q$. Let's determine the electric fields in this situation by making use of Gauss' law.

(a) First, sketch this physical situation. (b) What is the type of symmetry exhibited in this problem? What does the symmetry tell you about the direction and magnitude of the electric field in this problem? How is this symmetry used in the application of Gauss' law? What type of Gaussian surface is appropriate to this situation?

(c) What is the electric field for $r < a$? (d) What is the electric field for $a < r < b_1$? (e) What is the electric field for $b_1 < r < b_2$? What does this imply about the total charge on the inside surface of the metal shell? (f) What is the electric field for $r > b_2$? (g) Plot the electric field component as a function of r on a graph. Mark a, b_1, and b_2 for $b_1 \approx b_2 \approx 2a$, and plot the field approximately to scale. (h) Determine the electric charge densities on each of the three metal surfaces and show that they lead to the correct electric field magnitudes just outside those surfaces.

Answers

(b) This problem exhibits spherical symmetry; everything is symmetric about the center of the sphere. This symmetry implies that the direction of the electric field must be in the radial direction—that is, along a line from the center of the sphere to the point at which the field is being calculated. The direction of the electric field at a point with position vector $\vec{r}$ must be along a radial unit vector, either $\hat{r}$ or $-\hat{r}$; the field cannot have a component in any other direction. Also, the electric field strength must be a function of just r. Thus, all points with the same value of r must have the same magnitude of the electric field. We'll write the electric field at such points as $E_r(r) \, \hat{r}$.

Applying Gauss' law in this situation involves using spherical Gaussian surfaces centered at the center of the sphere. The electric flux through a Gaussian surface of radius r is $4\pi r^2 E_r(r)$.

(c) There is no electric field inside the sphere because that region is a conducting material. Recall that in an electrostatic situation (no moving charges) the electric field must be zero in a conductor.

(d) The total charge inside a Gaussian surface of radius r with $a < r < b_1$ is just Q, the net charge on the metal sphere. Gauss' law in this region gives

$$4\pi r^2 E_r(r) = \frac{Q}{\varepsilon_0},$$

so

$$E_r(r) = \frac{Q}{4\pi\varepsilon_0 r^2}.$$

Thus, the electric field is

$$E_r(r)\,\hat{\mathbf{r}} = \frac{Q}{4\pi\varepsilon_0 r^2}\,\hat{\mathbf{r}}.$$

(e) The region $b_1 < r < b_2$ is in the metal shell (a conducting material) and must have zero electric field. Gauss' law then implies that the total charge inside a Gaussian surface through this region must be zero. Since the metal sphere has a charge of Q, the inside surface (at $r = b_1$) of the metal shell must have a charge of $-Q$.

(f) The total charge of the metal sphere and the metal shell is $Q - 2Q = -Q$, so Gauss' law in the region outside the metal shell gives

$$4\pi r^2 E_r(r) = -\frac{Q}{\varepsilon_0},$$

so

$$E_r(r) = -\frac{Q}{4\pi\varepsilon_0 r^2}.$$

Thus, the electric field is

$$E_r(r)\,\hat{\mathbf{r}} = -\frac{Q}{4\pi\varepsilon_0 r^2}\,\hat{\mathbf{r}}.$$

(g) Plot $E = 0$ for $0 < r < a$, $E > 0$ for $a < r < b_1$, $E = 0$ for $b_1 < r < b_2$, and $E < 0$ for $b_2 < r$. Set $E(b_1) \approx \frac{1}{4}E(a)$ and $E(b_2) \approx -E(b_1)$.

(h) On the surface at $r = a$, the total charge is Q. Since the total surface area of the sphere is $4\pi a^2$, the charge density at that surface is $\sigma = Q/4\pi a^2$. That means the electric field just outside the sphere is $\sigma/\varepsilon_0 = Q/4\pi\varepsilon_0 a^2$, which agrees with the result of part (d) for $r = a$.

On the surface at $r = b_1$, the total charge is $-Q$ and the surface area is $4\pi b_1^2$. Therefore, $\sigma = -Q/4\pi b_1^2$, which leads to $\sigma/\varepsilon_0 = -Q/4\pi\varepsilon_0 b_1^2$. This agrees with the result of part (d), since at $r = b_1$ the field E_r is then $-\sigma/\varepsilon_0 = +Q/4\pi\varepsilon_0 b_1^2$.

On the surface at $r = b_2$, the surface density is $\sigma = -Q/4\pi b_2^2$ and the component of the electric field there is $\sigma/\varepsilon_0 = -Q/4\pi\varepsilon_0 b_2^2$. This matches the value of E_r at $r = b_2$ that we found in part (f).

25
Electric Potential

ADDITIONAL SAMPLE PROBLEMS

Sample Problem 25-7

The nucleus of a hydrogen atom consists of a single proton, which can be treated as a particle (or point charge).

(a) With the electric potential equal to zero at infinite distance, what is the electric potential V due to the proton at a radial distance $r = 2.12 \times 10^{-10}$ m from it?

SOLUTION: The **Key Idea** here is that, because we can treat the proton as a particle, the electric potential V it produces at a point at distance r is given by Eq. 25-26,

$$V = \frac{1}{4\pi\varepsilon_0}\frac{q}{r}.$$

Here charge q is e ($= 1.6 \times 10^{-19}$ C). Substituting this and the given value for r, we find

$$V = \frac{(8.99 \times 10^9\ \text{N}\cdot\text{m}^2/\text{C}^2)(1.60 \times 10^{-19}\ \text{C})}{2.12 \times 10^{-10}\ \text{m}}$$

$$= 6.78\ \text{V}. \qquad \text{(Answer)}$$

(b) What is the electric potential energy U in electron-volts of an electron at the given distance from the nucleus? (The potential en-

ergy is actually that of the electron–proton system—the hydrogen atom.)

SOLUTION: The **Key Idea** here is that when a particle of charge q is located at a point where the electric potential due to other charges is V, the electric potential energy U is given by Eq. 25-5 ($V = U/q$). Using the electron's charge $-e$, we find

$$U = qV = (-1.60 \times 10^{-19}\ \text{C})(6.78\ \text{V})$$

$$= -1.0848 \times 10^{-18}\ \text{J} = -6.78\ \text{eV}. \qquad \text{(Answer)}$$

(c) If the electron moves closer to the proton, does the electric potential energy increase or decrease?

SOLUTION: The **Key Ideas** of parts (a) and (b) apply here also. As the electron moves closer to the proton, the electric potential V due to the proton at the electron's position increases (because r decreases). Thus, the value of V in part (b) increases. Because the electron is negatively charged, this means that the value of U becomes more negative. Hence, the potential energy U of the electron (that is, of the system or atom) decreases.

Sample Problem 25-8

The potential at the center of a uniformly charged circular disk of radius $R = 3.5$ cm is $V_0 = 550$ V, relative to zero potential at an infinite distance.

(a) What is the total charge q on the disk?

SOLUTION: The **Key Idea** here is that the potential V at any point z on the central axis of the disk is related to the surface charge density σ of the disk by Eq. 25-37. Since $z = 0$ at the center of the disk, we get

$$V_0 = \frac{\sigma R}{2\varepsilon_0},$$

from which

$$\sigma = \frac{2\varepsilon_0 V_0}{R}. \qquad (25\text{-}44)$$

The total charge q on the disk is $\sigma(\pi R^2)$, so from Eq. 25-44 we have

$$q = \sigma(\pi R^2) = 2\pi\varepsilon_0 R V_0$$

$$= (2\pi)(8.85 \times 10^{-12}\ \text{C}^2/\text{N}\cdot\text{m}^2)(0.035\ \text{m})(550\ \text{V})$$

$$= 1.1 \times 10^{-9}\ \text{C} = 1.1\ \text{nC}, \qquad \text{(Answer)}$$

in which we use Eq. 25-9 to write $1\ \text{V} = 1\ \text{J/C} = 1\ \text{N}\cdot\text{m/C}$.

(b) What is the potential at a point on the axis of the disk a distance $z = 5.0R$ from the center of the disk?

SOLUTION: Using the **Key Idea** of part (a) and substituting $z = 5.0R$ into Eq. 25-37 lead to

$$V = \frac{\sigma}{2\varepsilon_0}[\sqrt{(5.0R)^2 + R^2} - 5.0R].$$

Substituting σ from Eq. 25-44 then yields

$$V = \frac{V_0}{R}(\sqrt{26R^2} - 5.0R) = V_0(\sqrt{26} - 5.0)$$

$$= (550\ \text{V})(0.099) = 54\ \text{V}. \qquad \text{(Answer)}$$

Sample Problem 25-9

An alpha particle (which consists of two protons and two neutrons) passes through the region of electron orbits in a gold atom, moving directly toward the gold nucleus, which has 79 protons and 118 neutrons. The alpha particle slows and then comes to a momentary stop, at a center-to-center separation r of 9.23 fm, before it begins to move back along its original path (Fig. 25-44). (Because the gold nucleus is much more massive than the alpha particle, we can assume the gold nucleus does not move.) What was the kinetic energy K of the alpha particle when it was initially far away (hence external to the gold atom)? Assume that the only force acting between the alpha particle and the gold nucleus is the (electrostatic) Coulomb force.

SOLUTION: The **Key Idea** here is that during the entire process, the mechanical energy of the *alpha particle + gold atom* system is conserved. When the alpha particle is outside the atom, the electric potential energy of the system is zero, because the atom has an equal number of electrons and protons, is thus electrically neutral, and so does not produce an external electric field. However, once the alpha particle has passed through the region of electron orbits on its way toward the nucleus, it is acted on by a repulsive Coulomb force on its protons from those in the nucleus. (The neutrons, being electrically neutral, do not participate in producing this force. The electrons, now being outside the location of the alpha particle, act like a uniformly charged spherical shell, which produces no internal force.)

As the alpha particle slows because of the repulsive force, its kinetic energy is transferred to electric potential energy of the system. The transfer is complete when the alpha particle momentarily stops. Using the principle of conservation of mechanical energy, we can equate the initial kinetic energy K of the alpha particle to the electric potential energy U of the system at the instant the alpha particle stops:

$$K = U. \qquad (25\text{-}45)$$

By substituting the right side of Eq. 25-43 for U, with $q_1 = 2e$, $q_2 = 79e$ (in which e is the elementary charge, 1.60×10^{-19} C), and $r = 9.23$ fm, we can rewrite Eq. 25-45 as

$$\begin{aligned}
K &= \frac{1}{4\pi\varepsilon_0} \frac{(2e)(79e)}{9.23 \text{ fm}} \\
&= \frac{(8.99 \times 10^9 \text{ N} \cdot \text{m}^2/\text{C}^2)(158)(1.60 \times 10^{-19} \text{ C})^2}{9.23 \times 10^{-15} \text{ m}} \\
&= 3.94 \times 10^{-12} \text{ J} = 24.6 \text{ MeV}. \qquad \text{(Answer)}
\end{aligned}$$

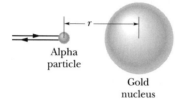

Fig. 25-44 Sample Problem 25-9. An alpha particle, traveling head-on toward the center of a gold nucleus, comes to a momentary stop (at which time all its kinetic energy has been transferred to electric potential energy) and then reverses its path.

QUESTIONS

11. (a) In Fig. 25-3a, does the electric potential increase toward the right or toward the left? (b) If adjacent equipotential surfaces differ by 10 V and the rightmost one is at an electric potential of -100 V, what is the electric potential of the leftmost one? If we move an electron toward the right, is the work done on the electron by (c) our force and (d) the electric field positive or negative?

12. Figure 25-45 shows a thin, uniformly charged rod and three points at the same distance d from the rod. Without calculation, rank the magnitude of the electric potential the rod produces at those three points, greatest first.

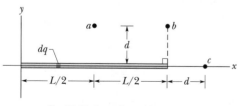

Fig. 25-45 Questions 12 and 13.

13. *Organizing question:* Figure 25-45 shows a rod of uniform positive charge density λ lying along an x axis; it also shows a differ-ential charge element dq. The net electric potential V at point c due to the rod is to be evaluated, with the potential taken to be zero at infinite distance. In the expression

$$dV = \frac{1}{4\pi\varepsilon_0} \frac{dq}{r},$$

what should be substituted for (a) dq and (b) r to prepare for the required integration? (c) What are the limits of the integration? (d) If V is to be expressed in terms of the charge Q on the rod, what should be substituted for the charge density?

14. Figure 25-46 shows two situations in which we move an electron in from an infinite distance to a point midway between two

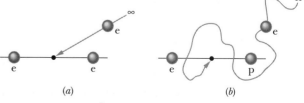

Fig. 25-46 Question 14.

charged particles (either proton or electron) that are fixed in place. In each situation, is the work done on the incoming electron *by the electric field* positive, negative, or zero?

15. The electric potential at the xyz coordinates (2 m, 0.5 m, 0.2 m) is given by $V = 2x - 3y + 4z$. Rank the magnitudes of the electric field components E_x, E_y, and E_z there, greatest first.

16. Figure 25-47 shows three systems of particles of charge $+q$ or $-q$, forming one equilateral and two isosceles triangles with edge lengths of either d, $2d$, or $d/2$. (a) Rank the systems according to their electric potential energy, most positive first. (b) How much work must we do to assemble the system of Fig. 25-47*b* if the particles are initially infinitely far apart?

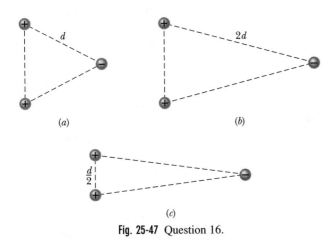

(a)   (b)

(c)

Fig. 25-47 Question 16.

17. (a) If the alpha particle of Sample Problem 25-9 had a smaller initial kinetic energy than the calculated 24.6 MeV, would it momentarily stop farther from, closer to, or at the same distance of 9.23 fm from the gold nucleus? (Again, assume only Coulomb forces act between the alpha particle and the gold nucleus.) (b) If, instead, the alpha particle is replaced by a single proton with the same kinetic energy of 24.6 MeV, will the proton momentarily stop at the same distance of 9.23 fm from the gold nucleus, farther from it, or closer to it?

18. We have seen that, inside a hollow conductor, you are shielded from the fields of outside charges. If you are *outside* a hollow conductor that contains charges, are you shielded from the fields of those charges?

19. Three isolated, empty, spherical shells of the same radius have the following uniform charges: shell A, $+q$; shell B, $+2q$; shell C, $+3q$. Set the electric potential to be zero at an infinite distance from the shells. Then rank the shells, greatest first, according to (a) the electric potential at the surface of the shell, (b) the electric potential at the center of the shell, (c) the electric field magnitude at the surface of the shell, and (d) the electric field magnitude at the center of the shell.

20. Repeat Question 19 but with the electric potential now set to zero at the center of each shell.

21. A proton is either released at rest or launched with a certain velocity in a uniform electric field. Which of the graphs in Fig. 25-48 could possibly show how the kinetic energy of the proton changes during the proton's motion?

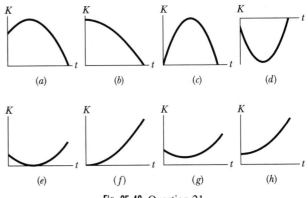

(a)   (b)   (c)   (d)

(e)   (f)   (g)   (h)

Fig. 25-48 Question 21.

22. In Fig. 25-49, a particle is to be released at rest at point A and then is to be accelerated directly through point B by an electric field. The potential difference between points A and B is 100 V. Which point should be at higher electric potential if the particle is (a) an electron, (b) a positron (the positively charged antiparticle of the electron), (c) a proton, and (d) an alpha particle (a nucleus of two protons and two neutrons)? (e) For each situation and without a calculator or written calculation, determine the particle's kinetic energy in electron-volts as the particle reaches point B.

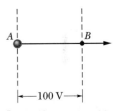

Fig. 25-49 Question 22.

23. Figures 25-50*a* through *c* are graphs of electric potential V ver-

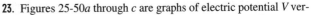

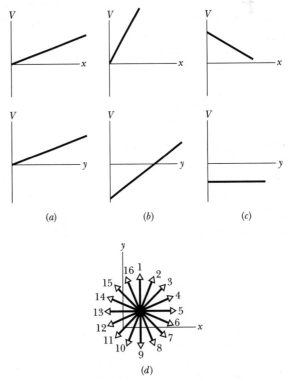

(a)   (b)   (c)

(d)

Fig. 25-50 Question 23.

sus *x* and *y* in a certain region for three situations. (In Fig. 25-50*a*, the plots are identical.) Figure 25-50*d* gives 16 choices of electric field in the region. In choices 3, 7, 11, and 15, the electric field makes an angle of 45° with the *x* axis. Which of the 16 choices best gives the direction of the electric field corresponding to (a) Fig. 25-50*a*, (b) Fig. 25-50*b*, and (c) Fig. 25-50*c*?

24. Figure 25-51 shows three spherical shells in separate situations, with each shell having the same uniform positive net charge. Points 1, 4, and 7 are at the same radial distances from the centers of their respective shells; so are points 2, 5, and 8; and so are points 3, 6, and 9. With the electric potential taken equal to zero at an infinite distance, rank the numbered points according to the electric potential at those points, greatest first.

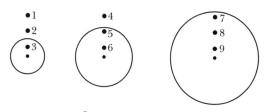

Fig. 25-51 Question 24.

DISCUSSION QUESTIONS

25. Engineers and scientists often define the potential of Earth's surface ("ground") to be zero. If, instead, they defined it to be +100 V, what effect would the change have on measured values of (a) potentials and (b) potential differences?

26. What would happen to you if you were on an electrically isolated stand and your potential was increased by 10 kV with respect to the ground?

27. In what situations is the electron-volt a more convenient unit of energy than the joule?

28. How would a proton-volt compare with an electron-volt? The mass of a proton is 1840 times the mass of an electron.

29. Do electrons tend to move toward regions of high potential or of low potential? Why?

30. Why is it possible to shield a room against electrical forces but not against gravitational forces?

31. Suppose that Earth's surface had a net charge that was not zero.

Why would it still be possible to adopt that surface as a standard reference point of potential and to assign the potential $V = 0$ to it?

32. Does the potential of a positively charged isolated conductor have to be positive? Give an example to prove your point.

33. An electrical worker was accidentally electrocuted and a newspaper account reported: "He accidentally touched a high-voltage cable and 20 000 V of electricity surged through his body." Criticize this statement.

34. Advice to mountaineers caught in lightning and thunderstorms is to (a) get rapidly off peaks and ridges and (b) put both feet together and crouch in the open, only the feet touching the ground. What is the basis for this good advice?

35. If $\vec{E}$ equals zero at a given point, must V equal zero at that point? Give some examples to prove your answer.

36. If you know only $\vec{E}$ at a given point, can you calculate V at that point? If not, what other information do you need?

37. Is the uniformly charged, nonconducting disk of Section 25-8 an equipotential surface? Explain.

38. Distinguish between potential difference and difference of potential energy. Give statements in which each term is used properly.

39. Ions and free electrons act like condensation centers; water drops form around them when they are airborne. Explain why.

40. If V is constant throughout a given region of space, what can you say about $\vec{E}$ in that region?

41. How can you ensure that the electric potential in a given region of space will have the same value throughout that region?

42. Devise an arrangement of three point charges, separated by finite distances, that has zero electric potential energy.

43. We have seen (Section 25-11) that the potential inside a conductor is the same as that on its surface. (a) What if the conductor is irregularly shaped and has an irregularly shaped cavity inside? (b) What if the cavity has a small "worm hole" connecting it to the outside? (c) What if the cavity is closed but has a point charge suspended within it? For each situation, discuss the potential within the conducting material and at different points within the cavity.

44. An isolated, spherical, conducting shell has a negative charge. What will happen if a positively charged metal object is placed in contact with the interior surface of the shell? Discuss the three cases in which the magnitude of the positive charge is (a) less than, (b) equal to, and (c) greater than the magnitude of the negative charge.

EXERCISES & PROBLEMS

58. Figure 25-52 shows a ring of outer radius R and inner radius $r = 0.200R$; the ring has a uniform surface charge density σ. With $V = 0$ at infinity, find an expression for the electric potential at point P on the central axis of the ring, at a distance $z = 2.00R$ from the center of the ring.

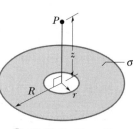

Fig. 25-52 Problem 58.

59. Suppose that the negative charge in a copper one-cent coin were removed to a very large distance from Earth—perhaps to a distant galaxy—and that the positive charge were distributed uniformly over Earth's surface. By how much would the electric potential at Earth's surface change? (See Sample Problem 22-7 in this book.)

60. (a) If an isolated conducting sphere 10 cm in radius has a net charge of 4.0 μC, and $V = 0$ at infinity, what is the potential on the surface of the sphere? (b) Can this situation actually occur,

given that the air around the sphere undergoes electrical breakdown when the field exceeds 3.0 MV/m?

61. Much of the material making up Saturn's rings is in the form of tiny dust grains having radii on the order of 10^{-6} m. These grains are located in a region containing a dilute ionized gas, and they pick up excess electrons. As an approximation, suppose each grain is spherical, with radius $R = 1.0 \times 10^{-6}$ m. How many electrons would one grain have to pick up to have a potential of -400 V on its surface (taking $V = 0$ at infinity)?

62. In Fig. 25-53, a particle with charge $+5q$ is brought in from infinity to the position shown; the other two particles in the figure are fixed in place. What is the ratio of the potential energy of this three-particle system to that of the original two-particle system?

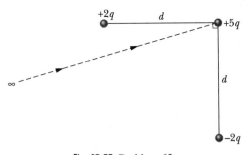

Fig. 25-53 Problem 62.

63. A charge of 1.50×10^{-8} C lies on an isolated metal sphere of radius 16.0 cm. With $V = 0$ at infinity, what is the electric potential at points on the sphere's surface?

64. In a certain situation, the electric potential varies along the x axis as shown in the graph of Fig. 25-54. For each of the intervals ab, bc, cd, de, ef, fg, and gh, determine the x component of the electric field, and then plot E_x versus x. (Ignore behavior at the interval end points.)

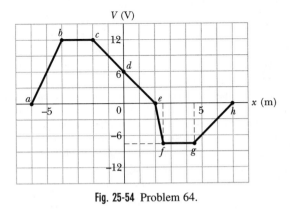

Fig. 25-54 Problem 64.

65. Starting from Eq. 25-30, find the electric field due to a dipole at a point on the dipole axis.

66. A disk like that of Fig. 25-14 has radius $R = 2.20$ cm. Its surface charge density is 1.50×10^{-6} C/m² from $r = 0$ to $R/2$ and 8.00×10^{-7} C/m² from $r = R/2$ to R. (a) What is the total charge on the disk? (b) With $V = 0$ at infinity, what is the electric potential at a point on the central axis of the disk, at a distance $z = R/2$ from the center of the disk?

67. Two infinite lines of charge are parallel to and in the same plane with the z axis. One, of charge per unit length $+\lambda$, is a distance a to the right of this axis. The other, of charge per unit length $-\lambda$, is a distance a to the left of this axis. Sketch some of the equipotential surfaces due to this arrangement.

68. In the Millikan oil-drop experiment (see Section 23-8), a uniform electric field of 1.92×10^5 N/C is maintained in the region between two plates separated by 1.50 cm. Find the potential difference between the plates.

69. In Fig. 25-55, what is the net electric potential at the origin due to the circular arc of charge $+Q$ and the two particles of charges $+4Q$ and $-2Q$? Take the radius of curvature of the arc as $R = 2.00$ m, and the center of curvature as the origin.

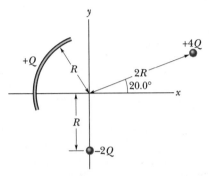

Fig. 25-55 Problem 69.

70. In Fig. 25-56, three long parallel lines of charge, with the linear charge densities shown, extend perpendicular to the page in both directions. Sketch some electric field lines; also sketch the cross sections in the plane of the figure of some equipotential surfaces.

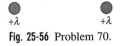

Fig. 25-56 Problem 70.

71. Figure 25-57 shows three circular arcs, each of radius R and total charge as indicated. What is the net electric potential at the center of curvature?

72. A long, solid, conducting cylinder has a radius of 2.0 cm. The electric field at the surface of the cylinder is 160 N/C, directed radially outward. Let A, B, and C be points that are 1.0 cm, 2.0 cm, and 5.0 cm, respectively, from the central axis of the cylinder. (a) What is the electric field at point C? What are the electric potential differences (b) $V_B - V_C$ and (c) $V_A - V_B$?

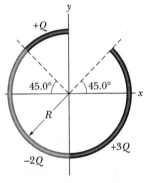

Fig. 25-57 Problem 71.

73. In Fig. 25-58 two charged particles are fixed in place on an x axis. If $d = 1.00$ m, what is the electric potential difference $V_B - V_A$ between points B and A?

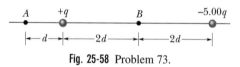

Fig. 25-58 Problem 73.

74. A nonuniform linear charge distribution given by $\lambda = bx$, where b is a constant, is located along an x axis from $x = 0$ to $x = 0.20$ m. If $b = 20$ nC/m^2, what is the electric potential (relative to a potential of zero at infinity) at (a) the origin and (b) the point $y = 0.15$ m on the y axis?

75. Consider a point charge $q = 1.5 \times 10^{-8}$ C, and take $V = 0$ at infinity. (a) What are the shape and dimensions of an equipotential surface having a potential of 30 V due to q alone? (b) Are surfaces whose potentials differ by a constant amount (1.0 V, say) evenly spaced?

76. Point charges of equal magnitudes (25 nC) and opposite signs are placed on diagonally opposite corners of a 60 cm × 80 cm rectangle. Point A is the corner of this rectangle nearest the positive charge, and point B is the corner of this rectangle nearest the negative charge. Determine the potential difference $V_B - V_A$.

77. A graph of the x component of the electric field as a function of x in a region of space is shown in Fig. 25-59. The y and z components of the electric field are zero in this region. If the electric potential at the origin is 10 V, (a) what is the electric potential at $x = 2.0$ m, (b) what is the greatest positive value of the electric potential for points on the x axis for which $0 \le x \le 6.0$ m, and (c) for what value of x is the electric potential zero?

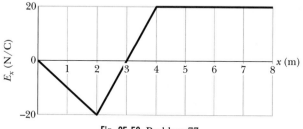

Fig. 25-59 Problem 77.

78. Two uniformly charged, infinite, nonconducting planes are parallel to the yz plane and positioned at $x = -50$ cm and $x = +50$ cm. The charge densities on the planes are -50 nC/m^2 and $+25$ nC/m^2, respectively. What is the magnitude of the potential difference between the origin and the point on the x axis at $x = +80$ cm? (*Hint:* Use Gauss' law for planar symmetry to determine the electric field in each region of space; see Section 24-8.)

79. The electric field in a region of space has the components $E_y = E_z = 0$ and $E_x = (4.00$ N/C$)x$. Point A is on the y axis at $y = 3.0$ m, and point B is on the x axis at $x = 4.0$ m. What is the potential difference $V_B - V_A$?

80. Two charges $q = +2.0$ μC are fixed in space a distance $d = 2.0$ cm apart, as shown in Fig. 25-60. (a) With $V = 0$ at infinity, what is the electric potential at point C? (b) You bring a third charge $q = +2.0$ μC from infinity to C. How much work must you do? (c) What is the potential energy U of the three-charge configuration when the third charge is in place?

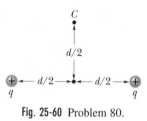

Fig. 25-60 Problem 80.

81. In the quark model of fundamental particles, a proton is composed of three quarks: two "up" quarks, each having charge $+2e/3$, and one "down" quark, having charge $-e/3$. Suppose that the three quarks are equidistant from one another. Take the distance to be 1.32×10^{-15} m and calculate (a) the electric potential energy of the subsystem of two "up" quarks and (b) the total electric potential energy of the three-quark system.

82. (a) A proton of kinetic energy 4.80 MeV travels head-on toward a lead nucleus. Assuming that the proton does not penetrate the nucleus and the only force between the proton and the nucleus is the Coulomb force, calculate the smallest center-to-center separation that occurs between the proton and the nucleus when the proton momentarily stops. (b) If the proton were replaced with an alpha particle of the same initial kinetic energy, how would the smallest center-to-center separation compare with that in (a)?

83. The charges and coordinates of two point charges located in the xy plane are: $q_1 = +3.0 \times 10^{-6}$ C, $x = +3.5$ cm, $y = +0.50$ cm; and $q_2 = -4.0 \times 10^{-6}$ C, $x = -2.0$ cm, $y = +1.5$ cm. How much work must be done to locate these charges at their given positions, starting from infinite separation?

84. A net charge of $+16$ μC is uniformly distributed on a thin circular ring that lies in an xy plane with its center at the origin. The radius of the ring is 3.0 cm. If point A is at the origin and point B is on the z axis at $z = 4.0$ cm, what is $V_B - V_A$?

85. What is the magnitude of the electric field at the point $(3\hat{i} - 2\hat{j} + 4\hat{k})$ m if the electric potential is given by $V = 2xyz^2$, where V is in volts and x, y, and z are in meters?

86. Suppose that N electrons can be placed in either of two configurations. In the first configuration they are all placed on the circumference of a narrow ring of radius R and are uniformly distributed so that the distance between adjacent electrons is the same everywhere; in the second configuration $N - 1$ of the electrons are placed on the ring as before and one electron is placed in the center of the ring. (a) For which configuration is the electrostatic potential energy of the system less? Consider all integer values of N from 2 to 15. (b) What is the smallest value of N for which the second configuration is less energetic than the first? (c) For the value of N found in (b), consider any one circumference electron—call it e_0. How many other circumference electrons are closer to e_0 than the central electron is?

CLUSTERED PROBLEMS

87. In Fig. 25-61a, a particle of charge q and mass m is released from rest at a point midway between two large conducting plates that are vertical and separated by distance D. As the particle falls due to the gravitational force on it, it moves rightward because of a potential difference ΔV between the plates. Just as the particle runs into a plate, (a) what distance d has it traveled downward and (b) what is its speed v? Evaluate (c) distance d and (d) speed v using the values $q = 0.10$ nC, $m = 1.0$ μg, $D = 10$ cm, and $\Delta V = 10$ V.

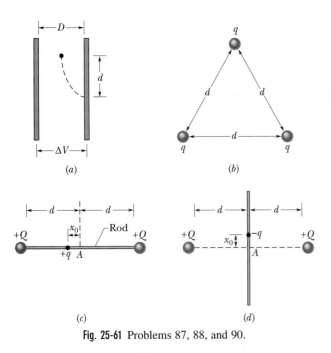

(a)

(b)

(c)

(d)

Fig. 25-61 Problems 87, 88, and 90.

88. In Fig. 25-61b, three particles with the same charge q and same mass m are initially fixed in place to form an equilateral triangle with edge lengths d. (a) If the particles are released simultaneously,

what are their speeds when they have traveled a large distance (effectively an infinite distance) from each other? (Measure the speeds in the original rest frame of the particles.)

Suppose, instead, the particles are released one at a time: The first one is released, and then, when that first one is at a large distance, a second one is released, and then, when that second one is at a large distance, the last one is released. What then are the final speeds of (b) the first particle, (c) the second particle, and (d) the last particle?

89. Two small spheres of radius r and mass m are released from rest when their centers are separated by distance $D \gg r$. Sphere 1 has charge Q; sphere 2 has charge $2Q$. What are the final speeds of (a) sphere 1 and (b) sphere 2? (Let the speeds be measured in the original rest frame of the particles.) Suppose, instead, that sphere 2 has charge $-2Q$. What then are the speeds of (c) sphere 1 and (d) sphere 2 just as the spheres meet? Next suppose that the collision is elastic. What then are the speeds of (e) sphere 1 and (f) sphere 2 when the separation between the spheres is again D?

90. In Fig. 25-61c, two particles of identical charge $+Q$ are fixed in place at opposite ends of an insulated rod of length $2d$. A third particle, of charge $+q$ and mass m, can slide along the rod. The particle is displaced from the rod's midpoint (point A) by distance x_0 and then released from rest. (a) Find an expression for the velocity of the particle as it passes back through point A. (b) Simplify the expression using the assumption that $x_0 \ll d$. (c) Evaluate your simplified expression using the values $Q = 1.0$ μC, $m = 1.0$ μg, $q = 1.0$ nC, $x_0 = 0.10$ mm, and $d = 1.0$ cm.

Now the particles of charge $+Q$ are kept in place while the rod is rotated by 90° about its midpoint (Fig. 25-61d). Let the third particle be of charge $-q$; it is displaced from point A by distance x_0 and then released from rest. (d) Repeat parts (a) and (b) for this arrangement.

26
Capacitance

Sample Problem 26-7

The plates of a parallel-plate capacitor are separated by a distance $d = 1.0$ mm. What must be the plate area if the capacitance is to be 1.0 F?

SOLUTION: The **Key Idea** here is that, because this is a parallel-plate capacitor, we can relate capacitance C, plate area A, and plate separation d via Eq. 26-9 ($C = \varepsilon_0 A/d$). It gives us

$$A = \frac{Cd}{\varepsilon_0} = \frac{(1.0 \text{ F})(1.0 \times 10^{-3} \text{ m})}{8.85 \times 10^{-12} \text{ F/m}}$$
$$= 1.1 \times 10^8 \text{ m}^2. \qquad \text{(Answer)}$$

This is the area of a square more than 10 km on edge. The farad is indeed a large unit. Modern technology, however, has permitted the construction of 1 F capacitors of very modest size. These "supercaps" are used as backup voltage sources for computers; they can maintain the computer memory for up to 30 days in case of power failure.

Sample Problem 26-8

What is the potential energy of the two-capacitor system in Sample Problem 26-3, before and after switch S in Fig. 26-10 is closed?

SOLUTION: The **Key Idea** here is that we can find the energy U stored in a capacitor if we know the capacitance C and either the charge q on the capacitor or the electric potential V to which the capacitor is charged. Here we know V, so we can use Eq. 26-22. Initially, only capacitor C_1 is charged and has a potential energy; its potential difference is $V_0 = 6.30$ V. Thus, from Eq. 26-22, the initial potential energy is

$$U_i = \tfrac{1}{2}C_1 V_0^2 = (\tfrac{1}{2})(3.55 \times 10^{-6} \text{ F})(6.30 \text{ V})^2$$
$$= 7.045 \times 10^{-5} \text{ J} = 70.5 \text{ } \mu\text{J}. \qquad \text{(Answer)}$$

After the switch has been closed, the capacitors come to the same final potential difference $V = 1.79$ V. The final potential energy is then

$$U_f = \tfrac{1}{2}C_1 V^2 + \tfrac{1}{2}C_2 V^2 = \tfrac{1}{2}(C_1 + C_2)V^2$$
$$= (\tfrac{1}{2})(3.55 \times 10^{-6} \text{ F} + 8.95 \times 10^{-6} \text{ F})(1.79 \text{ V})^2$$
$$= 2.00 \times 10^{-5} \text{ J} = 20.0 \text{ } \mu\text{J}. \qquad \text{(Answer)}$$

Thus, $U_f < U_i$, by about 72%.

This is not a violation of the principle of energy conservation. The "missing" energy appears as thermal energy in the connecting wires (as we shall discuss in Chapter 27) and as radiated energy.

Sample Problem 26-9

Let us surround the isolated conducting sphere of Sample Problem 26-4 with an imaginary spherical surface of radius R_i that is concentric with the sphere. What is R_i (in terms of the sphere's radius R) if the imaginary surface encloses 0.50 of the total energy stored in the electric field of the sphere?

SOLUTION: A **Key Idea** here is that to find the energy enclosed by the imaginary spherical surface, we can use the energy density u for points enclosed by that surface. We can write the density as

$$u = \frac{dU}{dV}, \qquad (26\text{-}36)$$

where dU is the differential amount of energy stored in the differential volume dV. Take dV to be the volume of a thin spherical shell of radius r and thickness dr, concentric with the sphere. Then the volume of the shell is the product of the shell's surface area $4\pi r^2$ and its thickness dr, or

$$dV = 4\pi r^2 \, dr. \qquad (26\text{-}37)$$

Solving Eq. 26-36 for dU and substituting for dV from Eq. 26-37 and for u from Eq. 26-23 ($u = \tfrac{1}{2}\varepsilon_0 E^2$), we find

$$dU = (\tfrac{1}{2}\varepsilon_0 E^2)(4\pi r^2 \, dr). \qquad (26\text{-}38)$$

From Section 24-6, we know that inside a conducting sphere ($r < R$), E is zero. Thus, the stored energy is also zero there. From Eq. 24-15, the electric field outside a sphere with uniform charge q is

$$E = \frac{1}{4\pi\varepsilon_0}\frac{q}{r^2}.$$

Substituting this into Eq. 26-38 gives us

$$dU = \frac{1}{8\pi\varepsilon_0}\frac{q^2}{r^2}\,dr \qquad (26\text{-}39)$$

for all points outside the sphere.

To find the energy enclosed by the imaginary surface, we integrate both sides of Eq. 26-39. On the right side, the limits of integration are $r = R$ (at the sphere's surface) and $r = R_i$ (at the imaginary surface). We obtain

$$\int dU = \int_R^{R_i}\frac{1}{8\pi\varepsilon_0}\frac{q^2}{r^2}\,dr$$

and

$$U = -\frac{q^2}{8\pi\varepsilon_0}\left(\frac{1}{R_i} - \frac{1}{R}\right). \qquad (26\text{-}40)$$

We want

$$U = 0.50U_{\text{tot}}. \qquad (26\text{-}41)$$

From Sample Problem 26-4a, with a slight change in notation, we know that the total energy U_{tot} of the sphere's electric field is

$$U_{\text{tot}} = \frac{q^2}{8\pi\varepsilon_0 R}. \qquad (26\text{-}42)$$

Substituting Eq. 26-40 for U and Eq. 26-42 for U_{tot} into Eq. 26-41 and solving for R_i, we find

$$R_i = 2R. \qquad \text{(Answer)}$$

Thus, 0.50 of the stored energy is enclosed by an imaginary spherical surface with a radius that is twice the radius of the conducting sphere.

QUESTIONS

12. Two capacitors are wired to a battery. (a) In which arrangement, parallel or series, is the potential difference across each capacitor the same and the same as that across the equivalent capacitance? (b) In which is the charge on each capacitor the same and the same as that on the equivalent capacitance?

13. *Organizing question:* Figure 26-37a shows a circuit with three capacitors, and Fig. 26-37b shows the circuit with their equivalent capacitor, of capacitance C_{eq}, which has a charge of 60 μC. Without a calculator (and, if possible, without written calculation), find (a) the charge on and (b) the voltage across capacitor 3 and then (c) the charge on and (d) the voltage across capacitor 1.

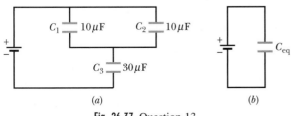

(a)   (b)

Fig. 26-37 Question 13.

14. In Fig. 26-38, the capacitors are initially uncharged. When the switch is thrown to position A, does the battery (of potential difference V) then begin to charge (a) capacitor 1 and (b) capacitor 2? When equilibrium is reached, what is the potential across (c) capacitor 1 and (d) capacitor 2? (e) When, later, the switch is thrown to position B, is capacitor 2 then charged by the battery or by capacitor 1? When equilibrium is again reached, (f) is the potential across capacitor 1 greater than, less than, or equal to that across capacitor 2 and (g) is it greater than, less than, or equal to the battery's potential difference?

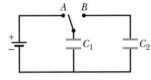

Fig. 26-38 Questions 14 and 15.

15. *Organizing question:* In Question 14 and Fig. 26-38, take $C_1 = 10$ μF and $C_2 = 20$ μF. (a) Using these data, set up an equation that gives q_1 in terms of q_2, where q_1 is the charge on capacitor 1 and q_2 is the charge on capacitor 2 when the capacitors reach equilibrium. (b) Then set up an equation that gives q_2 in terms of the initial charge q_0 stored on capacitor 1 by the battery.

16. An oil-filled parallel-plate capacitor was designed to have a capacitance C and to operate safely at or below a certain potential difference V_{max} without undergoing breakdown. However, the design is flawed and the capacitor occasionally breaks down. What can be done to redesign the capacitor, keeping C and V_{max} unchanged and using the same dielectric (the oil)?

17. (a) In Sample Problem 26-2, is the potential difference across capacitor 2 more than, less than, or equal to that across capacitor 1? (b) Is the charge on capacitor 2 more than, less than, or equal to that on capacitor 1?

18. In Sample Problem 26-3, if we increase the capacitance of capacitor 2, do the following increase, decrease, or remain the same: (a) the final potential difference across each capacitor, and the share of q_0 received by (b) capacitor 1 and (c) capacitor 2?

19. Figure 26-39 shows three circuits with identical capacitors. The

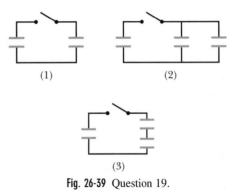

(1)   (2)

(3)

Fig. 26-39 Question 19.

capacitors at the left in the circuits initially have identical charges q_0; the capacitors at the right are initially uncharged. As in Sample Problem 26-3, we transfer charge from the charged capacitor by closing the switch. Without written calculation, rank the three circuits according to (a) the equivalent capacitance of the capacitors at the right and (b) the final (equilibrium) charge on the initially charged capacitor, greatest first.

20. Figure 26-40 shows a circuit containing a battery and three parallel-plate capacitors with identical plate separations (filled with air). The capacitors lie along an x axis, and a graph of the electric potential V along that axis is shown. (a) Which point at the battery, A or B, is at lower electric potential? (b) Is the plate at the right of capacitor 3 positively or negatively charged? Rank the capacitors according to (c) the charge on them, (d) their capacitance, (e) their plate area, and (f) the magnitude of the electric field between their plates, greatest first. (g) What is the direction of that electric field on the x axis?

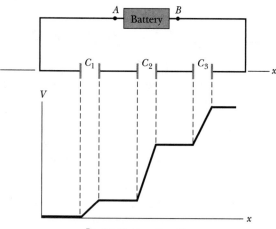

Fig. 26-40 Question 20.

21. A parallel-plate capacitor is connected to a battery of electric potential difference V. If the plate separation is decreased, do the following quantities increase, decrease, or remain the same: (a) the capacitor's capacitance, (b) the potential difference across the capacitor, (c) the charge on the capacitor, (d) the energy stored by the capacitor, (e) the magnitude of the electric field between the plates, and (f) the energy density of that electric field?

DISCUSSION QUESTIONS

22. A capacitor is connected across a battery. (a) Why does each plate receive a charge of exactly the same magnitude? (b) Is this true even if the plates are of different sizes?

23. Can there be a potential difference between two adjacent conductors that have the same amount of excess positive charge?

24. In Fig. 26-2, suppose that the two bodies are nonconductors, the charge being distributed arbitrarily over their surfaces. (a) Would Eq. 26-1 ($q = CV$) hold, with C independent of the charge arrangement? (b) How would you define V in this case?

25. You are given a parallel-plate capacitor with square plates of area A and separation d, in a vacuum. What is the qualitative effect of each of the following on the capacitance? (a) Reduce d. (b) Put a slab of copper between the plates, touching neither plate. (c) Double the area of both plates. (d) Double the area of one plate only. (e) Slide the plates parallel to each other so that the area of overlap is, say, 50% of its original value. (f) Double the potential difference between the plates. (g) Tilt one plate so that the separation remains d at one end but is $\frac{1}{2}d$ at the other.

26. Figure 26-41 shows two isolated, *distant* conductors, each of which has a certain capacitance. Suppose you join these conductors with a fine wire. How do you calculate the capacitance of the combination? In joining the conductors with the wire, have you connected them in parallel or in series?

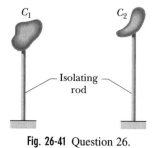

Fig. 26-41 Question 26.

27. Capacitors often are stored with a wire connected across their terminals. Why is this done?

28. If you did not neglect the fringing of the electric field lines in a parallel-plate capacitor, would you calculate a higher or a lower capacitance?

29. Two circular copper disks are facing each other a certain distance apart. In what ways could you reduce the capacitance of this combination?

30. Would you expect the dielectric constant of a material to vary with temperature? If so, how? Does whether or not the molecules have permanent dipole moments matter here?

31. Discuss similarities and differences when (a) a dielectric slab and (b) a conducting slab are inserted between the plates of a parallel-plate capacitor. Assume the slab thicknesses to be one-half the plate separation.

32. A dielectric object in a nonuniform electric field experiences a net force. Why is there no net force if the field is uniform?

33. A stream of tap water can be deflected if a charged rod is brought close to the stream. Explain why.

34. Water has a high dielectric constant. Why isn't it used ordinarily as a dielectric material in capacitors?

35. A parallel-plate capacitor is charged by using a battery, which is then disconnected. A dielectric slab is then slipped completely between the plates. Describe qualitatively what happens to the charge, the capacitance, the potential difference, the electric field, the stored energy, and the slab.

36. While a parallel-plate capacitor remains connected to a battery, a dielectric slab is slipped between the plates. Describe qualitatively what happens to the charge, the capacitance, the potential difference, the electric field, and the stored energy. Is work required to insert the slab?

EXERCISES & PROBLEMS

49. Some burn victims are treated while lying in a closed chamber into which oxygen-enriched air is introduced under pressure. Such a victim is passed into the chamber on a stretcher, which then rests on insulating supports inside the metal framework of the chamber. Following treatment, the stretcher and patient are pulled out of the chamber and onto a trolley, to be rolled away. In at least two reported cases, a fire developed within the stretcher soon after it was pulled out of the chamber, at the end that was last to emerge. Investigating engineers reasoned that the fire was due to an electric spark inside the stretcher. The motion of the patient and activity of the hospital staff could have produced an electrostatic charge on the patient and stretcher. This charge would then induce a charge of the same magnitude but opposite sign on the metal framework of the chamber, just below the stretcher. Thus, the stretcher and that framework would effectively form a parallel-plate capacitor.

The engineers suspected that a spark between the two plates of this capacitor caused the fire. However, to do so, the capacitor would have to have had an electric potential difference V of at least 2000 V and an energy U of at least 0.20 mJ. During a test with the stretcher inside the chamber, the potential difference of the capacitor was found to be (only) 600 V, and the capacitance was 250 pF. (a) What then was U? (b) Could a spark have started a fire then?

Surprisingly, as the stretcher was pulled out of the chamber, with the top plate of the capacitor sliding past the lower plate, the capacitance was reduced without any reduction in the charge on the plates. When the capacitance was reduced to 25 pF, what were (c) V and (d) U? (e) Could a spark have started a fire as the stretcher was pulled out of the chamber?

50. As a safety engineer, you must evaluate the practice of storing flammable conducting liquids in nonconducting containers. The company supplying a certain liquid has been using a squat, cylindrical plastic container of radius $r = 0.20$ m and filling it to height $h = 10$ cm, which is not the container's full interior height (Fig. 26-42). Your investigation reveals that during handling at the company, the exterior surface of the container commonly acquires a negative charge density of magnitude 2.0 $\mu C/m^2$ (approximately uniform). Because the liquid is a conducting material, the charge on the container induces charge separation within the liquid. (a) How much negative charge is induced in the center of the liquid's bulk? (b) Assume the capacitance of the central portion of the liquid relative to ground is 35 pF. What is the potential energy associated with the negative charge in that effective capacitor? (c) If a spark occurs between the ground and the central portion of the liquid (through the venting port), the potential energy can be fed into the spark. The minimum spark energy needed to ignite the liquid is 10 mJ. In this situation, can a spark ignite the liquid?

51. *Plastic food wrap and failing parachutes.* Figure 26-43 shows, in cross section, two layers of plastic food wrap with a thin inter-

Fig. 26-43 Problem 51.

mediate layer of air of thickness x. The (uniform) surface charge density is $+15\ \mu C/m^2$ for the top layer and $-15\ \mu C/m^2$ for the bottom layer. Both layers have area A. In the air gap between the layers, what are (a) the magnitude of the electric field and (b) the electric energy density? (c) Find an expression for the electric energy U in the air gap. (d) Using Eq. 8-20 ($F = -dU/dx$), find an expression for the magnitude of the force needed to separate the plastic layers by an additional distance dx. (e) Then find the force per unit area required to separate the layers. (This is a typical force per unit area needed to separate plastic food wrap, which clings to itself because of the electrostatic charge put on it during its production and packaging.)

When a parachute is deployed, the nylon layers, which are close together in the packed parachute, must be separated by passing air currents. Electrostatic charge on adjacent layers can fight against this separation. When the parachute descent is slow, the force per unit area on the nylon sheets from the air currents might be as little as 6.0 N/m^2. (f) In that case, and assuming a situation like that in Fig. 26-43, what surface charge density will prevent the nylon layers from separating? (Such surface charge densities can easily be found in packed parachutes.)

52. In Fig. 26-44, $C_1 = 2.0\ \mu F$, $C_2 = 16\ \mu F$, and $C_3 = C_4 = 8.0\ \mu F$. Switch S is first thrown to the left until capacitor 1 reaches equilibrium. Then the switch is thrown to the right. When equilibrium is again reached, (a) how much charge is on capacitor 2 and (b) what is the potential across capacitor 2?

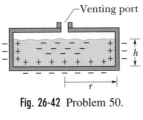

Fig. 26-42 Problem 50.

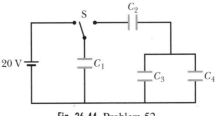

Fig. 26-44 Problem 52.

53. A certain parallel-plate capacitor is filled with a dielectric for which $\kappa = 5.5$. The area of each plate is 0.034 m^2, and the plates are separated by 2.0 mm. The capacitor will fail (short out and burn up) if the electric field between the plates exceeds 200 kN/C. What is the maximum energy that can be stored in the capacitor?

54. The capacitors in Fig. 26-45 each have capacitance 10 μF. What are the charges on (a) capacitor 1 and (b) capacitor 2?

55. Two sheets of aluminum foil have the same area, a separation of 1.0 mm, and (together) a capacitance of 10 pF, and they are charged to 12 V. (a) Calculate the area of each sheet. The separation

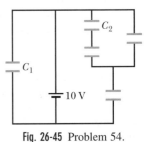

Fig. 26-45 Problem 54.

is now decreased by 0.10 mm with the charge held constant. (b) What is the new capacitance? (c) By how much does the potential difference change? Explain how a microphone might be constructed using this principle.

56. The capacitances of the four capacitors shown in Fig. 26-46 are given in terms of a certain quantity C. (a) If $C = 50\ \mu F$, what is the equivalent capacitance of the four-capacitor segment between points A and B? (*Hint:* First imagine that a battery is connected between those two points; then reduce the circuit to an equivalent capacitance.) (b) Repeat the question for points A and D.

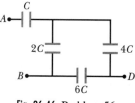

Fig. 26-46 Problem 56.

57. Figure 26-47 shows two identical capacitors with capacitance C in a circuit with two (ideal) diodes D. (An ideal diode has the property that positive charge flows through it only in the direction of the arrow and negative charge flows through it only in the opposite direction.) A 100 V battery is connected across the input terminals, first with terminal a connected to the positive battery terminal and later with terminal b connected there. In each case, what is the potential difference across the output terminals?

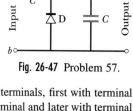

Fig. 26-47 Problem 57.

58. Figure 26-48 shows a variable "air gap" capacitor of the type used in manually tuned radios. Alternate plates are connected together; one group of plates is fixed in position, and the other group is capable of rotation. Consider a capacitor of n plates of alternating polarity, each plate having area A and separated from adjacent plates by a distance d. Show that this capacitor has a maximum capacitance of

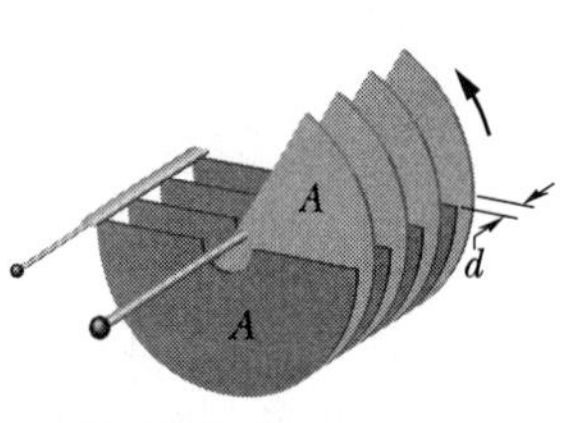

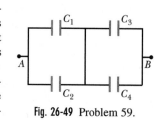

Fig. 26-48 Problem 58.

$$C = \frac{(n-1)\varepsilon_0 A}{d}.$$

59. Figure 26-49 shows a four-capacitor arrangement that is connected to a larger circuit at points A and B. The capacitances are $C_1 = 10\ \mu F$ and $C_2 = C_3 = C_4 = 20\ \mu F$. The charge on capacitor 1 is 30 μC. What is the magnitude of the potential difference $V_A - V_B$?

Fig. 26-49 Problem 59.

60. A capacitor of capacitance $C_1 = 6.00\ \mu F$ is connected in series with a capacitor of capacitance $C_2 = 4.00\ \mu F$, and a potential difference of 200 V is applied across the pair. (a) Calculate the equivalent capacitance. (b) What is the charge on each capacitor? (c) What is the potential difference across each capacitor?

61. Repeat Exercise 60 for the same two capacitors but with them now connected in parallel.

62. A capacitor of unknown capacitance C is charged to 100 V and connected across an initially uncharged 60 μF capacitor. If the final potential difference across the 60 μF capacitor is 40 V, what is C?

63. You have several 2.0 μF capacitors, each capable of withstanding 200 V without undergoing electrical breakdown (in which they conduct charge instead of storing it). How would you assemble a combination having an equivalent capacitance of (a) 0.40 μF or (b) 1.2 μF, each combination capable of withstanding 1000 V?

64. (a) Calculate the energy density of the electric field at distance r from the center of an electron at rest. (b) If the electron is assumed to be an infinitesimal point, what does this calculation yield for the energy density in the limit as $r \to 0$?

65. Three capacitors are initially uncharged and connected as shown in Fig. 26-50. If no capacitor can withstand a potential difference of more than 100 V without failure, (a) what is the magnitude of the maximum potential difference that can exist between points A and B and (b) what is the maximum energy that can be stored in the three-capacitor arrangement?

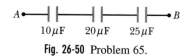

Fig. 26-50 Problem 65.

66. In Fig. 26-51, two parallel-plate capacitors A and B are connected in parallel across a 600 V battery. Each plate has area 80.0 cm²; the plate separations are 3.00 mm. Capacitor A is filled with air; capacitor B is filled with a dielectric of dielectric constant $\kappa = 2.60$. Find the magnitude of the electric field within (a) the dielectric of capacitor B and (b) the air of capacitor A. What are the free charge densities σ on the higher-potential plate of (c) capacitor A and (d) capacitor B? (e) What is the induced charge density σ' on the top surface of the dielectric?

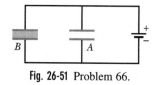

Fig. 26-51 Problem 66.

67. (a) A potential difference of 300 V is applied to a series connection of two capacitors, of capacitances $C_1 = 2.0\ \mu F$ and $C_2 = 8.0\ \mu F$. What are the charge on and the potential difference across each capacitor? (b) The charged capacitors are disconnected from each other and from the battery. They are then reconnected, positive plate to positive plate and negative plate to negative plate, with no external voltage being applied. What are the charge and the potential difference for each now? (c) Suppose the charged capacitors in (a) are reconnected with plates of *opposite* sign together. What then are the steady-state charge and potential difference for each?

68. A certain capacitor is charged to a potential difference V. If you wish to increase its stored energy by 10%, by what percentage should you increase V?

69. In Fig. 26-52, $C_1 = C_2 = C_3 = C_4 = 2.00\ \mu F$. (a) What is the charge on capacitor 1? (b) What is the voltage across capacitor 4?

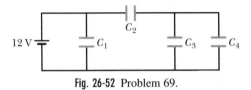

Fig. 26-52 Problem 69.

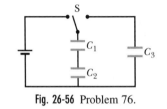

Fig. 26-56 Problem 76.

70. A parallel-plate air-filled capacitor has a capacitance of 130 pF. (a) What is the stored energy if the applied potential difference is 56.0 V? (b) With these data, can you calculate the energy density for points between the plates? Explain.

71. In Fig. 26-53, $C_1 = C_2 = C_4 = 8.0$ μF, and $C_3 = 4.0$ μF. What is the potential across capacitor 4?

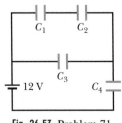

Fig. 26-53 Problem 71.

72. For making a parallel-plate capacitor, you have available two plates of copper, a sheet of mica (thickness = 0.10 mm, $\kappa = 5.4$), a sheet of glass (thickness = 2.0 mm, $\kappa = 7.0$), and a slab of paraffin (thickness = 1.0 cm, $\kappa = 2.0$). To obtain the largest capacitance, which sheet should you place between the copper plates?

73. In Fig. 26-54, (a) what is the equivalent capacitance of the six capacitors and (b) what is the net charge stored on them?

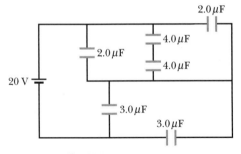

Fig. 26-54 Problem 73.

74. A slab of copper of thickness b is thrust into a parallel-plate capacitor of plate area A, as shown in Fig. 26-55; it is exactly halfway between the plates. (a) What is the capacitance after the slab is introduced? (b) If a charge q is maintained on the plates, what is the ratio of the stored energy before to that after the slab is inserted? (c) How much work is done on the slab as it is inserted? Is the slab sucked in or must it be pushed in?

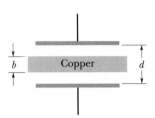

Fig. 26-55 Problems 74 and 75.

75. Repeat Problem 74, assuming that the potential difference, rather than the charge, is held constant.

76. In Fig. 26-56, $C_1 = C_2 = C_3 = 2.00$ μF. Switch S is first thrown to the left to charge capacitors 1 and 2 to 48 μC. Then the switch is thrown to the right. When equilibrium is reached, how much charge is on capacitor 3?

77. You have been assigned to design a transportable capacitor that can store 250 kJ of energy. You decide on a parallel-plate type with dielectric. (a) What is the minimum capacitor volume possible if you use thin plates and a dielectric whose dielectric strength is listed in Table 26-1? (b) Modern high-performance capacitors that can store 250 kJ have volumes of 0.0870 m^3. Assuming that the dielectric used has the same dielectric strength as in (a), what must be its dielectric constant?

78. A parallel-plate capacitor has plates of area 0.12 m^2 and a separation of 1.2 cm. A battery charges the plates to a potential difference of 120 V and is then disconnected. A dielectric slab of thickness 4.0 mm and dielectric constant 4.8 is then placed symmetrically between the plates. (a) What is the capacitance before the slab is inserted? (b) What is the capacitance with the slab in place? (c) What is the free charge q before and after the slab is inserted? What is the magnitude of the electric field (d) in the space between the plates and dielectric and (e) in the dielectric itself? (f) With the slab in place, what is the potential difference across the plates? (g) How much external work is involved in the process of inserting the slab?

79. In the capacitor of Sample Problem 26-6 (Fig. 26-17), (a) what fraction of the energy is stored in the air gaps? (b) What fraction is stored in the slab?

80. (a) Three capacitors are connected in parallel. Each has plate area A and plate spacing d. What must be the spacing of a single capacitor of plate area A if its capacitance equals that of the parallel combination? (b) What must be the spacing of an equivalent capacitor if the three capacitors are connected in series?

81. In Fig. 26-57, $C_1 = C_2 = 2.00$ μF, and $C_3 = 4.00$ μF. The switch is first thrown leftward until capacitor 1 reaches equilibrium, then it is thrown rightward. When equilibrium is again reached, what is the charge on capacitor 3?

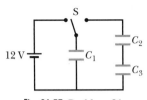

Fig. 26-57 Problem 81.

CLUSTERED PROBLEMS

82. A 40.0 μF capacitor 1 is charged to a potential difference of 10.0 V by a battery. Then the battery is removed, capacitor 1 is connected to the uncharged 30.0 μF capacitor 2 in Fig. 26-58a, and switch S is closed. When the system comes to equilibrium, what are (a) the charge on capacitor 1, (b) the charge on capacitor 2, (c) the potential difference across capacitor 1, and (d) the potential difference across capacitor 2? (e) What is the change in the potential energy of the two-capacitor system between the time the switch is closed and the time equilibrium is reached?

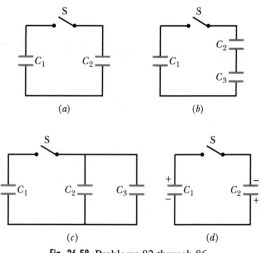

(a)   (b)

(c)   (d)

Fig. 26-58 Problems 82 through 86.

83. Repeat Problem 82 with the following change: Capacitor 2 is initially charged to 20.0 V instead of being uncharged.

84. A 40.0 μF capacitor 1 is charged to a potential difference of 10.0 V by a battery. Then the battery is removed, capacitor 1 is

connected to the uncharged 15.0 μF capacitor 2 and the uncharged 30.0 μF capacitor 3 in Fig. 26-58b, and switch S is closed. When the system comes to equilibrium, what are the charges on (a) capacitor 1, (b) capacitor 2, and (c) capacitor 3? Also, what are the potential differences across (d) capacitor 1, (e) capacitor 2, and (f) capacitor 3?

85. A 40.0 μF capacitor 1 is charged to a potential difference of 10.0 V by a battery. Then the battery is removed, capacitor 1 is connected to the uncharged 30.0 μF capacitor 2 and the uncharged 15.0 μF capacitor 3 in Fig. 26-58c, and switch S is closed. When the system comes to equilibrium, what are the charges on (a) capacitor 1, (b) capacitor 2, and (c) capacitor 3? Also, what are the potential differences across (d) capacitor 1, (e) capacitor 2, and (f) capacitor 3?

86. A 40.0 μF capacitor 1 and a 15.0 μF capacitor 2 are connected in parallel across a 10.0 V battery until equilibrium is established. Then the capacitors are disconnected from the battery and their connections with each other are reversed, with a switch S between them on one side (Fig. 26-58d). The switch is then closed. When the capacitors reach equilibrium, (a) how much charge has flowed through the switch and (b) what is the potential difference across either capacitor?

27
Current and Resistance

Sample Problem 27-7

Consider a strip of silicon that has a rectangular cross section with width $w = 3.2$ mm and height $h = 250$ μm, and through which there is a uniform current i of 5.2 mA. The silicon strip is a semiconductor that has been doped with a controlled phosphorus impurity. The doping has the effect of greatly increasing n, the number of charge carriers per unit volume, as compared with the value for pure silicon. In this case, $n = 1.5 \times 10^{23}$ m^{-3}.

(a) What is the current density in the strip?

SOLUTION: The **Key Idea** here is that current density J and current i are related via the cross-sectional area of the conductor. Also, because here the current is uniform over the cross-sectional area A, they are related by Eq. 27-5 ($J = i/A$), which gives us

$$J = \frac{i}{wh} = \frac{5.2 \times 10^{-3}\ \text{A}}{(3.2 \times 10^{-3}\ \text{m})(250 \times 10^{-6}\ \text{m})}$$
$$= 6500\ \text{A/m}^2. \qquad \text{(Answer)}$$

(b) What is the drift speed in the strip?

SOLUTION: The **Key Idea** here is that the drift speed is related to the current density J and the number n of conduction electrons per unit volume. Here, Eq. 27-7 gives us

$$v_d = \frac{J}{ne} = \frac{6500\ \text{A/m}^2}{(1.5 \times 10^{23}\ \text{m}^{-3})(1.60 \times 10^{-19}\ \text{C})}$$
$$= 0.27\ \text{m/s} = 27\ \text{cm/s}. \qquad \text{(Answer)}$$

Note that the current density (6500 A/m^2) for this semiconductor turns out to be comparable to the current density for the copper conductor in Sample Problem 27-3:

$$J = \frac{i}{\pi r^2} = \frac{17 \times 10^{-3}\ \text{A}}{2.54 \times 10^{-6}\ \text{m}^2} = 6700\ \text{A/m}^2.$$

That is, the rate at which charge flows through a unit area is about the same for the two devices. Yet, the drift speed (0.27 m/s) in the semiconductor, as found here, is *much* greater than the drift speed (4.9×10^{-7} m/s) in the copper conductor, as found in Sample Problem 27-3.

If you recheck the calculations, you will see that this large difference in drift speeds occurs because the number n of charge carriers per unit volume is much smaller in the semiconductor. Thus, if the current densities are to be comparable, the fewer conduction electrons in the semiconductor must move much faster than the electrons in the copper conductor.

Sample Problem 27-8

(a) What is the magnitude of the electric field that is applied to the copper wire of Sample Problem 27-3?

SOLUTION: We need two **Key Ideas** here: (1) We can relate the magnitude E of the electric field to the current density J with Eq. 27-11 ($E = \rho J$). (2) Because the current density is uniform, we can then relate J, the given current i ($= 17$ mA), and the wire's cross-sectional area A ($= 2.54 \times 10^{-6}$ m^2) via Eq. 27-5 ($J = i/A$). Putting these two ideas together and using the value of ρ for copper in Table 27-1, we find

$$E = \rho \frac{i}{A} = (1.69 \times 10^{-8}\ \Omega \cdot \text{m}) \frac{17 \times 10^{-3}\ \text{A}}{2.54 \times 10^{-6}\ \text{m}^2}$$
$$= 1.1 \times 10^{-4}\ \text{V/m}. \qquad \text{(Answer)}$$

(b) What is the magnitude of the electric field in the silicon semiconductor of Sample Problem 27-7? Because of the particular doping, that silicon is an *n*-type silicon.

SOLUTION: Here again the **Key Idea** is to relate E and J with Eq. 27-11. From Sample Problem 27-7 we know that $J = 6500$ A/m^2, and from Table 27-1 we see that $\rho = 8.7 \times 10^{-4}$ $\Omega \cdot$ m for *n*-type silicon. Thus, from Eq. 27-11,

$$E = \rho J = (8.7 \times 10^{-4}\ \Omega \cdot \text{m})(6500\ \text{A/m}^2)$$
$$= 5.7\ \text{V/m}. \qquad \text{(Answer)}$$

Note that the applied electric field in the semiconductor is much greater than that in the copper conductor. If you recheck the calculations, you will find that this difference is required by the large difference in the resistivities of the two devices. The need for a much greater electric field in the semiconductor is consistent with the need for a much greater drift speed that we found in Sample Problem 27-7. If the current densities are to be comparable in the two devices, the electric field applied to the semiconductor must be greater, to make possible the acceleration of the electrons to a greater drift speed.

Sample Problem 27-9

A wire of length $L = 2.35$ m and diameter $d = 1.63$ mm carries a current i of 1.24 A. The wire dissipates electrical energy at the rate P of 48.5 mW. Of what is the wire made?

SOLUTION: The **Key Idea** here is that we can identify the material by its resistivity, which we can find from the rate P at which energy is dissipated in the wire. From Eqs. 27-16 and 27-22 we have

$$P = i^2R = \frac{i^2\rho L}{A} = \frac{4i^2\rho L}{\pi d^2},$$

in which $A\ (=\frac{1}{4}\pi d^2)$ is the cross-sectional area of the wire. Solving for ρ, the resistivity of the material of which the wire is made, yields

$$\rho = \frac{\pi P d^2}{4i^2L} = \frac{(\pi)(48.5 \times 10^{-3}\ \text{W})(1.63 \times 10^{-3}\ \text{m})^2}{(4)(1.24\ \text{A})^2(2.35\ \text{m})}$$

$$= 2.80 \times 10^{-8}\ \Omega \cdot \text{m}.$$

Inspection of Table 27-1 tells us that the material is aluminum.

QUESTIONS

11. *Organizing question:* The following table gives the current density $J(r)$ in four wires, each with a radius R. J is in amperes per square meter, and radial distance r is in meters. For each wire, (a) give the units of the constant (a, b, c, or d) in the expression for $J(r)$ and (b) set up an equation to find the current in the wire between $r = 0$ and $r = R/2$.

Wire	Current Density
A	$J = ar$
B	$J = br^2$
C	$J = cr^3$
D	$J = d$

12. Figure 27-24 gives, for three wires of radius R, the current density $J(r)$ versus radius r, as measured from the center of a circular cross section through the wire. The wires are all made from the same material. Rank the wires according to the magnitude of the electric field (a) at the center, (b) halfway to the surface, and (c) at the surface, greatest first.

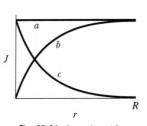

Fig. 27-24 Question 12.

13. *Organizing question:* the following list gives data for three situations in which a current is in a wire. For each situation, set up an equation, complete with known data, to find the energy dissipated in the wire in 2 min.

Situation 1: potential difference = 12 V, resistance = 2.0 Ω

Situation 2: current = 4.0 A, resistance = 3.0 Ω

Situation 3: potential difference = 12 V, current = 4.0 A

14. Figure 27-25 displays the circular cross section of a wire and three thin rings concentric with the wire. The rings have the same radial width dr, and the wire has a uniform current density. Rank the rings according to the amount of current passing through them, greatest first.

Fig. 27-25 Question 14.

15. In Fig. 27-26*a*, battery B_1 is recharging battery B_2. The current through B_2 and the potential across B_2 may be (a) 3 A and 4 V, (b) 2 A and 5 V, or (c) 6 A and 2 V. Rank these pairs of values according to the rate at which electric energy is transferred from B_1 to B_2, greatest first.

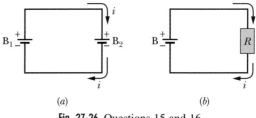

(a)   *(b)*

Fig. 27-26 Questions 15 and 16.

16. In three situations, a battery B and a resistance R are connected as in Fig. 27-26*b*. The values of R and the current through the resistance in the three situations are (a) 4 Ω and 2 A, (b) 3 Ω and 3 A, and (c) 3 Ω and 2 A. Rank the situations according to the rate at which electrical energy is transferred to thermal energy in the resistance, greatest first.

17. Is the filament resistance lower or higher in a 500 W lightbulb than in a 100 W bulb? (The same potential difference is applied to them when they are lighted.)

18. Figure 27-27 shows, as a function of time, the energy dissipated by current in a resistor. Rank the three lettered time periods according to (a) the current through the resistor and (b) the rate of dissipation in the resistor, greatest first.

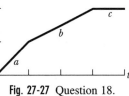

Fig. 27-27 Question 18.

DISCUSSION QUESTIONS

19. What conclusions can you draw by applying Eq. 27-4 to a closed surface through which a number of wires pass in random directions, carrying steady currents of different magnitudes?

20. In our convention for the direction of current arrows (a) would it have been more convenient, or even possible, to have assumed all charge carriers to be negative? (b) Would it have been more convenient, or even possible, to have labeled the electron as positive and the proton as negative?

21. List in tabular form similarities and differences between the flow of charge along a conductor, the flow of water through a horizontal pipe, and the conduction of heat through a slab. Consider such ideas as what causes the flow, what opposes it, what particles (if any) participate, and the units in which the flow may be measured.

22. Explain in your own words why we can have $\vec{E} \neq 0$ inside a conductor in this chapter, whereas we took $\vec{E} = 0$ for granted in Section 24-6.

23. Let a battery be connected to a copper cube at two corners defining a diagonal of the cube. Pass a hypothetical plane completely through the cube, tilted at an arbitrary angle. (a) Is the current i through the plane independent of the position and orientation of the plane? (b) Is there any position and orientation of the plane for which $\vec{J}$ is a constant in magnitude, direction, or both? (c) Does Eq. 27-4 hold for all orientations of the plane? (d) Does Eq. 27-4 hold for a closed surface of arbitrary shape, which may or may not lie entirely within the cube?

24. A potential difference V is applied to a copper wire of diameter d and length L. What is the effect on the electron drift speed of (a) doubling V, (b) doubling L, and (c) doubling d?

25. Why is it not possible to measure the drift speed for electrons by timing their travel along a conductor?

26. A potential difference V is applied to a circular cylinder of carbon by clamping it between circular copper electrodes, as in Fig. 27-28. Discuss the difficulty of calculating the resistance of the carbon cylinder using the relation $R = \rho L/A$.

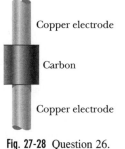

Fig. 27-28 Question 26.

27. How would you measure the resistance of a pretzel-shaped metal block? Give specific details to clarify the concept.

28. Sliding across the seat of an automobile can generate potentials of several thousand volts. Why isn't the sliding person electrocuted?

29. Discuss the difficulties of testing whether the filament of a lightbulb obeys Ohm's law.

30. How does the relation $V = iR$ apply to resistors that do *not* obey Ohm's law?

31. A fuse in an electrical circuit is a wire that is designed to melt, and thereby open the circuit, if the current exceeds a predetermined value. What are some characteristics of an ideal fuse wire?

32. Why does an incandescent lightbulb grow dimmer with use?

33. "The character and quality of our daily lives are influenced greatly by devices that do not obey Ohm's law." What can you say in support of this claim?

34. From a student's paper: "The relationship $R = V/i$ tells us that the resistance of a conductor is directly proportional to the potential difference applied to it." What do you think of this proposition?

35. Carbon has a negative temperature coefficient of resistivity. This means that its resistivity drops as its temperature increases. Would its resistivity disappear entirely at some high enough temperature?

36. What special characteristics must heating wire have?

37. Equation 27-22 ($P = i^2 R$) seems to suggest that the rate of energy dissipation in a resistor is reduced if the resistance is reduced. Equation 27-23 ($P = V^2/R$) seems to suggest just the opposite. How do you reconcile this apparent paradox?

38. Why do electric power companies reduce voltage during times of heavy demand? What is being saved?

39. Five wires of the same length and diameter are connected in turn between two points maintained at a constant potential difference. Will thermal energy be developed at the faster rate in the wire of (a) the smallest or (b) the largest resistance?

EXERCISES & PROBLEMS

46. Solar activity can cause wide-scale motions of ions in Earth's upper atmosphere. Those motions produce currents (called *telluric currents*) below ground, which set up electric potentials along Earth's surface. By measuring these potentials, engineers and geophysicists can determine what kinds of structures lie below the surface. As one example, Fig. 27-29 shows a cross section through the ground where a layer of wet porous rock with a resistivity ρ of $100\ \Omega \cdot m$ and a nonuniform thickness d overlays thick granite with a much greater resistivity. Assume that the current is confined to the porous rock and is directed along the x axis shown. Suppose also that the potential difference ΔV along the x axis is measured over a distance of $L = 20$ m in two regions: in region A, $\Delta V = 40$ μV; in region B, $\Delta V = 60$ μV. What then are the current densities in (a) region A and (b) region B?

Imagine a cross-sectional area perpendicular to the current; let the width (into and out of the page) be 1.0 m and the height be equal to the thickness d of the wet porous rock. (c) What is the current through such a cross-sectional area below region A, where $d = 3.8$ m? (d) Assuming that the current continues unchanged through another cross-sectional area of 1.0 m width below region B, find the thickness d of the porous rock there. (e) Sketch a graph of ΔV versus x directly on Fig. 27-29, and note how the variation in the granite's height can be inferred even if no actual d value is known.

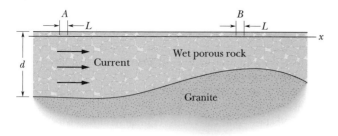

Fig. 27-29 Problem 46.

47. Figure 27-30a shows an overhead sketch of a section of the Trans Alaska Pipeline System, a petroleum pipeline with a length of 1000 km. In the figure, the pipeline extends through region A, where the ground resistivity is 500 $\Omega \cdot$ m, and region B, where the ground resistivity is 1000 $\Omega \cdot$ m. Currents i through the ground (the telluric currents of Problem 46) happen to be parallel to the pipeline in Fig. 27-30a. (a) Sketch the current streamlines through regions A and B (as in Fig. 27-4). The total resistance of the pipeline for its 1000 km length is 6.0 Ω. In a cross-sectional plane perpendicular to its length, the area of the pipe wall is 0.010 m². (b) What is the resistivity of the pipe?

Because the resistivity of the pipe is low compared to that of the ground, currents from the ground can enter the pipe, where they cause corrosion and possible leakage. Engineers monitor the currents to anticipate such damage. Assume that the currents are constant everywhere. Then the potential difference for any pair of points along the pipe must equal that for a pair of points along the adjacent ground. In Fig. 27-30a, points a and b along the pipe are separated by 1.0 km and have a potential difference of 8.0 mV. What are (c) the current in the pipe and (d) the current density in the adjacent ground? Points c and d along the pipe are separated by 1.0 km and have a potential difference of 9.5 mV. What are (e) the current in the pipe and (f) the current density in the adjacent ground? Is current entering or leaving the pipeline (g) where the pipeline moves into the left side of region B and (h) where it exits the right side of region B?

Figure 27-30b shows the pipeline in a region of uniform resistivity. The current along the pipe from e to f is 1.0 A. What are the currents in the sections of pipe (i) from g to h (that section is perpendicular to the ground currents) and (j) from f to g? Is current entering or leaving the pipeline at (k) point f and (l) point g?

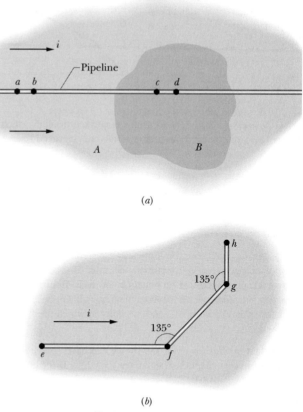

Fig. 27-30 Problem 47.

48. A cylindrical copper rod of length L and cross-sectional area A is re-formed to twice its original length with no change in volume. (a) Find the new cross-sectional area. (b) The resistance between its ends was R; what is it now?

49. A potential difference V is applied to a wire of cross sectional area A, length L, and resistivity ρ. You want to change the applied potential difference and stretch the wire so that the energy dissipation rate is multiplied by 30 and the current is multiplied by 4. What should be the wire's new length and area?

50. An aluminum rod with a square cross section is 1.3 m long and 5.2 mm on edge. (a) What is the resistance between its ends? (b) What must be the diameter of a cylindrical copper rod of length 1.3 m if its resistance is to be the same as that of the aluminum rod?

51. The copper windings of a motor have a resistance of 50 Ω at 20°C when the motor is idle. After the motor has run for several hours, the resistance rises to 58 Ω. What is the temperature of the windings now? Ignore changes in the dimensions of the windings. (Use Table 27-1.)

52. Using data taken from Fig. 27-11c, plot the resistance of the pn junction diode as a function of applied potential difference.

53. A 4.0-cm-long caterpillar crawls in the direction of electron drift along a 5.2-mm-diameter bare copper wire that carries a current of 12 A. (a) What is the potential difference between the two ends of the caterpillar? (b) Is its tail positive or negative compared to its head? (c) How much time would the caterpillar take to crawl 1.0 cm if it crawls at the drift speed of the electrons in the wire?

54. In Fig. 27-31, a resistance coil, wired to an external battery, is placed inside a thermally insulated cylinder fitted with a frictionless piston and containing an ideal gas. A current $i = 240$ mA exists in the coil, which has a resistance $R = 550$ Ω. At what speed v must the piston, of mass $m = 12$ kg, move upward to keep the temperature of the gas unchanged?

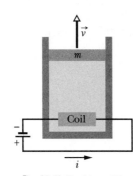

Fig. 27-31 Problem 54.

55. A certain brand of hot-dog cooker works by applying a potential difference of 120 V across opposite ends of the hot dog and allowing it to cook by means of the thermal energy produced. If the current is 10 A, 60 kJ is required to cook a hot dog, and the rate at which energy is supplied is unchanged, how long will it take to cook three hot dogs simultaneously?

56. The current density in a wire is uniform and equal to 2.0×10^6 A/m^2, the wire's length is 5.0 m, and the density of conduction electrons is 8.5×10^{28}/m^3. How long does an electron take (on the average) to travel the length of the wire?

57. A *pn* junction is formed from two different semiconducting materials in the form of identical cylinders with radius 0.165 mm, as depicted in Fig. 27-32. In one application, 3.50×10^{15} electrons per second flow across the junction from the n to the p side while 2.25×10^{15} *holes* per second flow from the p to the n side. (A hole acts like a particle with charge $+1.60 \times 10^{-19}$ C.) What are (a) the total current and (b) the current density?

Fig. 27-32 Problem 57.

58. A cylindrical metal rod is 1.60 m long and 5.50 mm in diameter. The resistance between its two ends (at 20°C) is 1.09×10^{-3} Ω. (a) What is the material? (b) A round disk, 2.00 cm in diameter and 1.00 mm thick, is formed of the same material. What is the resistance between the round faces, assuming that each face is an equipotential surface?

59. The headlights of a moving car require about 10 A from the 12 V alternator, which is driven by the engine. Assume the alternator is 80% efficient (its output electrical power is 80% of its input mechanical power), and calculate the horsepower the engine must supply to run the lights.

60. A potential difference of 1.20 V will be applied to a 33.0 m length of 18-gauge copper wire (diameter = 0.0400 in.). Calculate (a) the current, (b) the magnitude of the current density, (c) the magnitude of the electric field within the wire, and (d) the rate at which thermal energy will appear in the wire.

61. Copper and aluminum are being considered for a high-voltage transmission line that must carry a current of 60.0 A. The resistance per unit length is to be 0.150 Ω/km. Compute for each choice of cable material (a) the current density and (b) the mass per meter of the cable. The densities of copper and aluminum are 8960 and 2700 kg/m^3, respectively.

62. A 500 W heating unit is designed to operate with an applied potential difference of 115 V. (a) By what percentage will its heat output drop if the applied potential difference drops to 110 V? Assume no change in resistance. (b) If you took the variation of resistance with temperature into account, would the actual drop in heat output be larger or smaller than that calculated in (a)?

63. A straight conducting wire with a diameter of 1.0 mm carries a current of 2.0 A that is produced by a uniform electric field of 5.3 V/m inside the wire. What is the resistivity of the wire's material?

64. The resistance of a resistor is measured at several temperatures, as shown in the table. Enter the data in your graphing calculator and perform a linear regression fit of R versus T. Have your calculator graph the results of the linear regression fit; using the TRACE capability of the calculator (and perhaps the parameters of the fit), find the value of the resistance (a) at 20°C and (b) at 0°C. (c) Find the temperature coefficient *of resistance* (instead of resistivity) with a reference temperature of 20°C. (d) Find the temperature coefficient of resistance with a reference temperature of 0°C. (e) Find the resistance of the resistor at 265°C.

T, °C	50	100	150	200	250	300
R, Ω	139	171	203	234	266	298

65. A 200-m-long copper wire connects points A and B. The electric potential at point B is 50 V less than that at point A. If the resistivity of copper is 1.7×10^{-8} Ω · m, what are the magnitude and direction of the current density $\vec{J}$ in the wire?

66. The current density J in a certain wire with a circular cross section of radius $R = 2.00$ mm is given by $J = (3.0 \times 10^8)r$, with J in amperes per square meter and radial distance r in meters. What is the current through the outer section bounded by $r = R/2$ and $r = R$?

Clustered Problems

67. Wire A and wire B are made from the same material. Wire A has twice the diameter and half the length of wire B and a resistance of 8.0 Ω. (a) What is the resistance of wire B? (b) If the wires have equal currents, what is the ratio J_A/J_B of their current densities?

68. Wire C and wire D are made from different materials. The resistivity and diameter of wire C are 2.0×10^{-6} Ω · m and 1.00 mm, and those of wire D are 1.0×10^{-6} Ω · m and 0.50 mm. The wires are joined as shown in Fig. 27-33, and a current of 2.0 A is set up in them. What are the electric potential differences between (a) points 1 and 2 and (b) points 2 and 3? What are the rates at which energy is dissipated between (c) points 1 and 2 and (d) points 2 and 3?

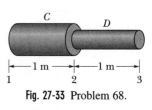

Fig. 27-33 Problem 68.

69. A certain length L of wire C in Problem 68 has a resistance of 8.0 Ω at 300 K. At 400 K it has a resistance of 10 Ω. (a) What is its resistance at 600 K? The material comprising wire D in Problem 68 has a temperature coefficient of resistivity that is twice that of the material in wire C. (b) What is the resistance of wire D at 600 K if its length is equal to L?

70. The tungsten filament in a certain lightbulb has a resistance of 2.0 Ω when the filament is at room temperature (300 K) with no current in it. When 12 V is applied to the filament, the current in the filament dissipates energy at the rate of 10 W. What then is the approximate temperature of the filament?

28

Circuits

Sample Problem 28-6

In Sample Problem 28-1, we found that the potential difference across the terminals of battery 1 in Fig. 28-6a is 3.84 V and the current i in the circuit is 0.2396 A. We also saw that this current i is a recharging flow through battery 2, against the emf of that battery. What is the potential difference across the terminals of battery 2, and is that difference greater than, less than, or equal to the emf (2.1 V) of the battery?

SOLUTION: The **Key Idea** is that because battery 2 is a real battery, with an internal resistance r_2 (= 1.8 Ω), the potential difference across its terminals is not equal to its emf. To find the potential difference, we mentally move from one terminal to the other, keep-

ing track of potential changes. Let us start at point c (the negative terminal of battery 2) and travel through battery 2 to point a (the positive terminal). We find

$$V_c + ir_2 + \mathscr{E}_2 = V_a$$

or
$$\begin{aligned} V_a - V_c &= ir_2 + \mathscr{E}_2 \\ &= (0.2396 \text{ A})(1.8 \ \Omega) + 2.1 \text{ V} \\ &= +2.5 \text{ V}. \qquad \text{(Answer)} \end{aligned}$$

The potential difference (2.5 V) between the terminals of the recharging battery is greater than the emf (2.1 V) of the battery.

Sample Problem 28-7

(a) In Sample Problem 28-5, a capacitor of capacitance C discharges through a resistor of resistance R. At what (time-dependent) rate P_r is thermal energy produced in the resistor during the discharging process?

SOLUTION: Two **Key Ideas** are required to find P_r. First, the rate at which thermal energy is produced due to a resistance is related to the current in the resistance according to Eq. 27-22 ($P = i^2R$). Second, the current here is decreasing during the discharging of the capacitor according to Eq. 28-37. Putting these two ideas together gives us

$$\begin{aligned} P_R = i^2R &= \left[-\frac{q_0}{RC} e^{-t/RC} \right]^2 R \\ &= \frac{q_0^2}{RC^2} e^{-2t/RC}. \qquad \text{(Answer)} \end{aligned}$$

(b) At what (time-dependent) rate P_C is stored energy lost by the capacitor during the discharging process?

SOLUTION: From Sample Problem 28-5b, we know that the energy U stored in the capacitor decreases during the discharge from its initial amount U_0 according to

$$U = U_0 e^{-2t/RC}.$$

Thus, the **Key Idea** here is that the rate P_C at which this energy is lost by the capacitor is the time derivative dU/dt:

$$P_C = \frac{dU}{dt} = \frac{d}{dt} (U_0 e^{-2t/RC}) = -\frac{2U_0}{RC} e^{-2t/RC}.$$

The initial charge on the capacitor is q_0, so by Eq. 26-21 the initial stored energy U_0 is $q_0^2/2C$. Substituting that for U_0 gives us

$$P_C = -\frac{q_0^2}{RC^2} e^{-2t/RC}. \qquad \text{(Answer)}$$

Note that $P_C + P_R = 0$. In words, stored energy lost by the capacitor is transferred completely to thermal energy of the resistor.

Sample Problem 28-8

The circuit in Fig. 28-45 consists of an ideal battery with emf $\mathscr{E} =$ 12 V, two resistors with resistances $R_1 = 4.0 \ \Omega$ and $R_2 = 6.0 \ \Omega$, and an initially uncharged capacitor with capacitance $C = 6.0 \ \mu$F. The circuit is completed when switch S is closed at time $t = 0$.

(a) At time $t = 2.0\tau$, what is the potential difference across the capacitor?

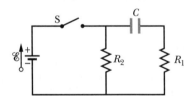

Fig. 28-45 Sample Problem 28-8. When switch S is closed, the circuit is complete and the battery begins to charge the capacitor.

SOLUTION: The **Key Idea** here is that once the switch is closed, the capacitor is charged through resistance R_1 by the battery's emf, which is connected across them just as in Fig. 28-13. (Resistance R_2 does not change this fact.) Thus, the potential across the capacitor changes with time according to Eq. 28-32,

$$V_C = \mathscr{E}(1 - e^{-t/RC}),$$

except that here the resistance is R_1 rather than R. Substituting $t = 2.0\tau = 2.0R_1C$ and given data, we then find that

$$V_C = (12 \text{ V})(1 - e^{-2.0R_1C/R_1C})$$
$$= (12 \text{ V})(1 - e^{-2.0}) = 10 \text{ V}. \qquad \text{(Answer)}$$

(b) At time $t = 2.0\tau$, what are the potential differences V_{R_1} and V_{R_2} across the two resistors? Do those potential differences increase, decrease, or remain the same while the capacitor is being charged?

SOLUTION: The **Key Idea** here is that we can apply the loop rule to the circuit of Fig. 28-45 to find the potential differences across the resistors. For the potential difference across R_1, we apply the rule

to the big loop so as to include R_1. Moving clockwise from the negative terminal of the battery, up in potential across the battery, then down in potential across the capacitor and resistance R_1, we find

$$\mathscr{E} - V_C - V_{R_1} = 0. \qquad (28\text{-}42)$$

We now know that at $t = 2.0\tau$, the potential difference V_C is equal to 10 V. Substituting this and $\mathscr{E} = 12$ V into Eq. 28-42 yields

$$V_{R_1} = 2.0 \text{ V}. \qquad \text{(Answer)}$$

During the charging of the capacitor, the battery's emf $\mathscr{E}$ is constant and the potential difference V_C across the capacitor increases. By writing Eq. 28-42 as $V_{R_1} = \mathscr{E} - V_C$, we see that V_{R_1} must decrease during the charging process.

If we apply the loop rule to the left-hand loop of Fig. 28-45, again clockwise from the negative terminal, we find

$$\mathscr{E} - V_{R_2} = 0$$

and

$$V_{R_2} = \mathscr{E} = 12 \text{ V}. \qquad \text{(Answer)}$$

Thus, V_{R_2} does not change during the charging process.

QUESTIONS

12. In Fig. 28-46, the four graphs are intended to give the electric potential around a closed, one-loop circuit containing an ideal battery and one or more resistors. (a) Which, if any, are physically possible? What is wrong with any wrong graph? (b) For any physically possible graph, which part of the graph corresponds to the battery and how many resistors are in the circuit?

and rank the segments according to the magnitude of the average electric field in them, greatest first. (b) Now assume that $R_1 > R_2$ and then again rank the segments. (c) What is the direction of the electric field along the x axis?

14. What is the equivalent resistance of three resistors, each of resistance R, if they are connected to an ideal battery (a) in series with one another and (b) in parallel with one another? (c) Is the potential difference across the series arrangement greater than, less than, or equal to that across the parallel arrangement?

15. *Organizing question:* In this chapter we have seen three basic techniques for solving for the current through a resistor located in a circuit of batteries and resistors: (1) solving a single-loop equation, (2) simplifying the circuit by finding the equivalent resistance of resistors in parallel or in series, and (3) solving two loop equations simultaneously. Which of these techniques is best for finding the current through resistor R_1 in the five circuits of Fig. 28-48?

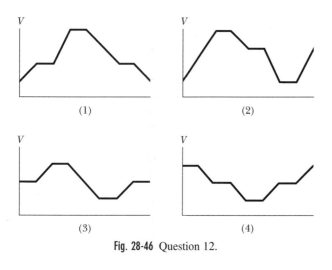

(1)   (2)

(3)   (4)

Fig. 28-46 Question 12.

13. In Fig. 28-47, a circuit consists of a battery and two uniform resistors, and the section lying along an x axis is divided into five segments of equal lengths. (a) Assume that $R_1 = R_2$

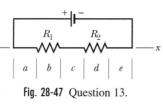

Fig. 28-47 Question 13.

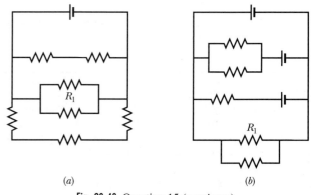

(a)   (b)

Fig. 28-48 Question 15 (*continues*).

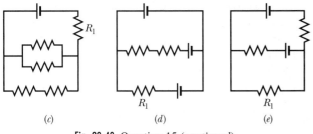

Fig. **28-48** Question 15 (*continued*).

16. Two resistors are wired to a battery. (a) In which arrangement, parallel or series, are the potential differences across each resistor and across the equivalent resistance all equal? (b) In which arrangement are the currents through each resistor and through the equivalent resistance all equal?

17. (a) In Fig. 28-49, when the branch with R_2 is added as indicated, does the rate at which electric energy is transferred to thermal energy in R_1 increase, decrease, or stay the same? (b) Does the rate at which electric energy is supplied by the battery increase, decrease, or stay the same? (c) Repeat (a) and (b) if, instead, R_2 is added in series with R_1.

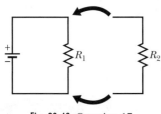

Fig. **28-49** Question 17.

18. Without written calculation, determine the potential difference across each capacitor in Fig. 28-50.

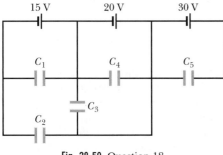

Fig. **28-50** Question 18.

19. *Organizing question:* The switch in Fig. 28-51 is closed at time $t = 0$. (a) Thereafter, what is the current through R_3 in terms of the currents through R_1 and R_2? (b) What is the relation between the charge q on the capacitor and the current through R_2? Using those relations and data given in the figure, set up loop equations for (c) the big loop and (d) the left-hand loop. (e) To produce a single differential equation for charge q, what step should now be taken? When the circuit finally reaches equilibrium, what are (f) the current through R_2 and (g) the potential difference across the capaci-

tor? (*Hint:* No calculator or even written calculation is needed for (f) and (g).)

The switch is reopened at $t = 0$ (we reset our reference clock). (h) What now is the relation between the charge q on the capacitor and the current through R_2? (i) What is the relation between the currents through R_2 and R_3? (j) Using these relations and known data, set up a loop equation for the discharge of the capacitor.

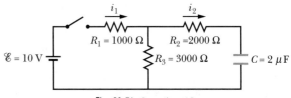

Fig. **28-51** Question 19.

20. After the switch in Fig. 28-13 is closed on point *a*, there is current *i* through resistance *R*. Figure 28-52 indicates that current for four sets of values of *R* and capacitance *C*: (1) R_0 and C_0, (2) $2R_0$ and C_0, (3) R_0 and $2C_0$, (4) $2R_0$ and $2C_0$. Which set goes with which curve?

21. *Organizing question:* The switch in the circuit of Fig. 28-53 is closed at time $t = 0$. Set up a loop equation, complete with known values, to find the charge q on the capacitor when the current in the resistor has dropped to 5 mA.

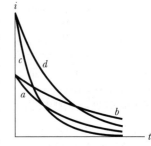

Fig. **28-52** Question 20.

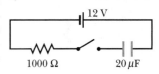

Fig. **28-53** Question 21.

22. The five sections of circuit in Fig. 28-54 are to be connected, in turn, to the same 12 V battery via a switch as in Fig. 28-13. The resistors are all identical; so are the capacitors. Rank the sections

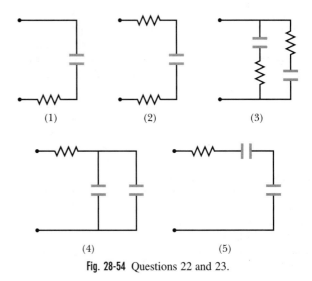

Fig. **28-54** Questions 22 and 23.

according to the time required for the capacitors to reach 50% of their final potential, greatest first.

23. Rank the five sections of Question 22 according to the potential across any resistor in the section when the potential across any capacitor in the section reaches 4 V, greatest first.

DISCUSSION QUESTIONS

24. Discuss the changes that would occur if, in Fig. 28-2, we increased the mass m by such an amount that the "motor" reversed direction and became a "generator"—that is, a seat of emf.

25. Devise a method for measuring the emf and the internal resistance of a battery.

26. How could you calculate $V_b - V_a$ in Fig. 28-4a by following a path from a to b that does not lie in the conducting circuit?

27. A 120 V bulb that operates at 25 W glows at normal brightness when connected across a bank of batteries. A 120 V bulb that operates at 500 W glows only dimly when connected across the same bank. How could this happen?

28. Under what circumstances can the potential difference between the terminals of a battery exceed the battery's emf?

29. The loop rule is based on the principle of conservation of energy, and the junction rule on the principle of conservation of charge. Explain how these rules are based on these principles.

30. Under what circumstances would you want to connect resistors in parallel? In series?

31. Under what circumstances would you want to connect batteries in parallel? In series?

32. What is the difference between emf and potential difference?

33. Do the junction and loop rules apply in a circuit containing a capacitor?

34. Show that the product RC in Eqs. 28-30 and 28-31 has the dimensions of time—that is, that 1 second = 1 ohm × 1 farad.

35. When a capacitor, a resistor, and a battery are connected in series, the charge that the capacitor stores is unaffected by the resistance of the resistor. What purpose then is served by the resistor?

36. Explain why, in Sample Problem 28-5, the energy falls to half its initial value more rapidly than does the charge.

37. A capacitor is connected across the terminals of a battery. Does the amount of charge that eventually appears on the capacitor plates depend on the internal resistance of the battery?

38. Devise a method whereby an RC circuit can be used to measure very high resistances.

39. In Fig. 28-13 suppose that switch S is closed on a. Explain why, in view of the fact that the negative terminal of the battery is not connected to resistance R, the initial current in R should be $\mathcal{E}/R$, as Eq. 28-31 predicts.

EXERCISES & PROBLEMS

57. When the tires of a car roll along pavement, a separation of charge occurs at the contact regions with the pavement—electrons move up onto the tire from the pavement, leaving trails of positive charge behind the car. As a result, the car can become highly charged. When the car stops moving, it discharges through the tires as a capacitor can discharge through a resistor. If a person outside the car attempts to touch the car or its occupants before the car is discharged, a painful spark may occur. Worse, if fuel vapors are present, the spark could cause a fire or an explosion.

Consider a just-stopped car with an effective capacitance $C = 500$ pF and an initial electric potential $V_0 = -30.0$ kV relative to ground. The charge on the car discharges through the *equivalent* resistance of the four tires (Fig. 28-55). In general, tires range in resistance R from 100 MΩ to 100 GΩ. (a) For that resistance range, determine the range of the time constant τ for the discharge circuit. (b) What is the initial (stored) energy associated with the charge on the car? (c) For the given resistance range, what is the range of time needed for the stored energy to decrease to 50 mJ, which is probably less energy than is needed to cause an explosion?

(d) In pit stops for professional racing, the pit crew feverishly runs to the car to fuel it and make any necessary changes. Given that haste, should the tires on a race car be of high or low resistance? (e) If that resistance requirement cannot be met, what else could be done to avoid sparks and the possibility of a fire or an explosion?

58. Because electrostatic discharges can easily ruin electronic equipment such as computers, electrical engineers worry about such discharges between workers and electronic equipment in

Fig. 28-55 Problem 57.

workplaces. Suppose that, while you are sitting in a chair, charge separation between your clothing and the chair puts you at a potential of 200 V, with the capacitance between you and the chair at 150 pF. When you stand up, the increased separation between your body and the chair decreases the capacitance to 10 pF. (a) What then is the potential of your body? That potential is reduced over time, as the charge on you drains through your body and shoes (you are a capacitor discharging through a resistance). Assume that the resistance along that route is 300 GΩ. If you touch an electrical component while your potential is greater than 100 V, you could ruin the component. (b) How long must you wait until your potential reaches the safe level of 100 V?

If you wear a conducting wrist strap that is connected to ground, your potential does not increase as much when you stand up; you also discharge more rapidly because the resistance through the grounding connection is much less than through your body and shoes. (c) Suppose that when you stand up, your potential is

1400 V and the chair-to-you capacitance is 10 pF. What resistance in that wrist-strap grounding connection will allow you to discharge to 100 V in 0.30 s, which is less time than you would need to reach for, say, your computer?

59. A gasoline gauge for an automobile is shown schematically in Fig. 28-56. The indicator (on the dashboard) has a resistance of 10 Ω. The tank unit is simply a float connected to a variable resistor whose resistance is 140 Ω when the tank is empty, is 20 Ω when the tank is full, and varies linearly with the volume of gasoline. Find the current in the circuit when the tank is (a) empty, (b) half-full, and (c) full. Treat the battery as ideal.

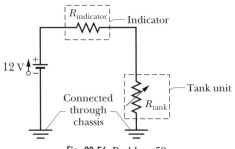

Fig. 28-56 Problem 59.

60. Wires A and B, having equal lengths of 40.0 m and equal diameters of 2.60 mm, are connected in series. A potential difference of 60.0 V is applied between the ends of the composite wire. The resistances of the wires are 0.127 and 0.729 Ω, respectively. Determine (a) the current density in each wire and (b) the potential difference across each wire. (c) Identify the wire materials. (See Table 27-1.)

61. A temperature-stable resistor is made by connecting a resistor made of silicon in series with one made of iron. If the required total resistance is 1000 Ω in a wide temperature range around 20°C, what should be the resistances of the two resistors? (See Table 27-1.)

62. In Fig. 28-5a, calculate the potential difference across R_2, assuming $\mathscr{E} = 12$ V, $R_1 = 3.0$ Ω, $R_2 = 4.0$ Ω, and $R_3 = 5.0$ Ω.

63. A total resistance of 3.00 Ω is to be produced by connecting an unknown resistance to a 12.0 Ω resistance. What must be the value of the unknown resistance, and should it be connected in series or in parallel?

64. A circuit containing five resistors connected to a battery with a 12.0 V emf is shown in Fig. 28-57. What is the potential difference across the 5.0 Ω resistor?

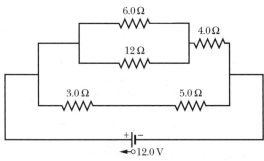

Fig. 28-57 Problem 64.

65. A group of N identical batteries of emf $\mathscr{E}$ and internal resistance r may be connected all in series (Fig. 28-58a) or all in parallel (Fig. 28-58b) and then across a resistor R. Show that both arrangements will give the same current in R if $R = r$.

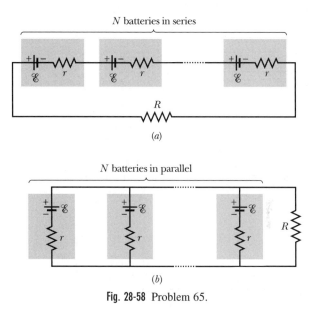

Fig. 28-58 Problem 65.

66. For sensitive manual control of current in a circuit, you can use a parallel combination of variable resistors of the sliding contact type, as in Fig. 28-59. (Moving the contact changes how much resistance is in the circuit.) Suppose the full resistance R_1 of resistor A is 20 times the full resistance R_2 of resistor B. (a) What procedure should be used to adjust the current i to the desired value? (b) Why is the parallel combination better than a single variable resistor?

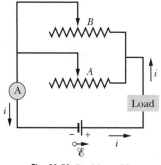

Fig. 28-59 Problem 66.

67. Two resistors R_1 and R_2 may be connected either in series or in parallel across an ideal battery with emf $\mathscr{E}$. We desire the rate of energy dissipation of the parallel combination to be five times that of the series combination. If $R_1 = 100$ Ω, what is R_2? (*Hint:* There are two answers.)

68. (a) If points a and b in Fig. 28-41 are connected by a wire of resistance r, show that the current in the wire is

$$i = \frac{\mathscr{E}(R_s - R_x)}{(R + 2r)(R_s + R_x) + 2R_sR_x},$$

where $\mathscr{E}$ is the emf of the ideal battery and $R = R_1 - R_2$. Assume that R_0 equals zero. (b) Is this formula consistent with the result of Problem 43?

69. In Fig. 28-60, the current through the 4.0 Ω resistor is indicated. What is the emf of the (ideal) battery?

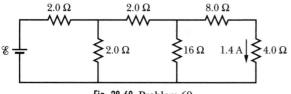

Fig. 28-60 Problem 69.

70. The switch in Fig. 28-61 is left closed for a long time so that the steady state is reached. Then at time $t = 0$ the switch is opened. What is the current through the 15 kΩ resistor at $t = 4.00$ ms?

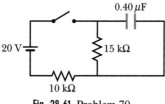

Fig. 28-61 Problem 70.

71. (a) What are the size and direction of current i_1 in Fig. 28-62? (Can you answer this making only mental calculations?) (b) At what rate is the battery supplying energy?

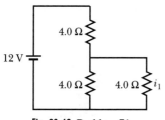

Fig. 28-62 Problem 71.

72. Figure 28-63 shows a portion of a circuit. What is the size of current i_1?

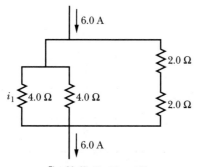

Fig. 28-63 Problem 72.

73. What are the potentials (a) V_1 and (b) V_2 at the points indicated in Fig. 28-64, where each resistance is 2.0 Ω? (The symbol at the upper right indicates that the circuit is grounded there; the potential is defined to be zero at that point.)

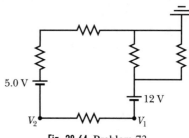

Fig. 28-64 Problem 73.

74. Each of the six real batteries in Fig. 28-65 has an emf of 20 V and a resistance of 4.0 Ω. (a) What is the current through the 4.0 Ω resistor? (b) What is the potential difference across each battery? (c) What is the power of each battery? (d) At what rate does each battery transfer energy to internal thermal energy?

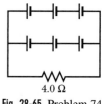

Fig. 28-65 Problem 74.

75. (a) How much work does an ideal battery with a 12.0 V emf do on an electron that passes through the battery from the positive to the negative terminal? (b) If 3.4×10^{18} electrons pass through each second, what is the power of the battery?

76. A 10-km-long underground cable extends east to west and consists of two parallel wires, each of which has resistance 13 Ω/km. A short develops at distance x from the west end when a conducting path of resistance R connects the wires (Fig. 28-66). The resistance of the wires and the short is then 100 Ω when the measurement is made from the east end, and 200 Ω when it is made from the west end. What are (a) x and (b) R?

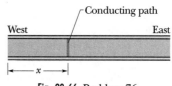

Fig. 28-66 Problem 76.

77. Thermal energy is to be generated in a 0.10 Ω resistor at the rate of 10 W by connecting the resistor to a battery whose emf is 1.5 V. (a) What potential difference must exist across the resistor? (b) What must be the internal resistance of the battery?

78. A battery of emf $\mathscr{E} = 2.00$ V and internal resistance $r = 0.500$ Ω is driving a motor. The motor is lifting a 2.00 N mass at constant speed $v = 0.500$ m/s. Assuming no energy losses, find (a) the current i in the battery–motor circuit and (b) the potential difference V across the terminals of the motor. (c) Discuss the fact that there are two solutions to this problem.

79. Energy is supplied by a device of emf $\mathscr{E}$ to a transmission line with a total resistance R. Find the ratio of the energy dissipation rate in the line for $\mathscr{E} = 110\,000$ V to the energy dissipation rate for $\mathscr{E} = 110$ V, assuming energy is supplied at the same rate for the two cases.

80. In Fig. 28-67, find the equivalent resistance between points (a) A and B, (b) A and C, and (c) B and C. (*Hint:* Imagine that a battery is connected between points A and C.)

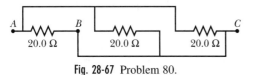

Fig. 28-67 Problem 80.

81. A 120 V power line is protected by a 15 A fuse. What is the maximum number of 500 W lamps that can be simultaneously operated in parallel on this line without "blowing" the fuse because of an excess of current?

82. Figure 28-68 shows a battery connected across a uniform resistor R_0. A sliding contact can move across the resistor from $x = 0$ at the left to $x = 10$ cm at the right. Moving the contact changes how much resistance is to the left of the contact and how much is to the right. Find an equation for the rate at which energy is dissipated in resistor R as a function of x. Plot the function for $\mathscr{E} = 50$ V, $R = 2000$ Ω, and $R_0 = 100$ Ω.

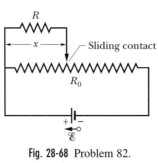

Fig. 28-68 Problem 82.

83. A three-way 120 V lamp bulb that contains two filaments is rated for 100-200-300 W. One filament burns out. Afterward, the bulb operates at the same intensity (dissipates energy at the same rate) on its lowest and its highest switch positions but does not operate at all on the middle position. (a) How are the two filaments wired to the three switch positions? (b) Calculate the resistances of the filaments.

84. What current, in terms of $\mathscr{E}$ and R, does the ammeter in Fig. 28-69 read? Assume that it has zero resistance and that the battery is ideal.

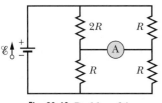

Fig. 28-69 Problem 84.

85. In Fig. 28-12, assume that $\mathscr{E} = 5.0$ V, $r = 2.0$ Ω, $R_1 = 5.0$ Ω, and $R_2 = 4.0$ Ω. If the ammeter resistance R_A is 0.10 Ω, what percent error does it introduce into the measurement of the current? Assume that the voltmeter is not present.

86. The circuit of Fig. 28-70 shows a capacitor, two ideal batteries, two resistors, and a switch S. Initially S has been open for a long time. If it is then closed for a long time, by how much does the charge on the capacitor change? Assume $C = 10$ μF, $\mathscr{E}_1 = 1.0$ V, $\mathscr{E}_2 = 3.0$ V, $R_1 = 0.20$ Ω, and $R_2 = 0.40$ Ω.

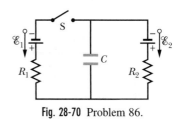

Fig. 28-70 Problem 86.

87. In the circuit of Fig. 28-71, what value of R will result in no current through the 20.0 V battery?

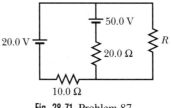

Fig. 28-71 Problem 87.

88. The capacitor in Fig. 28-72 is uncharged when the switch is closed. What are the initial currents through (a) the 10 kΩ resistor and (b) the 20 kΩ resistor? (c) What is the current through the 10 kΩ resistor a very long time after the switch is closed?

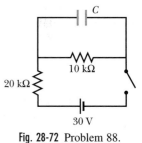

Fig. 28-72 Problem 88.

89. What are the sizes and directions of currents (a) i_1 and (b) i_2 in Fig. 28-73? (See if you can answer this using only mental calculation.) At what rates is energy being transferred at (c) the 16 V battery and (d) the 8.0 V battery, and for each, is energy being supplied or absorbed? The batteries are ideal.

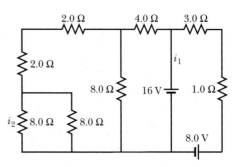

Fig. 28-73 Problem 89.

90. What are the size and direction of current i_1 in Fig. 28-74? (*Hint:* This can be answered by using only mental calculation.)

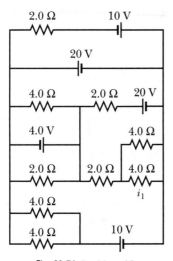

Fig. 28-74 Problem 90.

91. Considering all possible values of R in Fig. 28-75, find the maximum possible rate at which energy can be supplied by the battery.

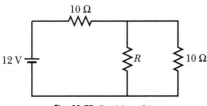

Fig. 28-75 Problem 91.

92. What are the size and direction of current i in Fig. 28-76, where all resistances are 4.0 Ω and all batteries have an emf of 10 V? (*Hint:* This can be answered using only mental calculation.)

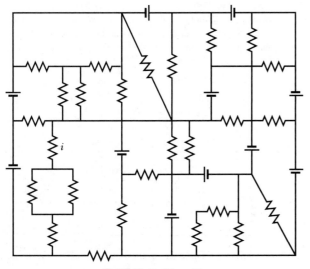

Fig. 28-76 Problem 92.

Clustered Problems

93. In Fig. 28-77a, $\mathscr{E} = 6.00$ V, $R_1 = 100$ Ω, $R_2 = 300$ Ω, and $R_3 = 600$ Ω. (a) What is the equivalent resistance between points A and B? (b) What is the electric potential across R_1? (c) What is the current through R_3?

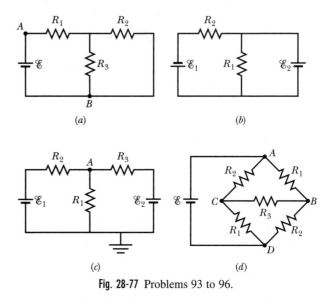

Fig. 28-77 Problems 93 to 96.

94. In Fig. 28-77b, $\mathscr{E}_1 = 6.00$ V (note battery orientation), $\mathscr{E}_2 = 12.0$ V, $R_1 = 200$ Ω, and $R_2 = 100$ Ω. What is the current (magnitude and direction) through (a) R_1, (b) R_2, and (c) the 12 V battery?

95. In Fig. 28-77c, $\mathscr{E}_1 = 6.00$ V, $\mathscr{E}_2 = 12.0$ V, $R_1 = 100$ Ω, $R_2 = 200$ Ω, and $R_3 = 300$ Ω. One point of the circuit is grounded, as indicated by the standard symbol for grounding; the electric potential of that point is defined to be zero. Grounding a point in a circuit does not alter the electric potential differences around the circuit and thus does not alter the currents in the circuit. What is the current (magnitude and direction) through (a) R_1, (b) R_2, and (c) R_3? (d) What is the electric potential at point A?

96. In Fig. 28-77d, $\mathscr{E} = 12.0$ V, $R_1 = 2000$ Ω, $R_2 = 3000$ Ω, and $R_3 = 4000$ Ω. What are the potential differences (a) $V_A - V_B$, (b) $V_B - V_C$, (c) $V_C - V_D$, and (d) $V_A - V_C$?

Magnetic Fields

ADDITIONAL SAMPLE PROBLEMS

Sample Problem 29-9

Figure 29-41 shows a proton traveling through a region of uniform electric and magnetic fields. The electric field $\vec{E}$ is directed parallel to the y axis and has a magnitude of 2000 V/m; the magnetic field $\vec{B}$ is directed out of the figure (in the positive direction of the z axis, which is not drawn) and has a magnitude of 0.50 T. At the instant the proton's velocity vector is in the positive direction of the x axis as shown, find the electric force, the magnetic force, and the net force on the proton for proton speeds of 2.0 m/s, 4000 m/s, and 10 000 m/s.

SOLUTION: Let us find the electric and magnetic forces individually and then vectorially add them to get the net force.

Electric force: From Eq. 23-28 ($\vec{F}_E = q\vec{E}$), we see that the electric force $\vec{F}_E$ is related only to the electric field $\vec{E}$ and the proton's charge q. A **Key Idea** here is that $\vec{F}_E$ *does not* depend on the proton's velocity $\vec{v}$. Thus, for each of the three given speeds, the magnitude of $\vec{F}_E$ is

$$F_E = qE = (1.6 \times 10^{-19}\ \text{C})(2000\ \text{V/m})$$
$$= 3.2 \times 10^{-16}\ \text{N}.$$

Because q is a positive charge, the direction of $\vec{F}_E$ is the direction of $\vec{E}$, which is the positive direction of y. Table 29-5 gives the y component of $\vec{F}_E$ for the three speeds; $\vec{F}_E$ has no x or z component.

Magnetic force: From Eq. 29-2 ($\vec{F}_B = q\vec{v} \times \vec{B}$), we see that the magnetic force $\vec{F}_B$ is related to the magnetic field $\vec{B}$, the proton's charge q, and its velocity $\vec{v}$. A **Key Idea** here is that $\vec{F}_B$ *does* depend on the proton's velocity $\vec{v}$. The magnitude of $\vec{F}_B$ is given by Eq. 29-3 as

$$F_B = |q|vB \sin \phi,$$

where ϕ, the angle between velocity $\vec{v}$ and magnetic field $\vec{B}$, is 90° in Fig. 29-41. For the proton speed of 2.0 m/s, we then have

TABLE 29-5 Results for the Situation of Fig. 29-41

Proton Speed (m/s)	$\vec{F}_E$ (N)	$\vec{F}_B$ (N)	$\vec{F}_{net}$ (N)
		y Components of	
2.0	3.2×10^{-16}	-1.6×10^{-19}	3.2×10^{-16}
4000	3.2×10^{-16}	-3.2×10^{-16}	0
10 000	3.2×10^{-16}	-8.0×10^{-16}	-4.8×10^{-16}

$$F_B = (1.6 \times 10^{-19}\ \text{C})(2.0\ \text{m/s})(0.50\ \text{T})(\sin 90°)$$
$$= 1.6 \times 10^{-19}\ \text{N}.$$

Using the right-hand rule of Fig. 29-2, we see that $\vec{F}_B$ is in the negative direction of the y axis and thus has the y component -1.6×10^{-19} N. You can show that the other two given proton speeds yield the other two y components of $\vec{F}_B$ listed in Table 29-5; $\vec{F}_B$ has no x or z component.

Net force: Because $\vec{F}_B$ and $\vec{F}_E$ are both directed along the y axis, the net force $\vec{F}_{net}$ on the proton at the instant shown in Fig. 29-41 is the algebraic sum of their y components; the sums are given in Table 29-5 for the three speeds. Note that when the proton speed v is small, $\vec{F}_E$ dominates $\vec{F}_{net}$ because the magnitude of $\vec{F}_B$, which depends on v, is small. When $v = 4000$ m/s, the magnitude of $\vec{F}_B$ is enough to match the magnitude of $\vec{F}_E$, and so $\vec{F}_{net} = 0$. Finally, when v is even greater, so is the magnitude of $\vec{F}_B$, which then dominates $\vec{F}_{net}$.

You can show that if $\vec{B}$ is reversed in Fig. 29-41, Table 29-5 must be replaced by Table 29-6 because $\vec{F}_B$ is then in the positive direction of y. Note that now $\vec{F}_{net}$ is not zero for $v = 4000$ m/s (and is not zero for any other speed).

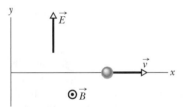

Fig. 29-41 Sample Problem 29-9. A proton travels through crossed electric and magnetic fields. At a certain instant, its velocity vector is directed along the x axis.

TABLE 29-6 Results When $\vec{B}$ in Fig. 29-41 Is Reversed

Proton Speed (m/s)	$\vec{F}_E$ (N)	$\vec{F}_B$ (N)	$\vec{F}_{net}$ (N)
		y Components of	
2.0	3.2×10^{-16}	1.6×10^{-19}	3.2×10^{-16}
4000	3.2×10^{-16}	3.2×10^{-16}	6.4×10^{-16}
10 000	3.2×10^{-16}	8.0×10^{-16}	11.2×10^{-16}

Sample Problem 29-10

Figure 29-42 shows a length of wire with a central semicircular arc, placed in a uniform magnetic field $\vec{B}$ that is directed out of the plane of the figure. If the wire carries a current i, what net magnetic force $\vec{F}$ acts on it?

SOLUTION: Because the wire carries a current in a magnetic field $\vec{B}$, magnetic forces $\vec{F}_B$ can act on the wire. One **Key Idea** here is that to find the net magnetic force on the wire, we can separately consider the forces on the two straight sections and the forces on the central arc.

   Straight sections: Both straight sections have a length vector $\vec{L}$ directed to the right. Because $\vec{B}$ is directed outward and current i is taken to be a positive quantity, Eq. 29-26 tells us that the magnetic forces on both sections ($\vec{F}_1$ and $\vec{F}_3$ in Fig. 29-42) act downward. From Eq. 29-27 with $\phi = 90°$, the magnitudes of those forces are

$$F_1 = F_3 = iLB. \tag{29-41}$$

   Central arc: The **Key Idea** here is that we cannot apply Eq. 29-26 to the central arc because the arc does not have a single direction for a length vector $\vec{L}$ along it. However, we can apply Eq. 29-28 ($d\vec{F}_B = i\,d\vec{L} \times \vec{B}$) to a segment of the arc with differential length dL and, thus, differential length vector $d\vec{L}$. No matter where the segment is located on the central arc, the differential magnetic force $d\vec{F}$ acting on it is directed radially toward point O, the center of the arc, as shown in Fig. 29-42.

   To find the net force $\vec{F}_2$ acting on the arc, we must sum (via integration) the differential forces $d\vec{F}$ acting on all the segments of the arc. However, another **Key Idea** greatly simplifies the summation—we can apply symmetry. Labeling the angle θ as shown in Fig. 29-42, we resolve the force $d\vec{F}$ on our original segment into a horizontal component $dF \cos \theta$ and a vertical component $dF \sin \theta$. Then we similarly resolve the force acting on a symmetrically opposite segment of the arc into a horizontal component and a vertical component. The horizontal force components on every such pair of segments are equal in magnitude and opposite in direction, so those components cancel each other, leaving only the vertical components. Thus, to find the vector sum of the forces $d\vec{F}$ on the arc, we need to sum only the vertical components $dF \sin \theta$, which are all directed downward. We can now write the magnitude of the net force $\vec{F}_2$ on the arc as

$$F_2 = \int_0^\pi dF \sin \theta. \tag{29-42}$$

   From Eq. 29-28, we see that, because $d\vec{L}$ and $\vec{B}$ are perpendicular,

$$dF = iB\,dL \sin \theta = iB\,dL \sin 90° = iB\,dL.$$

Substituting this into Eq. 29-42 yields

$$F_2 = \int_0^\pi iB\,dL \sin \theta, \tag{29-43}$$

which we cannot integrate until we eliminate θ or dL. Noting in Fig. 29-42 that arc segment dL is subtended by angle $d\theta$ in a circle of radius R, we write

$$dL = R\,d\theta,$$

substitute this into Eq. 29-43, and obtain, finally,

$$F_2 = \int_0^\pi (iBR\,d\theta) \sin \theta = iBR \int_0^\pi \sin \theta\,d\theta = 2iBR.$$

Thus, the net force on the entire wire acts downward and has the magnitude

$$F = F_1 + F_2 + F_3 = iLB + 2iBR + iLB$$
$$= 2iB(L + R). \tag{Answer}$$

Note that this force is equal to the force that would act on a straight wire of length $2(L + R)$. This would be true no matter what the shape of the central segment.

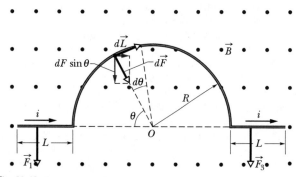

Fig. 29-42 Sample Problem 29-10. A wire segment carrying a current i is immersed in a magnetic field.

Sample Problem 29-11

Figure 29-43a shows a wire carrying a current $i = 6.0$ A in the positive direction of the x axis and lying in a *nonuniform* magnetic field given by $\vec{B} = (2.3 \text{ T/m})x\hat{i} + (2.0 \text{ T/m})x\hat{j}$, with $\vec{B}$ in teslas and x in meters. What is the net magnetic force $\vec{F}_B$ on the section of the wire between $x = 0$ and $x = 2.0$ m?

SOLUTION: Because the wire carries a current in a magnetic field $\vec{B}$, magnetic forces $\vec{F}_B$ can act on the wire. However, we cannot find $\vec{F}_B$ from Eq. 29-26 ($\vec{F}_B = i\vec{L} \times \vec{B}$) because $\vec{B}$ is not uniform along the wire and thus does not have a single value. The **Key Idea** here is that, instead, we must mentally divide the wire into differential lengths and use Eq. 29-28 to find the differential force $d\vec{F}_B$ on each length. Then we can sum these differential forces to find the net magnetic force $\vec{F}_B$ on the full section of wire.

   Figure 29-43b shows a differential length dx of the wire. Its length vector $d\vec{L}$ has magnitude dx and its direction is the positive

$$dF_B = i \, d\vec{L} \times \vec{B}$$
$$= i(dx \, \hat{i}) \times (2.3x\hat{i} + 2.0x\hat{j})$$
$$= i \, dx[2.3x(\hat{i} \times \hat{i}) + 2.0x(\hat{i} \times \hat{j})]$$
$$= i \, dx[0 + 2.0x\hat{k}] = 2.0ix \, dx \, \hat{k}, \qquad (29\text{-}45)$$

where the constant 2.0 has the unit teslas per meter. From this result we see that the magnetic force does not depend on the x component of $\vec{B}$ (because that component is parallel to the current). We also see that the magnetic force dF_B on length dx of the wire is in the positive direction of the z axis (out of the page in Fig. 29-43c) and has magnitude $dF_B = (2.0 \text{ T/m})ix \, dx$.

Because the direction of the force dF_B is the same for all the differential lengths dx of the wire, we can find the magnitude of the total force by summing all the differential force magnitudes dF_B. To do so, we integrate dF_B from $x = 0$ to $x = 2.0$ m and then substitute the given data. We get

$$F_B = \int dF_B = \int_0^{2.0 \text{ m}} (2.0 \text{ T/m})ix \, dx$$
$$= (2.0 \text{ T/m})i \left[\tfrac{1}{2}x^2 \right]_0^{2.0 \text{ m}} = (2.0 \text{ T/m})(6.0 \text{ A})(\tfrac{1}{2})(2.0 \text{ m})^2$$
$$= 24 \text{ T} \cdot \text{A} \cdot \text{m} = 24 \text{ N}. \qquad \text{(Answer)}$$

This force is directed along the positive direction of the z axis.

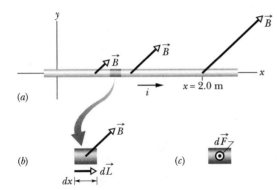

(a)

(b)

(c)

Fig. 29-43 Sample Problem 29-11. (a) A wire with current i lies in a nonuniform magnetic field $\vec{B}$. (b) An element of the wire, with differential length vector $d\vec{L}$ and length dx. (c) The differential force $d\vec{F}$ acting on the element of (b) due to the magnetic field; the force is directed out of the page.

direction of the x axis (which is the direction of the current). Thus, we can write this vector $d\vec{L}$ as

$$d\vec{L} = dx \, \hat{i}. \qquad (29\text{-}44)$$

(Be careful not to confuse the unit vector $\hat{i}$ with the current i.) Now, by Eq. 29-28, the differential force $d\vec{F}_B$ on the length dx of the wire is

QUESTIONS

11. *Organizing question:* Figure 29-44 shows a wire of length 2 m carrying a 4 A current along a y axis through a magnetic field. The magnetic field is given in the table for six situations (angles are measured counterclockwise from the positive direction of the x axis). For each situation, set up an equation, complete with known data, in order to find the magnitude of the net magnetic force on the 2.0 m length of wire. (Here, x and y are in meters.)

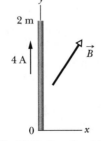

Fig. 29-44 Question 11.

Situation	Magnetic Field
a	$B = 3 \, \mu\text{T}$ at $60°$
b	$B = 3y \, \mu\text{T}$ at $30°$
c	$\vec{B} = (2\hat{i} + 3y\hat{j}) \, \mu\text{T}$
d	$\vec{B} = (2\hat{i} + 3x\hat{j}) \, \mu\text{T}$
e	$\vec{B} = (2y\hat{i} + 3\hat{j}) \, \mu\text{T}$
f	$\vec{B} = (2x^2\hat{i} + 3y^2\hat{j}) \, \mu\text{T}$

12. Figure 29-45 shows four directions for the velocity vector $\vec{v}$ of a negatively charged particle moving at angle θ to a uniform magnetic field $\vec{B}$. (a) Rank the directions according to the magnitude of the magnetic force on the particle, greatest first. (b) Which directions give a magnetic force out of the plane of the page?

13. The dead-quiet "caterpillar drive" for submarines in the movie *The Hunt for Red October* is based on a *magnetohydrodynamic* (MHD) drive; as the ship moves forward, seawater flows through multiple channels in a structure built around the rear of the hull. Figure 29-46 shows the essentials of a channel. Magnets, positioned along opposite sides of the channel with opposite poles facing each other, create a magnetic field within the channel. Electrodes (not shown) create an electric field across the channel. The

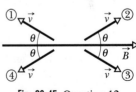

Fig. 29-45 Question 12.

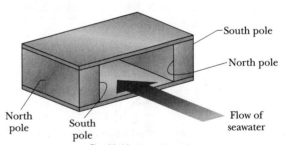

Fig. 29-46 Question 13.

electric field drives a current across the channel and through the water; the magnetic force on the current propels the water toward the rear of the channel, thus propelling the ship forward. In Fig. 29-46, should the electric field be directed upward, downward, leftward, rightward, frontward, or rearward?

14. Figure 29-47 shows four views of a horseshoe magnet and a straight wire in which electrons are flowing out of the page, perpendicular to the plane of the magnet. In which case will the magnetic force on the wire be directed toward the top of the page?

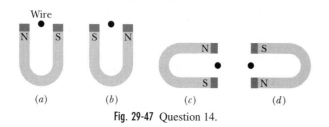

Fig. 29-47 Question 14.

15. A wire carries a current i in the negative x direction, through a magnetic field $\vec{B}$. Without written calculation, rank the following choices for $\vec{B}$ according to the magnitudes of the magnetic forces they exert on the wire, greatest first: $\vec{B}_1 = 2\hat{i} + 3\hat{j}$, $\vec{B}_2 = 4\hat{i} - 3\hat{j}$, $\vec{B}_3 = 6\hat{i} + 3\hat{k}$, and $\vec{B}_4 = -8\hat{i} - 3\hat{k}$.

16. Figure 29-48 shows the cross section of a solid conductor carrying a current perpendicular to the page. (a) Which pair of the four terminals (a, b, c, d) should be used to measure the Hall voltage if the magnetic field is in the positive direction of the x axis, the charge carriers are negative, and they move out of the page? Which terminal of the pair is at the higher potential? (b) Repeat for a magnetic field in the negative direction of the y axis and positive charge carriers moving out of the page. (c) Discuss the situation if the magnetic field is in the positive z direction.

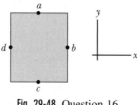

Fig. 29-48 Question 16.

17. An airplane flies due west over Massachusetts, where Earth's magnetic field is directed downward and to the north. (a) On which wing, left or right, are some of the conduction electrons moved to the wingtip by the magnetic force on them? (b) Which wingtip gets the conduction electrons if the flight is eastward?

18. Figure 29-49 gives snapshots for three situations in which a positively charged particle passes through a uniform magnetic field $\vec{B}$. The velocities $\vec{v}$ of the particle differ in orientation in the three snapshots but not in magnitude. Rank the situations according to (a) the period, (b) the frequency, and (c) the pitch of the particle's motion, greatest first.

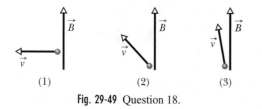

Fig. 29-49 Question 18.

19. Figure 29-50 shows the path of an electron in a region of uniform magnetic field. The path consists of two straight sections, each between a pair of uniformly charged plates, and two half-circles. Which plate is at the higher electric potential in (a) the top pair of plates and (b) the bottom pair? (c) What is the direction of the magnetic field?

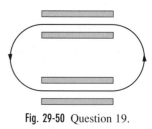

Fig. 29-50 Question 19.

20. The potential energies associated with four orientations of a magnetic dipole in a magnetic field are (1) $-6U_0$, (2) $8U_0$, (3) $6U_0$, and (4) $-7U_0$, where U_0 is positive. Rank the orientations according to (a) the angle between the directions of the magnetic dipole moment $\vec{\mu}$ and the magnetic field $\vec{B}$ and (b) the magnitude of the torque on the magnetic dipole, greatest first.

DISCUSSION QUESTIONS

21. Of the three vectors in the equation $\vec{F}_B = q\vec{v} \times \vec{B}$, which pairs are always perpendicular? Which may have any angle between them?

22. Why do we not simply define the direction of the magnetic field $\vec{B}$ to be the direction of the magnetic force that acts on a moving charge?

23. Imagine that you are sitting in a room with your back to the back wall and that a beam of electrons, traveling horizontally from the back wall toward the front wall, is deflected to your right. What is the direction of the uniform magnetic field that exists in the room?

24. How could we show that the forces between two magnets are not electrostatic forces?

25. If an electron is not deflected in passing through a certain region of space, can we be sure that there is no magnetic field in that region?

26. If a moving electron is deflected in passing through a certain region of space, can we be sure that a magnetic field exists in that region?

27. A beam of electrons can be deflected either by an electric field or by a magnetic field. Is one method better than the other? Is one method in any sense easier?

28. A charged particle passes through a magnetic field and is deflected. This means that a force acted on it and changed its momentum. Where there is a force, there must be a third-law force pair. On what object does the second force of the pair act?

29. Imagine the room in which you are seated is filled with a uniform magnetic field with $\vec{B}$ directed vertically downward. At the center of the room two electrons are suddenly projected horizontally with the same initial speed but in opposite directions. (a) Describe their motions. (b) Describe the motions if, instead, one particle is an electron and the other is a positron. (The particles will gradually slow down as they collide with molecules of the air in the room.)

30. In Fig. 29-3 why are the electron and positron tracks spirals? In other words, why does the radius of curvature change in the uniform magnetic field in which the chamber is immersed?

31. What are the primary functions of (a) the electric field and (b) the magnetic field in a cyclotron?

32. What central fact makes the operation of a conventional cyclotron possible? Ignore relativistic considerations.

33. A bare copper wire emerges from one wall of a room, crosses the room, and disappears into the opposite wall. You are told that there is a steady current in the wire. How can you find its direction? Describe as many ways as you can think of. You may use any reasonable piece of equipment, but you may not cut the wire.

34. In Section 29-7, we see that a magnetic field $\vec{B}$ exerts a force on the conduction electrons in, say, a cooper wire carrying a current i. We have tacitly assumed that this same force acts on the conductor itself. Are there some missing steps in this argument? If so, supply them.

35. A current in a magnetic field experiences a force. Therefore, it should be possible to pump a conducting liquid by sending a current through the liquid (in an appropriate direction) and letting the liquid pass through a magnetic field. Design such a pump. This principle is used to pump liquid sodium (a conductor, but highly corrosive) in some nuclear reactors, where it is used as a coolant. What advantages would such a pump have?

36. A conductor, even though it is carrying a current, has zero net charge. Why then does a magnetic field exert a force on it?

37. You wish to modify a galvanometer (see Sample Problem 29-7) to make it into a (a) an ammeter and (b) a voltmeter. What do you need to do in each case?

38. A rectangular current loop is in an arbitrary orientation in an external magnetic field. How much work is required to rotate the loop completely about an axis perpendicular to its plane?

39. Equation 29-37 ($\vec{\tau} = \vec{\mu} \times \vec{B}$) shows that there is no torque on a current loop in a magnetic field if the angle between the axis of the loop and the field is (a) 0° or (b) 180°. Discuss the nature of the equilibrium (that is, is it stable, neutral, or unstable?) for these two positions.

40. The work required to turn a current loop end-for-end in an external magnetic field is $2\mu B$. Does this result hold no matter what the original orientation of the loop was?

41. Imagine that the room in which you are seated is filled with a uniform magnetic field with $\vec{B}$ pointing vertically upward. A circular loop of wire has its plane horizontal. For what direction of current in the loop, as viewed from above, will the loop be in stable equilibrium with respect to forces and torques of magnetic origin?

42. The torque exerted by a magnetic field on a magnetic dipole can be used to measure the strength of that magnetic field. For an accurate measurement, does it matter whether the dipole moment is small or not? Recall that, in the case of measurement of an electric field, the test charge was to be as small as possible so as not to disturb the source of the field.

43. You are given a smooth sphere the size of a Ping-Pong ball and told that it contains a magnetic dipole. What experiments would you carry out to find the magnitude and direction of its magnetic dipole moment?

EXERCISES & PROBLEMS

60. A wire lying along an x axis from $x = 0$ to $x = 1.00$ m carries a current of 3.00 A in the positive x direction. It is immersed in a magnetic field given by

$$\vec{B} = (4.00 \text{ T/m}^2)x^2\hat{i} - (0.600 \text{ T/m}^2)x^2\hat{j}.$$

In unit-vector notation, what is the magnetic force on the wire?

61. In Fig. 29-51, a conducting rectangular solid moves at constant velocity $\vec{v} = (20.0 \text{ m/s})\hat{i}$ through a uniform magnetic field $\vec{B} = (30.0 \text{ mT})\hat{j}$. What are (a) the resulting electric field within the solid, in unit-vector notation, and (b) the resulting potential difference across the solid?

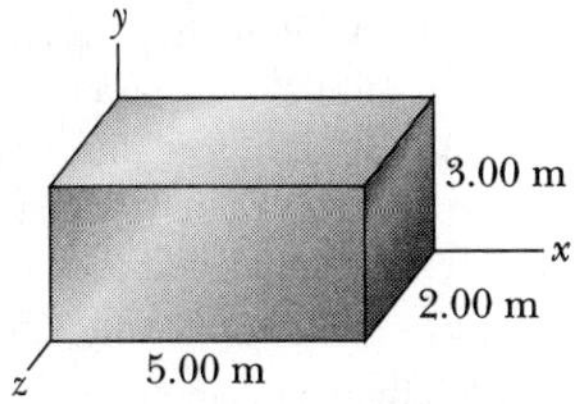

Fig. 29-51 Problem 61.

62. In Fig. 29-52, a particle moves along a circle in a region of uniform magnetic field of magnitude $B = 4.00$ mT. The particle is either a proton or an electron (you must decide which). It experiences a magnetic force of magnitude 3.20×10^{-15} N. What are (a) the particle's speed, (b) the radius of the circle, and (c) the period of the motion?

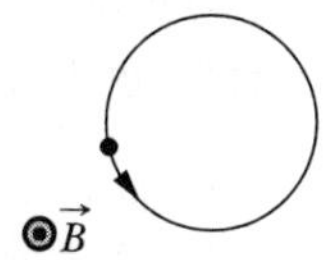

Fig. 29-52 Problem 62.

63. An electron moves through a region of uniform magnetic field of magnitude 60 μT, directed along the positive direction of an x axis. The electron has a velocity of $(32\hat{i} + 40\hat{j})$ km/s as it enters the field. What are (a) the radius of the helical path taken by the electron and (b) the pitch of that path? (c) To an observer looking into the magnetic field region from the entrance point of the electron, does the electron spiral clockwise or counterclockwise as it moves deeper into the region?

64. A particle of mass 10 g and charge 80 μC moves through a uniform magnetic field, in a region where the free-fall acceleration is $-9.8\hat{j}$ m/s^2. The velocity of the particle is a constant $20\hat{i}$ km/s, which is perpendicular to the magnetic field. What, then, is the magnetic field?

65. A particle of mass 6.0 g moves at 4.0 km/s in an xy plane, in a region with a uniform magnetic field given by $5.0\hat{i}$ mT. At one instant, when the particle's velocity is directed 37° counterclockwise from the positive direction of the x axis, the magnetic force on the particle is $0.48\hat{k}$ N. What is the particle's charge?

66. A proton, a deuteron ($q = +e, m = 2.0$ u), and an alpha particle ($q = +2e, m = 4.0$ u), accelerated through the same potential difference, enter a region of uniform magnetic field $\vec{B}$, moving perpendicular to $\vec{B}$. (a) Compare their kinetic energies. If the radius of the proton's circular path is 10 cm, what are the radii of (b) the deuteron's path and (c) the alpha particle's path?

67. A physicist is designing a cyclotron to accelerate protons to one-tenth the speed of light. The magnet used will produce a field of magnitude 1.4 T. Calculate (a) the required radius of the cyclotron dees and (b) the corresponding oscillator frequency. Relativity considerations are not significant.

68. A metal wire of mass m slides without friction on two horizontal rails spaced a distance d apart, as in Fig. 29-53. The track lies in a vertical uniform magnetic field $\vec{B}$. There is a constant current i through generator G, along one rail, across the wire, and back down the other rail. Find the speed and direction of the wire's motion as a function of time t, assuming it to be stationary at $t = 0$.

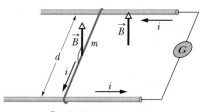

Fig. 29-53 Problem 68.

69. In a Hall-effect experiment, a current of 3.0 A sent lengthwise through a conductor 1.0 cm wide, 4.0 cm long, and 10 μm thick produces a transverse (across the width) Hall potential difference of 10 μV when a magnetic field of 1.5 T is passed perpendicularly through the thickness of the conductor. From these data, find (a) the drift velocity of the charge carriers and (b) the number density of charge carriers. (c) Show on a diagram the polarity of the Hall potential difference with assumed current and magnetic field directions, assuming also that the charge carriers are electrons.

70. In Fig. 29-54, an electron moves at speed $v = 100$ m/s along a straight line toward the right through uniform electric and magnetic fields. The magnetic field $\vec{B}$ is directed into the

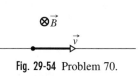

Fig. 29-54 Problem 70.

page and has magnitude 5.00 T. What are the magnitude and direction of the electric field?

71. At one instant a proton has velocity

$$\vec{v} = (-2.00\hat{i} + 4.00\hat{j} - 6.00\hat{k}) \text{ m/s}$$

in a uniform magnetic field

$$\vec{B} = (2.00\hat{i} - 4.00\hat{j} + 8.00\hat{k}) \text{ mT}.$$

At that instant, what are (a) the magnetic force $\vec{F}$ on the proton, in unit-vector notation, (b) the angle between $\vec{v}$ and $\vec{F}$, and (c) the angle between $\vec{v}$ and $\vec{B}$?

72. The bent wire shown in Fig. 29-55 lies in a uniform magnetic field. The two straight sections of the wire each have length 2.0 m, and the wire carries a current of 2.0 A. What is the net magnetic force on the wire in unit-vector notation if the magnetic field is given by (a) 4.0$\hat{k}$ T and (b) 4.0$\hat{i}$ T?

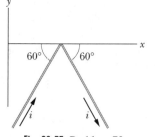

Fig. 29-55 Problem 72.

73. A 5.0 μC particle moves through a region containing the magnetic field $-20\hat{i}$ mT and the electric field $300\hat{j}$ V/m. At one instant the velocity of the particle is $(17\hat{i} - 11\hat{j} + 7.0\hat{k})$ km/s. At that instant, what is the net electromagnetic force (the sum of the electric and magnetic forces) on the particle?

74. A particle with charge 2.0 C moves through a uniform magnetic field. At one instant its velocity is $(2.0\hat{i} + 4.0\hat{j} + 6.0\hat{k})$ m/s and the magnetic force on it is $(4.0\hat{i} - 20\hat{j} + 12\hat{k})$ N. The x and y components of the magnetic field are equal. What is $\vec{B}$?

75. (a) In a magnetic field with magnitude $B = 0.50$ T, for what path radius will an electron circulate at 10% the speed of light? (b) What will be its kinetic energy in electron-volts?

76. Two singly ionized atoms have the same charge q but masses that differ by a small amount Δm. They are introduced into the mass spectrometer described in Sample Problem 29-3. (a) Calculate the difference in mass in terms of V, q, m (of either), B, and the distance Δx between the spots they produce on the photographic plate. (b) Calculate Δx for a beam of singly ionized chlorine atoms of masses 35 and 37 u if $V = 7.3$ kV and $B = 0.50$ T.

77. Figure 29-56 shows a wire of arbitrary shape carrying a current i between points a and b. The wire lies in a plane that is perpendicular to a uniform magnetic field $\vec{B}$. (a) Prove that the force on the wire is the same as that on a straight wire carrying a current i directly from a to b. (*Hint:* Replace the wire with a series of "steps" parallel and perpendicular to the straight line joining a and b.) (b) Prove that the force on the wire becomes zero when points a and b are brought together so that the wire is a complete loop whose plane is perpendicular to field $\vec{B}$.

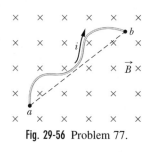

Fig. 29-56 Problem 77.

78. A long, rigid conductor, lying along an x axis, carries a current of 5.0 A in the negative x direction. A magnetic field $\vec{B}$ is present, given by $\vec{B} = 3.0\hat{i} + 8.0x^2\hat{j}$, with x in meters and $\vec{B}$ in milliteslas. Find, in unit-vector notation, the force on the 2.0 m segment of the conductor that lies between $x = 1.0$ m and $x = 3.0$ m.

Tutorial Problem

79. In this problem, we try to determine the electric and magnetic fields at a point by observing the net force on a charged particle at that point. The particle has a mass of 1.80×10^{-25} kg and an electric charge of $+4e$. We use a Cartesian coordinate system chosen with the y axis directed upward and the origin on the ground. (a) If the particle is placed at rest at a point A for which $\vec{r} = (2.45 \text{ m})\hat{j}$, what is the gravitational force on the particle? Express the force in unit-vector notation.

(b) Suppose that in addition to the gravitational force there is a force $+(7.31 \times 10^{-14} \text{ N})\hat{i}$ on the particle when it is at rest. Compare the magnitude of this force with that of the gravitational force in part (a). What fundamental interaction is responsible for this force? Name the field providing the force and determine its value in unit-vector notation.

(c) Now suppose that the same particle is sent through point A with velocity $\vec{v} = (4.12 \times 10^6 \text{ m/s})\hat{i}$ and is found to experience an additional force $-(5.43 \times 10^{-14} \text{ N})\hat{k}$. What can you say about the magnetic field at point A? Explain your reasoning in complete sentences. (d) Devise an experiment that might enable you, with the information already gathered, to determine the complete magnetic field vector. Explain the experiment in complete sentences.

(e) Now consider a new physical situation. If a proton moves through a uniform magnetic field $\vec{B}$ with a velocity of

$$\vec{v} = (3.0 \times 10^6 \text{ m/s})\hat{i} - (4.0 \times 10^6 \text{ m/s})\hat{j},$$

it experiences a magnetic force $\vec{F}_B = (3.2 \times 10^{-15} \text{ N})\hat{k}$. If, instead, it moves through the field with a velocity along $\hat{k}$, the magnetic force on it is parallel to $\hat{j}$. Determine $\vec{B}$, explaining your solution logically in full sentences (not with only a string of formulas and a few words).

Answers

(a) The gravitational force is

$$\vec{F}_{\text{grav}} = -mg\hat{j} = -(1.80 \times 10^{-25} \text{ kg})(9.80 \text{ m/s}^2)\hat{j}$$
$$= -(1.76 \times 10^{-24} \text{ N})\hat{j}.$$

(b) This force has a magnitude greater than 10^{10} times that of the gravitational force. Presumably an electric or magnetic interaction is responsible for this force. Since the particle is at rest, it must be an electric field that produces this force. From the relation $\vec{F} = q\vec{E}$ we can determine the electric field:

$$\vec{E} = \frac{\vec{F}}{q} = \frac{(7.31 \times 10^{-14} \text{ N})\hat{j}}{(4)(1.602 \times 10^{-19} \text{ C})}$$
$$= (1.14 \times 10^5 \text{ N/C})\hat{j}.$$

The electric field is directed upward from Earth's surface, in the same direction as the force on the particle, because the particle has a positive electric charge.

(c) First, since the velocity is along $\hat{i}$ and the force is along $\hat{k}$, there clearly must be a component of the magnetic field along $\hat{j}$. From Eq. 29-2,

$$F_z = q(v_x B_y - v_y B_x) = q v_x B_y,$$

because $v_y = 0$. Thus, we see that

$$B_y = +\left(\frac{F_z}{q v_x}\right)$$
$$= \frac{-5.43 \times 10^{-14} \text{ N}}{(4)(1.602 \times 10^{-19} \text{ C})(4.12 \times 10^6 \text{ m/s})}$$
$$= -0.0206 \text{ T}.$$

We also know that there cannot be a component of $\vec{B}$ along $\hat{k}$, for it would have shown up as a component of force along $\hat{j}$. However, we can't learn anything about the component of the magnetic field along $\hat{i}$, because that is also the direction of $\vec{v}$ and such a component wouldn't make any contribution to the magnetic force.

(d) We can determine the component of $\vec{B}$ along $\hat{i}$ if we send the particle through point A with a velocity along the $\hat{j}$ direction (or the $\hat{k}$ direction or any direction other than exactly along $\hat{i}$). We then find a component of the magnetic force along $\hat{k}$ that depends on B_x, and that allows us to determine B_x.

(e) The magnetic forces in the two situations must both be perpendicular to the magnetic field, since $\vec{F}_B = q\vec{v} \times \vec{B}$. In one case the force is parallel to $\hat{k}$ and in the other case the force is parallel to $\hat{j}$, so $\vec{B}$ must be perpendicular to both. That means $\vec{B}$ is along $\pm\hat{i}$. Knowing this, we can make use of the numerical information. In doing so, we note that v_x has no effect on the magnetic force since it is the component of the velocity parallel to the magnetic field; so only v_y needs to be taken into account.

We know that $\vec{F}_B = (3.2 \times 10^{-15} \text{ N})\hat{k}$ in the first case. We also see that

$$\vec{F}_B = e(\vec{v} \times B_x\hat{i})$$
$$= (1.60 \times 10^{-19} \text{ C})(-4.0 \times 10^6 \text{ m/s})B_x(\hat{j} \times \hat{i})$$
$$= (6.4 \times 10^{-13} \text{ C} \cdot \text{m/s})B_x\hat{k}.$$

Therefore,

$$B_x = \frac{3.2 \times 10^{-15} \text{ N}}{6.4 \times 10^{-13} \text{ C} \cdot \text{m/s}} = 5.0 \times 10^{-3} \text{ T} = 5.0 \text{ mT}$$

and

$$\vec{B} = (5.0 \text{ mT})\hat{i}.$$

Magnetic Fields Due to Currents

Sample Problem 30-5

Figure 30-58 shows a wire consisting of two very long straight sections and a circular arc of central angle $\phi = 120°$ and radius $R = 2.0$ cm. The wire carries a current of 4.0 mA. What is the magnetic field $\vec{B}$ at point C, which is at the center of the arc?

SOLUTION: One **Key Idea** here is that we can find the magnetic field $\vec{B}$ at point C by applying the Biot–Savart law of Eq. 30-5 to the wire. A second **Key Idea** is that this application can be simplified by evaluating $\vec{B}$ separately for the three distinguishable sections of the wire—namely, (1) the horizontal straight section, (2) the tilted straight section, and (3) the circular arc.

Straight sections. Approximating both sections as semi-infinite straight wires and applying the Biot–Savart law to them lead to Eq. 30-9 for the magnitudes of the magnetic fields $\vec{B}_1$ and $\vec{B}_2$ they produce at point C. The perpendicular distances between point C and the near ends of the straight sections are both R. Thus, Eq. 30-9 gives us

$$B_1 = \frac{\mu_0 i}{4\pi R} \quad \text{and} \quad B_2 = \frac{\mu_0 i}{4\pi R}.$$

To find the direction of $\vec{B}_1$, we apply the right-hand rule displayed in Fig. 30-4. Mentally grasp section 1 of the wire with your right hand and with your thumb in the direction of the current. The di-

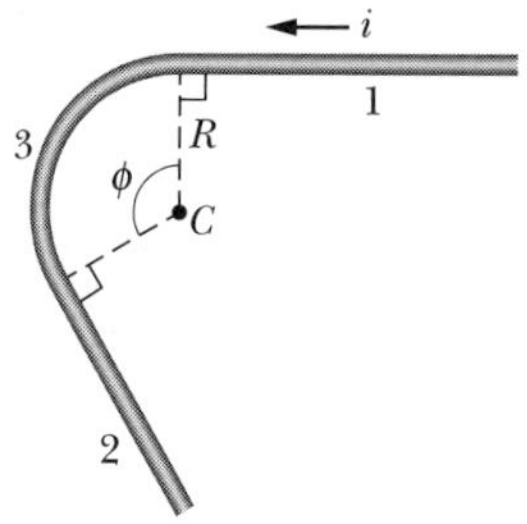

Fig. 30-58 Sample Problem 30-5. A wire carrying current i consists of two very long, straight sections and a circular arc of central angle ϕ and radius R.

rection in which your fingers curl around the wire indicates the direction of the magnetic field lines around the wire. In the region of point C (below section 1), your fingertips curl out of the plane of Fig. 30-58. Thus, $\vec{B}_1$ is directed out of that plane. You can similarly show that $\vec{B}_2$ is also directed out of the page at C.

Circular arc. The **Key Idea** here is that application of the Biot–Savart law to the circular arc leads to Eq. 30-11 for the magnitude B_3 of the magnetic field at C. Here the central angle ϕ of the arc is $120° = 2\pi/3$ rad, so Eq. 30-11 gives us

$$B_3 = \frac{\mu_0 i \phi}{4\pi R} = \frac{\mu_0 i (2\pi/3)}{4\pi R} = \frac{2\pi}{3} \frac{\mu_0 i}{4\pi R}.$$

(For convenience, we leave B_3 in the same form as we found for B_1 and B_2.) To find the direction of $\vec{B}_3$, we again apply the right-hand rule displayed in Fig. 30-4, and again we find that the field at point C is directed out of the page.

Net field. Generally, to find the net magnetic field of two or more individual magnetic fields, we must vectorially add the individual fields, not simply add their magnitudes. Here, however, all three sections of the wire produce magnetic fields directed out of the page at point C. Thus, the net magnetic field $\vec{B}$ at point C is directed out of the page and we can find its magnitude as a simple sum:

$$
\begin{aligned}
B &= B_1 + B_2 + B_3 \\
&= \frac{\mu_0 i}{4\pi R} + \frac{\mu_0 i}{4\pi R} + \frac{2\pi}{3} \frac{\mu_0 i}{4\pi R} \\
&= \frac{\mu_0 i}{4\pi R}\left(2 + \frac{2\pi}{3}\right) \\
&= \frac{(4\pi \times 10^{-7}\ \text{T} \cdot \text{m/A})(0.0040\ \text{A})}{4\pi(0.020\ \text{m})}\ (4.09) \\
&= 8.2 \times 10^{-8}\ \text{T}. \qquad\qquad \text{(Answer)}
\end{aligned}
$$

Sample Problem 30-6

Two long parallel wires a distance $2d$ apart carry equal currents i in opposite directions, as shown in Fig. 30-59a. Derive an expression for $B(x)$, the magnitude of the net magnetic field for points on the x axis and between the wires.

SOLUTION: One **Key Idea** here is that the net magnetic field $\vec{B}$ at any point near the two wires is the vector sum of the magnetic fields due to the currents in the two wires. A second **Key Idea** is that we can find the magnetic field due to any current by applying the Biot–

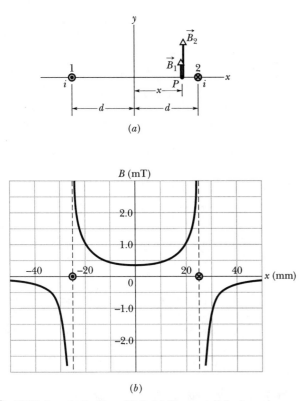

(a)

(b)

Fig. 30-59 Sample Problem 30-6. (a) Two parallel wires carry currents of the same magnitude in opposite directions (out of and into the page). At points between the wires, such as P, the magnetic fields due to the separate currents have the same direction. (b) A plot of $B(x)$ for $i = 25$ A and a wire separation of 50 mm.

Savart law to the current. For points near a long straight wire, that law leads to Eq. 30-6 ($B = \mu_0 i/2\pi R$).

Let us consider an arbitrary point P on the x axis, as shown in Fig. 30-59a; the point is at distance $d + x$ from wire 1 at the left, and distance $d - x$ from wire 2 at the right. Thus, Eq. 30-6

tells us that at point P the currents i in those wires produce magnetic fields $\vec{B}_1$ and $\vec{B}_2$ with magnitudes

$$B_1(x) = \frac{\mu_0 i}{2\pi(d + x)} \quad \text{and} \quad B_2(x) = \frac{\mu_0 i}{2\pi(d - x)}.$$

We want to combine $\vec{B}_1(x)$ and $\vec{B}_2(x)$ to find their vector sum, which is the net field $\vec{B}(x)$. To find the directions of $\vec{B}_1(x)$ and $\vec{B}_2(x)$, we apply the right-hand rule of Fig. 30-4 to each of the currents in Fig. 30-59a. For wire 1, with current out of the page, we mentally grasp the wire with the right hand, with the thumb pointing out of the page. Then the direction in which our fingers curl around the wire indicates the direction of the magnetic field lines around the wire. Near point P (between the two wires), our fingertips would point up the page. (Recall that the magnetic field at a point near a long straight current must be perpendicular to a radial line connecting the point and the current.) Thus, $\vec{B}_1(x)$ is directed up the page, in the positive direction of the y axis.

Repeating this analysis for the current in wire 2, we find that $\vec{B}_2(x)$ is also directed in the positive direction of the y axis. Thus, a **Key Idea** here is that because the two fields have the same direction, we can find the magnitude of their sum by simply adding their magnitudes:

$$B(x) = B_1(x) + B_2(x) = \frac{\mu_0 i}{2\pi(d + x)} + \frac{\mu_0 i}{2\pi(d - x)}$$

$$= \frac{\mu_0 i d}{\pi(d^2 - x^2)}. \qquad \text{(Answer)} \quad (30\text{-}33)$$

Inspection of this relation shows that between the wires (1) $B(x)$ is symmetric about the origin ($x = 0$); (2) $B(x)$ has its minimum value ($= \mu_0 i/\pi d$) at this point; and (3) $B(x) \rightarrow \infty$ as $x \rightarrow \pm d$. At $x = \pm d$, the point P in Fig. 30-59a is within the wires on their axes. Our derivation of Eq. 30-6, however, is valid only for points outside the wires, so Eq. 30-33 holds only up to the surface of the wires.

Figure 30-59b plots Eq. 30-33 for $i = 25$ A and $2d = 50$ mm. We leave it as an exercise to show what the plot suggests: that Eq. 30-33 holds also for points beyond the wires—that is, for points with $|x| > d$.

Sample Problem 30-7

Figure 30-60a shows, in horizontal cross section, three wires that are meant to carry current from a lightning rod on top of a house to the ground in case lightning strikes the rod. The wires are parallel, have length $L = 4.0$ m, and are spaced $r = 5.0$ mm apart. Assume that, during a strike, the current in each wire is $i = 5000$ A. What are the magnitude and direction of the net force on each wire due to the currents in the other two wires?

SOLUTION: One **Key Idea** here is that a magnetic force acts on a wire that carries current through a magnetic field set up by another current. Each current in Fig. 30-60a is in the magnetic field set up by two other currents. Thus, to find the net magnetic force on one of the wires, we must vectorially add the two magnetic forces acting on it. A second **Key Idea** is that because the wires are long, straight, and have parallel currents, the magnetic forces tend to pull them

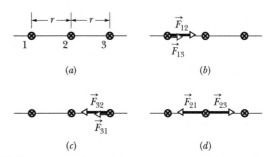

(a)

(b)

(c)

(d)

Fig. 30-60 Sample Problem 30-7. (a) Three long parallel wires carry identical currents i into the plane of the page. (b) The forces acting on wire 1 due to the currents in wires 2 and 3. (c) The forces acting on wire 3 due to the currents in wires 1 and 2. (d) The forces acting on wire 2 due to the currents in wires 1 and 3.

together, with a force magnitude given by Eq. 30-15. Let us apply this second idea to each wire, in turn.

Wire 1: The magnetic force $\vec{F}_{12}$ on wire 1 due to the current in wire 2 is directed rightward, pulling wire 1 toward wire 2 (Fig. 30-60*b*). Similarly, the magnetic force $\vec{F}_{13}$ on wire 1 due to the current in wire 3 is directed rightward, pulling wire 1 toward wire 3. Because these forces have the same directions, we can simply add their magnitudes to find the magnitude of the net force on wire 1. From Eq. 30-15, those magnitudes are

$$F_{12} = \frac{\mu_0 L i^2}{2\pi r} \quad \text{and} \quad F_{13} = \frac{\mu_0 L i^2}{2\pi(2r)},$$

where *r* is the distance between wires 1 and 2, and *2r* is the distance between wires 1 and 3. Thus, the net force $\vec{F}_1$ on wire 1 is directed to the right and has the magnitude

$$F_1 = F_{12} + F_{13}$$
$$= \frac{\mu_0 L i^2}{2\pi r}\left(1 + \frac{1}{2}\right) = \frac{\mu_0 L i^2}{2\pi r}\,(1.50)$$
$$= \frac{(4\pi \times 10^{-7}\,\text{T}\cdot\text{m/A})(4.0\,\text{m})(5000\,\text{A})^2}{(2\pi)(0.0050\,\text{m})}\,(1.50)$$
$$= 6000\,\text{N}. \qquad \text{(Answer)}$$

Wire 3: Using similar arguments, you can show that the two forces on wire 3 are as shown in Fig. 30-60*c* and that the net force $\vec{F}_3$ is directed leftward and has the magnitude

$$F_3 = 6000\,\text{N}. \qquad \text{(Answer)}$$

Wire 2: The magnetic force $\vec{F}_{21}$ on wire 2 due to the current in wire 1 is directed leftward, toward wire 1 (Fig. 30-60*d*). Similarly, the magnetic force $\vec{F}_{23}$ on wire 2 due to the current in wire 3 is directed rightward, toward wire 3. To find the magnitudes of $\vec{F}_{21}$ and $\vec{F}_{23}$, we again use Eq. 30-15, this time substituting *r* for the distance *d* for both forces:

$$F_{21} = \frac{\mu_0 L i^2}{2\pi r} \quad \text{and} \quad F_{23} = \frac{\mu_0 L i^2}{2\pi r}.$$

Thus, forces $\vec{F}_{21}$ and $\vec{F}_{23}$ have the same magnitudes. Because they have opposite directions, this means they cancel and the net force on wire 2 is

$$\vec{F}_2 = 0. \qquad \text{(Answer)}$$

When bundles of closely spaced wires carry the large currents typical of a lightning strike, the magnetic forces among the wires can collapse the bundle; then the intense heating due to the currents can fuse the wires together.

QUESTIONS

11. *Organizing question:* Figure 30-61 shows five situations in which a wire's current *i* produces a magnetic field $\vec{B}$ at a point *P*. For each situation, we have seen how the magnitude *B* can be determined immediately or by means of a formula derived from the Biot–Savart law. Give that immediate result or the formula. The situations are (a) wire in circular arc, *P* at center of curvature; (b) straight wire of finite length, *P* along an extension of the wire; (c) straight wire of semi-infinite length, *P* along an extension of the wire; (d) straight wire of infinite length, *P* off the wire; and (e) straight wire of semi-infinite length, *P* aligned with the end but off the wire.

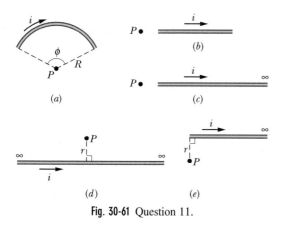

Fig. 30-61 Question 11.

12. Figure 30-62 shows a wire with current *i*, a differential length element *ds* of the wire (at the origin), and five numbered points.

The points are in the *xy* plane, at equal distances from element *ds*. A straight line through the element connects points 2 and 4; another straight line through the element connects points 3 and 5. (a) Rank the points according to the magnitude of the magnetic field $d\vec{B}$ produced there by the current in element *ds*, greatest first. (b) What is the direction of that field at each point?

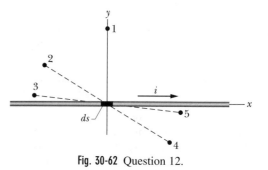

Fig. 30-62 Question 12.

13. *Organizing question:* In Fig. 30-63, two long straight wires (shown in cross section) carry currents i_1 and i_2 either directly into or directly out of the page and are at equal distances from a *y* axis. The currents set up a net magnetic field $\vec{B}$ at a point *P* on the *y* axis; an example of a possible field is shown. Directions around *P* are divided into four regions, *K*, *L*, *M*, and *N*, by the dashed lines. (Note the two perpendicular symbols in the figure.) For the example shown, $\vec{B}$ is directed into angular region *K*. What directions (into the page or out of the page) are required of the currents if field $\vec{B}$ is to be in angular region (a) *K*, (b) *L*, (c) *M*, and (d) *N*?

Suppose that the currents can range between 0 and 5 A. (e) For each of the four numbered borders between regions, give the currents in the wires, using either 0 or 5 A, that would result in $\vec{B}$ being directed along that border. (f) For each region, determine whether we should increase or decrease which current (consider first i_1 and then i_2) to cause a $\vec{B}$ vector in that region to rotate counterclockwise.

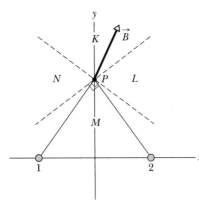

Fig. 30-63 Question 13.

14. In Fig. 30-64, three long wires, with identical currents either directly into or directly out of the page, form three partial squares. Rank the squares according to the magnitude of the net magnetic field produced by the currents at the (empty) upper right corner of the square, greatest first.

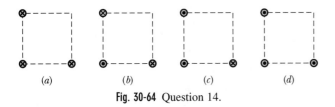

Fig. 30-64 Question 14.

15. In Fig. 30-65, a messy loop of wire is placed on a slick table with points a and b fixed in place. If a current is then sent through the wire, will the wire be pushed outward into an arc or will it be pulled inward?

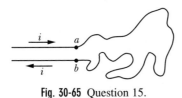

Fig. 30-65 Question 15.

16. Figure 30-66 gives, as functions of radial distance r, the magnitude B of the magnetic field inside and outside four wires (a, b, c, and d) carrying currents that are uniformly distributed across the cross sections of the wires. Overlapping portions of the plots are indicated by double labels. Rank the wires according to (a) their radii, (b) the magnitudes of the magnetic fields on their surfaces,

and (c) the values of their currents, greatest first. (d) Is the current density in wire a greater than, less than, or equal to that in wire c?

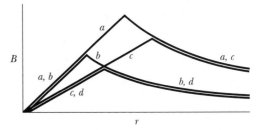

Fig. 30-66 Question 16.

17. *Organizing question:* Figure 30-67 shows a circular region of radius R in which a current i is directed out of the page. It also shows two Amperian loops (of radii r_1 and r_2) that form circles concentric with the circular region. Set up the right side of Ampere's law (Eq. 30-16) for loops 1 and 2 for the following situations, where either the total current i, the current density J, or the current i_{enc} encircled by a loop at radius r is given. (Worry about worrisome subscripts.)

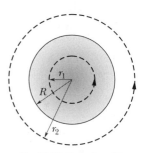

Fig. 30-67 Question 17.

Situation	Current or Current Density
a	$i = 2$ A, uniform
b	$J = 3$ A/m², uniform
c	$i_{enc} = (4 \text{ A})\left(\dfrac{r}{R}\right)$
d	$i_{enc} = (5 \text{ A})\left(\dfrac{r}{R}\right)^2$
e	$J = (6 \text{ A/m}^2)\left(1 - \dfrac{r}{R}\right)$

18. Figure 30-68 represents a snapshot of the velocity vectors of four electrons near a wire carrying current i. The four velocities have the same magnitude; velocity $\vec{v}_2$ is directed into the page. Particles 1 and 2 are at the same distance from the wire, as are particles 3 and 4. Rank the particles according to the magnitudes of the magnetic forces on them due to current i, greatest first.

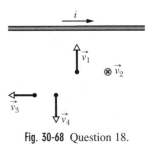

Fig. 30-68 Question 18.

19. The position vector of a particle moving around a circle of radius r is $\vec{r}$. What is the value of $\oint \vec{r} \cdot d\vec{s}$ around the circle?

DISCUSSION QUESTIONS

20. A beam of protons emerges from a cyclotron. Do these particles cause a magnetic field?

21. Discuss analogies and differences between Coulomb's law and the Biot–Savart law.

22. Consider a magnetic field line. Is the magnitude of $\vec{B}$ constant or variable along such a line? Can you give an example of each case?

23. In electronic equipment, wires that carry equal but opposite currents are often twisted together to reduce their magnetic effect at distant points. Why is this effective?

24. Drifting electrons constitute the current in a wire, and a magnetic field is associated with this current. What current and magnetic field would be measured by an observer moving along with the drifting electrons?

25. Consider two charges, first (a) of the same sign and then (b) of opposite signs, that are moving along separate parallel paths with the same velocity. Compare the directions of the mutual electric and magnetic forces in each case.

26. Two long parallel conductors carry equal currents i in the same direction. Sketch roughly the net magnetic field lines due to the two currents. Does your figure suggest an attraction between the wires?

27. A current is sent through a vertical spring from whose lower end an object is hanging. What will happen?

28. Two long straight wires pass near one another perpendicularly. The wires are free to move. Describe what happens when currents are sent through both of them.

29. Apply Ampere's law qualitatively to the three Amperian loops shown in Fig. 30-69.

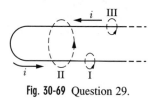

Fig. 30-69 Question 29.

30. Discuss analogies and differences between Gauss' law and Ampere's law.

31. A steady longitudinal uniform current is set up in a long copper tube. Is there a magnetic field (a) inside and/or (b) outside the tube?

32. A long straight wire of radius R carries a steady current i. How does the magnetic field generated by this current depend on R? Consider points both outside and inside the wire.

33. Two long solenoids are nested on the same axis, as Fig. 30-70 shows. They carry identical currents but in opposite directions. If there is no magnetic field inside the inner solenoid, what can you say about n, the number of turns per unit length, for the two solenoids? Which one, if either, has the larger value?

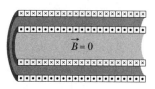

Fig. 30-70 Question 33.

34. A steady current is set up in a cubical network of resistive wires, connected as in Fig. 30-71. Use symmetry arguments to show that the magnetic field at the center of the cube is zero.

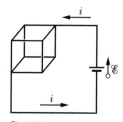

Fig. 30-71 Question 34.

EXERCISES & PROBLEMS

59. Figure 30-72a shows a cross section of a hollow cylindrical conductor of radii a and b, carrying a uniformly distributed current i. (a) Show that the magnetic field magnitude $B(r)$ for the radial distance r in the range $b < r < a$ is given by

$$B = \frac{\mu_0 i}{2\pi(a^2 - b^2)} \frac{r^2 - b^2}{r}.$$

(b) Show that when $r = a$, this equation gives the magnetic field magnitude B at the surface of a long straight wire carrying current i; when $r = b$, it gives zero magnetic field; and when $b = 0$, it gives the magnetic field inside a solid conductor of radius a carrying current i. (c) Assume that $a = 2.0$ cm, $b = 1.8$ cm, and $i = 100$ A, and plot $B(r)$ for the range $0 < r < 6$ cm.

60. Figure 30-72b shows a cross section of a long conductor of a type called a coaxial cable and gives its radii (a, b, c). Equal but opposite currents i are uniformly distributed in the two conductors. Derive expressions for $B(r)$ with radial distance r in the ranges (a) $r < c$, (b) $c < r < b$, (c) $b < r < a$, and (d) $r > a$. (e) Test these expressions for all the special cases that occur to you. (f) Assume that $a = 2.0$ cm, $b = 1.8$ cm, $c = 0.40$ cm, and $i = 120$ A and plot the function $B(r)$ over the range $0 < r < 3$ cm.

61. A long hairpin is formed by bending a very long wire as shown in Fig. 30-73. If the wire carries a 10 A current, what are the direction and magnitude of $\vec{B}$ at (a) point a and (b) point b midway between the wires? Take $R = 5.0$ mm and the distance between a and b to be *much* larger than R (each straight section is "infinite").

Fig. 30-73 Problem 61.

62. Figure 30-74 shows a 3.0 cm segment of wire, centered at the origin, carrying a current of 2.0 A in the positive y direction (as part of some complete circuit). To calculate the magnitude of the

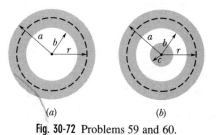

Fig. 30-72 Problems 59 and 60.

magnetic field $\vec{B}$ produced by the segment at a point several meters from the origin, one may use the Biot–Savart law in the form $B = (\mu_0/4\pi)i \, \Delta s \, (\sin \theta)/r^2$, in which $\Delta s = 3.0$ cm. This is because r and θ are essentially constant over the segment. Calculate $\vec{B}$ (in unit-vector notation) at the following (x, y, z) coordinates: (a) $(0, 0, 5.0$ m$)$, (b) $(0, 6.0$ m, $0)$, (c) $(7.0$ m, 7.0 m, $0)$, (d) $(-3.0$ m, -4.0 m, $0)$.

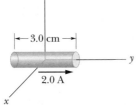

Fig. 30-74 Problem 62.

63. Two long wires a distance d apart carry equal antiparallel currents i, as in Fig. 30-75. (a) Show that the magnitude of the magnetic field at point P, which is equidistant from the wires, is given by

$$B = \frac{2\mu_0 id}{\pi(4R^2 + d^2)}.$$

(b) What is the direction of $\vec{B}$?

64. Assume that both currents in Fig. 30-59a (Sample Problem 30-6 in this supplement) are in the same direction, out of the plane of the figure. (a) Show that the magnitude of the net magnetic field at points on the x axis and between the wires is given by

$$B(x) = \frac{\mu_0 ix}{\pi(x^2 - d^2)}.$$

(b) Assume that $i = 10$ A and $d = 2.0$ cm in Fig. 30-59a, and plot $B(x)$ for the range -2 cm $< x < 2$ cm. Assume that the wire diameters are negligible.

65. Three long wires all lie in an xy plane parallel to the x axis. They are spaced equally, 10 cm apart. The two outer wires each carry a current of 5.0 A in the positive x direction. What is the magnitude of the force on a 3.0 m section of either of the outer wires if the current in the center wire is (a) 3.2 A in the positive x direction and (b) 3.2 A in the negative x direction?

66. A long thin wire carries an unknown current. Coaxial with the wire is a long, thin, cylindrical conducting surface that carries a current of 30 mA. The cylindrical surface has a radius of 3.0 mm. If the magnitude of the magnetic field at a point 5.0 mm from the wire is 1.0 μT, what is the current in the wire?

67. In a particular region there is a uniform current density of 15 A/m^2 in the positive z direction. What is the value of $\oint \vec{B} \cdot d\vec{s}$ when that line integral is taken along the three straight-line segments from $(4d, 0, 0)$ to $(4d, 3d, 0)$ to $(0, 0, 0)$ to $(4d, 0, 0)$, where $d = 20$ cm?

68. Two infinitely long wires carry equal currents i. Each follows a 90° arc on the circumference of the same circle of radius R, in the configuration shown in Fig. 30-76. Show that $\vec{B}$ at the center of the circle is the same as the field $\vec{B}$ a distance R below an infinite straight wire carrying a current i to the left.

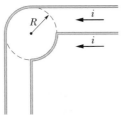

Fig. 30-76 Problem 68.

69. A 10-gauge bare copper wire (2.6 mm in diameter) can carry a current of 50 A without overheating. For this current, what is the magnitude of the magnetic field at the surface of the wire?

70. Each of two long, straight, parallel wires, 10 cm apart, carries a current of 100 A. Figure 30-77 shows a cross section, with the wires running perpendicular to the page and point P lying on the perpendicular bisector of the line between the wires. Find the magnitude and direction of the magnetic field at P when the current in the left-hand wire is out of the page and the current in the right-hand wire is (a) out of the page and (b) into the page.

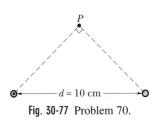

Fig. 30-77 Problem 70.

71. Figure 30-78 shows two current segments. In the upper segment, an arc of radius 4.0 cm subtends an angle of 120° with center P. The lower segment includes a larger semicircle of radius 5.0 cm, also with center P. (a) If $I = 0.40$ A, what is the net magnetic field at point P due to these current segments? (b) If the direction of the current in the semicircle is reversed, what now is the net magnetic field at point P?

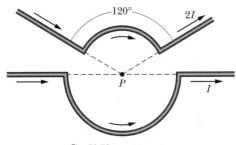

Fig. 30-78 Problem 71.

72. Two long parallel wires lie in an xy plane. One wire lies along the line $y = 10.0$ cm and carries a current of 6.00 A in the positive x direction. The other wire lies along the line $y = 5.00$ cm and carries a current of 10.0 A in the positive x direction. (a) What is the resulting magnetic field at the origin? (b) For what value of y is the resulting magnetic field zero? (c) If the 6.00 A current is reversed so that it is now in the negative x direction, for what value of y is the resulting magnetic field now zero?

73. A long, hollow, cylindrical conductor (inner radius $= 2.0$ mm, outer radius $= 4.0$ mm) carries a current of 24 A distributed uniformly across its cross section. A long thin wire that is coaxial with the cylinder carries a current of 24 A in the opposite direction. What are the magnitudes of the magnetic fields (a) 1.0 mm, (b) 3.0 mm, and (c) 5.0 mm from the central axis of the wire and cylinder?

74. Show that if the thickness of a toroid is very small compared to its radius of curvature (a very skinny toroid), then Eq. 30-26 for the field inside a toroid reduces to Eq. 30-25 for the field inside a solenoid. Explain why this result is to be expected.

75. A long straight wire carries a current of 50 A. An electron, traveling at 1.0×10^7 m/s, is 5.0 cm from the wire. What force acts on the electron if the electron velocity is directed (a) toward the wire, (b) parallel to the wire in the direction of the current, and (c) perpendicular to the two directions defined by (a) and (b)?

76. The magnitude of the magnetic field 88.0 cm from the axis of a long straight wire is 7.30 μT. What is the current in the wire?

77. For the wires in Sample Problem 30-6 in this supplement, show that Eq. 30-33 holds for points beyond the wires—that is, for points with $|x| > d$.

78. Three long wires are parallel to a z axis and each carries a current of 10 A in the positive z direction. Their points of intersection with the xy plane form an equilateral triangle with sides of 50 cm, as shown in Fig. 30-79. A fourth wire (wire b) passes through the midpoint of the base of the triangle and is parallel to the other three wires. What must be the current in wire b for the net magnetic force on wire a to be zero?

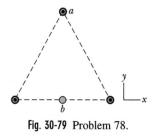

Fig. 30-79 Problem 78.

79. A long wire is known to have a radius greater than 4.0 mm and to carry a current that is uniformly distributed over its cross section. The magnitude of the magnetic field due to that current is 0.28 mT at a point 4.0 mm from the axis of the wire, and 0.20 mT at a point 10 mm from the axis of the wire. What is the radius of the wire?

80. Figure 30-80 shows five very long wires (in cross section) that carry currents directly into or out of the page, as indicated. The wires are uniformly spaced, 0.50 m apart; the currents are $i_1 = 2.00$ A, $i_2 = 4.00$ A, $i_3 = 0.25$ A, $i_4 = 4.00$ A, and $i_5 = 2.00$ A. What is the magnitude of the net force per unit length acting on the central wire due to the currents in the other wires?

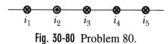

Fig. 30-80 Problem 80.

Tutorial Problem

81. Let's consider an infinitely long wire that is a solid cylinder of radius a. The wire carries a current i; we will make the simplifying assumption that this current is uniformly distributed through the interior of the wire. (a) Make two cross-sectional sketches of the wire, one along the axis and one perpendicular to the axis. Mark the direction of the current on each diagram. (b) What is the magnitude of the current density (current per unit area) in the wire? (c) For a problem with cylindrical symmetry, it is most appropriate to write the magnetic field as a function of cylindrical coordinates r, θ, and z, where r and θ are polar coordinates in the xy plane and z is the coordinate along the cylinder axis. Explain in complete sentences why the magnetic field would not be expected to depend on θ or z, but might depend on r. (d) What do you expect the magnetic field lines to look like inside and outside the wire? Sketch them on your diagram of part (a), showing the correct direction of the field. (e) The appropriate unit vectors to use in a problem with cylindrical symmetry are $\hat{r}$ (lying in the xy plane, directed straight out from the z axis), $\hat{\theta}$ (lying in the xy plane, perpendicular to $\hat{r}$, and pointing in the counterclockwise direction

of increasing θ), and $\hat{z}$ (along the z axis; $\hat{z}$ is equivalent to $\hat{k}$). In terms of these unit vectors, what is the direction of the current in the wire?

(f) Name and describe in a sentence the fundamental principle of physics that can be used to determine the magnitude of the magnetic field. (g) Use this principle to determine the magnetic field in the regions $r < a$ (inside the wire) and $r > a$ (outside the wire). Explain clearly any path or surface you use. (h) Graph the magnitude of $\vec{B}$ as a function of r.

Answers

(b) The wire has radius a and cross-sectional area πa^2, so its current density $\vec{J}$ has magnitude $J = i/\pi a^2$.

(c) All values of z are equivalent because the wire is infinitely long, so the magnetic field should not depend on z. Also, all directions from the wire (all values of θ) are equivalent, so the magnetic field should not depend on θ. However, clearly, all values of r are not equivalent, so we expect that the magnetic field will depend on r.

(d) The magnetic field lines should be circles centered on the axis of the wire. From the right-hand rule, if the current in the wire is directed up out of the plane of the paper, the field lines will be counterclockwise. That the field lines are circles follows from the cylindrical symmetry of the wire.

(e) The current in the wire is along $\hat{z}$, by our choice of direction for $\hat{z}$.

(f) This is the highly symmetric type of situation in which Ampere's law can be used. That law says that the line integral of the magnetic field vector around a closed path equals $\mu_0 i_{enc}$, where i_{enc} is the net current through the path. In symbols,

$$\oint \vec{B} \cdot d\vec{s} = \mu_0 i_{enc}.$$

(g) The appropriate Amperian path to use is a circle of radius r lying in the xy plane and concentric with the wire. Along this path the magnetic field is parallel to $d\vec{s}$. For $r < a$,

$$\oint \vec{B}(r) \cdot d\vec{s} = \mu_0 i_{enc}$$

becomes

$$2\pi r B = \mu_0 \left(\frac{i}{\pi a^2} \right) \pi r^2,$$

so

$$B(r) = \frac{\mu_0 i r}{2 \pi a^2}.$$

For $a < r$,

$$\oint \vec{B}(r) \cdot d\vec{s} = \mu_0 i_{enc}$$

becomes

$$2\pi r B = \mu_0 i,$$

so

$$B(r) = \frac{\mu_0 i}{2 \pi r}.$$

31

Induction and Inductance

ADDITIONAL SAMPLE PROBLEMS

Sample Problem 31-10

Figure 31-63a shows a rectangular conducting loop of resistance R, width L, and length b being pulled at constant speed v through a region of width d in which a uniform magnetic field $\vec{B}$ is produced by an electromagnet. Let $L = 40$ mm, $b = 10$ cm, $d = 15$ cm, $R = 1.6\ \Omega$, $B = 2.0$ T, and $v = 1.0$ m/s.

(a) Plot the magnetic flux Φ_B through the loop as a function of the position x of the right side of the loop.

SOLUTION: The **Key Idea** here is that the amount of magnetic flux Φ_B through the loop depends on how much area of the loop is in

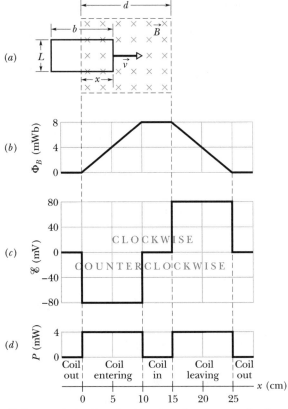

Fig. 31-63 Sample Problem 31-10. (a) A closed conducting loop is pulled at constant velocity $\vec{v}$ completely through a magnetic field. (b) The flux through the loop as a function of the position x of the right side of the loop. (c) The induced emf as a function of x. (d) The rate at which thermal energy appears in the loop as a function of x.

magnetic field $\vec{B}$. Thus, the flux through the loop changes as the loop moves into and then out of the field. The flux is zero when the loop is not in the field; it is BLb (= 8 mWb) when the loop is entirely in the field; it is BLx when the loop is entering the field; and then it is $BL[b - (x - d)]$ when the loop is leaving the field. These results lead to the plot of Fig. 31-63b, which you should verify.

(b) Plot the induced emf as a function of the position of the loop. Indicate the directions of the induced emf.

SOLUTION: To find the magnitude of the emf $\mathscr{E}$ induced in the loop, we use the **Key Idea** that $\mathscr{E}$ is related to the rate $d\Phi_B/dt$ at which the magnetic flux through the loop changes, according to Faraday's law in the form of Eq. 31-6. We can rewrite that equation to include the speed v (= dx/dt) at which the loop moves through the field, as follows:

$$\mathscr{E} = -\frac{d\Phi_B}{dt} = -\frac{d\Phi_B}{dx}\frac{dx}{dt} = -\frac{d\Phi_B}{dx}v. \qquad (31\text{-}72)$$

Note that $d\Phi_B/dx$ is the slope of the curve of Fig. 31-63b. Thus, the magnitude of the induced emf $\mathscr{E}$ is the product of that slope and the constant speed v for each point x along the loop's path. For example, we can find the slope of the segment between $x = 0$ and $x = 10$ cm as

$$\text{slope} = \frac{\Delta\Phi_B}{\Delta x} = \frac{8.0\ \text{mWb} - 0}{10\ \text{cm} - 0} = 0.80\ \text{mWb/cm} = 0.080\ \text{Wb/m}.$$

Then at $x = 5$ cm, Eq. 31-72 tells us that the induced emf $\mathscr{E}$ is

$$\mathscr{E} = -(\text{slope})v = -(0.080\ \text{Wb/m})(1.0\ \text{m/s})$$
$$= -0.080\ \text{V} = -80\ \text{mV}.$$

Similar calculations yield the curve shown in Fig. 31-63c.

To determine the direction of the induced emf, the **Key Idea** is to apply Lenz's law to the loop of Fig. 31-63a. When the loop is entering the magnetic field, a magnetic flux through the loop and into the page is increasing. Lenz's law tells us that the induced magnetic field must oppose that increase, meaning that the induced field must be directed out of the page. A curled–straight right-hand rule tells us this induced field requires a counterclockwise current in the loop. When the loop is leaving the magnetic field, the flux through it is decreasing, the induced magnetic field is into the page, and so the current in the loop is clockwise. In Fig. 31-63c, a counterclockwise emf is plotted as a negative value, a clockwise emf as

221

a positive value. There is *no* emf when the loop is either entirely out of the field or entirely in it because, in these two situations, the flux through the loop is not changing.

(c) Plot the rate of production of thermal energy in the loop as a function of the position of the loop.

SOLUTION: One **Key Idea** here is that because the loop is complete and conducting, the induced emf $\mathscr{E}$ can drive a current i around it. A second **Key Idea** is that because the loop has resistance R, electric energy of that current is transferred to thermal energy at a rate P, according to Eq. 31-16 ($P = i^2R$). Thus, substituting $i = \mathscr{E}/R$ in Eq. 31-16 gives us

$$P = i^2R = \frac{\mathscr{E}^2}{R}.$$

We can calculate P for any value of x by squaring the ordinate of the curve of Fig. 31-63c for that value and dividing by R, being careful of powers of 10 in the units. The result is plotted in Fig. 31-63d. Note that thermal energy is produced only when the loop is entering or leaving the magnetic field.

In practice, the external magnetic field $\vec{B}$ cannot drop sharply to zero at its boundary but must approach zero smoothly. The result would be a rounding of the corners of the curves plotted in Fig. 31-63.

Sample Problem 31-11

Figure 31-64 shows a cross section, in the plane of the page, of a toroid of N turns like that in Fig. 30-20a but of rectangular cross section; its dimensions are as indicated.

(a) What is its inductance L?

SOLUTION: The first **Key Idea** here is to apply the definition of inductance L given in Eq. 31-30 ($L = N\Phi_B/i$) to a toroid. To do this, we need an expression for the magnetic flux Φ_B inside the toroid due to the current i through it. From Eq. 30-26, we already know the magnitude B of the magnetic field within the toroid (also due to i):

$$B = \frac{\mu_0 iN}{2\pi r}, \qquad (31\text{-}73)$$

where r is the distance from the center of the toroid. This equation holds regardless of the shape or dimensions of the toroid's cross section.

A second **Key Idea** is that because B is *not* uniform over the cross section, we cannot use Eq. 31-4 ($\Phi_B = BA$) to find the flux Φ, but instead must use Eq. 31-3,

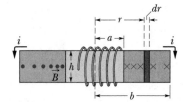

Fig. 31-64 Sample Problem 31-11. A cross section of a toroid, showing the current in the windings and the associated magnetic field. See Fig. 30-20. The nonuniform magnetic field within the toroid is represented by nonuniformly spaced dots and ×s.

$$\Phi_B = \int \vec{B} \cdot d\vec{A}. \qquad (31\text{-}74)$$

The direction of $\vec{B}$ is everywhere perpendicular to the cross section, as shown in Fig. 31-64; $\vec{B}$ is thus parallel to the differential cross-sectional area vector $d\vec{A}$, so the dot product in Eq. 31-74 gives $B\,dA$. For the differential area dA, we can use the area $h\,dr$ of the strip shown in Fig. 31-64. Substituting these quantities and Eq. 31-73 into Eq. 31-74 and integrating from $r = a$ to $r = b$ yield

$$\Phi_B = \int_a^b Bh\,dr = \int_a^b \frac{\mu_0 iN}{2\pi r} h\,dr$$

$$= \frac{\mu_0 iNh}{2\pi} \int_a^b \frac{dr}{r} = \frac{\mu_0 iNh}{2\pi} \ln\frac{b}{a}.$$

Equation 31-30 then gives us

$$L = \frac{N\Phi_B}{i} = \frac{N}{i}\frac{\mu_0 iNh}{2\pi} \ln\frac{b}{a},$$

so

$$L = \frac{\mu_0 N^2 h}{2\pi} \ln\frac{b}{a}. \qquad (\text{Answer}) \quad (31\text{-}75)$$

(b) The toroid shown in Fig. 31-64 has $N = 1250$ turns, $a = 52$ mm, $b = 95$ mm, and $h = 13$ mm. What is its inductance?

SOLUTION: From Eq. 31-75

$$L = \frac{\mu_0 N^2 h}{2\pi} \ln\frac{b}{a}$$

$$= \frac{(4\pi \times 10^{-7}\ \text{H/m})(1250)^2(13 \times 10^{-3}\ \text{m})}{2\pi} \ln\frac{95\ \text{mm}}{52\ \text{mm}}$$

$$= 2.45 \times 10^{-3}\ \text{H} \approx 2.5\ \text{mH}. \qquad (\text{Answer})$$

Sample Problem 31-12

A 3.56 H inductor is placed in series with a 12.8 Ω resistor, and an ideal battery of emf 3.24 V is suddenly applied across the RL combination at time $t = 0$.

(a) At $t = 0.278$ s (which is one inductive time constant) after the

emf is applied, what is the rate P_{emf} at which energy is being delivered by the battery?

SOLUTION: One **Key Idea** here is that the rate at which the battery delivers energy is the product of the emf and the current at any time t, according to Eq. 28-14 ($P_{emf} = i\mathcal{E}$). A second **Key Idea** is that the current increases with time according to Eq. 31-43,

$$i = \frac{\mathcal{E}}{R}(1 - e^{-t/\tau_L}).$$

At $t = \tau_L$, this yields

$$i = \frac{3.24\ V}{12.8\ \Omega}(1 - e^{-1}) = 0.1600\ A.$$

Thus, the rate at which the battery delivers energy is

$$P_{emf} = i\mathcal{E} = (0.1600\ A)(3.24\ V)$$
$$= 0.5184\ W \approx 518\ mW. \qquad \text{(Answer)}$$

(b) At $t = 0.278$ s, at what rate P_B is energy being stored in the magnetic field of the inductor?

SOLUTION: The **Key Idea** here is this: The rate at which the energy U_B of an inductor's magnetic field can change at any instant depends on both the current i through the inductor and the rate di/dt at which that current is changing *at that instant,* all according to Eq. 31-50 ($dU_B/dt = Li\ di/dt$). From part (a), we know that at $t = 0.278$ s, the current is 0.1600 A. To find di/dt at that time, we first differentiate Eq. 31-43 for the current:

$$\frac{di}{dt} = \frac{d}{dt}\left[\frac{\mathcal{E}}{R}(1 - e^{-t/\tau_L})\right] = \frac{\mathcal{E}}{R\tau_L}(e^{-t/\tau_L}) = \frac{\mathcal{E}}{L}(e^{-t/\tau_L}).$$

Then we evaluate this equation at $t = 0.278$ s by substituting $t = \tau_L$ and other data, obtaining

$$\frac{di}{dt} = \frac{3.24\ V}{3.56\ H}e^{-1} = 0.3348\ A/s.$$

Now from Eq. 31-50 the desired rate is

$$P_B = \frac{dU_B}{dt} = Li\frac{di}{dt}$$
$$= (3.56\ H)(0.1600\ A)(0.3348\ A/s)$$
$$= 0.1907\ W \approx 191\ mW. \qquad \text{(Answer)}$$

(c) At $t = 0.278$ s, at what rate P_R is energy appearing as thermal energy in the resistor?

SOLUTION: One way to answer this question is to assume that the battery, inductor, and resistor form a closed system. Then we can apply the **Key Idea** that the total energy of the system must be conserved. Energy is being transferred *from* the battery at rate P_{emf}, *to* the magnetic field at rate P_B, and *to* thermal energy at rate P_R. Because the total energy is conserved, these rates are related by

$$P_{emf} = P_B + P_R.$$

Thus, $\quad P_R = P_{emf} - P_B = 0.5184\ W - 0.1907\ W$
$$= 0.3277\ W \approx 328\ mW. \qquad \text{(Answer)}$$

We can verify this result (and help justify our assumption that the system here is closed) by using Eq. 27-22 ($P = i^2R$). At $t = 0.278$ s, the current i is 0.1600 A. Thus, Eq. 27-22 yields

$$P_R = i^2R = (0.1600\ A)^2(12.8\ \Omega)$$
$$= 0.3277\ W \approx 328\ mW. \qquad \text{(Answer)}$$

QUESTIONS

11. *Organizing question:* In Fig. 31-65, a wire loop forms a rectangle of height H and width W. The loop lies in a magnetic field that does not vary with time, is directed out of the page, and has a magnitude B. We wish to find the magnetic flux Φ_B through the loop for three choices of $B(x, y)$:
(a) $B = ax$, (b) $B = by$, and (c) $B = cxy$, where a, b, and c are constants. For each choice, which of the following expressions can be used to find Φ_B?

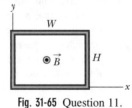

Fig. 31-65 Question 11.

(1) BHW
(2) $\int BH\ dx$
(3) $\int BW\ dx$
(4) $\int BH\ dy$
(5) $\int BW\ dy$
(6) $\int B\ dx\ dy$

(d) For each choice, is the emf induced around the loop clockwise, counterclockwise, or nonexistent?

12. In Fig. 31-22, the current in coil 1 is given, in three situations, by (1) $i_1 = 3\cos(4t)$, (2) $i_1 = 10\cos(t)$, and (3) $i_1 = 5\cos(2t)$, with i_1 in amperes and t in seconds. For the three situations, rank (a) the mutual inductance of the coils and (b) the magnitude of the maximum emf appearing in coil 2 due to i_1, greatest first.

13. *Organizing question:* Figure 31-66 shows a circular region of radius R in which a magnetic flux Φ_B is directed out of the page.

It also shows two integration paths (of radii r_1 and r_2) that form circles concentric with the circular region. Set up the right side of Faraday's law (Eq. 31-22) for paths 1 and 2 for the following situations, where either the total flux Φ_B, the flux $\Phi_{B,enc}$ encircled by a path of radius r, or the magnitude B of the associated magnetic field is given. Neglect the minus sign in Eq. 31-22.

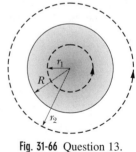

Fig. 31-66 Question 13.

Perform any differentiation that may be required, but not any integration (and be warned of menacing subscripts).

Situation	Flux or Field
a	$\Phi_B = (4\ T \cdot m^2/s)t$, uniform
b	$B = (2\ T/s)t$, uniform
c	$B = (3\ T)\left(\dfrac{r}{R}\right)$
d	$\Phi_{B,enc} = (5\ T \cdot m^2/s)\left(\dfrac{r}{R}\right)t$
e	$B = (4\ T/s)\left(1 - \dfrac{r}{R}\right)t$

14. Figure 31-67 shows three circuits with identical batteries, inductors, and resistors. Rank the circuits, greatest first, according to the current through the resistor labeled R (a) long after the switch is closed, (b) just after the switch is reopened a long time later, and (c) long after it is reopened.

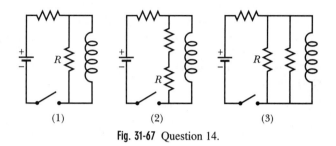

Fig. 31-67 Question 14.

15. In Fig. 31-17, assume that the (ideal) battery has an emf of 5 V. After the switch is thrown to position a, consider the inductor at the stages when the potential difference across the resistor is (1) 1 V, (2) 2 V, and (3) 3 V. Rank those stages according to (a) the rate at which the current is changing, (b) the total flux through the inductor, and (c) the emf across the inductor, greatest first.

16. Suppose that in Question 15 and Fig. 31-17 the current has reached its equilibrium value when the switch is thrown to position b. Consider the inductor at the stages when the potential difference across the resistor is (1) 1 V, (2) 2 V, and (3) 3 V. Rank those stages according to (a) the rate at which the current is changing, (b) the total flux through the inductor, and (c) the emf across the inductor, greatest first.

17. *Organizing question:* The switch in Fig. 31-68 is closed at time $t = 0$. (a) Thereafter, what is the current through R_3 in terms of the currents through R_1 and R_2? Using that relation and given data, set up loop equations for (b) the left-hand loop and (c) the big loop. (d) To produce a single differential equation from these results, written in one unknown current, what step should now be taken and which current is then in the differential equation?

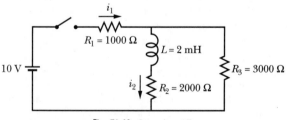

Fig. 31-68 Question 17.

18. Here are three sets of values for the emf $\mathscr{E}$ of the battery and the potential difference V_R across the resistor in the circuit of Fig. 31-18 at different times after the current begins to increase: (a) 12 V and 3 V, (b) 24 V and 16 V, (c) 18 V and 10 V. Rank the sets according to the potential difference across the inductor at those times, greatest first.

19. The number of turns per unit length, current, and cross-sectional area for three solenoids of the same length are given in the following table. Rank the solenoids according to (a) their inductance and (b) the flux through each turn, greatest first.

Solenoid	Turns per Unit Length	Current	Area
1	$2n_1$	i_1	$2A_1$
2	n_1	$2i_1$	A_1
3	n_1	i_1	$4A_1$

20. If the variable resistance R in the left-hand circuit of Fig. 31-69 is increased at a steady rate, is the current induced in the right-hand loop clockwise or counterclockwise?

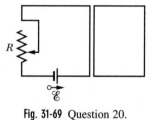

Fig. 31-69 Question 20.

21. In Fig. 31-70, a circular loop is moved at a constant velocity through regions where uniform magnetic fields of the same magnitude are directed into or out of the page. (The field is zero outside the dashed lines.) At which of the seven indicated loop positions is the emf induced in the loop (a) clockwise, (b) counterclockwise, and (c) zero?

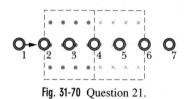

Fig. 31-70 Question 21.

22. In Fig. 31-71, for three situations, magnetic fields are directed through rectangular wire loops that have areas of either A_0 or $A_0/2$. In each situation, the field is perpendicular to the plane of the loop and is increasing in magnitude at the rate of 5 mT/s. At the instant shown, the magnitudes of the fields are either B_0, $2B_0$, or $3B_0$. (a) At that instant, rank the loops according to the magnitude of the emf induced in them, greatest first. (b) For each situation, what is the direction of the emf induced in the loop?

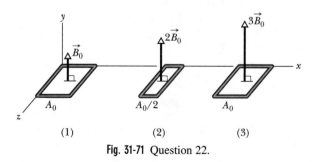

Fig. 31-71 Question 22.

23. Figure 31-72 shows three situations in which a wire loop lies partially in a magnetic field. The magnitude of the field is either increasing or decreasing, as indicated. In each situation, a battery is part of the loop. In which situations are the induced emf and the battery emf in the same direction along the loop?

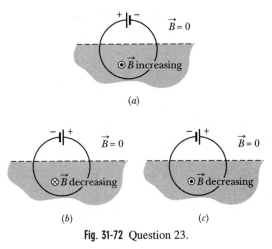

Fig. 31-72 Question 23.

24. In Fig. 31-73, a wire loop has been bent so that it has three segments: segment ab (a quarter circle), bc (a square corner), and ca (straight). Here are three choices for a magnetic field through the loop:

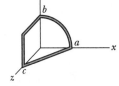

Fig. 31-73 Question 24.

(1) $\vec{B}_1 = 3\hat{i} + 7\hat{j} - 5t\hat{k}$,

(2) $\vec{B}_2 = 5t\hat{i} - 4\hat{j} - 15\hat{k}$,

(3) $\vec{B}_3 = 2\hat{i} - 5t\hat{j} - 12\hat{k}$,

where $\vec{B}$ is in milliteslas and t is in seconds. Without written calculation, rank the choices according to (a) the work done per unit charge in setting up the induced current and (b) that induced current, greatest first. (c) For each choice, what is the direction of the induced current in the figure?

25. Figure 31-74 gives four situations (similar to that in Fig. 31-10) in which we pull rectangular wire loops out of identical magnetic fields (directed into the page) at the same constant speed. The loops have edge lengths of either L or $2L$, as drawn. Rank the situations according to (a) the magnitude of the force required of us and (b) the rate at which energy is transferred from us to thermal energy of the loop, greatest first.

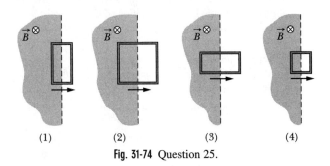

Fig. 31-74 Question 25.

26. In Fig. 31-12b, is the direction of the eddy currents in the plate clockwise or counterclockwise when the plate is (a) entering the magnetic field from the left (as shown), (b) leaving the field toward the right, (c) entering the field from the right, and (d) leaving the field toward the left?

DISCUSSION QUESTIONS

27. Are induced emfs and currents different in any way from emfs and currents provided by a battery connected to a conducting loop?

28. Is the magnitude of the voltage induced in a coil through which a magnet moves affected by the strength of the magnet? If so, explain how.

29. Explain the difference between a magnetic field $\vec{B}$ and the flux Φ_B of a magnetic field. Are they vectors or scalars? In what units may each be expressed? How are these units related? Are either or both (or neither) properties of a given point in space?

30. Can a charged particle at rest be set in motion by the action of a magnetic field? If not, why not? If so, how?

31. You drop a bar magnet along the central axis of a long copper tube. Describe the motion of the magnet and the energy interchanges involved. Neglect air resistance.

32. You are playing with a metal loop, moving it back and forth in a magnetic field. How can you tell, without detailed inspection, whether or not the loop has a narrow saw cut across it, making it an incomplete loop?

33. Figure 31-75 shows an inclined wooden track that passes, for part of its length, through a strong magnetic field. You roll a copper coin down the track. Describe the motion of the coin as it rolls from the top of the track to the bottom.

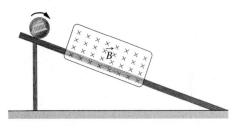

Fig. 31-75 Question 33.

34. Figure 31-76 shows a copper ring hung from a ceiling by two threads. Describe in detail how you might most effectively use a bar magnet to get this ring to swing back and forth.

35. A bar magnet moves inside a long solenoid, along its central axis. Is an emf induced in the solenoid? Explain your answer.

Fig. 31-76 Question 34.

36. Two conducting loops face each other a distance d apart (Fig. 31-77). An observer sights along their common axis from left to

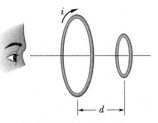

Fig. 31-77 Question 36.

right. If a clockwise current i is suddenly established in the larger loop, by a battery not shown, (a) what is the direction of the induced current in the smaller loop, and (b) what is the direction of the force (if any) that acts on the smaller loop?

37. The north pole of a magnet is moved away from a copper ring, as shown in Fig. 31-78. In the part of the ring farthest from the reader, what is the direction of the current?

38. In Fig. 31-79, a short solenoid carrying a steady current is moving toward a conducting loop. What is the direction of the current induced in the loop as one sights toward it as shown?

Fig. 31-78 Question 37.

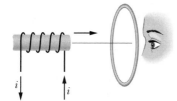

Fig. 31-79 Question 38.

39. What is the direction of the induced current through resistance R in Fig. 31-80 (a) immediately after switch S is closed, (b) some time after switch S is closed, and (c) immediately after switch S is opened? (d) When switch S is held closed, from which end of the longer coil do magnetic field lines emerge? This is the effective north pole of the coil.
(e) How do the conduction electrons in the coil containing R know about the flux within the longer coil? What really gets those electrons moving?

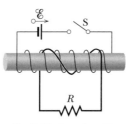

Fig. 31-80 Question 39.

40. In Faraday's law of induction, does the induced emf depend on the resistance in the circuit? If so, how?

41. Suppose that the direction of induced emfs was governed by what we can call the Antilenz law: The induced current will appear in such a direction that it aids the change that produced it. Design a machine based on this law that would make a lot of money for you. (Alas, the Antilenz law is *false*.)

42. The loop of wire shown in Fig. 31-81 rotates with constant angular speed about the x axis. A uniform magnetic field $\vec{B}$, whose direction is that of the positive y axis, is present. For what portions of the rotation is the induced current in the PQ side of the loop (a) from P to Q, (b) from Q to P, and (c) zero? (d) Repeat (a)

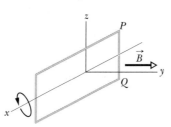

Fig. 31-81 Question 42.

through (c) for rotation that is reversed from that shown in the figure.

43. The conducting loop shown in cross section in Fig. 31-82 is removed from the permanent magnet by pulling it vertically upward. (a) What is the direction of the induced current in the loop? (b) Is a force required to remove the loop? (c) Does the total amount of thermal energy produced in removing the loop depend on the time taken to remove it?

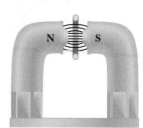

Fig. 31-82 Question 43.

44. A planar closed loop is placed in a uniform magnetic field. In what ways can the loop be moved without inducing an emf? Consider both translations and rotations.

45. *Electromagnetic shielding.* In Fig. 31-83, a conducting sheet lies in a plane perpendicular to a magnetic field $\vec{B}$. (a) If $\vec{B}$ suddenly changes, the full change in $\vec{B}$ is not immediately detected at points near P. Explain. (b) If the resistivity of the sheet is zero, the change is never detected at P. Explain. (c) If $\vec{B}$ changes periodically at high frequency and the conductor is made of a material of low resistivity, the region near P is almost completely shielded from the changes in flux. Explain. (d) Why is such a conducting sheet not useful as a shield from static magnetic fields?

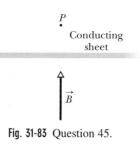

Fig. 31-83 Question 45.

46. (a) In Fig. 31-13*b*, must the circle of radius r be a conducting loop in order that $\vec{E}$ and $\mathcal{E}$ be present? (b) If the circle of radius r were not concentric with the field (moved slightly to the left, say), would $\mathcal{E}$ change? Would the configuration of $\vec{E}$ around the circle change? (c) For a concentric circle of radius r, with $r > R$, does an emf exist? Do electric fields then exist?

47. A copper ring and a wooden ring of the same dimensions are placed so that there is the same changing magnetic flux through each. Compare the induced electric fields in the two rings.

48. Explain how a long straight wire can show self-induction effects. How would you go about looking for them?

49. If the same magnetic flux passes through each turn of a coil, the inductance of the coil may be computed from Eq. 31-30 ($L = N\Phi_B/i$). How might we compute L for a coil for which this assumption is not valid?

50. Show that the dimensions of the two expressions for L, Eq. 31-30 ($N\Phi_B/i$) and Eq. 31-37 [$\mathcal{E}_L/(di/dt)$] are the same.

51. You want to wind a coil so that it has resistance but approximately no inductance. How would you do it?

52. Is the inductance per unit length for a solenoid near its center the same as, less than, or greater than the inductance per unit length near its ends? Justify your answer.

53. Explain why the self-inductance of a coaxial cable is expected to increase when the radius of the outer conductor is increased, the radius of the inner conductor remaining fixed.

54. A steady current is set up in a coil with a very large inductive time constant. When the current is interrupted with a switch, a heavy arc tends to appear at the switch blades. Explain why. (*Note:* Interrupting currents in highly inductive circuits can be destructive and dangerous.)

55. Suppose that you connect an ideal (that is, approximately resistanceless) coil across an ideal (again, approximately resistanceless) battery. You might think that, because there is no resistance in the circuit, the current would jump at once to a very large value. On the other hand, you might think that, because the inductive time constant ($= L/R$) is extremely large, the current would rise very slowly, if at all. What actually happens?

56. In an *RL* circuit like that of Fig. 31-18, can the self-induced emf ever be greater than the battery emf?

57. In an *RL* circuit like that of Fig. 31-18, is the current in the resistor always the same as the current in the inductor?

58. In the circuit of Fig. 31-17, the self-induced emf is a maximum at the instant the switch is closed on *a*. How can this be, considering that there is no current in the inductor at this instant?

59. The switch in Fig. 31-17, having been closed on *a* for a long time, is thrown to *b*. What happens to the energy that is stored in the inductor?

60. A coil has a (measured) inductance *L* and a (measured) resistance *R*. Is its inductive time constant necessarily given by $\tau_L = L/R$? Bear in mind that we derived that equation (see Fig. 31-17) for a situation in which the inductive and resistive elements are separated. Discuss.

61. Figure 31-19*a* and Fig. 28-14*b* (multiplying by *R* does not change the shape) are plots of $V_R(t)$ for, respectively, an *RL* circuit and an *RC* circuit. Why are these two curves so different? Account for each in terms of physical processes going on in the appropriate circuits.

62. Two solenoids, *A* and *B*, have the same diameter and length and contain only one layer of copper windings, with adjacent turns touching, the insulation thickness being negligible. Solenoid *A* contains many turns of fine wire, and solenoid *B* contains fewer turns of heavier wire. (a) Which solenoid has the greater self-inductance? (b) Which solenoid has the greater inductive time constant? Justify your answers.

63. Can you make an argument based on the manipulation of bar magnets to suggest that energy may be stored in a magnetic field? Explain.

64. Draw all the formal analogies that you can think of between a parallel-plate capacitor (for electric fields) and a long solenoid (for magnetic fields).

65. In each of the following operations energy is expended. Some of this energy is returnable (can be reconverted) into electric energy that can be made to do useful work, and some becomes unavailable for useful work or is wasted in other ways. In which of the following cases will there be the *least* percentage of returnable electric energy: (a) charging a capacitor, (b) charging a storage battery, (c) sending a current through a resistor, (d) setting up a magnetic field, (e) moving a conductor in a magnetic field? Explain.

66. The current in a solenoid is reversed. What changes does this make in the magnetic field $\vec{B}$ and the energy density u_B at various points along the solenoid axis?

67. Commercial devices such as motors and generators that are involved in the transformation of energy between electrical and mechanical forms involve magnetic rather than electrostatic fields. Why should this be so?

68. A heavy current is passed, clockwise, through both coils shown in Fig. 31-84. *Q* is the horizontal midpoint of the long coil whose ends are *P* and *S*. The horizontal midpoint *R* of the short coil is originally located a distance *x* from *Q*. Describe the subsequent motion of point *R*.

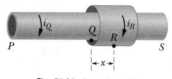

Fig. 31-84 Question 68.

69. In a case of mutual induction, such as in Fig. 31-22, is self-induction also present? Discuss.

70. You are given two similar flat circular coils of *N* turns each. The centers of the coils are maintained a fixed distance apart. For what orientation will their mutual inductance *M* be the greatest? For what orientation will it be the least? Why?

71. A circular coil of *N* turns surrounds a long solenoid. Is the mutual inductance greater when the coil is near the center of the solenoid or when it is near one end? Justify your answer.

72. A long cylinder is wound from left to right with one layer of wire, giving it *n* turns per unit length with a self-inductance of L_1, as in Fig. 31-85*a*. If the winding is now continued, in the same *sense* but returning from right to left, as in Fig. 31-85*b*, so as to give a second layer also of *n* turns per unit length, what then is the value of the self-inductance? Explain.

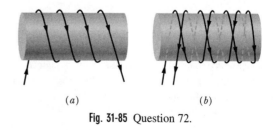

(*a*)　　　　　　　　(*b*)

Fig. 31-85 Question 72.

EXERCISES & PROBLEMS

76. Switch S in Fig. 31-86 is closed for $t < 0$ and is opened at $t = 0$. When current i_1 through L_1 and current i_2 through L_2 are *first* equal to each other, what is their common value? (The resistors have the same resistance R.)

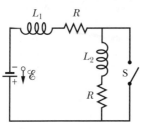

Fig. 31-86 Problem 76.

77. A coil with an inductance of 2.0 H and a resistance of 10 Ω is suddenly connected to a resistanceless battery with $\mathscr{E} = 100$ V. (a) What is the equilibrium current? (b) How much energy is stored in the magnetic field when this current exists in the coil?

78. How long would it take, following the removal of the battery, for the potential difference across the resistor in an RL circuit (with $L = 2.00$ H, $R = 3.00$ Ω) to decay to 10.0% of its initial value?

79. (a) For the toroid of Sample Problem 31-11b (in this supplement), find an expression for the energy density as a function of the radial distance r from the center. (b) By integrating the energy density over the volume of the toroid, calculate the total energy stored in the field of the toroid; assume $i = 0.500$ A. (c) Using Eq. 31-51, evaluate the energy stored in the toroid directly from the inductance and compare it with your answer in (b).

80. A long thin solenoid can be bent into a ring to form a toroid. Show that if the solenoid is long and thin enough, the equation for the inductance of a toroid (Eq. 31-75 in Sample Problem 31-11, in this supplement) is equivalent to that for a solenoid of the appropriate length (Eq. 31-32).

81. Figure 31-87 shows a circuit consisting of a battery, a resistance, and an inductance. When the switch is closed, how long does the current take to build up to 2.0 mA?

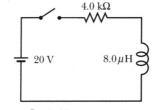

Fig. 31-87 Problem 81.

82. A long solenoid with a radius of 25 mm has 100 turns/cm. A single loop of wire of radius 5.0 cm is placed around the solenoid, the central axes of the loop and the solenoid coinciding. In 10 ms the current in the solenoid is reduced from 1.0 A to 0.50 A at a uniform rate. What emf appears in the loop?

83. A switch is closed at time $t = 0$ to connect a 10 V emf to a series combination of a 10 Ω resistor and a 10 mH inductor. What is the energy stored in the inductor at $t = 2.0$ ms?

84. A square wire loop 20 cm on a side, with resistance 20 mΩ, has its plane normal to a uniform magnetic field of magnitude $B = 2.0$ T. If you pull two opposite sides of the loop away from each other, the other two sides automatically draw toward each other, reducing the area enclosed by the loop. If the area is reduced to zero in time $\Delta t = 0.20$ s, what are (a) the average emf and (b) the average current induced in the loop during Δt?

85. In the circuit shown in Fig. 31-88, $\mathscr{E} = 12$ V. The switch has been open for a long time before it is closed at $t = 0$. At what rate is the current in the inductor changing (a) immediately after the switch is closed and (b) when the current in the battery is 0.50 A? (c) What is the current in the battery when the circuit reaches its steady-state condition?

86. The flux linkage through a certain coil of 0.75 Ω resistance would be 26 mWb if there were a current of 5.5 A in it. (a) Calculate the inductance of the coil. (b) If a 6.0 V battery were suddenly connected across the coil, how long would it take for the current to rise from 0 to 2.5 A?

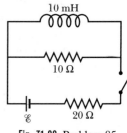

Fig. 31-88 Problem 85.

87. The magnetic energy stored in a certain inductor is 25.0 mJ when the current in it is 60.0 mA. (a) Calculate its inductance. (b) What current is required for the stored magnetic energy to be four times as great?

88. A long cylindrical solenoid with 100 turns/cm has a radius of 1.6 cm. Assume that the magnetic field it produces is parallel to its axis and is uniform in its interior. (a) What is its inductance per meter of length? (b) If the current changes at the rate 13 A/s, what emf is induced per meter?

89. Calculate the energy needed to produce, in a cube that is 10 cm on edge, (a) a uniform electric field of 100 kV/m and (b) a uniform magnetic field of 1.0 T. (Both these large fields are readily available in the laboratory.) (c) From these answers, determine which type of field can store greater amounts of energy.

90. Figure 31-89 shows a uniform magnetic field $\vec{B}$ confined to a cylindrical volume of radius R. The magnitude of $\vec{B}$ is decreasing at a constant rate of 10 mT/s. What are the instantaneous accelerations (direction and magnitude) experienced by an electron placed at a, at b, and at c? Assume $r = 5.0$ cm.

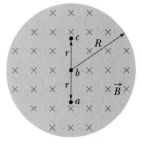

Fig. 31-89 Problem 90.

91. The switch in Fig. 31-90 is suddenly closed at time $t = 0$. At $t = 5.0$ s, the current is 1.0 A. What is inductance L?

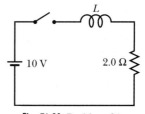

Fig. 31-90 Problem 91.

92. In the circuit shown in Fig. 31-91, $\mathscr{E} = 6.0$ V. (a) At what rate is the current in the 0.30 H inductor changing immediately after the switch is closed? (b) What is the current in the 0.30 H inductor when the circuit reaches its steady-state condition?

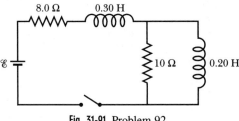

Fig. 31-91 Problem 92.

93. At a certain location in the southern hemisphere, Earth's magnetic field has a magnitude of 42 μT and is directed upward at 57° to the vertical. Calculate the flux through a horizontal surface of area 2.5 m²; see Fig. 31-92, in which area vector $\vec{A}$ has arbitrarily been chosen to be upward.

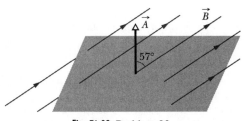

Fig. 31-92 Problem 93.

94. A circular loop (radius = 14 cm) of wire is placed in a magnetic field that makes an angle of 30° with the normal to the plane of the loop. The magnitude of this field increases at a constant rate from 30 mT to 60 mT in 15 ms. If the loop has a resistance of 5.0 Ω, what is the magnitude of the current induced in the loop when the field is 50 mT?

95. In Fig. 31-93, the fuse in the upper branch is an ideal 3.0 A fuse. It has zero resistance as long as the current through it remains less than 3.0 A. If the current reaches 3.0 A, it "blows" and thereafter has infinite resistance. Switch S is closed at time $t = 0$. (a) When does the fuse blow? (*Hint:* Equation 31-43 does not apply. Rethink Eq. 31-41.) (b) Sketch a graph of the current i through the inductor as a function of time. Mark the time at which the fuse blows.

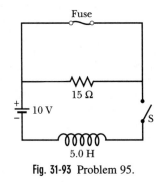

Fig. 31-93 Problem 95.

96. At $t = 0$ a battery is connected across an inductor and a resistor that are connected in series. The table gives the measured potential difference across the inductor as a function of time following the connection of the battery. Find (a) the emf of the battery and (b) the time constant of the circuit.

t (ms)	V_L (V)	t (ms)	V_L (V)
1.0	18.2	5.0	5.98
2.0	13.8	6.0	4.53
3.0	10.4	7.0	3.43
4.0	7.90	8.0	2.60

97. (a) What is the energy density of Earth's magnetic field, which has a magnitude of 50 μT? (b) Assuming this density to be relatively constant over distances small compared with Earth's radius and neglecting variations near the magnetic poles, how much energy would be stored between Earth's surface and a spherical shell 16 km above the surface?

98. A solenoid is wound with a single layer of insulated copper wire (of diameter 2.5 mm) and is 4.0 cm in diameter and 2.0 m long. (a) How many turns are on the solenoid? (b) What is the inductance per meter for the solenoid near its center? Assume that adjacent wires touch and that insulation thickness is negligible.

99. At time $t = 0$, a 45 V potential difference is suddenly applied to the leads of a coil with inductance $L = 50$ mH and resistance $R = 180$ Ω. At what rate is the current through the coil increasing at $t = 1.2$ ms?

100. In Fig. 31-94 a conducting rod of mass m and length L slides without friction on two long horizontal rails. A uniform vertical magnetic field $\vec{B}$ fills the region in which the rod is free to move. The generator G supplies a constant current i directed as shown. (a) Find the velocity of the rod as a function of time, assuming it to be at rest at $t = 0$. The generator is now replaced with a battery that supplies a constant emf $\mathscr{E}$. (b) Show that the velocity of the rod now approaches a constant terminal value $\vec{v}$ and give its magnitude and direction. (c) What is the current in the rod when this terminal velocity is reached? (d) Analyze this situation and that with the generator from the point of view of energy transfers.

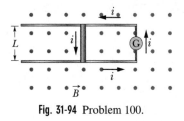

Fig. 31-94 Problem 100.

101. Figure 31-95a shows a wire in the shape of a circle enclosing an area of 3.0 m². The resistance of the wire is 9.0 Ω. The wire is

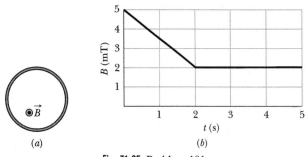

Fig. 31-95 Problem 101.

immersed in a uniform magnetic field that is directed out of the page; the magnitude of the field for $t \geq 0$ is shown in Fig. 31-95*b*. What are the magnitudes and directions of the currents induced in the wire at the times (a) 0.50 s, (b) 1.5 s, (c) 3.0 s, and (d) 4.0 s?

102. A toroid having a 5.00 cm square cross section and an inside radius of 15.0 cm has 500 turns of wire and carries a current of 0.800 A. What is the magnetic flux through the cross section?

103. In the circuit of Fig. 31-96, the switch has been open for a long time when it is closed at time $t = 0$. What are (a) the current through the battery and (b) the rate at which that current is changing immediately after the switch is closed? What are (c) the current and (d) its rate of change at time $t = 3.0$ μs? What are (e) the current and (f) its rate of change a long time later?

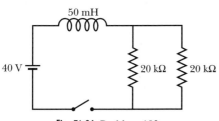

Fig. 31-96 Problem 103.

104. Figure 31-97 shows a closed loop of wire that consists of a pair of equal semicircles, of radius 3.7 cm, lying in mutually perpendicular planes. The loop was formed by folding a plane circular loop along a diameter until the two halves became perpendicular. A uniform magnetic field $\vec{B}$ of magnitude 76 mT is directed perpendicular to the

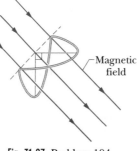

Fig. 31-97 Problem 104.

fold diameter and makes equal angles (of 45°) with the planes of the semicircles. The magnetic field is reduced to zero at a uniform rate during a time interval of 4.5 ms. Determine the magnitude of the induced emf and the direction of the induced current in the loop during this interval.

105. A 50-turn circular coil (radius = 15 cm) with a total resistance of 4.0 Ω is placed in a uniform magnetic field directed perpendicularly to the plane of the coil. The magnitude of this field varies with time according to $B = A \sin(\omega t)$, where $A = 80$ μT and $\omega = 50\pi$ rad/s. What is the magnitude of the current induced in the coil at $t = 20$ ms?

Clustered Problems

In the problems in this cluster, a wire (called a slider*) is forced to move at constant velocity $\vec{v}$ (of magnitude 12.0 m/s) along two long rails (see Fig. 31-98). A resistor of resistance $R = 5.00$ Ω connects the rails at the ends opposite the direction of motion; the rails have negligible resistance. A uniform magnetic field $\vec{B}$, directed into the page in the figures, exists throughout the region. The resistor, rails, and slider form a complete conducting loop.*

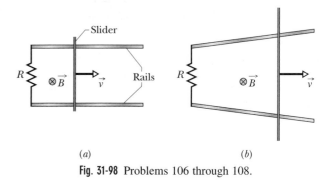

(*a*)   (*b*)

Fig. 31-98 Problems 106 through 108.

106. See this cluster's setup. In Fig. 31-98*a*, the magnitude of the magnetic field is a constant 200 mT. The rails are parallel, with a separation of 30.0 cm. What are the magnitude and direction of the current in the loop if (a) the slider has negligible resistance and (b) the slider has a resistance of 2.00 Ω?

107. See this cluster's setup. In Fig. 31-98*b*, the magnitude of the magnetic field is a constant 200 mT. The slider, which has negligible resistance, begins its motion at $t = 0$ at the resistor. There the rails are separated by 30 cm; their separation increases by 20.0 cm with every 1.00 m of distance from the resistor. (a) What are the magnitude and direction of the current in the loop when the slider reaches the point where the rail separation is 50 cm? (b) Write an equation for the current as a function of time t.

108. See this cluster's setup. In Fig. 31-98*a*, the slider begins at the resistor at time $t = 0$. The magnitude of the magnetic field is 200 mT just then but is increasing at a constant rate of 50.0 mT/s. The resistance of the slider is negligible; the rails are parallel, with a separation of 30 cm. What are the magnitude and direction of the current in the loop at (a) $t = 0$ and (b) $t = 1.00$ s? (c) Write an equation for the current as a function of time.

32
Magnetism of Matter; Maxwell's Equations

Sample Problem 32-5

In Tucson, Arizona, in 1964, the north pole of a compass needle pointed 13° east of geographic north, and the north pole of a dip-meter needle pointed downward, 59° below the horizontal. The

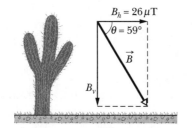

Fig. 32-29 Sample Problem 32-5. Earth's magnetic field, along with its horizontal and vertical components, at Tucson, Arizona, in 1964.

horizontal component B_h of Earth's magnetic field $\vec{B}$ in Tucson had a magnitude of 26 μT. What was the magnitude B of the field then, in units of gauss? (Earth's field is often measured in gauss.)

SOLUTION: The **Key Ideas** here are these:

1. The compass, held level, indicates the *horizontal* direction to the magnetic north pole; let that direction and a vertical line (at the location of the compass) define a vertical plane.

2. The dip meter, positioned in that vertical plane, then indicates that the direction of the magnetic field $\vec{B}$ is 59° below the horizontal, as shown in Fig. 32-29.

From the figure we then have

$$B = \frac{B_h}{\cos \theta} = \frac{26 \ \mu\text{T}}{\cos 59°} = 50 \ \mu\text{T} = 0.50 \text{ gauss.} \quad \text{(Answer)}$$

Sample Problem 32-6

If the compass needle of Sample Problem 32-2 is jarred slightly from its (horizontal) north–south equilibrium position, it oscillates about that position. If the period of oscillation is 2.2 s, what is the horizontal component of the local magnetic field?

SOLUTION: One **Key Idea** here is that the compass needle is a magnet with a magnetic dipole moment $\vec{\mu}$, which is directed along the needle's length, from its south pole to its north pole. A second **Key Idea** is that when the horizontal needle is jarred from its equilibrium position, Earth's magnetic field $\vec{B}$ produces a torque on the needle, about the needle's pivot axis. However, because the needle is free to rotate only horizontally, only the horizontal component B_h of Earth's field produces a torque to rotate the needle back toward its equilibrium position. From Eq. 29-36 ($\tau = \mu B \sin \theta$), we can write this torque as

$$\tau = -\mu B_h \sin \theta, \quad (32\text{-}46)$$

in which the minus sign indicates that τ opposes the angular displacement θ. Because the rotation angle is small, we may write $\sin \theta \approx \theta$ so that

$$\tau = -\mu B_h \theta. \quad (32\text{-}47)$$

Because μ and B_h are both constant, Eq. 32-47 tells us that the restoring torque is proportional to the negative of the angular displacement. This kind of relation is the hallmark of angular simple harmonic motion, as we saw in Section 16-5. From Eqs. 16-22 and 16-23, the period of oscillation may then be written as

$$T = 2\pi \sqrt{\frac{I}{\mu B_h}},$$

which yields

$$B_h = \frac{I}{\mu} \left(\frac{2\pi}{T} \right)^2, \quad (32\text{-}48)$$

where I is the rotational inertia of the needle. Approximating the needle as being a uniform thin rod, we use Table 11-2e to find

$$I = \frac{mL^2}{12} = \frac{(1.185 \times 10^{-4} \text{ kg})(0.030 \text{ m})^2}{12}$$

$$= 8.888 \times 10^{-9} \text{ kg} \cdot \text{m}^2.$$

Substituting this value, the value we obtained for μ, and the given value for T into Eq. 32-48, we find

$$B_h = \frac{8.888 \times 10^{-9} \text{ kg} \cdot \text{m}^2}{2.682 \times 10^{-3} \text{ J/T}} \left(\frac{2\pi}{2.2 \text{ s}}\right)^2$$

$$= 2.7 \times 10^{-5} \text{ T}, \qquad \text{(Answer)}$$

which is approximately the value we used in Sample Problem 32-5 for Tucson. Thus, even with an inexpensive compass, we can measure a local magnetic field by jarring the needle and timing its oscillations.

Sample Problem 32-7

An area $A = 0.020$ m^2 lies in an xy plane and is fully within a uniform electric field $\vec{E}$ directed along the z axis. The component E of the field along the z axis varies sinusoidally with time t as shown in Fig. 32-30.

(a) Find an expression for the displacement current $i_d(t)$ through area A.

SOLUTION: One **Key Idea** here is that the displacement current through area A is related to the rate $d\Phi_E/dt$ at which the electric flux through the area changes with time, according to Eq. 32-34 ($i_d = \varepsilon_0 d\Phi_E/dt$). A second **Key Idea** is that, because the electric field is uniform and directed perpendicular to area A, the electric flux Φ_E through A is EA. Thus, Eq. 32-34 gives us

$$i_d = \varepsilon_0 \frac{d\Phi_E}{dt} = \varepsilon_0 \frac{d(EA)}{dt}.$$

Here $E(t)$ varies with time and A is constant, so we can write

$$i_d = \varepsilon_0 A \frac{dE}{dt}. \qquad (32\text{-}49)$$

To evaluate dE/dt, we can first write the sinusoidal variation of $E(t)$ in the style of the simple harmonic motion of Eq. 16-3:

$$E = E_m \cos(\omega t + \phi), \qquad (32\text{-}50)$$

where E_m is the oscillation amplitude, ω is the angular frequency, and ϕ is the phase constant. From Fig. 32-30, we see that $\phi = 0$ (because $E(t)$ has its maximum positive value at time $t = 0$). We don't know ω, but we can easily measure the period T of the motion

in Fig. 32-30; so we use Eq. 16-5 ($\omega = 2\pi/T$) to replace ω in Eq. 32-50. With these two changes, Eq. 32-50 becomes

$$E = E_m \cos\frac{2\pi}{T}t. \qquad (32\text{-}51)$$

Substituting this into Eq. 32-49, we have

$$i_d = \varepsilon_0 A \frac{d}{dt}\left(E_m \cos\frac{2\pi}{T}t\right)$$

$$= -\frac{2\pi\varepsilon_0 A E_m}{T}\sin\frac{2\pi}{T}t. \qquad (32\text{-}52)$$

From Fig. 32-30, we see that $E_m = 800$ kN/C (or 800×10^3 N/C) and $T = 4.0$ μs (or 4.0×10^{-6} s). Substituting these values and the given $A = 0.020$ m^2 into Eq. 32-52, we find that

$$i_d = -(0.222 \text{ A}) \sin[(1.5708 \times 10^6 \text{ s}^{-1})t]$$

$$\approx -(220 \text{ mA}) \sin[(1.6 \times 10^6 \text{ s}^{-1})t]. \qquad \text{(Answer)} \quad (32\text{-}53)$$

(b) For $t \geq 0$, at what times is the displacement current i_d equal to zero?

SOLUTION: A **Key Idea** here is that the displacement current i_d is directly related (via Eq. 32-49) to the rate dE/dt at which the electric field changes. Thus, i_d must be zero when dE/dt is zero. The value of dE/dt is the slope of the curve in Fig. 32-30; that slope is zero at times $t = 0$ μs, 2 μs, 4 μs, ... Thus, for $t \geq 0$,

$$i_d = 0 \qquad \text{at } t = 0 \text{ } \mu\text{s, 2 } \mu\text{s, 4}\mu\text{s, ...} \quad \text{(Answer)}$$

We can also answer this question by using Eq. 32-53. From it, we can see that, for $t \geq 0$, the displacement current i_d is zero when

$$\sin(1.5708 \times 10^6 \text{ s}^{-1})t = 0.$$

This condition occurs when

$$(1.5708 \times 10^6 \text{ s}^{-1})t = n\pi \qquad \text{where } n = 0, 1, 2, \text{ . . . .}$$

Solving for t and substituting the values of n then give us

$$t = \frac{n\pi}{1.5708 \times 10^6 \text{ s}^{-1}} = 0 \text{ } \mu\text{s, 2 } \mu\text{s, 4 } \mu\text{s, ...} \quad \text{(Answer)}$$

From Fig. 32-30 or Eq. 32-53, you can also show that i_d has its maximum magnitude at times $t = 1$ μs, 3 μs, 5 μs, ...

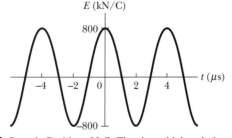

Fig. 32-30 Sample Problem 32-7. The sinusoidal variation of a uniform electric field $\vec{E}$ in a certain region.

QUESTIONS

12. Figure 32-31 shows four steel bars; three are permanent magnets. One of the poles is indicated. Through experiment we find that ends a and d attract each other, ends c and f repel, ends e and h attract, and ends a and h attract. (a) Which ends are north poles? (b) Which bar is not a magnet?

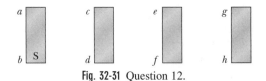

Fig. 32-31 Question 12.

13. In Figure 32-14b, $\vec{E}$ is directed into the page and is increasing in magnitude. Is the direction of the magnetic field $\vec{B}$ clockwise or counterclockwise if, instead, $\vec{E}$ is out of the page and (a) increasing and (b) decreasing? (c) What is the direction of $\vec{B}$ if $\vec{E}$ is out of the page and not changing?

14. Figure 32-32 shows a face-on view of one of the two square plates of a parallel-plate capacitor, as well as four loops that are located between the plates. The capacitor is being discharged. (a) Neglecting fringing of the magnetic field, rank the loops according to the magnitude of $\oint \vec{B} \cdot d\vec{s}$ along them, greatest first. (b) Along which loop, if any, is the angle between the directions of $\vec{B}$ and $d\vec{s}$ constant (so that their dot product can easily be evaluated)? (c) Along which loop, if any, is B constant (so that B can be brought in front of the integral sign in Eq. 32-27)?

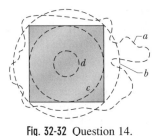

Fig. 32-32 Question 14.

15. In Fig. 32-33, three orientations of a compass needle in a uniform magnetic field $\vec{B}$ are given, with the north and south ends of the needle labeled. The first two orientations involve the same angle θ_0. The needle can pivot about its mounting pin at its midpoint. (a) Rank the orientations according to the torque on the needle due to the magnetic field, greatest first. (b) For each orientation, the needle is released from rest. Rank the orientations according to the maximum kinetic energy the needle has during its rotation (assuming that the friction between needle and mounting pin is negligible), greatest first.

16. Figure 32-34 shows a section of a long straight wire through which there is a uniform current i in the positive direction of an x axis. Point a is just above the wire, in the xy plane. Assume that the current is constant. (a) What is the direction of the magnetic field $\vec{B}$ at point a due to current i? (b) What is the direction of the electric field $\vec{E}$ responsible for the current? (c) Is the electric field constant or is it varying with time?

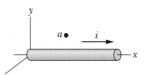

Fig. 32-34 Question 16.

Now assume that current i is increasing. (d) Is the magnitude of the electric field through the wire increasing, decreasing, or remaining the same? (e) What is the direction of $d\vec{E}/dt$? (f) What is the direction of the displacement current i_d associated with $d\vec{E}/dt$? (g) What is the direction of the magnetic field $\vec{B}_d$ at point a due to i_d, and is this direction the same as or opposite to that of the magnetic field $\vec{B}$ due to current i?

Finally, assume that current i is decreasing. (h) Are the magnetic fields at point a due to i and i_d in the same direction or in opposite directions?

17. *Organizing question:* Figure 32-35 shows a circular area of radius R in which an electric flux Φ_E is directed out of the page. It also shows two integration paths (of radii r_1 and r_2) that form circles concentric with the circular region. Set up the right side of the Ampere–Maxwell law (Eq. 32-29) for paths 1 and 2 for the following situations, where either total flux Φ_E, the flux $\Phi_{E,\text{enc}}$ encircled by a path of radius r, or the magnitude E of the associated electric field is given. Perform any differentiation that may be required, but not any integration (and be wary of pesky subscripts).

Situation	Flux or Field
a	$\Phi_E = (4 \text{ V} \cdot \text{m/s})t$, uniform
b	$E = (2 \text{ V/m} \cdot \text{s})t$, uniform
c	$E = (3 \text{ V/m})\left(\dfrac{r}{R}\right)$
d	$\Phi_{E,\text{enc}} = (5 \text{ V} \cdot \text{m/s})\left(\dfrac{r}{R}\right)t$
e	$E = (4 \text{ V/m} \cdot \text{s})\left(1 - \dfrac{r}{R}\right)t$

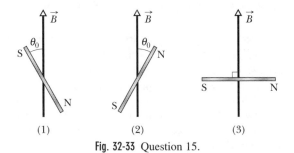

Fig. 32-33 Question 15.

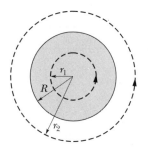

Fig. 32-35 Questions 17 and 21.

18. The Curie temperature of iron is 1043 K. If, instead, it were (a) 380 K or (b) 200 K, would VCRs and audiotape players work?

19. When three magnetic dipoles are near each other, they each experience the magnetic field of the other two, and the three-dipole system has a certain potential energy. Figure 32-36 shows two arrangements

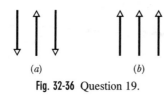

Fig. 32-36 Question 19.

in which three magnetic dipoles are side by side. Each dipole has the same magnitude of magnetic dipole moment, and the spacings between adjacent dipoles are equal. In which arrangement is the potential energy of the three-dipole system greater?

20. Figure 32-37 represents three rectangular samples of a ferromagnetic material in which the magnetic dipoles of the domains have been directed out of the page (encircled dot) by a very strong applied field B_0. In each sample, an island domain still has its magnetic field directed into the page (encircled ×). Sample 1 is one (pure) crystal. The other samples contain impurities collected along lines; domains cannot easily spread across such lines.

The applied field is now to be reversed and its magnitude kept moderate. The change causes the island domain to grow. (a) Rank the three samples according to the success of that growth, greatest growth first. Ferromagnetic materials in which the magnetic dipoles are easily changed are said to be *magnetically soft;* when the changes are difficult, requiring strong applied fields, the materials are said to be *magnetically hard.* (b) Of the three samples, which is the most magnetically hard?

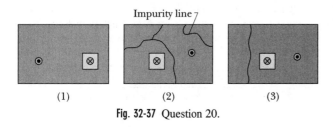

Fig. 32-37 Question 20.

21. *Organizing question:* Figure 32-35 shows a circular area of radius R in which a displacement current i_d is directed out of the page. It also shows two integration paths (of radii r_1 and r_2) that form circles concentric with the circular region. Set up the right side of the Ampere–Maxwell law (Eq. 32-29) for paths 1 and 2 for the following situations, where either the total displacement current i_d, the displacement current $i_{d,enc}$ encircled by a path of radius r, or the displacement current density J_d is given. (Be alert to irksome subscripts.)

Situation	Current or Current Density
a	$i_d = 2$ A, uniform
b	$J_d = 3$ A/m^2, uniform
c	$i_{d,enc} = (5\ \text{A})\left(\dfrac{r}{R}\right)$
d	$J_d = (6\ \text{A/m}^2)\left(1 - \dfrac{r}{R}\right)$

DISCUSSION QUESTIONS

22. Two iron bars appear identical. One is a magnet and one is not. How can you tell them apart? You are not permitted to suspend either bar as a compass needle or to use any other apparatus.

23. Two iron bars always attract, no matter the combination in which their ends are brought near each other. Can you conclude that one of the bars must be unmagnetized?

24. How can you determine the polarity of an unlabeled magnet?

25. Must all permanent magnets have north and south poles? Consider shapes other than the bar or horseshoe magnet, like the circular refrigerator magnet.

26. Starting with dipoles $\vec{a}$ and $\vec{e}$ in the positions and orientations shown in Fig. 32-38, with $\vec{a}$ fixed but $\vec{e}$ free to rotate, what happens (a) if $\vec{a}$ is an electric dipole and $\vec{e}$ is a magnetic dipole, (b) if $\vec{a}$ and $\vec{e}$ are both magnetic dipoles, (c) if $\vec{a}$ and $\vec{e}$ are both electric dipoles? Answer the same questions for $\vec{e}$ fixed and $\vec{a}$ free to rotate.

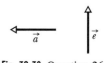

Fig. 32-38 Question 26.

27. How might the magnetic dipole moment of Earth be measured?

28. Give three reasons for believing that the flux Φ_B of Earth's magnetic field is greater through the boundaries of Alaska than through those of Texas.

29. You are a manufacturer of compasses. (a) Describe ways in which you might magnetize the needles. (b) The end of the needle that points north is usually painted a characteristic color. Without suspending the needle in Earth's field, how might you find out which end of the needle to paint? (c) Is the painted end a north or a south magnetic pole?

30. Would you expect the magnetization at saturation for a paramagnetic substance to be very much different from that for a saturated ferromagnetic substance of about the same size? Why or why not?

31. The magnetization induced in a given diamagnetic sphere by a given external magnetic field does not vary with temperature, in contrast to the situation in paramagnetism. Explain this behavior in terms of our discussion of the origin of diamagnetism.

32. Explain why a magnet attracts an unmagnetized iron object such as a nail.

33. Does any net force or torque act on (a) an unmagnetized iron bar or (b) a permanent bar magnet when either is placed in a uniform magnetic field?

34. A nail is placed at rest on a smooth tabletop near a strong magnet. It is released and is attracted to the magnet. What is the source of the kinetic energy the nail has just before it strikes the magnet?

35. Compare the magnetization curves for a paramagnetic substance (Fig. 32-9) and for a ferromagnetic substance (Fig. 32-11). What would a similar curve for a diamagnetic substance look like?

36. A "friend" borrows your favorite compass and paints the entire needle red. You discover this when you are lost in a cave and have with you two flashlights, a few meters of wire, and (of course) this book. How might you discover which end of your compass needle is the north-seeking end?

37. A Rowland ring is being supplied with a constant current. What happens to the magnetic induction in the ring if a small slot is cut out of the ring, leaving an air gap?

38. How could you magnetize an iron bar if Earth were the only magnet around?

39. How would you go about shielding a certain volume of space from constant external magnetic fields? If you think it can't be done, explain why.

40. In your own words, explain why Faraday's law of induction can be interpreted by saying: "A changing magnetic field generates an electric field."

41. If a uniform flux Φ_E through a plane circular ring decreases with time, is the induced magnetic field (as viewed along the direction of $\vec{E}$) clockwise or counterclockwise?

42. Why is it so easy to show that "a changing magnetic field produces an electric field" but so hard to show in a simple way that "a changing electric field produces a magnetic field"?

43. In Fig. 32-14b consider a circle with radius $r > R$. How can a magnetic field be induced around this circle, as computed in Sample Problem 32-3c? After all, there is no electric field at the location of this circle and $dE/dt = 0$ there.

44. In Fig. 32-14a,b what is the direction of the displacement current i_d? In this same figure, can you find a rule relating the directions (a) of $\vec{B}$ and $\vec{E}$ and (b) of $\vec{B}$ and $d\vec{E}/dt$?

45. What advantages are there in calling $\varepsilon_0 d\Phi_E/dt$ in the Ampere–Maxwell law a displacement current?

46. Can a displacement current be measured with an ammeter? Explain.

47. Why are the magnetic effects of conduction currents in wires so easy to detect but the magnetic effects of displacement currents in capacitors so hard to detect?

48. In Table 32-1 there are three kinds of apparent lack of symmetry in Maxwell's equations. (a) The quantities ε_0 and/or μ_0 appear in the first and fourth equations but not in the second and third. (b) There is a minus sign in the third but no minus sign in the fourth. (c) There are missing "magnetic pole terms" in the second and third. Which of these represent a genuine lack of symmetry? If magnetic monopoles were discovered, how would you rewrite these equations to include them? (Hint: Let p be the magnetic pole strength.)

EXERCISES & PROBLEMS

39. A silver wire has resistivity $\rho = 1.62 \times 10^{-8}\ \Omega \cdot m$ and a cross-sectional area of 5.00 mm². The current in the wire is uniform and changing at the rate of 2000 A/s when the current is 100 A. (a) What is the (uniform) electric field in the wire when the current in the wire is 100 A? (b) What is the displacement current in the wire at that time? (c) What is the ratio of the magnetic field due to the displacement current to that due to the current at a distance r from the wire?

40. A parallel-plate capacitor with circular plates of radius R is being discharged by a current of 6.0 A. (a) At what distances from the central axis is the induced magnetic field equal to 75% of the maximum value of that field? (b) What is the maximum value of that field if $R = 0.040$ m?

41. The capacitor in Fig. 32-39 with circular plates of radius $R = 18.0$ cm is connected to a source of emf $\mathcal{E} = \mathcal{E}_m \sin \omega t$, where $\mathcal{E}_m = 220$ V and $\omega = 130$ rad/s. The maximum value of the displacement current is $i_d = 7.60\ \mu A$. Neglect fringing of the electric field at the edges of the plates. (a) What is the maximum value of the current i in the circuit? (b) What is the maximum value of $d\Phi_E/dt$, where Φ_E is the electric flux through the region between the plates? (c) What is the separation d between the plates? (d) Find the maximum value of the magnitude of $\vec{B}$ between the plates at a distance $r = 11.0$ cm from the center.

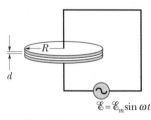

Fig. 32-39 Problem 41.

42. Two wires, parallel to the z axis and a distance 4r apart, carry equal currents i in opposite directions, as shown in Fig. 32-40. A circular cylinder of radius r and length L has its axis on the z axis, midway between the wires. Use Gauss' law for magnetism to calculate the net outward magnetic flux through the half of the cylindrical surface above the x axis. (Hint: Find the flux through that portion of the xz plane that is within the cylinder.)

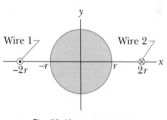

Fig. 32-40 Problem 42.

43. A sample of the paramagnetic salt to which the magnetization curve of Fig. 32-9 applies is immersed in a uniform magnetic field of 2.0 T. At what temperature will the degree of magnetic saturation of the sample be (a) 50% and (b) 90%?

44. If an electron in an atom has orbital angular momentum with m_l values limited by ± 3, how many values of (a) $L_{orb,z}$ and (b) $\mu_{orb,z}$ can it have? In terms of h, m, and e, what are the greatest and least allowed magnitudes for (c) $L_{orb,z}$ and (d) $\mu_{orb,z}$? (e) What is the greatest allowed magnitude for the z component of its *net* angular momentum (orbital plus spin)? (f) How many values (signs included) are allowed for the z component of its net angular momentum?

45. A parallel-plate capacitor, with circular plates of radius $R = 16$ mm and gap width $d = 5.0$ mm, has a uniform electric field between the plates. Starting at time $t = 0$, the potential difference between the plates is $V = (100\ V)e^{-t/\tau}$, where the time constant

$\tau = 12$ ms. At the radial distance $r = 0.80R$ from the capacitor axis within the gap, what is the magnetic field (a) as a function of time for $t \geq 0$ and (b) at time $t = 3\tau$?

46. A magnetic flux of 7.0 mW is directed outward through the flat bottom face of the closed surface shown in Fig. 32-41. Along the flat top face (which has a radius of 4.2 cm) there is a magnetic field $\vec{B}$ of magnitude 0.40 T directed perpendicular to the face. What is the magnetic flux (magnitude and direction) through the curved part of the surface?

Fig. 32-41 Problem 46.

47. A parallel-plate capacitor with circular plates is being charged. Consider a circular loop centered on the central axis and located between the plates. If the loop's radius of 3.00 cm is greater than the plate radius, then what is the displacement current between the plates when the magnetic field along the loop has magnitude 2.00 μT?

48. Figure 32-42 gives the variation of an electric field that is perpendicular to a circular area of 2.0 m². During the time period shown, what is the greatest displacement current through the area?

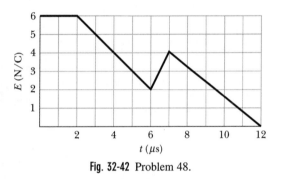

Fig. 32-42 Problem 48.

49. Consider a solid containing N atoms per unit volume, each atom having a magnetic dipole moment $\vec{\mu}$. Suppose the direction of $\vec{\mu}$ can be only parallel or antiparallel to an externally applied magnetic field $\vec{B}$ (this will be the case if $\vec{\mu}$ is due to the spin of a single electron). According to statistical mechanics, the probability of an atom being in a state with energy U is proportional to $e^{-U/kT}$, where T is the temperature and k is Boltzmann's constant. Thus, because energy U is $-\vec{\mu} \cdot \vec{B}$, the fraction of atoms whose dipole moment is parallel to $\vec{B}$ is proportional to $e^{\mu B/kT}$ and the fraction of atoms whose dipole moment is antiparallel to $\vec{B}$ is proportional to $e^{-\mu B/kT}$. (a) Show that the magnetization of this solid is $M = N\mu \tanh(\mu B/kT)$. Here tanh is the hyperbolic tangent function: $\tanh(x) = (e^x - e^{-x})/(e^x + e^{-x})$. (b) Show that the result given in (a) reduces to $M = N\mu^2 B/kT$ for $\mu B \ll kT$. (c) Show that the result of (a) reduces to $M = N\mu$ for $\mu B \gg kT$. (d) Show that both (b) and (c) agree qualitatively with Fig. 32-9.

50. In the lowest energy state of the hydrogen atom, the most probable distance of the single electron from the central proton (the nucleus) is 5.2×10^{-11} m. (a) Compute the magnitude of the proton's electric field at that distance. The component $\mu_{s,z}$ of the proton's spin magnetic dipole moment measured on a z axis is 1.4×10^{-26} J/T. (b) Compute the magnitude of the proton's magnetic field at the distance 5.2×10^{-11} m on the z axis. (*Hint:* Use Eq. 30-29.) (c) What is the ratio of the spin magnetic dipole moment of the electron to that of the proton?

51. A magnetic compass has its needle of mass 0.050 kg and length 4.0 cm aligned with the horizontal component of Earth's magnetic field at a place where that component has the value $B_h = 16$ μT. After the compass is given a momentary gentle shake, the needle oscillates with angular frequency $\omega = 45$ rad/s. Assuming that the needle is a uniform thin rod mounted at its center, find its magnetic dipole moment.

52. A parallel-plate capacitor with circular plates of radius R is being discharged. The displacement current through a central circular area, parallel to the plates and with radius $R/2$, is 2.0 A. What is the discharging current?

53. What are the measured components of the orbital magnetic dipole moment of an electron with (a) $m_l = 3$ and (b) $m_l = -4$?

54. Figure 32-43 shows two charged plates with plate area 4.0×10^{-2} m². The electric field $\vec{E}$ between them is drawn for time $t = 0$, and the field magnitude is given by

$$E = 4.0 \times 10^5 \text{ V/m} - (6.0 \times 10^4 \text{ V/m} \cdot \text{s})t,$$

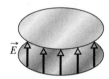
Fig. 32-43 Problem 54.

with positive values corresponding to a field up the page and negative values down the page. In the figure, is the induced magnetic field clockwise or counterclockwise at (a) $t = 0$ and (b) $t = 100$ s? What is the magnitude of the displacement current between the plates at (c) $t = 0$ and (d) $t = 100$ s?

55. Earth has a magnetic dipole moment of 8.0×10^{22} J/T. (a) What current would have to be produced in a single turn of wire extending around Earth at its geomagnetic equator if we wished to set up such a dipole? Could such an arrangement be used to cancel out Earth's magnetism (b) at points in space well above Earth's surface or (c) on Earth's surface?

56. A charge q is distributed uniformly around a thin ring of radius r. The ring is rotating about an axis through its center and perpendicular to its plane, at an angular speed ω. (a) Show that the magnetic moment due to the rotating charge is

$$\mu = \tfrac{1}{2}q\omega r^2.$$

(b) What is the direction of this magnetic moment if the charge is positive?

57. Figure 32-44a is a one-axis graph along which two of the allowed energy values (*levels*) of an atom are plotted. When the atom

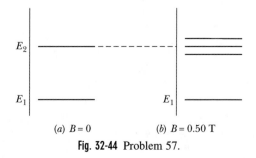
Fig. 32-44 Problem 57.

is placed in a magnetic field of 0.50 T, the graph changes to that of Fig. 32-44b because of the energy associated with $\vec{\mu}_{orb} \cdot \vec{B}$. (We neglect $\vec{\mu}_s$.) Level E_1 is unchanged, but level E_2 splits into a (closely spaced) triplet of levels. What are the allowed values of m_l associated with (a) energy level E_1 and (b) energy level E_2? (c) On the graph, what is the spacing between the triplet levels?

58. A Rowland ring is formed of ferromagnetic material. It is circular in cross section, with an inner radius of 5.0 cm and an outer radius of 6.0 cm, and is wound with 400 turns of wire. (a) What current must be set up in the windings to attain a toroidal field $B_0 = 0.20$ mT? (b) A secondary coil wound around the toroid has 50 turns and resistance 8.0 Ω. If, for this value of B_0, we have $B_M = 800B_0$, how much charge moves through the secondary coil when the current in the toroid windings is turned on?

59. A magnetic field $\vec{B}$ is directed along a z axis. The energy difference between parallel and antiparallel alignments of the z component of an electron's spin magnetic moment with $\vec{B}$ is 6.0×10^{-25} J. What is the magnitude of $\vec{B}$?

60. In Fig. 32-45, a parallel-plate capacitor is being discharged by a current $i = 5.0$ A. The plates are square with edge length $L = 8.0$ mm. (a) What is the rate at which the electric field between the plates is changing? (b) What is the value of $\int \vec{B} \cdot d\vec{s}$ around the dashed path, where $H = 2.0$ mm and $W = 3.0$ mm?

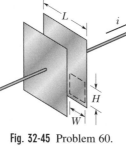

Fig. 32-45 Problem 60.

Tutorial Problem

61. Maxwell's equations are written in the textbook in what is called their *integral form,* in which line and surface integrals of electric and magnetic fields are related to various quantities enclosed by associated surfaces and volumes, respectively. Maxwell's equations can also be written at single points in space in what is called their *differential form,* which involves partial derivatives. In a vacuum these equations are:

Gauss' law: $\vec{\nabla} \cdot \vec{E} = 0$

Gauss' law for magnetism: $\vec{\nabla} \cdot \vec{B} = 0$

Faraday's law: $\vec{\nabla} \times \vec{E} = -\dfrac{\partial \vec{B}}{\partial t}$

Ampere–Maxwell law: $\vec{\nabla} \times \vec{B} = \mu_0 \varepsilon_0 \dfrac{\partial \vec{E}}{\partial t}$

In these expressions, $\vec{\nabla}$ (*del,* or *nabla*) represents the vector differential operator

$$\vec{\nabla} \equiv \hat{i}\left(\frac{\partial}{\partial x}\right) + \hat{j}\left(\frac{\partial}{\partial y}\right) + \hat{k}\left(\frac{\partial}{\partial z}\right),$$

which is treated just like a vector but has to have some function on the right side on which it can operate. For example, its dot product with a vector is a scalar function called the *divergence* (abbreviated div):

$$\text{div } \vec{B} \equiv \vec{\nabla} \cdot \vec{B} = \frac{\partial B_x}{\partial x} + \frac{\partial B_y}{\partial y} + \frac{\partial B_z}{\partial z}.$$

By Gauss' law for magnetism, this quantity must be zero for any real magnetic field, which greatly restricts the possible functional form of the magnetic field.

Similarly, the vector cross product of nabla with a vector gives another vector called the curl of that vector:

$$\text{curl } \vec{E} \equiv \vec{\nabla} \times \vec{E}$$
$$= \hat{i}\left(\frac{\partial E_z}{\partial y} - \frac{\partial E_y}{\partial z}\right) + \hat{j}\left(\frac{\partial E_x}{\partial z} - \frac{\partial E_z}{\partial x}\right) + \hat{k}\left(\frac{\partial E_y}{\partial x} - \frac{\partial E_x}{\partial y}\right).$$

Thus, the differential form of Faraday's law says that the curl of the electric field vector must equal the negative of the partial time derivative of the magnetic field. This is equivalent to three separate equations for the three Cartesian coordinates; for example, the equation for the x components in Faraday's law gives

$$\left(\frac{\partial E_z}{\partial y} - \frac{\partial E_y}{\partial z}\right) = -\frac{\partial B_x}{\partial t}.$$

(a) Maxwell's treatise on electromagnetism was written before vectors were used in physics, so he had more than four equations. When written in terms of components, how many separate (scalar) differential equations are there in Maxwell's equations?

Now let's consider a plane electromagnetic wave moving in the positive x direction with speed of light c. That the wave is a plane wave means that its behavior (its electric and magnetic fields) is the same at all points in the yz plane. In other words, the electric and magnetic fields have no dependence on y or z, only on x and t. Thus, we expect the electric and magnetic fields to be expressible as $\vec{E}(x, t)$ and $\vec{B}(x, t)$. (b) Explain why the electric and magnetic fields of this plane electromagnetic wave must actually be functions of $x - ct$.

Now let's take the plane electromagnetic wave to be a sinusoidal wave with angular wave number k. Let's write its electric and magnetic fields in the form

$$\vec{E}(x - ct) = (E_{x0}\hat{i} + E_{y0}\hat{j} + E_{z0}\hat{k}) \sin k(x - ct)$$

and $\quad \vec{B}(x - ct) = (B_{x0}\hat{i} + B_{y0}\hat{j} + B_{z0}\hat{k}) \sin k(x - ct + \phi),$

where we are allowing the possibilities that both fields have components along all three axes and the magnetic field has a different phase constant. (c) Explain how Gauss' law predicts that the electric field of the wave is transverse (in other words, E_{x0} is zero) and how Gauss' law for magnetism predicts that the magnetic field of the wave is transverse.

From here on, suppose that $E_{z0} = 0$. (d) Explain how our freedom to choose a coordinate system makes this possible, without loss of generality. (e) Apply Faraday's law to show that (1) the magnetic field must be perpendicular to the electric field (i.e., $B_{y0} = 0$), (2) there is no phase difference between the electric and magnetic fields (i.e., $\phi = 0$), and (3) the magnetic and electric field magnitudes are related by $B_{z0} = E_{y0}/c$.

If you have successfully proved the results of previous parts, you should now have

$$\vec{E}(x - ct) = [E_{y0} \sin k(x - ct)] \hat{j}$$

and
$$\vec{B}(x - ct) = \left[\left(\frac{E_{y0}}{c}\right)\sin k(x - ct)\right]\hat{k}.$$

(f) Substitute these expressions into the Ampere–Maxwell law. What new equation results from this substitution?

Which of Maxwell's equations can be used to prove the following characteristics of the plane electromagnetic wave: (g) The electric field is transverse, (h) the square of the speed of propagation of the waves is equal to $1/\mu_0\varepsilon_0$, (i) there is no phase difference between the electric and magnetic fields, (j) the magnetic and electric field magnitudes are related by $cB = E$, (k) the magnetic field is transverse, (l) the electric and magnetic fields are perpendicular to each other? The grand finale! (m) Draw the coordinate axes for this wave, and sketch the appearance at $t = 0$ of the electric and magnetic field vectors along the x axis.

Answers

(a) There are eight: one each for Gauss' laws and three each for the other two.

(b) For the wave to be a wave propagating with speed c in the positive x direction, the wave function must be a function of $x - ct$ if the values of the fields at some point x are to have shifted to a new value of x that is greater by ct.

(c) Gauss' law states that, in a vacuum, $\partial E_x/\partial x + \partial E_y/\partial y + \partial E_z/\partial z = 0$. Since there is no y or z dependence to the wave here, this law simply reduces to $\partial E_x/\partial x = 0$, but that statement can be true only if $E_{x0} = 0$ (not a nonzero constant because that wouldn't be a wave).

Gauss' law for magnetism states that, in a vacuum, $\partial B_x/\partial x + \partial B_y/\partial y + \partial B_z/\partial z = 0$. Since there is no y or z dependence to the wave, this law reduces to $\partial B_x/\partial x = 0$, which can be true only if $B_{x0} = 0$.

Thus, the electric and magnetic fields can have components along only the y and z directions; in other words, they can have only components that are transverse to the direction of propagation of the wave, which is along x.

(d) Since we are free to choose any directions we want for two of our coordinate axes (the third is then fixed by our necessity to have a right-hand coordinate system), we can simply choose to have the y axis along the direction of the electric field. (We previously chose the x direction along the direction of propagation of the wave.) There is no loss of generality. This choice does not affect anything physical; it affects only the numerical values of the components. This choice obviously simplifies our expression for the electric field, since it has only one component. If we are lucky, it may not complicate the expression for the magnetic field.

(e) Faraday's law for an electric field in the y direction and a magnetic field in the y and z directions becomes (since there is no y or z dependence to the fields)
$$\hat{k}\left(\frac{\partial E_y}{\partial x}\right) = -\hat{j}\left(\frac{\partial B_y}{\partial t}\right) - \hat{k}\left(\frac{\partial B_z}{\partial t}\right).$$

(1) The y components of this vector equation tell us that $0 = \partial B_y/\partial t$, which can be true only with $B_{y0} = 0$. Thus, the magnetic field is only in the z direction; that is, it is perpendicular to the electric field (as well as to the direction of propagation). (2) The z components give $\partial E_y/\partial x = \partial B_z/\partial t$, or
$$kE_{y0}\cos k(x - ct) = -[-kcB_{z0}\cos k(x - ct - \phi)].$$

This equation can be satisfied at all times only if $\phi = 0$, which means that there is no phase difference between the electric and magnetic fields. (3) The equation then shows that
$$kE_{y0} = kcB_{z0},$$

which means that $E_{y0} = cB_{z0}$, or $B_{z0} = E_{y0}/c$.

(f) Substituting into the Ampere–Maxwell law, we find
$$\hat{j}\left(-\frac{\partial B_z}{\partial x}\right) = \mu_0\varepsilon_0\left(\frac{\partial E_y}{\partial t}\right)\hat{j}.$$

Thus, we have
$$-\frac{\partial B_z}{\partial x} = \mu_0\varepsilon_0\left(\frac{\partial E_y}{\partial t}\right),$$

which leads to
$$-k\left(\frac{E_{y0}}{c}\right)\cos k(x - ct) = \mu_0\varepsilon_0(-kcE_{y0})\cos k(x - ct),$$

in which most of the terms cancel out to give
$$\frac{1}{c} = \mu_0\varepsilon_0 c$$

or
$$c = \sqrt{\frac{1}{\mu_0\varepsilon_0}}.$$

This shows how the speed c of electromagnetic waves is related to the permittivity and permeability of a vacuum.

(g) Gauss' law (for electricity);

(h) Ampere–Maxwell law;

(i) Faraday's law of induction;

(j) Faraday's law of induction;

(k) Gauss' law for magnetism;

(l) Faraday's law of induction.

(m) See Fig. 34-5 in the textbook.

Electromagnetic Oscillations and Alternating Current

Sample Problem 33-10

An LC oscillator, with $L = 12$ mH and $C = 1.7$ μF, begins to oscillate at time $t = 0$, when the capacitor has its maximum charge. The resistance R in the circuit is negligible.

(a) What is the first time t_1 at which the electromagnetic energy of the oscillator is shared equally between the electric field of the capacitor and the magnetic field of the inductor?

SOLUTION: One **Key Idea** here is that because R is negligible, we can assume that the electromagnetic energy of the circuit is conserved as it is transferred back and forth between the capacitor and the inductor. A second **Key Idea** is that the energy $U_E(t)$ stored in the electric field at any time can be written as given in Eq. 33-16,

$$U_E = \frac{Q^2}{2C} \cos^2(\omega t + \phi). \tag{33-83}$$

Because the capacitor has its maximum charge at $t = 0$, its electric field is maximum then and so is U_E. This means that the cosine function in Eq. 33-83 is maximum at $t = 0$, and so ϕ must be zero. Substituting 0 for ϕ and replacing $Q^2/2C$ with $U_{E,\max}$, we can rewrite Eq. 33-83 as

$$U_E = U_{E,\max} \cos^2 \omega t. \tag{33-84}$$

We want time t_1 when the capacitor has half the total energy of the circuit—that is, when $U_E = 0.5U_{E,\max}$. Substituting this requirement into Eq. 33-84 leads to

$$\cos^2 \omega t_1 = 0.5$$

or

$$\cos \omega t_1 = \sqrt{0.5}.$$

Substituting for ω from Eq. 33-4 ($\omega = 1/\sqrt{LC}$) and then taking the inverse cosine of both sides yield

$$\frac{t_1}{\sqrt{LC}} = \cos^{-1} \sqrt{0.5} = 0.785 \text{ rad}. \tag{33-85}$$

Solving Eq. 33-85 for t_1, we then have

$$
\begin{aligned}
t_1 &= (0.785 \text{ rad})\sqrt{LC} \\
&= (0.785 \text{ rad})\sqrt{(0.012 \text{ H})(1.7 \times 10^{-6} \text{ F})} \\
&= 1.12 \times 10^{-4} \text{ s}. \quad \text{(Answer)}
\end{aligned}
$$

(b) At what next time t_2 is U_E again equal to $0.5U_{E,\max}$?

SOLUTION: The **Key Idea** here is that U_E reaches its maximum value every time the electric field in the capacitor reaches its maximum magnitude, which occurs twice during each full oscillation of the circuit (see Fig. 33-1). Thus, $U_{E,\max}$ occurs twice during the oscillation period T, or once every $T/2$. The half-maximum $0.5U_{E,\max}$ occurs as U_E is increasing to its maximum value and then again as U_E is decreasing from that value. Thus, $0.5U_{E,\max}$ occurs four times during the oscillation period T, or once every $\frac{1}{4}T$. If U_E reaches its first half-maximum at time t_1, then it will reach its next half-maximum at time

$$t_2 = t_1 + \tfrac{1}{4}T. \tag{33-86}$$

From Eq. 16-5 ($\omega = 2\pi/T$) and Eq. 33-4 ($\omega = 1/\sqrt{LC}$), we find

$$
\begin{aligned}
T &= 2\pi\sqrt{LC} \\
&= 2\pi\sqrt{(0.012 \text{ H})(1.7 \times 10^{-6} \text{ F})} \\
&= 8.974 \times 10^{-4} \text{ s}.
\end{aligned}
$$

If we then substitute this for T and 1.12×10^{-4} s for t_1, Eq. 33-86 becomes

$$
\begin{aligned}
t_2 &= 1.12 \times 10^{-4} \text{ s} + \tfrac{1}{4}(8.974 \times 10^{-4} \text{ s}) \\
&= 3.4 \times 10^{-4} \text{ s}. \quad \text{(Answer)}
\end{aligned}
$$

Tactic 2: *Memory Devices for RLC Resonance Curves*

A resonance curve like those in Fig. 33-13 is a rich source of information about driven series RLC circuits (in which a sinusoidal emf source drives a resistance, inductance, and capacitance that are connected in series, as in Fig. 33-7). The features of such a curve might be remembered better with Fig. 33-33, which shows a resonance curve transformed into Resonance Hill—that is, the hill

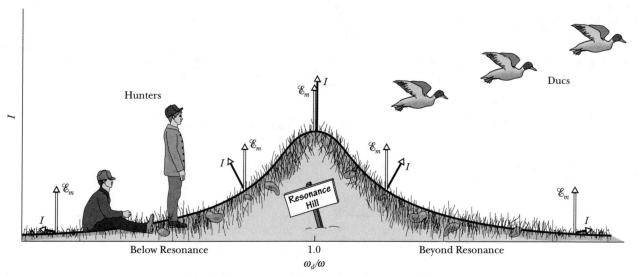

Fig. 33-33 Problem-Solving Tactic 2. Resonance Hill.

has the shape of a resonance curve and its peak is located at $\omega_d/\omega = 1.0$. The features of Resonance Hill are these:

1. "Ducs" (for inductance) are shown on the right side of the hill to indicate the side of a resonance curve for which circuits are more inductive than capacitive ($X_L > X_C$). The ducs are beyond the peak, that is, in the region where ω_d/ω is greater than 1.0 (where the driving angular frequency ω_d is greater than the natural angular frequency ω of the circuit).

2. Hunters, with way-cool L. L. Bean caps (for capacitance), are shown on the left side of the hill. They and their caps indicate the side of a resonance curve for which circuits are more capacitive than inductive ($X_C > X_L$). The hunters are below the peak—that is, in the region where ω_d/ω is less than 1.0 (where the driving angular frequency ω_d is less than the natural angular frequency ω of the circuit).

3. The ducs *rise* to the *right,* indicating that if the inductance of a circuit is *increased,* the point representing the circuit on a resonance curve moves to the *right.* Thus, with the increase in L, the ratio ω_d/ω becomes greater because ω_d is unchanged but $\omega (= 1/\sqrt{LC})$ is decreased.

4. The hunters also *rise* to the *right* (the standing hunter is to the right of the sitting hunter). This indicates that if the capacitance of a circuit is *increased,* the point representing the circuit on a resonance curve moves to the *right.* Thus, with the increase in C, the ratio ω_d/ω becomes greater because ω_d is unchanged but ω is decreased.

5. "Emf stalks" $\mathscr{E}_m$ grow directly upward on Resonance Hill and sprout "eye thorns" I at various angles. Their arrangements on the hill indicate the arrangements of $\mathscr{E}_m$ and I in phasor diagrams for various driven series *RLC* circuits, as in Fig. 33-12.

   Beyond (to the right of) the peak, the eye thorns sprout to the right, indicating that I lags $\mathscr{E}_m$ in a phasor diagram for a circuit represented in that region.

   Below (to the left of) the peak, the eye thorns sprout to the left, indicating that I leads $\mathscr{E}_m$ in a phasor diagram for a circuit represented in that region.

   At the peak, the eye thorn sprouts directly upward, indicating that I is aligned with $\mathscr{E}_m$ in a phasor diagram for a circuit at resonance.

6. The length of the eye thorns I is greatest at the peak of Resonance Hill (which is why the hunters are not there) and progressively less at greater distances from the peak. This indicates that the amplitude I of the current in a driven series *RLC* circuit is greatest when the circuit is at resonance, and progressively less the further the circuit is from resonance. Also, from Eq. 33-62 ($I = \mathscr{E}_m/Z$), we know that $Z = \mathscr{E}_m/I$. Thus, impedance Z is least when the circuit is at resonance, and progressively greater the further the circuit is from resonance.

7. The angle between an emf stalk $\mathscr{E}_m$ and its eye thorn I represents the phase constant ϕ of the current. That constant is positive on the right (positive) side of the hill, negative on the left side of the hill, and zero at the top of the hill. Also, the size of the phase constant ϕ is progressively greater the further a circuit is from the resonance peak. At great distances to the right from the peak, ϕ approaches $+90°$ (but cannot exceed that limiting value—the eye thorns cannot grow into the ground). Similarly, at great distances to the left from the peak, ϕ approaches $-90°$.

An example: Let us put Resonance Hill to work. If a particular series *RLC* circuit is driven with an angular frequency ω_d that is somewhat greater than its natural frequency ω, can you see the following from Fig. 33-33 without any calculation?

1. The circuit is represented by a point to the right side of the resonance-curve peak.

2. The circuit is more inductive than capacitive ($X_L > X_C$).

3. The current amplitude I is less than it would be if the circuit were at resonance, and the impedance Z of the circuit is greater than it would then be.

4. The current in the circuit lags the driving emf.

5. The phase constant ϕ for the current is positive and less than $+90°$.

Can you also see that if we increase either L or C (or both) in the circuit, the following occur?

1. The circuit moves farther to the right on the resonance curve and thus further from resonance.

2. The current amplitude I decreases, and the impedance Z increases.

3. The phase constant ϕ for the current becomes more positive (but still is less than $+90°$), and the current in the circuit lags the driving emf even more than it did previously.

Sample Problem 33-11

A series *RLC* circuit is driven by an alternating-current generator, as in Fig. 33-7. The amplitude $\mathcal{E}_m$ of the alternating emf is 12.0 V, and the resistance R is 40 Ω. At time $t = 0$, phasors $\mathcal{E}_m$ and I have the orientations shown in Fig. 33-34a. At time $t = 10.0$ ms, the phasors have rotated 120° and have the orientations shown in Fig. 33-34b.

(a) Is the circuit in resonance with the generator, more capacitive than inductive, or more inductive than capacitive? What is the phase constant ϕ of the current i in the circuit? Is the driving angular frequency ω_d equal to, less than, or greater than the natural angular frequency ω of the circuit?

SOLUTION: One **Key Idea** here is to apply the mnemonic "*ELI* positively is the *ICE* man" to the circuit. From Fig. 33-34a or b, we see that phasor I leads phasor $\mathcal{E}_m$ in their rotation. Thus, the circuit is an *ICE* circuit, which means that the circuit is more capacitive than inductive and that phase constant ϕ is negative. We see from Fig. 33-34 that the angle between phasors $\mathcal{E}_m$ and I is 30°. Thus, ϕ must be

$$\phi = -30° = -\pi/6 \text{ rad} = -0.524 \text{ rad.} \quad \text{(Answer)}$$

An alternative **Key Idea** is to apply Resonance Hill (Fig. 33-33). Since Fig. 33-34a or b indicates that phasor I leads phasor $\mathcal{E}_m$, we know the circuit is represented by a point on the left side of the hill. There the hunter "caps" tell us that the circuit is more capacitive than inductive. There, also, the ratio ω_d/ω is less than 1.0, so ω_d must be less than ω. Finally, on the left side of the hill, the phase constant is negative; thus, from Fig. 33-34, we again see that $\phi = -30° = -\pi/6$ rad.

(b) What is driving angular frequency ω_d?

SOLUTION: The **Key Idea** here is that the phasors shown in Fig. 33-34 rotate with an angular frequency equal to the driving angular frequency ω_d. Because the phasors rotate 120° (= $2\pi/3$ rad) in 10 ms (= 0.010 s), we have

$$\omega_d = \frac{2\pi/3 \text{ rad}}{0.010 \text{ s}} = 209 \text{ rad/s.} \quad \text{(Answer)}$$

(c) What is the current amplitude I?

SOLUTION: The **Key Idea** here is that the current amplitude I is related to the given emf amplitude $\mathcal{E}_m$ by Eq. 33-62,

$$I = \frac{\mathcal{E}_m}{Z}, \quad (33\text{-}87)$$

in which the impedance Z of the circuit is

$$Z = \sqrt{R^2 + (X_L - X_C)^2}. \quad (33\text{-}88)$$

Impedance X_L is given by Eq. 33-49 ($X_L = \omega_d L$), and impedance X_C is given by Eq. 33-39 ($X_C = 1/\omega_d C$). Because we do not know the inductance L and capacitance C, evaluating X_L and X_C individually is not possible, and thus Eq. 33-88 seems useless.

However, in a second **Key Idea** we note that only the difference $X_L - X_C$ of the impedances, and not the individual values, is needed in Eq. 33-88. Moreover, that difference also appears in Eq. 33-65 for the phase constant ϕ:

$$\tan \phi = \frac{X_L - X_C}{R}. \quad (33\text{-}89)$$

Here R is given as 40 Ω, and we know from part (a) that ϕ is $-30°$. Solving Eq. 33-89 for $X_L - X_C$ and substituting these values give us

$$X_L - X_C = R \tan \phi = (40 \ \Omega) \tan(-30°) = -23.09 \ \Omega.$$

Next, substituting this into Eq. 33-88 leads to

$$Z = \sqrt{(40 \ \Omega)^2 + (-23.09 \ \Omega)^2} = 46.186 \ \Omega.$$

Substituting this value into Eq. 33-87 now yields

$$I = \frac{12.0 \text{ V}}{46.186 \ \Omega} = 0.2598 \text{ A} \approx 260 \text{ mA.} \quad \text{(Answer)}$$

(d) Write an equation for the current $i(t)$ in the circuit as a function of time.

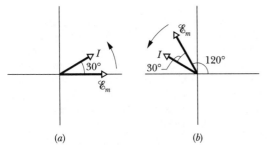

(a)   (b)

Fig. 33-34 Sample Problem 33-11. The emf $\mathcal{E}_m$ and current I phasors for a driven series *RLC* circuit at times (a) $t = 0$ and (b) $t = 10.0$ ms.

SOLUTION: The **Key Idea** here is that the current $i(t)$ has the form of Eq. 33-29:

$$i = I \sin(\omega_d t - \phi).$$

Substituting $I = 260$ mA from part (c), $\omega_d = 209$ rad/s from part (b), and $\phi = -0.524$ rad from part (a), we have

$$i = (260 \text{ mA}) \sin[(209 \text{ rad/s})t + 0.524 \text{ rad}].   \text{(Answer)}$$

QUESTIONS

13. The following table gives, for three series *RLC* circuits, the amplitude $\mathcal{E}_m$ of the driving emf and the values of R, L, and C. Without written calculation, rank the circuits according to (a) the amplitude I of the current at resonance and (b) the angular frequency at resonance, greatest first.

Circuit	$\mathcal{E}_m$ (V)	R (Ω)	L (mH)	C (μF)
1	25	5.0	200	10
2	60	12	100	5.0
3	80	10	300	10

14. Suppose that for a particular driving angular frequency, the emf leads the current in a series *RLC* circuit. If you decrease the driving angular frequency slightly, do (a) the phase constant and (b) the current amplitude increase, decrease, or remain the same?

15. The driving angular frequency in a certain series *RLC* circuit is less than the natural angular frequency of the circuit. (a) Is the phase constant ϕ positive, negative, or zero? (b) Does the current lead or lag the emf?

16. A system for transmitting electric power to residential users is diagrammed in Fig. 33-35. For each of the two transformers, determine (a) whether it should be a step-up or step-down transformer and (b) whether the number of turns in the secondary is greater than or less than that in the primary.

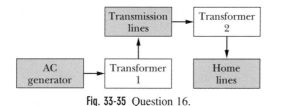

Fig. 33-35 Question 16.

17. (a) Does the phasor diagram of Fig. 33-36 correspond to an alternating emf source connected to a resistor, a capacitor, or an inductor? (b) If the angular speed of the phasors is increased, does the length of the current phasor increase or decrease when the scale of the diagram is maintained?

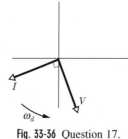

Fig. 33-36 Question 17.

18. (a) Does the graph of Fig. 33-37a correspond to an alternating emf source connected to a resistor, a capacitor, or an inductor? (b) If the driving frequency of the emf source is increased, which

one of the other four graphs in Fig. 33-37 (all to the same scale as Fig. 33-37a) then best represents the relation between $i(t)$ and $v(t)$? (c) Which one is the best representation if the driving frequency is decreased?

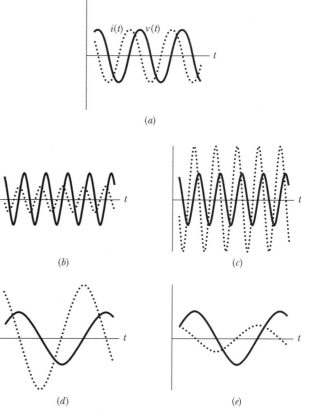

Fig. 33-37 Question 18.

19. Suppose that the graph of current i for a driven series *RLC* circuit is initially that in Fig. 33-12b. You wish to change the graph first to that in Fig. 33-12f and then to that in Fig. 33-12d by varying one of the following circuit parameters: (a) the driving frequency f_d, (b) the inductance L, (c) the capacitance C, or (d) the resistance R. For each parameter, determine whether varying that parameter can actually produce the desired change and, if yes, decide whether the value of the parameter should be increased or decreased to do so.

20. Figure 33-38 represents four stages in the rotation of a loop in a uniform magnetic field, as in Fig. 33-6, except that here field $\vec{B}$ is directed up the page. One side of the loop is drawn with a darker line so that we can follow the rotation. In (a) the loop is tipped downward toward us (the near side is below the horizontal); in (b) it is tipped upward (the near side is above the horizontal); in (c) it is tipped downward; and in (d) it is tipped upward. In each stage, is the direction of the current induced in the side drawn with a darker line rightward or leftward?

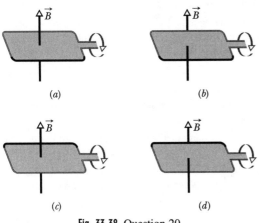

Fig. 33-38 Question 20.

21. Figure 33-39a gives the potential $v_L(t)$ across an inductor due to an alternating emf source, and Fig. 33-39b gives the phasor diagram for that potential at four times. (a) Which of the lettered points in Fig. 33-39a corresponds to which of the numbered phasor orientations in Fig. 33-39b? If the frequency of the emf source is increased without any change in the amplitude of the source's emf, do the following increase, decrease, or remain the same: (b) the amplitude of the potential across the inductor, (c) the rotation rate of the phasor for that potential, (d) the reactance of the inductor, and (e) the amplitude of the current through the inductor?

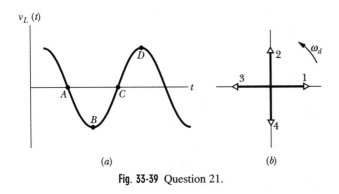

Fig. 33-39 Question 21.

22. Figure 33-40 shows the current i and driving emf $\mathscr{E}$ for four series RLC circuits having the same I_{rms} and the same $\mathscr{E}_{\text{rms}}$. The phase difference (in radians) between the two curves for each circuit is indicated. (a) Rank the circuits according to their power factors, greatest first. Then rank them according to the rates at which energy is dissipated in (b) the resistor, (c) the inductor, and (d) the capacitor, greatest first.

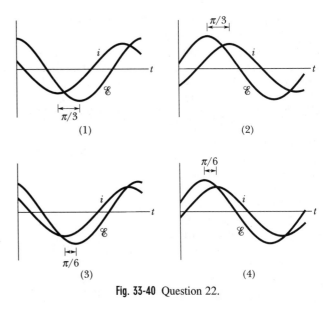

Fig. 33-40 Question 22.

23. For each of the curves of $q(t)$ in Fig. 33-41 for an LC circuit, determine the least positive phase constant ϕ required in Eq. 33-12 to produce the curve.

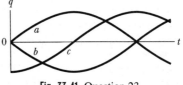

Fig. 33-41 Question 23.

24. What is the least positive phase constant ϕ in Eq. 33-17 that will give the curve of Fig. 33-42 for the magnetic field energy U_B in an LC circuit?

Fig. 33-42 Question 24.

DISCUSSION QUESTIONS

25. Why doesn't the LC circuit of Fig. 33-1 simply stop oscillating when the capacitor has been completely discharged?

26. How might you start an LC circuit into oscillation with its initial condition being represented by Fig. 33-1c? Devise a switching scheme to bring this about.

27. The lower curve (b) in Fig. 33-2 is proportional to the derivative of the upper curve (a). Explain why.

28. In an oscillating LC circuit, assumed resistanceless, what determines (a) the frequency and (b) the amplitude of the oscillations?

29. In connection with Figs. 33-1c and 33-1g, explain how there can be a current in the inductor even though there is no charge on the capacitor.

30. In Fig. 33-1, what changes are required to cause the oscillations to proceed counterclockwise around the figure?

31. What constructional difficulties would you encounter if you tried to build an LC circuit of the type shown in Fig. 33-1 to oscillate (a) at 0.01 Hz or (b) at 10^{10} Hz?

32. Two inductances L_1 and L_2 and two capacitances C_1 and C_2 can be connected in series as in Fig. 33-43a or as in Fig. 33-43b. Are the frequencies of the two oscillating circuits equal? Consider the cases (a) $C_1 = C_2$, $L_1 = L_2$ and (b) $C_1 \neq C_2$, $L_1 \neq L_2$.

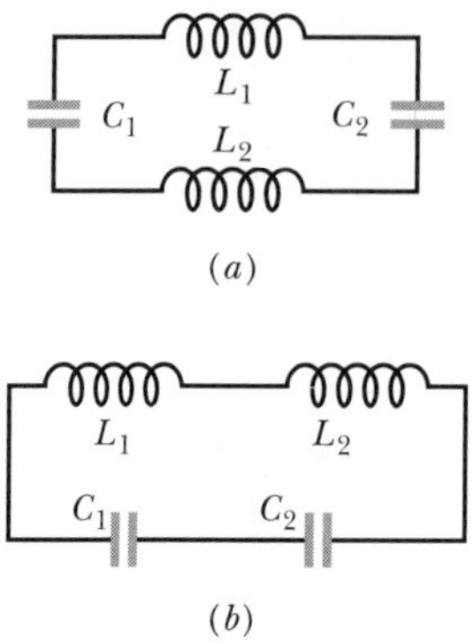

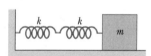

(a)

(b)

Fig. 33-43 Question 32.

33. In the mechanical analog of the oscillating LC circuit, what mechanical quantity corresponds to potential difference?

34. Discuss the assertion that the resonance curves of Fig. 33-13 and Fig. 16-17 cannot truly be compared because the former is a plot of the current amplitude I and the latter of the displacement amplitude x_m. Are these "corresponding" quantities? Does it make any difference if our purpose is only to exhibit the resonance phenomenon?

35. Two identical springs are joined and connected to an object with mass m, the arrangement being free to oscillate on a horizontal frictionless surface as in Fig. 33-44. Sketch the electromagnetic analog of this mechanical oscillating system.

Fig. 33-44 Question 35.

36. In the circuit of Fig. 33-7, why is it safe to assume that (a) the alternating current of Eq. 33-29 has the same angular frequency ω as the alternating emf of Eq. 33-28, and (b) the phase angle ϕ in Eq. 33-29 does not vary with time? What would happen if either of these (true) statements were false?

37. How does a phasor differ from a vector? We know that emfs, potential differences, and currents are not vector quantities. How then can we justify constructions such as Fig. 33-11?

38. Would any of the discussion of Section 33-8 be invalid if the phasor diagrams were to rotate in the clockwise direction, rather than the counterclockwise direction that we assumed? Explain.

39. Suppose that, in a series RLC circuit, the frequency of the applied voltage were changed continuously from a very low value to a very high value. How would the phase constant change?

40. Does it seem intuitively reasonable that a capacitive reactance ($= 1/\omega C$) should vary inversely with angular frequency, whereas an inductive reactance ($= \omega L$) varies directly with this quantity?

41. During World War II, at a large research laboratory in the United States, an alternating-current generator was located a mile or so from the laboratory building it served. A technician increased the speed of the generator to compensate for what he called "the loss of frequency along the transmission line" connecting the generator with the laboratory building. Comment on this reasoning.

42. Discuss in your own words what it means to say that an alternating current "leads" or "lags" an alternating emf.

43. Suppose, as stated in Section 33-9, that a given circuit is "more inductive than capacitive"—that is, that $X_L > X_C$. (a) Does this mean, for a fixed angular frequency, that L is relatively "large" and C is relatively "small," or that L and C are both relatively "large"? (b) For fixed values of L and C does this mean that ω is relatively "large" or relatively "small"?

44. How could you determine, in a series RLC circuit, whether the frequency of an applied emf is above or below resonance?

45. What is wrong with this statement: "If $X_L > X_C$, then we must have $L > 1/C$"?

46. How, if at all, must Kirchhoff's rules (the loop and junction rules) for direct-current circuits be modified when they are applied to alternating-current circuits?

47. Do the loop and junction rules apply to multiloop alternating-current circuits as well as to multiloop direct-current circuits?

48. In Sample Problem 33-8, what would be the effect on P_{avg} if you increased (a) R, (b) C, and (c) L? How would ϕ in Eq. 33-76 change in these three cases?

49. Do commercial power station engineers like to use a low power factor or a high one, or does it make any difference to them? Between what values can the power factor range? What determines the power factor; is it characteristic of the generator, of the transmission line, of the circuit to which the transmission line is connected, or of some combination of these?

50. Can the instantaneous power of a source of alternating emf ever be negative? Can the power factor ever be negative? If so, explain the meaning of these negative values.

51. In a driven series RLC circuit the emf leads the current for a particular frequency of operation. You now lower the driving frequency slightly. Does the total impedance of the circuit increase, decrease, or stay the same? Explain.

52. If you know only the power factor ($= \cos \phi$ in Eq. 33-76) for a given RLC circuit, can you tell whether or not the current is leading or lagging the applied alternating emf? If so, how? If not, why not?

53. What is the permissible range of values of the phase constant ϕ in Eq. 33-29? Of the power factor in Eq. 33-76?

54. Why is it useful to use rms notation for alternating currents and voltages?

55. You want to reduce your electric bill. Do you want a small or a large power factor, or does it make any difference? If it does, is there anything that you can do about it? Discuss.

56. In Eq. 33-76, is ϕ the phase angle between $\mathcal{E}(t)$ and $i(t)$ or between $\mathcal{E}_{rms}$ and i_{rms}? Explain.

57. A doorbell transformer is designed for a primary rms input of 120 V and a secondary rms output of 6 V. What would happen if the primary and secondary connections were accidentally interchanged during installation? Would you have to wait for someone to push the doorbell to find out? Discuss.

58. You are given a transformer enclosed in a wooden box, its primary and secondary terminals being available at two opposite faces of the box. How could you find its turns ratio without opening the box?

59. In the transformer of Fig. 33-15, with the secondary circuit open, what is the phase relationship between (a) the driving emf and the primary current, (b) the driving emf and the magnetic field in the transformer core, and (c) the primary current and the magnetic field in the transformer core?

EXERCISES & PROBLEMS

66. An electric motor connected to a 120 V, 60.0 Hz ac outlet does mechanical work at the rate of 0.100 hp (1 hp = 746 W). If it draws an rms current of 0.650 A, what is its effective resistance, relative to power transfer? Is this the same as the resistance of its coils, as measured with an ohmmeter with the motor disconnected from the outlet?

67. An electric motor has an effective resistance of 32.0 Ω and an inductive reactance of 45.0 Ω when working under load. The rms voltage across the alternating source is 420 V. Calculate the rms current.

68. Show mathematically, rather than graphically as shown in Fig. 33-14b, that the average value of $\sin^2(\omega t - \phi)$ over an integer number of half-cycles is $\frac{1}{2}$.

69. For a sinusoidally driven series RLC circuit, show that over one complete cycle with period T (a) the energy stored in the capacitor does not change; (b) the energy stored in the inductor does not change; (c) the driving emf device supplies energy $(\frac{1}{2}T)\mathcal{E}_m I \cos \phi$; and (d) the resistor dissipates energy $(\frac{1}{2}T)RI^2$. (e) Show that the quantities found in (c) and (d) are equal.

70. An ac generator provides emf to a resistive load in a remote factory over a two-cable transmission line. At the factory a step-down transformer reduces the voltage from its (rms) transmission value V_t to a much lower value that is safe and convenient for use in the factory. The transmission line resistance is 0.30 Ω/cable, and the power of the generator is 250 kW. Calculate the voltage decrease along the transmission line and the rate at which energy is dissipated in the line as thermal energy if (a) V_t = 80 kV, (b) V_t = 8.0 kV, and (c) V_t = 0.80 kV. Comment on the acceptability of each choice.

71. For a certain driven series RLC circuit, the maximum generator emf is 125 V and the maximum current is 3.20 A. If the current leads the generator emf by 0.982 rad, what are (a) the impedance and (b) the resistance of the circuit? (c) Is the circuit predominantly capacitive or inductive?

72. In a certain series RLC circuit being driven at a frequency of 60.0 Hz, the maximum voltage across the inductor is 2.00 times the maximum voltage across the resistor and 2.00 times the maximum voltage across the capacitor. (a) By what angle does the current lag the generator emf? (b) If the maximum generator emf is 30.0 V, what should be the resistance of the circuit to obtain a maximum current of 300 mA?

73. An LC circuit oscillates at a frequency of 10.4 kHz. (a) If the capacitance is 340 μF, what is the inductance? (b) If the maximum current is 7.20 mA, what is the total energy in the circuit? (c) What is the maximum charge on the capacitor?

74. An ac generator produces emf $\mathcal{E} = \mathcal{E}_m \sin(\omega_d t - \pi/4)$, where $\mathcal{E}_m$ = 30.0 V and ω_d = 350 rad/s. The current in the circuit attached to the generator is given by $i(t) = I \sin(\omega_d t + \pi/4)$, where I = 620 mA. (a) At what time after t = 0 does the generator emf first reach a maximum? (b) At what time after t = 0 does the current first reach a maximum? (c) The circuit contains a single element other than the generator. Is it a capacitor, an inductor, or a resistor? Justify your answer. (d) What is the value of the capacitance, inductance, or resistance, as the case may be?

75. A series RLC circuit has a resonant frequency of 6.00 kHz. When it is driven at 8.00 kHz, it has an impedance of 1.00 kΩ and a phase constant of 45°. What are the values of (a) R, (b) L, and (c) C for this circuit?

76. A series RLC circuit is driven in such a way that the maximum voltage across the inductor is 1.50 times the maximum voltage across the capacitor and 2.00 times the maximum voltage across the resistor. (a) What is ϕ for the circuit? (b) Is the circuit inductive, capacitive, or in resonance? The resistance is 49.9 Ω, and the current amplitude is 200 mA. (c) What is the amplitude of the driving emf?

77. A 45.0 μF capacitor and a 200 Ω resistor are connected in series with an ac source with a voltage amplitude V_s of 100 V. The frequency f of the source can be varied from 0 to 100 Hz. (a) Write an equation for the capacitive reactance X_C. (b) Simultaneously plot the resistance R, the capacitive reactance X_C, and the impedance Z versus f for the range 0 < f < 100 Hz. (c) From the plots, determine the value of f for which $X_C = R$.

78. (a) For the situation of Problem 77, simultaneously plot the voltage V_C across the capacitor, the voltage V_R across the resistor, and the (constant) voltage amplitude V_s across the source versus f for the range 0 < f < 100 Hz. (b) From the plots, determine the value of f for which $V_C = V_R$. (c) What is V_R at that frequency? (d) Determine the value of f for which $V_R = 0.50V_s$. (e) What is V_C at that frequency? (f) Determine the value of f for which $V_C = 0.50V_s$. (g) What is V_R at that frequency?

79. Consider the circuit shown in Fig. 33-45. With switch S_1 closed and the other two switches open, the circuit has a time constant τ_C (see Section 28-8). With switch S_2 closed and the other two switches open, the circuit has a time constant τ_L (see Section 31-9). With switch S_3 closed and the other two switches open, the circuit oscillates with a period T. Show that $T = 2\pi\sqrt{\tau_C \tau_L}$.

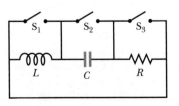

Fig. 33-45 Problem 79.

80. (a) By direct substitution of Eq. 33 25 into Eq. 33-24, show that $\omega' = \sqrt{(1/LC) - (R/2L)^2}$. (b) By what fraction does the frequency of oscillation shift when the resistance is increased from 0 to 100 Ω in a circuit with $L = 4.40$ H and $C = 7.30$ μF.

81. A 45.0 mH inductor has a reactance of 1.30 kΩ. (a) What is its operating frequency? (b) What is the capacitance of a capacitor with the same reactance at that frequency? (c) If the frequency is doubled, what are the new reactances of the inductor and capacitor?

82. A three-phase generator G produces electric power that is transmitted by means of three wires as shown in Fig. 33-46. The electric potentials (relative to a common reference level) of these wires are $V_1 = A \sin \omega_d t$ for wire 1, $V_2 = A \sin(\omega_d t - 120°)$ for wire 2, and $V_3 = A \sin(\omega_d t - 240°)$ for wire 3. Some types of heavy industrial equipment (for example, motors) have three terminals and are designed to be connected directly to these three wires. To use a more conventional two-terminal device (for example, a lightbulb), one connects it to any two of the three wires. Show that the potential difference between *any two* of the wires (a) oscillates sinusoidally with angular frequency ω_d and (b) has an amplitude of $A\sqrt{3}$.

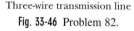

Three-wire transmission line

Fig. 33-46 Problem 82.

83. The ac generator in Fig. 33-47 supplies 120 V at 60.0 Hz. With the switch open as in the diagram, the current leads the generator emf by 20.0°. With the switch in position 1, the current lags the generator emf by 10.0°. When the switch is in position 2, the current is 2.00 A. Find the values of R, L, and C.

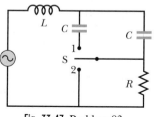

Fig. 33-47 Problem 83.

84. An *LC* oscillator has a 1.00 nF capacitance, a 3.00 V maximum potential on the capacitor, and a 1.73 mA maximum current. (a) What is that oscillator's period of oscillation? What are the maximum energies stored in (b) the capacitor and (c) the inductor? (d) What is the maximum rate at which the current changes? (e) What is the maximum rate at which the inductor gains energy?

85. (a) In an oscillating *LC* circuit, in terms of the maximum charge Q on the capacitor, what is the charge there when the energy in the electric field is 50.0% of that in the magnetic field? (b) What fraction of a period must elapse following the time the capacitor is fully charged for this condition to occur?

86. In an oscillating *LC* circuit, $L = 8.00$ mH and $C = 1.40$ μF. At time $t = 0$, the current is maximum at 12.0 mA. (a) What is the maximum charge on the capacitor during the oscillations? (b) At what times t is the rate of change of energy in the capacitor maximum? (c) What is that maximum rate of change?

87. A series *RLC* circuit is driven by a generator at frequency 1050 Hz. The inductance is 90 mH; the capacitance is 0.50 μF; and the phase constant has a magnitude of 60° (you should supply the appropriate sign for the angle). (a) What is the resistance? To increase the current amplitude in the circuit, should we increase or decrease (b) the driving frequency, (c) the inductance, and (d) the capacitance?

88. A series *RLC* circuit is driven by a generator at a frequency of 2000 Hz and an emf amplitude of 170 V. The inductance is 60 mH, the capacitance is 0.40 μF, and the resistance is 200 Ω. (a) What is the phase constant in radians? (b) What is the current amplitude?

89. When under load and operating at an rms voltage of 220 V, a certain electric motor draws an rms current of 3.00 A. It has a resistance of 24.0 Ω and no capacitive reactance. What is its inductive reactance?

90. A series connection of resistor ($R = 35.0$ Ω), capacitor ($C = 8.65$ μF), and inductor ($L = 50.0$ mH) is driven at 1200 Hz by an alternating source operating at an rms voltage of 112 V. (a) What is the rms current in the circuit? What are the rms voltages across (b) the resistor, (c) the capacitor, and (d) the inductor? What is the average rate at which energy is dissipated in (e) the resistor, (f) the capacitor, and (g) the inductor?

91. A 7.00 μF capacitor has an initial potential of 12 V when it is connected across an inductor. The combination then oscillates at a frequency of 715 Hz. What is the inductance of the inductor?

92. An *LC* oscillator consists of a 2.00 nF capacitor and a 2.00 mH inductor. The maximum voltage is 4.00 V. What are (a) the frequency of the oscillations, (b) the maximum current, (c) the maximum energy stored in the inductor, and (d) the maximum rate di/dt at which the current changes?

Tutorial Problems

93. Consider a series *RLC* circuit that is driven by an emf of rms amplitude 120 V and frequency f. Suppose the inductance is 20.0 mH, the resistance is 30.0 Ω, and the capacitance is 40.0 μF. (a) Sketch the circuit. Make a list of the quantities whose numerical values are provided here, giving each the correct symbol, and determine the resonant frequency of the circuit. (b) Several important physical quantities associated with an ac circuit have the unit of ohms. Determine all of them for the situation where $f_d = 60$ Hz.

(c) Determine the peak (maximum) current in the circuit and the phase constant. Write the complete numerical expressions for the emf $\mathcal{E}(t)$ and the current $i(t)$. (d) Determine the expression for the instantaneous power of the generator in this circuit. What is the average value of the power? Suppose the frequency of the emf were increased slightly. Determine qualitatively what would happen to each of the following quantities: (e) the inductive reactance, (f) the capacitive reactance, (g) the impedance, (h) the peak current, (i) the power factor, and (j) the average power. For what frequency of the generator will (k) the impedance be a minimum and (l) the average power be a maximum?

Answers

(a) See Fig. 33-7. $\mathcal{E}_{rms} = 120$ V, $L = 20.0$ mH, $R = 30.0$ Ω, and $C = 40.0$ μF. The resonant frequency of this circuit is

$$f = \frac{1}{2\pi\sqrt{LC}}$$

$$= \frac{1}{2\pi\sqrt{(20.0 \times 10^{-3} \text{ H})(40.0 \times 10^{-6} \text{ F})}}$$

$$= 178 \text{ Hz}.$$

(b) First, there is the resistance $R = 30\ \Omega$. Then there are the inductive reactance

$$X_L = \omega_d L = 2\pi f_d L = 2\pi(60 \text{ Hz})(20.0 \times 10^{-3} \text{ H}) = 7.54\ \Omega,$$

the capacitive reactance

$$X_C = \frac{1}{\omega_d C} = \frac{1}{2\pi f_d C} = \frac{1}{2\pi(60 \text{ Hz})(40.0 \times 10^{-6} \text{ F})} = 66.3\ \Omega,$$

and the impedance

$$Z = \sqrt{R^2 + (X_L - X_C)^2}$$
$$= \sqrt{(30.0\ \Omega)^2 + (7.54\ \Omega - 66.3\ \Omega)^2}$$
$$= 66.0\ \Omega.$$

(c) The peak current is

$$I = \frac{\mathscr{E}_m}{Z} = \frac{\sqrt{2}\mathscr{E}_{\text{rms}}}{Z} = \frac{\sqrt{2}(120 \text{ V})}{66.0\ \Omega} = 2.57 \text{ A}.$$

The phase constant is

$$\phi = \tan^{-1}\left(\frac{X_L - X_C}{R}\right)$$
$$= \tan^{-1}\left(\frac{7.54\ \Omega - 66.3\ \Omega}{30.0\ \Omega}\right)$$
$$= -63.0° = -1.10 \text{ rad}.$$

This minus sign means that the current leads the emf, which is consistent with the fact that the capacitive reactance has a greater magnitude than the inductive reactance. The emf is

$$\mathscr{E}(t) = \mathscr{E}_m \sin \omega t$$
$$= \sqrt{2}\mathscr{E}_{\text{rms}} \sin 2\pi f t$$
$$= \sqrt{2}(120 \text{ V}) \sin[2\pi(60 \text{ Hz})t]$$
$$= (170 \text{ V}) \sin(377 \text{ rad/s})t.$$

The current is

$$i(t) = I \sin(2\pi f t - \phi)$$
$$= (2.57 \text{ A}) \sin[2\pi(60 \text{ Hz})t - (-1.10 \text{ rad})]$$
$$= (2.57 \text{ A}) \sin[(377 \text{ rad/s})t + 1.10 \text{ rad}].$$

(d) The instantaneous power is equal to

$$i\mathscr{E} = (2.57 \text{ A}) \sin[(377 \text{ rad/s})t + 1.10 \text{ rad}]$$
$$\times (170 \text{ V}) \sin(377 \text{ rad/s})t$$
$$= (437 \text{ W}) \sin[(377 \text{ rad/s})t + 1.10 \text{ rad}]$$
$$\times \sin(377 \text{ rad/s})t.$$

The average value of the power is

$$I_{\text{rms}}\mathscr{E}_{\text{rms}} \cos \phi = \tfrac{1}{2} I\mathscr{E}_m \cos \phi$$
$$= (0.5)(2.57 \text{ A})(170 \text{ V}) \cos(-1.10 \text{ rad})$$
$$= 99.1 \text{ W}.$$

(e) Increasing the driving frequency always increases the inductive reactance of an ac circuit. **(f)** Increasing the driving frequency always decreases the capacitive reactance of an ac circuit. **(g)** Since the driving frequency is now closer to the resonant frequency, the impedance decreases. **(h)** Since the impedance decreases, the peak current increases. Since the driving frequency is now closer to the resonant frequency, **(i)** the power factor increases (it is now closer to its maximum value of 1) and **(j)** the average power increases. **(k)** The impedance is a minimum and **(l)** the average power is a maximum when $Z = R$, which occurs when $f_d = f = 178$ Hz.

94. In this problem we compare the electric currents and the energy dissipation when a resistor, a capacitor, and an inductor are, in turn, connected to an ac generator. Suppose that the ac generator produces an emf

$$\mathscr{E}(t) = (160 \text{ V}) \sin 2\pi(60 \text{ Hz})t.$$

For each of the following parts, (1) draw a diagram for the circuit; (2) write the basic equation for the circuit; (3) determine the electric current as a function of time and find the maximum value of the current; (4) make a plot of the emf and the electric current on the same diagram; and (5) describe what's happening in the circuit, as a function of time, from an energy point of view and find the maximum power of the source.

 (a) Suppose the generator is first connected to a 40 Ω resistor. (b) Next, suppose the generator is connected to a 40 μF capacitor (assumed to have zero resistance). (c) Next, suppose the generator is connected to a 40 mH inductor (assumed to have no resistance).

Answers

(a) (1) See Fig. 33-8a. (2) Kirchhoff's law gives the basic equation $\mathscr{E} - iR = 0$. (3) The electric current is

$$i(t) = \frac{\mathscr{E}}{R} = \frac{160 \text{ V}}{40\ \Omega} \sin 2\pi(60 \text{ Hz})t = (4.0 \text{ A}) \sin 2\pi(60 \text{ Hz})t.$$

The maximum current is 4.0 A. (4) See Fig. 33-8b, which shows that the emf $\mathscr{E}$ and the current i are exactly in phase. (5) From an energy point of view, the ac generator is supplying energy at a sinusoidally varying rate $i\mathscr{E} = i^2R$; that energy is being dissipated in the resistor. The source power is always positive, but varies from 0 to (4.0 A)(160 V) = 640 W.

(b) (1) See Fig. 33-9a. (2) Kirchhoff's law gives the basic equation $\mathscr{E} - q/C = 0$. (3) The charge q on the capacitor is

$$q = \mathscr{E}C$$
$$= (160 \text{ V})(40\ \mu\text{F}) \sin 2\pi(60 \text{ Hz})t$$
$$= (0.0064 \text{ A}) \sin 2\pi(60 \text{ Hz})t.$$

The electric current can be determined by differentiating this result with respect to t:

$$i(t) = \frac{dq}{dt} = 2\pi(60 \text{ Hz})(0.0064 \text{ A}) \cos 2\pi(60 \text{ Hz})t$$
$$= (2.4 \text{ A}) \cos 2\pi(60 \text{ Hz})t.$$

The maximum current is 2.4 A. (4) See Fig. 33-9b, which shows that the electric current i leads the emf $\mathscr{E}$. (5) From an energy point of view, the ac generator is supplying energy at the varying rate $i\mathscr{E}$. The energy shuttles between the source and the electric field of the capacitor. Part of the time the source power is positive, and part of the time it is negative.

(c) (1) See Fig. 33-10a. (2) Kirchhoff's law gives, as the basic equation, the differential equation $\mathscr{E} - L\, di/dt = 0$. (3) Integrating the equation $di/dt = \mathscr{E}/L$ gives the electric current as

$$i(t) = -\frac{160 \text{ V}}{2\pi(60 \text{ Hz})(40 \text{ mH})} \cos 2\pi(60 \text{ Hz})t$$
$$= -(10.6 \text{ A}) \cos 2\pi(60 \text{ Hz})t.$$

The maximum current is 10.6 A. (4) See Fig. 33-10b, which shows that the electric current i lags the emf $\mathscr{E}$. (5) From an energy point of view, the ac generator is supplying energy at the varying rate $i\mathscr{E}$. The energy shuttles between the source and the magnetic field of the inductor. The source power is positive for part of the cycle and negative for part of the cycle.

34
Electromagnetic Waves

Sample Problem 34-6

A submerged swimmer is looking directly upward through the air–water interface in a pool.

(a) Over what range of angles do rays reach the swimmer's eyes from light sources external to the water? Assume that the light is monochromatic and that the index of refraction of water is 1.33.

SOLUTION: One **Key Idea** here is that light reaches the swimmer's eyes from external sources by refracting through the air–water interface, in accordance with Eq. 34-44. Let us associate subscript 1 in that equation with air and subscript 2 with water. Then, with $n_1 = 1.00$ and $n_2 = 1.33$, we have $n_2 > n_1$. Thus, the refraction of light rays at the air–water interface bends the rays *toward* the normal, as in Fig. 34-18b. The bending of an arbitrary light ray, with angle of incidence θ_1 and angle of refraction θ_2, is shown in Fig. 34-56a; the swimmer's eyes are located at point E. Note that the refracted ray also makes an angle θ_2 with the vertical at E.

The value of θ_2 depends on the angle of incidence θ_1 according to Eq. 34-44:

$$\sin \theta_2 = \frac{n_1}{n_2} \sin \theta_1. \qquad (34\text{-}51)$$

Thus, to find the angles at which rays reach point E from external sources, we must consider the range of values of θ_1. That will give us the range of values of θ_2.

The least value of θ_1 is 0°, which is the value for an incident ray that is perpendicular to the air–water interface. For that value Eq. 34-51 gives us

$$\sin \theta_2 = \frac{1.00}{1.33} \sin 0° = 0$$

or $\qquad\qquad \theta_2 = 0°.$

Incident ray A in Fig. 34-56b shows this refraction situation: The incident ray is not bent; it reaches E along the vertical through E.

The maximum value of θ_1 is approximately 90°, which is the value for an incident ray that is almost parallel to the air–water interface. Equation 34-51 now gives us

$$\sin \theta_2 = \frac{1.00}{1.33} \sin 90° = 0.752$$

or $\qquad\qquad \theta_2 = 48.8°.$

In Fig. 34-56b, the two incident rays B show this refraction situation: These two rays are incident on the interface at the greatest

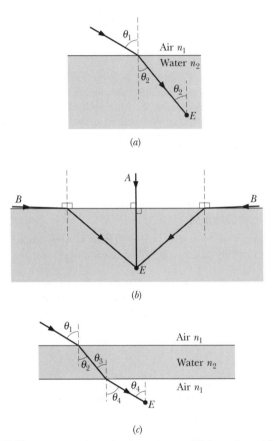

Fig. 34-56 Sample Problem 34-6. (*a*) A ray of light, refracted into water, reaches a swimmer's eyes located at point E. The angle of the ray at E is measured relative to the vertical through E.
(*b*) Ray A, perpendicular to the air–water interface, and rays B, almost parallel to the interface, reach point E. (*c*) The swimmer's eyes at point E are now in air that is trapped by a mask of transparent plastic. A ray reaching E refracts at two air–water interfaces.

possible angle (90°) but are 48.8° to the vertical when they reach the swimmer.

Figure 34-56b shows only one plane of the swimmer's field of view. However, if we rotate that plane about the vertical, we see that it then gives the swimmer's entire field of view. Thus, all the refracted rays reaching the swimmer from external sources are contained within a vertical cone that has its apex located at E and intersects the air–water interface in a circle directly above E. Moreover, the apex angle is

$$2\theta_2 = 97.6° \approx 98°.$$

The swimmer's entire view of the external world comes through the overhead circle, which acts as a personal window for the swimmer. It allows a range of vision from about 49° on one side of the vertical to about 49° on the other side.

(b) The swimmer now wears a swimming mask over his eyes; the thin flat layer of transparent plastic through which the swimmer looks is horizontal; and air fills the interior of the mask. Over what range of angles do rays now reach the swimmer's eyes from light sources external to the water? (Neglect refraction by the plastic faceplate of the mask; its inclusion would not alter the result.)

SOLUTION: The **Key Idea** here is that, as in part (a), the light can reach the observer's eyes only via refraction, except now the light must refract twice. Let us examine the arbitrary light ray shown in Fig. 34-56c. As before, the ray refracts from air into water, but now, to reach point E, it must refract a second time, from the water into the air held by the mask. Let the angle of incidence at the second

refraction be θ_3 and the angle of refraction be θ_4. Note that here the refracted ray makes the same angle θ_4 with the vertical at E.

Because the two air–water interfaces in Fig. 34-56c are parallel, $\theta_3 = \theta_2$. From Eq. 34-44, the angle of refraction θ_4 at the lower interface is given by

$$\sin\theta_4 = \frac{n_2}{n_1}\sin\theta_3.$$

First substituting θ_2 for θ_3 and then substituting for $\sin\theta_2$ from Eq. 34-51, we find

$$\sin\theta_4 = \frac{n_2}{n_1}\sin\theta_2 = \frac{n_2 n_1}{n_1 n_2}\sin\theta_1 = \sin\theta_1,$$

which (in this situation) gives us

$$\theta_4 = \theta_1.$$

In words, the arbitrary ray in Fig. 34-56c reaches point E traveling parallel to its original direction, as do all other rays reaching E from external sources. Thus, the range of those rays is from about 90° on one side of the vertical to about 90° on the other side. This means that with the mask, the swimmer sees the external world as if the water were not present, and not compressed into a 98° cone.

Sample Problem 34-7

(a) How can we use a glass plate with index of refraction $n = 1.57$ to produce fully polarized light from a beam of unpolarized light?

SOLUTION: The **Key Idea** here is that light reflected by the plate is fully polarized if the light is incident on the plate at the Brewster angle θ_B. To find that angle, we can use either Eq. 34-49 (the general equation) or Eq. 34-50 (when the reflecting material is in air, as here). Using Eq. 34-50, we find

$$\theta_B = \tan^{-1} n = \tan^{-1} 1.57 = 57.5°.$$

To produce polarized light, we would arrange for the unpolarized

beam to reflect from the glass plate with an angle of incidence of 57.5°.

(b) What is the angle of refraction of the light that enters the glass plate when the light is incident at the Brewster angle θ_B?

SOLUTION: The **Key Idea** here is that we can find the angle of refraction θ_r by using either the law of refraction of Eq. 34-44 or the simple relation of Eq. 34-48 ($\theta_B + \theta_r = 90°$) for light incident at the Brewster angle, as it is here. Choosing the latter, we find

$$\theta_r = 90° - \theta_B = 90° - 57.5° = 32.5°. \quad \text{(Answer)}$$

QUESTIONS

13. *Organizing question:* Which of the following four pairs of functions are properly written to describe the electric and magnetic components of an electromagnetic wave?

Pair 1:
$$E = E_m \sin\frac{2\pi}{\lambda}(x - ct)$$

$$B = \frac{E_m}{c}\sin\frac{2\pi}{\lambda}(x - ct)$$

Pair 2:
$$E = cB_m \sin\left(\frac{2\pi x}{\lambda} - \frac{2\pi t}{T}\right)$$

$$B = B_m \sin\left(\frac{2\pi x}{\lambda} - \frac{2\pi t}{T}\right)$$

Pair 3:
$$E = E_m \sin\omega\left(\frac{x}{c} - t\right)$$

$$B = \sqrt{\mu_0\varepsilon_0}\, E_m \sin\omega\left(\frac{x}{c} - t\right)$$

Pair 4:
$$E = \frac{B_m}{\sqrt{\mu_0\varepsilon_0}}\sin 2\pi\left(\frac{x}{\lambda} - ft\right)$$

$$B = B_m \sin 2\pi\left(\frac{x}{\lambda} - ft\right)$$

14. A particular point source of light emits isotropically, with an intensity I of 8 W/m² at a radial distance r of 1 m from the source. (a) Which of the curves in Fig. 34-57 best gives the intensity I of the light as a function of r? (b) If we place the source in sooty air, where the soot particles absorb the light they intercept, which of the curves then best gives $I(r)$?

15. In Fig. 34-58, three spherical dust particles suddenly experience a radiation pressure when a beam of laser light is turned on. The radii of the particles are R, $2R$, and $3R$; the particles have the same density and totally absorb the light they intercept. Rank the three particles according to (a) the radiation pressure on them, (b) the radiation force on them, (c) their masses, (d) the accelerations they experience due to the radiation force, and (e) the distance they travel in a certain time after the laser is turned on, all greatest first.

16. Light is sent through the two-sheet polarizing system shown in Fig. 34-59. If the ratio of the emerging intensity to the initial intensity is 0.7, is the light initially polarized or unpolarized?

17. *Organizing question:* Figure 34-60 represents six situations in which light is sent toward you through a polarizing sheet. The general arrangement is like that of Fig. 34-12, except here the view of the sheet is head-on. The initial state of polarization of the light (before it reaches the sheet) is represented by the arrows. The polarizing direction of the sheet is represented by a dashed line. (a) For each situation, determine whether the one-half rule or the cosine-squared rule applies. (b) For the latter, determine what angle should be used in the calculation of the intensity of the light transmitted by the sheet.

18. Three polarizing sheets are positioned like the two sheets in Fig. 34-59, and initially unpolarized light is sent into the system. How many different final intensities can be produced if the polarizing directions of the sheets are the following: one is parallel to the y axis, one is rotated 20° clockwise from the y axis about the light's line of travel, and one is rotated 20° in the opposite direction about that line of travel?

19. Initially unpolarized light is sent through the two-sheet polarizing system of Fig. 34-59. The emerging light is polarized 20° clockwise from the y axis and has half the initial intensity. What are the polarizing directions of the sheets?

20. Figure 34-61*a* shows an overhead view of a rectangular room with fully reflecting walls. The room length L and width W are both

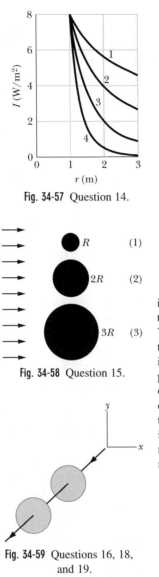

Fig. 34-57 Question 14.

Fig. 34-58 Question 15.

Fig. 34-59 Questions 16, 18, and 19.

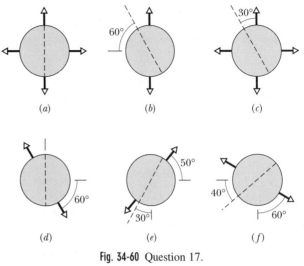

Fig. 34-60 Question 17.

integer numbers of units. You fire a laser from corner a at 45° to the walls. Small clay armadillos are located in the other corners. You are concerned as to whether your shot will hit one of those targets or yourself. Figure 34-61*b* shows a way to find out by drawing repeated reflections of the room and extending a straight light path through them. Which corner is hit depends on the ratio L/W, once that ratio is reduced to lowest terms. (For example, 4/2 reduces to 2/1.) Figure 34-61*b* is for $L/W = 2/1$. We see that the target in corner d is hit after one reflection. Determine which corner is hit for any (reduced) L/W in the form of (a) even number/odd number, (b) odd number/even number, and (c) odd number/odd number.

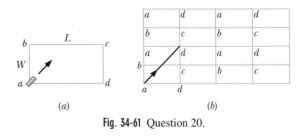

Fig. 34-61 Question 20.

21. Figure 34-62*a* depicts total internal reflection for light inside a material with an index of refraction n_1 when air is outside the material. A light ray reaching point A from within the shaded region at the left (such as the ray shown) fully reflects at that point and ends up in the shaded region at the right. The other two parts of Fig. 34-62 show similar situations for two other materials. Rank the indexes of refraction of the three materials, greatest first.

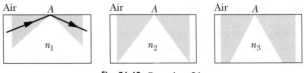

Fig. 34-62 Question 21.

22. In Figure 34-63, a light ray refracts from air into a plastic rod and then partially reflects (at an angle of 60°) and partially refracts at a point along the side of the rod. What is the value of angle θ?

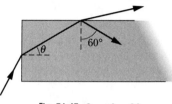

Fig. 34-63   Question 22.

23. A beam of white light in plastic is incident on the interface between the plastic and the external air. If the angle of incidence is equal to the critical angle for yellow light, does any of (a) the red light and (b) the blue light in the beam refract from the plastic into the air?

24. Sound waves traveling through water can be trapped by total internal reflection somewhat as light can be trapped in an optical fiber. Figure 34-64a depicts a channel in the ocean where sound waves are trapped in a certain depth range as they travel rightward. (The paths taken by two waves are represented by the two curved rays.) As explained for Eq. 18-3, the speed of sound in a material depends on the ratio of the material's bulk modulus B to its density ρ. Figure 34-64b gives four choices for how that ratio might vary with depth for water. Which choice best corresponds to the entrapment shown in Fig. 34-64a?

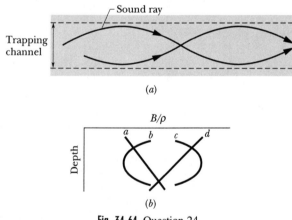

Fig. 34-64   Question 24.

25. Light in air is incident on a glass surface. In that situation, is the Brewster angle for green light greater than, less than, or the same as that for yellow light?

DISCUSSION QUESTIONS

26. Electromagnetic waves reach us from the farthest depths of space. From the information they carry, can we tell what the universe is like at the present moment? At any selected time in the past?

27. Comment on this definition of the limits of the spectrum of visible light, given by a physiologist: "The limits of the visible spectrum occur when the eye is no better adapted than any other organ of the body to serve as a detector."

28. List several ways in which radio waves differ from visible light waves. In what ways are they the same?

29. "Displacement currents are present in a traveling electromagnetic wave and we may associate the magnetic field component of the wave with these currents." Is this statement true? Discuss it in detail.

30. H. G. Wells, in his novel *The Invisible Man,* described a concoction that would render the person who drank it invisible. Give arguments to prove that a truly invisible person would be blind.

31. Can an electromagnetic wave be deflected by a magnetic field? By an electric field?

32. Why is Maxwell's modification of Ampere's law (that is, the term $\mu_0\varepsilon_0\, d\Phi_E/dt$ in Table 32-1) needed to understand the propagation of electromagnetic waves?

33. Can an object absorb light energy without having linear momentum transferred to it? If so, give an example. If not, explain why.

34. When you turn on a flashlight, does it experience any force associated with the emission of the light?

35. What is the relation, if any, between the intensity I of an electromagnetic wave and the magnitude S of its Poynting vector?

36. As we normally experience them, radio waves are almost always polarized and visible light is almost always unpolarized. Why should this be so?

37. You are given a number of polarizing sheets. Explain how you would use them to rotate the plane of polarization of a plane polarized wave through any given angle. How could you do it with the least energy loss?

38. Why do sunglasses made of polarizing materials have a marked advantage over those that simply depend on absorption effects? What disadvantages might they have?

39. Why cannot sound waves be polarized?

40. Unpolarized light falls on two polarizing sheets so oriented that no light is transmitted. If a third polarizing sheet is placed between them, can light be transmitted? If so, explain how.

41. Find a way to identify the polarizing direction of a polarizing sheet. No marks appear on the sheet.

42. Can you think of any "everyday" observation (that is, without experimental apparatus) to show that the speed of light is not infinite?

43. In a vacuum, does the speed of light depend on (a) the wavelength, (b) the frequency, (c) the intensity, (d) the state of polarization, (e) the speed of the source, or (f) the speed of the observer?

44. Explain how polarization by reflection could occur if the light were incident on the interface from the side with the higher index of refraction (glass to air, for example).

45. What is a plausible explanation for the observation that a street appears darker when wet than when dry?

EXERCISES & PROBLEMS

67. *Rainbow.* Figure 34-65 shows a light ray entering and then leaving a falling, spherical raindrop after one internal reflection (compare with Fig. 34-22). The ray is deviated (turned) from its original direction of travel by angular deviation θ_{dev}. (a) Show that θ_{dev} is

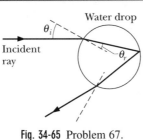

Fig. 34-65 Problem 67.

$$\theta_{dev} = 180° + 2\theta_i - 4\theta_r,$$

where θ_i is the angle of incidence of the ray on the drop and θ_r is the angle of refraction of the ray within the drop. (b) Using Snell's law, substitute for θ_r in terms of θ_i and the index of refraction n of the water. Then graph θ_{dev} versus θ_i for the range of possible θ_i values, but in doing so, replace n with a list: 1.331 for red light and 1.343 for blue light.

The red-light curve and the blue-light curve have different minima, which means that there is a different *angle of minimum deviation* for each color. The light of any given color that leaves the drop at that color's angle of minimum deviation is especially bright because rays bunch up at that angle. Thus, the bright red light leaves the drop at one angle and the bright blue light leaves it at another angle. The story is the same for other drops in your view (Fig. 34-22).

Determine the angles of minimum deviation from the curves of θ_{dev} for (c) red light and (d) blue light. Let us assume that these colors form the inner and outer edges of a rainbow. (e) What then is the angular width of the rainbow?

68. The *first-order* (or *primary*) *rainbow* described in Problem 67 is the type commonly seen in regions where rainbows appear. It is produced by light reflecting once inside the drops. More rare is the *second-order* (or *secondary*) *rainbow* that is produced by light reflecting twice inside the drops (Fig. 34-66a). (a) Show that the angular deviation of such light is

$$\theta_{dev} = (180°)k + 2\theta_i - 2(k + 1)\theta_r,$$

where k is the number of internal reflections. Using the procedure of Problem 67, find the angles of minimum deviation for (b) red light and (c) blue light in the second-order rainbow. (d) What is the angular width of that rainbow?

The *third-order rainbow* depends on three internal reflections (Fig. 34-66b). It probably occurs but cannot be seen because it is very faint and lies in the bright sky surrounding the direction of the Sun. What are the angles of minimum deviation for (e) the red light and (f) the blue light in this rainbow and (g) what is the rainbow's angular width?

69. A catfish is 2.00 m below the surface of a smooth lake. (a) What is the diameter of the circle on the surface through which the fish can see the world outside the water? (b) If the fish descends, does the diameter of the circle increase, decrease, or remain the same?

70. In Fig. 34-67, light that is polarized parallel to the y axis is sent into a system of two polarizing sheets. The fraction of the initial light intensity that emerges from the system is 0.200. What is the angle θ shown for the second sheet?

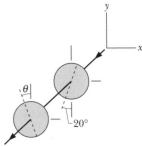

Fig. 34-67 Problem 70.

71. A light wave polarized parallel to a y axis is traveling in the negative direction of a z axis. The rms value of the electric field is 50.0 V/m and the wavelength is 250 nm. (a) Using this information, write an expression for the magnetic field component of the wave. Part of the wave is totally reflected by a chip in the shape of an equilateral triangle that is mounted on the z axis and is perpendicular to that axis. The triangle has an edge length of 2.00 μm, and the chip is fully illuminated on the side facing the light source. (b) What is the force on the chip due to the light?

72. What is the radiation force on a totally reflecting sail at a distance of 3.0×10^{11} m from the Sun if the sail is square with edge length 2.0 m and has its surface perpendicular to the direction of the sunlight?

73. The magnetic component of an electromagnetic wave has an rms value of 56.0 nT. (a) What is the amplitude of the electric component of the wave? (b) What is the intensity of the light?

74. In Fig. 34-1, verify that the uniform spaces between successive powers of 10 must be the same on the wavelength scale and on the frequency scale.

75. Start from Eqs. 34-11 and 34-17 and show that $E(x, t)$ and $B(x, t)$, the electric and magnetic field components of a plane traveling electromagnetic wave, must satisfy the "wave equations"

$$\frac{\partial^2 E}{\partial t^2} = c^2 \frac{\partial^2 E}{\partial x^2} \quad \text{and} \quad \frac{\partial^2 B}{\partial t^2} = c^2 \frac{\partial^2 B}{\partial x^2}.$$

76. Prove that the intensity of an electromagnetic wave is the product of the wave's energy density and its speed.

77. The average intensity of the solar radiation that strikes normally on a surface just outside Earth's atmosphere is 1.4 kW/m². (a) What radiation pressure is exerted on this surface, assuming complete absorption? (b) How does this pressure compare with Earth's sea-level atmospheric pressure, which is 1.0×10^5 Pa?

78. In Figure 34-68, two light rays pass from air through five transparent layers of plastic whose boundaries are parallel, whose indexes of refraction are as given, and whose thicknesses are unknown. The rays emerge back into air at the right. With respect to

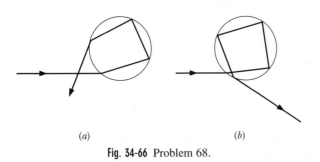

(a)   (b)

Fig. 34-66 Problem 68.

a normal to the last interface, what is the angle of (a) emerging ray *a* and (b) emerging ray *b*? (c) What are your answers if there is glass, with $n = 1.5$, instead of air on the left and right sides of the plastic layers? (*Hint:* Save yourself much time by first solving the problems algebraically.)

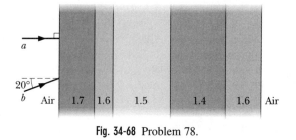

Fig. 34-68 Problem 78.

79. When the atmosphere is cold, moisture can form ice crystals of various shapes. If the atmosphere in the direction of the Sun happens to contain a sufficient number of ice crystals in the shape of flat hexagonal plates, a bright (perhaps colorful) region, called a *sun dog,* appears to the left or right of the Sun. A sun dog is formed by rays of sunlight that pass through the ice plates. These rays are parallel with one another when they arrive at Earth. The rays that pass through an ice plate are redirected by refraction, and those that pass through at the angle of minimum deviation ψ (shown from overhead in Fig. 34-69; see Problem 51) can form a sun dog. The sun dog can then be seen at an angle ψ away from the Sun. If the index of refraction of ice is 1.31, what is ψ?

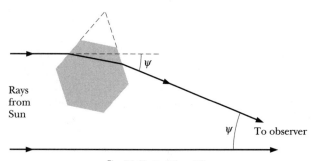

Fig. 34-69 Problem 79.

80. A point source of light is placed a distance *h* below the surface of a large deep lake. (a) Neglecting reflection at the surface except where it is total, show that the fraction *frac* of the light energy that escapes directly from the water surface is independent of *h* and is given by

$$frac = \tfrac{1}{2}(1 - \sqrt{1 - 1/n^2}),$$

where *n* is the index of refraction of the water. (b) Evaluate this fraction for $n = 1.33$.

81. In Fig. 34-70, light that is initially unpolarized is sent into a system of three polarizing sheets. What fraction of the initial light intensity is passed by the system?

82. The electric component of a beam of polarized light is

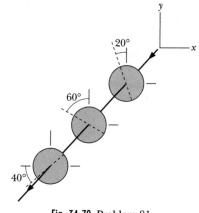

Fig. 34-70 Problem 81.

$$E_y = (5.00 \text{ V/m}) \sin[(1.00 \times 10^6 \text{ m}^{-1})z + \omega t].$$

(a) Write an expression for the magnetic field component of the wave, including a value for ω. What are (b) the wavelength, (c) the period, and (d) the intensity of this light? (e) Parallel to which axis does the magnetic field oscillate? (f) In which region of the electromagnetic spectrum is this wave?

83. In a region of space where gravitational forces can be neglected, a sphere is accelerated by a uniform light beam of intensity 6.0 mW/m². The sphere is totally absorbing and has a radius of 2.0 μm and a uniform density of 5.0×10^3 kg/m³. What is the magnitude of the sphere's acceleration due to the light?

84. A point source of light emits isotropically with a power of 200 W. What is the force due to the light on a totally absorbing sphere of radius 2.0 cm at a distance of 20 m from the source?

85. The rms value of the electric field in a certain light wave is 0.200 V/m. What is the amplitude of the associated magnetic field?

86. (a) Show that Eqs. 34-1 and 34-2 satisfy the wave equations displayed in Problem 75. (b) Show that any expressions of the form

$$E = E_m f(kx \pm \omega t) \quad \text{and} \quad B = B_m f(kx \pm \omega t),$$

where $f(kx \pm \omega t)$ denotes an arbitrary function, also satisfy these wave equations.

87. An electromagnetic wave is traveling in the negative *y* direction. At a particular position and time, the electric field is directed along positive *z* and has a magnitude of 100 V/m. What are the magnitude and direction of the magnetic field at that position and that time?

88. A helium–neon laser, radiating at 632.8 nm, has a power output of 3.0 mW and a full-angle beam divergence (see Exercise 13) of 0.17 mrad. (a) What is the intensity of the beam 40 m from the laser? (b) What is the power of a point source that provides this same intensity at the same distance?

89. Three polarizing sheets are stacked. The first and third are crossed; the one between has its polarizing direction at 45° to the polarizing directions of the other two. What fraction of the intensity of an originally unpolarized beam is transmitted by the stack?

90. A beam of light passes through an equilateral triangular prism that is oriented for minimum deviation (see Problem 51). The deviation is $\psi = 30.0°$. What is the index of refraction of the prism?

91. When red light in vacuum is incident at the Brewster angle on a certain glass slab, the angle of refraction is 32.0°. What are (a) the index of refraction of the glass and (b) the Brewster angle?

92. In the ray diagram of Fig. 34-71, where the angles are not drawn to scale, the ray is incident at the critical angle on the interface between materials 2 and 3. Angle $\phi = 60.0°$. Find (a) index of refraction n_3 and (b) angle θ. (c) If θ is decreased, is there refraction of light into material 3?

93. In Fig. 34-72, initially unpolarized light is sent toward a system of three polarizing sheets. What fraction of the initial light intensity emerges from the system?

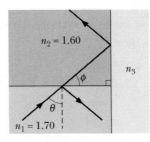

Fig. 34-71 Problem 92.

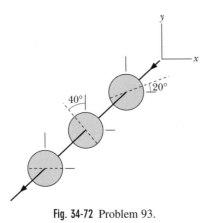

Fig. 34-72 Problem 93.

94. The magnetic component of a polarized wave of light is given by

$$B_x = (4.00 \ \mu T) \sin[ky + (2.00 \times 10^{15} \ s^{-1})t].$$

(a) In which direction does the wave travel, (b) parallel to which axis is it polarized, and (c) what is its intensity? (d) Write an expression for the electric field of the wave, including a value for the angular wave number. (e) What is the wavelength? (f) In which region of the electromagnetic spectrum is this wave?

95. A plane wave of light has an intensity of 1.00×10^4 W/m². What is the maximum strength of the magnetic field of the wave?

96. (a) The wavelength of the most energetic x rays produced when electrons are accelerated to a kinetic energy of 18 GeV in the Stanford Linear Accelerator and then slam into a solid target is 0.067 fm. What is the frequency of these x rays? (b) A VLF (very low frequency) radio wave has a frequency of only 30 Hz. What is its wavelength?

97. Earth's mean radius is 6.37×10^6 m and the mean Earth–Sun distance is 1.50×10^8 km. What fraction of the radiation emitted by the Sun is intercepted by the disk of Earth?

98. An optical fiber consists of a glass core (index of refraction n_1) surrounded by a coating (index of refraction $n_2 < n_1$). Suppose a beam of light enters one end of the fiber from air at an angle θ with the fiber axis as shown in Fig. 34-73. (a) Show that the greatest

possible value of θ for which a ray can travel down the fiber is $\theta = \sin^{-1}\sqrt{n_1^2 - n_2^2}$. (b) If the indexes of refraction of the glass and coating are 1.58 and 1.53, respectively, what is the greatest possible value of the incident angle θ?

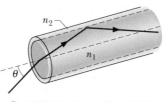

Fig. 34-73 Problems 98 and 99.

99. In an optical fiber (see Problem 98), different rays travel different paths along the fiber, leading to different travel times. This causes a light pulse to spread out as it travels along the fiber, resulting in information loss. The delay time should be minimized by the design of the fiber. Consider a ray that travels a distance L directly along a fiber axis and another that is repeatedly reflected, at the critical angle, as it travels to the same point as the first ray. (a) Show that the difference Δt in the times of arrival is

$$\Delta t = \frac{L}{c} \frac{n_1}{n_2} (n_1 - n_2),$$

where n_1 is the index of refraction of the glass core and n_2 is the index of refraction of the fiber coating. (b) Evaluate Δt for the fiber of Problem 98, with $L = 300$ m.

100. In Fig. 34-74a, light refracts from material 1 into a thin layer of material 2, crosses that layer, and then is incident at the critical angle on the interface between materials 2 and 3. (a) What is angle θ? (b) If θ is decreased, is there refraction of light into material 3?

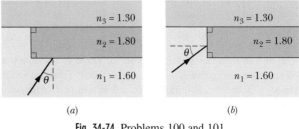

(a)   *(b)*

Fig. 34-74 Problems 100 and 101.

101. In Fig. 34-74b, light refracts from material 1 into a thin layer of material 2, crosses that layer, and then is incident at the critical angle on the interface between materials 2 and 3. (a) What is angle θ? (b) If θ is decreased, is there refraction of light into material 3?

Tutorial Problem

102. A plane electromagnetic wave with a wavelength of 200 nm is traveling in a vacuum in the positive x direction. Its magnetic field, whose maximum magnitude is 50 μT, is polarized parallel to the z axis. (a) Write the wave velocity as a vector. (b) Determine the following characteristics of the wave: its frequency f, its angular frequency ω, and its angular wave number k. What part of the electromagnetic spectrum is involved?

(c) Write a mathematical expression for the magnetic field of this wave. Take the phase constant to be zero, and remember that the magnetic field is a vector. (d) Similarly, write an expression for the electric field vector of this wave.

(e) Faraday's law applied to the electric and magnetic fields of a plane electromagnetic wave leads to the relation

$$\frac{\partial E_y}{\partial x} = -\frac{\partial B_z}{\partial t}$$

(Eq. 34-11, written in component notation). Verify that this equation is satisfied by the magnetic and electric fields derived in parts (c) and (d). (f) From his four electromagnetic equations, Maxwell derived the wave equations for electromagnetic waves as

$$\frac{\partial^2 \vec{E}}{\partial x^2} = \frac{1}{c^2}\frac{\partial^2 \vec{E}}{\partial t^2}$$

and

$$\frac{\partial^2 \vec{B}}{\partial x^2} = \frac{1}{c^2}\frac{\partial^2 \vec{B}}{\partial t^2}.$$

Verify that the electric field in this problem satisfies the wave equation for $\vec{E}$.

Answers

(a) Since the wave is traveling in the positive x direction, which is along the unit vector $\hat{i}$, the velocity of propagation of the wave is $\vec{v} = c\hat{i} = (3.00 \times 10^8 \text{ m/s})\hat{i}$.

(b) We have:

$$f = \frac{c}{\lambda} = \frac{3.00 \times 10^8 \text{ m/s}}{200 \times 10^{-9} \text{ m}} = 1.50 \times 10^{15} \text{ Hz}$$

$$\omega = 2\pi f = 2\pi(1.50 \times 10^{15} \text{ Hz}) = 9.42 \times 10^{15} \text{ rad/s}$$

$$k = \frac{2\pi}{\lambda} = \frac{2\pi}{200 \times 10^{-9} \text{ m}} = 3.14 \times 10^7 \text{ m}^{-1}.$$

Since visible light has wavelengths in the range of about 400 to 700 nm, a wavelength of 200 nm is shorter than the shortest visible light by a factor of 2, so it falls in the ultraviolet part of the spectrum.

(c) The magnetic field is parallel to the z axis and depends only on x (and time), so it can be written in the form

$$\vec{B}(x, t) = B_m \sin(kx - \omega t)\hat{k}.$$

Since we were told that $B_{\text{max}} = 50 \ \mu\text{T}$, and we determined the values of k and ω in part (b), we must have

$$\vec{B}(x, t) = (50 \ \mu\text{T}) \sin[(3.14 \times 10^7 \text{ m}^{-1})x$$
$$- (9.42 \times 10^{15} \text{ s}^{-1})t]\hat{k}.$$

(d) The maximum magnitude of the electric field is

$$E_m = cB_m = (3.00 \times 10^8 \text{ m/s})(50 \ \mu\text{T}) = 15 \text{ kV/m}.$$

Consequently, since the electric and magnetic fields are perpendicular to one another and are in phase,

$$\vec{E}(x, t) = (15 \text{ kV/m}) \sin[(3.14 \times 10^7 \text{ m}^{-1})x$$
$$- (9.42 \times 10^{15} \text{ s}^{-1})t]\hat{j}.$$

Note: This is $+\hat{j}$, not $-\hat{j}$, as can be determined by checking that $\vec{E}$, $\vec{B}$, and $\vec{v}$ are mutually perpendicular and form a right-handed coordinate system (in that order).

(e) First,

$$\frac{\partial E_y}{\partial x} = \frac{\partial}{\partial x}(15 \text{ kV/m}) \sin[(3.14 \times 10^7 \text{ m}^{-1})x$$
$$- (9.42 \times 10^{15} \text{ s}^{-1})t]$$
$$= (15 \text{ kV/m})(3.14 \times 10^7 \text{ m}^{-1}) \cos[(3.14 \times 10^7 \text{ m}^{-1})x$$
$$- (9.42 \times 10^{15} \text{ s}^{-1})t]$$
$$= (4.71 \times 10^{11} \text{ V/m}^2) \cos[(3.14 \times 10^7 \text{ m}^{-1})x$$
$$- (9.42 \times 10^{15} \text{ s}^{-1})t].$$

Next,

$$-\frac{\partial B_z}{\partial t} = -\frac{\partial}{\partial t}(50 \ \mu\text{T}) \sin[(3.14 \times 10^7 \text{ m}^{-1})x$$
$$- (9.42 \times 10^{15} \text{ s}^{-1})t]$$
$$= -(50 \ \mu\text{T})(-9.42 \times 10^{15} \text{ s}^{-1}) \cos[(3.14 \times 10^7 \text{ m}^{-1})x$$
$$- (9.42 \times 10^{15} \text{ s}^{-1})t]$$
$$= (4.71 \times 10^{11} \text{ V/m}^2) \cos[(3.14 \times 10^7 \text{ m}^{-1})x$$
$$- (9.42 \times 10^{15} \text{ s}^{-1})t].$$

Yes, these are equal.

(f) There is only one component to check—namely, E_y. We have already found

$$\frac{\partial E_y}{\partial x} = (4.71 \times 10^{11} \text{ V/m}^2) \cos[(3.14 \times 10^7 \text{ m}^{-1})x$$
$$- (9.42 \times 10^{15} \text{ s}^{-1})t].$$

Thus,

$$\frac{\partial^2 E_y}{\partial x^2} = \frac{\partial^2}{\partial x^2}(15 \text{ kV/m}) \sin[(3.14 \times 10^7 \text{ m}^{-1})x$$
$$- (9.42 \times 10^{15} \text{ s}^{-1})t]$$
$$= -(15 \text{ kV/m})(3.14 \times 10^7 \text{ m}^{-1})^2 \sin[(3.14 \times 10^7 \text{ m}^{-1})x$$
$$- (9.42 \times 10^{15} \text{ s}^{-1})t]$$
$$= -(1.48 \times 10^{19} \text{ V/m}^3) \sin[(3.14 \times 10^7 \text{ m}^{-1})x$$
$$- (9.42 \times 10^{15} \text{ s}^{-1})t].$$

Similarly,

$$\frac{\partial E_y}{\partial t} = \frac{\partial}{\partial t}(15 \text{ kV/m}) \sin[(3.14 \times 10^7 \text{ m}^{-1})x$$
$$- (9.42 \times 10^{15} \text{ s}^{-1})t]$$
$$= -(15 \text{ kV/m})(9.42 \times 10^{15} \text{ s}^{-1}) \cos[(3.14 \times 10^7 \text{ m}^{-1})x$$
$$- (9.42 \times 10^{15} \text{ s}^{-1})t]$$
$$= (1.41 \times 10^{20} \text{ V/m} \cdot \text{s}) \cos[(3.14 \times 10^7 \text{ m}^{-1})x$$
$$- (9.42 \times 10^{15} \text{ s}^{-1})t];$$

$$\frac{\partial^2 E_y}{\partial t^2} = \frac{\partial^2}{\partial t^2} (15 \text{ kV/m}) \sin[(3.14 \times 10^7 \text{ m}^{-1})x$$
$$- (9.42 \times 10^{15} \text{ s}^{-1})t]$$
$$= -(15 \text{ kV/m})(9.42 \times 10^{15} \text{ s}^{-1})^2 \sin[(3.14 \times 10^7 \text{ m}^{-1})x$$
$$- (9.42 \times 10^{15} \text{ s}^{-1})t]$$
$$= -(1.33 \times 10^{36} \text{ V/m} \cdot \text{s}^2) \sin[(3.14 \times 10^7 \text{ m}^{-1})x$$
$$- (9.42 \times 10^{15} \text{ s}^{-1})t].$$

Comparing, we see that

$$\frac{\partial^2 E_y}{\partial x^2} = (1.11 \times 10^{-17} \text{ s}^2/\text{m}^2)\left(\frac{\partial^2 E_y}{\partial t^2}\right) = \frac{1}{c^2}\left(\frac{\partial^2 E_y}{\partial t^2}\right),$$

which is the wave equation.

35

Images

Sample Problem 35-5

A basketball player with a height of 198 cm wants to see his entire height in a "full-length" mirror mounted on a wall. What is the least length the mirror must have?

SOLUTION: One **Key Idea** here is that if the player is to see his entire height in the mirror's image, light rays from his shoes and from

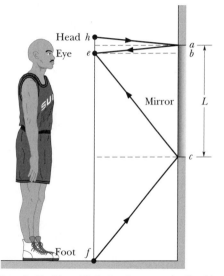

Fig. 35-36 Sample Problem 35-5. A mirror on a wall with the least possible length L that allows a basketball player to view his entire height in it.

the top of his head must be reflected to his eyes by the mirror. A second **Key Idea** is that any ray reflected by the mirror follows the rule of reflection: The ray's angle of reflection is equal to its angle of incidence.

In Fig. 35-36, the heights of the top of the player's head (h), his eyes (e), and the bottoms of his feet (f) are marked by dots. (Dot h has been drawn slightly too high for clarity.) The figure shows the paths followed by rays that leave his head and his feet and enter his eyes, reflecting from the mirror at points a and c, respectively, according to the rule of reflection. The mirror need occupy only the vertical distance L between those points.

From the geometry and Eq. 34-43,

$$ab = \tfrac{1}{2}he \quad \text{and} \quad bc = \tfrac{1}{2}ef,$$

so the required length of the mirror is

$$L = ab + bc = \tfrac{1}{2}(he + ef)$$
$$= (\tfrac{1}{2})(198 \text{ cm}) = 99 \text{ cm}. \qquad \text{(Answer)}$$

Thus, the mirror need be no longer than half the player's height, and this result is independent of his distance from the mirror. (If you have a full-length mirror available, you might experiment by taping newspaper over the portions of the mirror that do not contribute to your image. You will find that what you have left is just half your height. Mirrors that extend below point c in the figure just allow you to look at an image of the floor.)

Sample Problem 35-6

In Fig. 35-37a, a habanero seed O_1 is placed in front of two thin symmetrical coaxial lenses 1 and 2. The focal points for lens 1 are 3.0 cm from the lens; those for lens 2 are 6.0 cm from the lens. The lenses have separation $L = 8.5$ cm, and the seed is 9.0 cm to the left of lens 1.

(a) Where does the system of two lenses produce an image of the seed?

SOLUTION: This sample problem looks very much like Sample Problem 35-4, but we shall see some important and subtle differences. As in that earlier sample problem, we could locate the image produced by the system of lenses by tracing light rays from the seed

through the two lenses. However, the **Key Idea** here, as earlier, is that we can calculate the location of that image by working through the system in steps, lens by lens. We begin with the lens closer to the seed—lens 1. The image we seek is the "final one"—that is, image I_2 produced by lens 2.

Lens 1. Ignoring lens 2, we locate the image I_1 produced by lens 1 by applying Eq. 35-9 to lens 1 alone:

$$\frac{1}{p_1} + \frac{1}{i_1} = \frac{1}{f_1}. \qquad (35\text{-}27)$$

The object O_1 for lens 1 is the seed, which is 9.0 cm from the lens; thus, we substitute $+9.0$ cm for p_1. Another **Key Idea** here is that

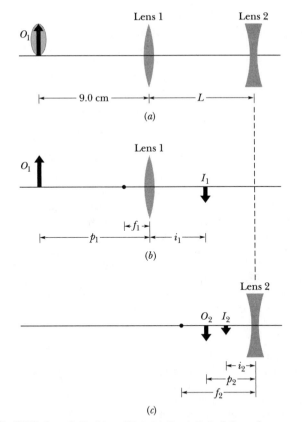

Fig. 35-37 Sample Problem 35-6. (*a*) Seed O_1 is 9.0 cm from a two-lens system with lens separation L. We use the arrow to orient the seed. (*b*) The image I_1 produced by lens 1 alone. (*c*) Image I_1 acts as object O_2 for lens 2 alone, which produces the final image I_2.

because lens 1 is a converging lens, its focal length is a positive quantity; thus, we substitute $+3.0$ cm for f_1. These two substitutions give us

$$\frac{1}{+9.0 \text{ cm}} + \frac{1}{i_1} = \frac{1}{+3.0 \text{ cm}},$$

which yields $i_1 = +4.5$ cm.

This tells us that the image I_1 is 4.5 cm from lens 1 and real. (We could have guessed that it is real by noting that the seed is outside the focal point of lens 1.) Since I_1 is real, it is on the opposite side of the lens from the seed (the object) and is inverted relative to the seed. These results differ from those of Sample Problem 35-4, and we must be careful to draw image I_1 on the correct side of lens 1, as is done in Fig. 35-37*b*.

Lens 2. One **Key Idea** for this next step of our solution is that we can treat image I_1 as an object O_2 for lens 2, and ignore lens 1. The object distance p_2 for this object is

$$p_2 = L - i_1 = 8.5 \text{ cm} - 4.5 \text{ cm} = 4.0 \text{ cm}.$$

(Because this object O_2 is on the same side of lens 2 as the source of light, this object distance p_2 is positive, as usual. If O_2 had turned out to be on the right side of lens 2, then p_2 would have been negative.)

Another **Key Idea** here is that because lens 2 is a diverging lens, its focal length is a negative quantity: $f_2 = -6.0$ cm. Substituting these values for p_2 and f_2 into Eq. 35-27 gives us

$$\frac{1}{+4.0 \text{ cm}} + \frac{1}{i_2} = \frac{1}{-6.0 \text{ cm}}$$

or
$$i_2 = -2.4 \text{ cm}. \qquad \text{(Answer)}$$

Thus, the "final" image I_2 produced by the system of two lenses is 2.4 cm from lens 2. The minus sign indicates that I_2 is virtual and on the same side of lens 2 as the object O_2 for that lens, as shown in Fig. 35-37*c*. I_2 turns out to be located between the two lenses. If you placed a card at that location, you would not see I_2 projected on the card; because I_2 is a virtual image, you could see it only by placing your eye in the path of the light rays emerging *at the right of lens 2*. I_2 would then appear to you to be on the opposite (left) side of lens 2, and 2.4 cm away from it.

(b) Seed O_1 has height $h = 2.0$ mm (measured perpendicular to the central axis of the lens system). What is the height of image I_2?

SOLUTION: One **Key Idea** here is that the height h' of image I_2 is the product of the magnitude of the system's overall lateral magnification M and the object's height h:

$$h' = |M|h. \qquad (35\text{-}28)$$

A second **Key Idea** is that M is the product of the individual lateral magnifications m_1 and m_2 of the lenses in the system, according to Eq. 35-11 ($M = m_1 m_2$). Thus, we can rewrite Eq. 35-28 as

$$h' = |m_1 m_2|h.$$

Substituting from Eq. 35-6 ($m = -i/p$) for each of the two lenses then yields

$$h' = \left| \left(-\frac{i_1}{p_1} \right) \left(-\frac{i_2}{p_2} \right) \right| h.$$

Finally, substituting known data leads to

$$h' = \left| \left(-\frac{+4.5 \text{ cm}}{+9.0 \text{ cm}} \right) \left(-\frac{-2.4 \text{ cm}}{+4.0 \text{ cm}} \right) \right| (2.0 \text{ mm})$$

$$= 0.60 \text{ mm}. \qquad \text{(Answer)}$$

QUESTIONS

11. *Organizing question:* In Fig. 35-38, stick figure O stands in front of a spherical mirror that is mounted within the boxed region; the central axis through the mirror is shown. The four stick figures I_1 to I_4 suggest general locations and orientations for the images that might be produced by the mirror. (The figures are only sketched in; neither their heights nor their distances from the mirror

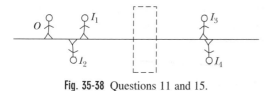

Fig. 35-38 Questions 11 and 15.

are drawn to scale.) (a) Which of the stick figures could not possibly represent images? Of the possible images, (b) which would be due to a concave mirror, (c) which would be due to a convex mirror, (d) which would be virtual, and (e) which would involve negative magnification?

12. (a) When the Sun is low over a perfectly calm body of water that stretches away from you to the horizon, you can see a single image of the Sun in the water via sunlight that reflects to you. If the Sun is 15° above the horizon, how far below the horizon is its image? (b) The scene differs when waves sweep over the water, because the tilted surfaces produced by the waves shift the image of the Sun upward toward the horizon or downward from it. The composite of all the images from the various changing tilts of the water surfaces results in a luminous *glitter path* of sunlight on the water. The shape of this luminous region might be similar to that shown in Fig. 35-39. Is the nearest point due to a surface that is tilted up toward you or up away from you? (*Hint:* Experiment with a flat mirror. Lay the mirror on a table so that in it you see the image of an object on the other side of the table. Then tilt the mirror's front edge or its back edge upward. Does the image move toward or away from you?)

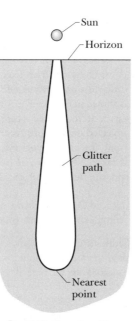

Fig. 35-39 Question 12.

13. *Organizing question:* Figure 35-40 shows three situations in which a stick figure stands in front of a spherical mirror; the focal points *F* are indicated. For each situation determine whether (a) the focal length, (b) the image distance, and (c) the lateral magnification are positive or negative. (*Hint:* If you filled out Table 35-1, this should be a snap, rather than a crackle or a pop.)

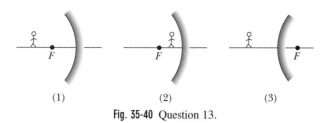

Fig. 35-40 Question 13.

14. A toy soldier is placed, in turn, in front of four mirrors, *A*, *B*, *C*, and *D*. The following table gives the object distances *p* and corresponding image distances *i*, all in centimeters. (a) Rank the mirrors according to the size of the (lateral) height of the image, greatest first. (b) Which mirrors produce an image of the toy soldier that could appear on a sheet of paper?

	A	B	C	D
p	2	4	2	6
i	4	-8	-6	2

15. *Organizing question:* In Fig. 35-38, stick figure *O* stands in front of a thin, symmetric lens that is mounted within the boxed region; the central axis through the lens is shown. The four stick figures I_1 to I_4 suggest general locations and orientations for the images that might be produced by the lens. (The figures are only sketched in; neither their height nor their distance from the lens is drawn to scale.) (a) Which of the stick figures could not possibly represent images? Of the possible images, (b) which would be due to a converging lens, (c) which would be due to a diverging lens, (d) which would be virtual, and (e) which would involve negative magnification?

16. Figure 35-41 shows a coordinate system in front of a flat mirror, with the *x* axis perpendicular to the mirror. Draw the image of the system in the mirror. (a) Which axis is reversed by the reflection? (b) If you face a mirror, is your image inverted (top for bottom)? Are your left and right reversed (as commonly believed)? (c) What then is reversed?

Fig. 35-41 Question 16.

17. *Organizing question:* Figure 35-42 shows three situations in which a stick figure stands in front of a symmetric thin lens; the focal points *F* are indicated. For each situation determine whether (a) the focal length, (b) the image distance, and (c) the lateral magnification are positive or negative. (*Hint:* If you have filled out Table 35-2, this should be a breeze, rather than a twister or a willy-nilly.)

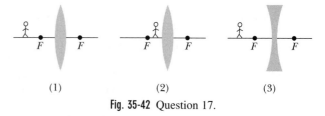

Fig. 35-42 Question 17.

18. (a) If an object is outside the focal point of a converging lens, does its image height increase, decrease, or stay the same when the object moves away from the lens? (*Hint:* Consider how the rays in Fig. 35-14*a* would change.) (b) If an object is in front of a diverging lens, does its image height increase, decrease, or stay the same when the object moves away from the lens? (*Hint:* Consider how the rays in Fig. 35-14*c* would change.)

19. As a penguin waddles toward a large converging lens, the image of the penguin, formed by the lens, moves as well. (a) If the penguin is outside the focal point, does the image move toward the lens or away from the lens? (b) Repeat part (a) for the penguin inside the focal point, waddling toward the lens.

20. A turnip sits before a thin converging lens, outside the focal point of the lens. The lens is filled with a transparent gel so that it is flexible; by squeezing its ends toward its center (as indicated in Fig. 35-43a), you can increase the curvature of its front and rear sides. (a) When you squeeze the lens, does the image of the turnip move toward or away from the lens? (b) Does the lateral height of the image increase, decrease, or stay the same? (c) Suppose that you must keep the image on a card at a certain distance behind the lens (Fig. 35-43b) while you move the turnip away from the lens. Must you increase or decrease your squeeze on the lens during the move?

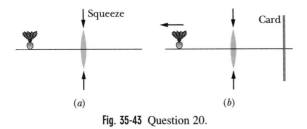

Fig. 35-43 Question 20.

DISCUSSION QUESTIONS

21. Design a periscope that makes use of total internal reflection. What are the advantages of such a device compared with one that uses mirrors?

22. Can you think of a system of mirrors that would let us see ourselves as others see us (without the so-called left–right reversal)? If so, draw it and prove your point by sketching some typical rays.

23. Devise a system of plane mirrors that will let you see the back of your head. Trace the rays to prove your point.

24. Can a virtual image be photographed by exposing film at the location of the image? Explain.

25. We have all seen TV pictures of a baseball game shot from a camera located somewhere behind second base. The pitcher and the batter are about 60 ft apart but they look much closer on the TV screen. Why are images viewed through a telephoto lens foreshortened in this way?

26. An unsymmetrical thin lens forms an image of a point object on its axis. Is the image location changed if the lens is reversed?

27. The *f-number* of a camera lens is its focal length divided by its aperture (effective diameter). Why is this useful to know in photography? How can the *f*-number of the lens be changed?

28. In William Golding's *Lord of the Flies* the character Piggy uses his eyeglasses to focus the Sun's rays and kindle a fire. Later, the boys abuse Piggy and break his glasses. He is unable to identify them at close range because he is nearsighted. Find the flaw in this narrative.

29. Explain the function of the objective lens of a microscope; why use an objective lens at all? Why not just use a very powerful simple magnifier?

30. Why do astronomers use optical telescopes in looking at the sky? After all, the stars are so far away that they still appear to be points of light, without any discernible detail.

EXERCISES & PROBLEMS

38. For a concave mirror, graph the lateral magnification m as a function of p/f, where p is the object distance and f is the (fixed) focal length.

39. Isaac Newton, having convinced himself (erroneously as it turned out) that chromatic aberration is an inherent property of refracting telescopes, invented the reflecting telescope, shown schematically in Fig. 35-44. He presented his second model of this telescope, with a magnifying power of 38, to the Royal Society (of England), which still has it. In Fig. 35-44 incident light falls, closely parallel to the telescope axis, on the objective mirror M. After reflection from small mirror M' (the figure is not to scale), the rays form a real, inverted image in the *focal plane* (the plane perpendicular to the line of sight, at focal point F). This image is then viewed through an eyepiece. (a) Show that the angular magnification m_θ for the device is given by Eq. 35-15:

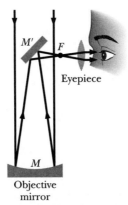

Fig. 35-44 Problem 39.

$$m_\theta = -f_{ob}/f_{ey},$$

where f_{ob} is the focal length of the objective mirror and f_{ey} is that of the eyepiece. (b) The 200 in. mirror in the reflecting telescope at Mt. Palomar in California has a focal length of 16.8 m. Estimate the size of the image formed by this mirror when the object is a meter stick 2.0 km away. Assume parallel incident rays. (c) The mirror of a different reflecting astronomical telescope has an effective radius of curvature of 10 m ("effective" because such mirrors are ground to a parabolic rather than a spherical shape, to eliminate spherical aberration defects). To give an angular magnification of 200, what must be the focal length of the eyepiece?

40. In Fig. 35-45, a box is somewhere at the left, on the central

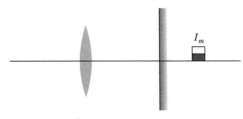

Fig. 35-45 Problem 40.

axis of the thin converging lens. The image I_m of the box produced by the plane mirror is 4.00 cm "inside" the mirror. The lens–mirror separation is 10.0 cm, and the focal length of the lens is 2.00 cm. (a) What is the distance between the box and the lens? Light reflected by the mirror travels back through the lens, which produces a final image of the box. (b) What is the distance between the lens and that final image?

41. An object is located to the left of a thin lens, on the central axis of the lens. The lens forms an image of the object with the same orientation as the object, and the image–object distance is 20 cm. What are the focal length of the lens and the lens–object distance if the image has (a) 2.0 times the height of the object and (b) 0.50 times the height of the object?

42. In Fig. 35-46, a fish watcher watches a fish through a 3.0-cm-thick glass wall of a fish tank. The watcher is level with the fish; the index of refraction of the glass is 8/5 and that of the water is 4/3. (a) To the fish, how far away does the watcher appear to be? (*Hint:* The watcher is the object. Light from that object passes through the wall's outside surface, which acts as a refracting surface. Find the image produced by that surface. Then treat that image as an object whose light passes through the wall's inside surface, which acts as another refracting surface. Find the image produced by that surface, and there is the answer.) (b) To the watcher, how far away does the fish appear to be?

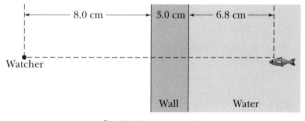

Fig. 35-46 Problem 42.

43. A small cup of green tea is positioned on the central axis of a spherical mirror. The lateral magnification of the cup is +0.250, and the distance between the mirror and its focal point is 2.00 cm. (a) What is the distance between the mirror and the image it produces? (b) Is the focal length positive or negative? (c) Is the image real or virtual?

44. In an eye that is *farsighted,* the eye focuses parallel rays so that the image I is formed behind the retina, as in Fig. 35-47a. In an eye that is *nearsighted,* the image I is formed in front of the retina, as in Fig. 35-47b. (a) How would you design a corrective lens for each eye defect? Make a ray diagram for each case. (b) If you need eyeglasses only for reading, are you nearsighted or are you

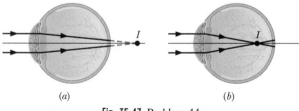

Fig. 35-47 Problem 44.

farsighted? (c) What is the function of bifocal glasses, in which the upper and lower parts have different focal lengths?

45. You have a supply of flat glass disks ($n = 1.5$) and a lens-grinding machine that can be set to grind a radius of curvature of either 40 cm or 60 cm. You are asked to prepare a set of six lenses like those shown in Fig. 35-48. What will be the focal length of each lens? Will the lens form a real or a virtual image of the Sun? (*Note:* Where you have a choice of radii of curvature, select the smaller one.)

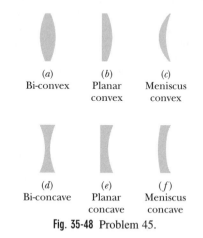

Fig. 35-48 Problem 45.

46. Light travels from point A to point B via reflection at point O on the surface of a mirror. Without using calculus, show that length AOB is a minimum when the angle of incidence θ is equal to the angle of reflection ϕ. (*Hint:* Consider the virtual image of A in the mirror.)

47. In Fig. 35-49a, a pea sits at a focal point of the first (nearer) thin diverging lens, 4.00 cm from that lens. The lenses are identical and separated by 10.0 cm, with a common central axis. (a) Where is the image of the pea produced by the second lens? (b) Is that image inverted or does it have the same orientation as the pea? (c) Is it real or virtual?

Fig. 35-49 Problems 47 and 48.

48. In Fig. 35-49b, a sand grain is 3.00 cm from the first (nearer) thin lens, on the central axis through the two symmetric lenses. The distance between focal point and lens is 4.00 cm for both lenses; the lenses are separated by 8.00 cm. (a) What is the distance between the second lens and the image that it produces of the sand grain? (b) Is the image to the left or right of the second lens? (c) Is it real or virtual? (d) Is it inverted from the sand grain?

49. A pepper seed is placed in front of a lens. The lateral magnification of the seed is +0.300. The absolute value of the lens's focal length is 40.0 cm. How far from the lens is the image?

50. One end of a long glass rod ($n = 1.5$) is a convex surface of radius 6.0 cm. An object is located in air along the axis of the rod, at a distance of 10 cm from the convex end. (a) How far apart are the object and the image formed by the glass rod? (b) Within what range of distances from the end of the rod must the object be located in order to produce a virtual image?

51. A grasshopper hops to a point on the central axis of a spherical mirror. The absolute magnitude of the mirror's focal length is 40.0 cm, and the lateral magnification of the grasshopper image produced by the mirror is +0.200. (a) Is the mirror convex or concave? (b) How far from the mirror is the grasshopper?

52. In Fig. 35-50, an object is placed a distance in front of a converging lens equal to twice the focal length f_1 of the lens. On the other side of the lens is a concave mirror of focal length f_2 separated from the lens by a distance $2(f_1 + f_2)$. (a) Find the location, type, orientation, and lateral magnification of the final image, as seen by an eye looking toward the mirror through the lens and just past (to one side of) the object. (b) Draw a ray diagram to locate the image.

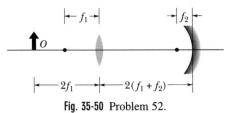

Fig. 35-50 Problem 52.

53. Using the lens maker's formula (Eq. 35-10), decide which of the thin lenses in Fig. 35-51 are converging and which are diverging for incident light rays that are parallel to the central axis of the lens.

54. A point object is 10 cm away from a plane mirror while the eye of an observer (with pupil diameter 5.0 mm) is 20 cm away. Assuming the eye and the object to be on the same line perpendicular to the mirror surface, find the area of the mirror used in observing the reflection of the point. (*Hint:* Adapt Fig. 35-4.)

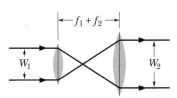

(a)   (b)   (c)   (d)

Fig. 35-51 Problem 53.

55. You are standing 1.0 m in front of a large shiny sphere that is 0.70 m in diameter. (a) How far from the surface of the sphere closest to you, and on which side of that surface, does your image appear to be? (b) If you are 2.0 m tall, how tall is your image? (c) Is the image inverted or erect with respect to you?

56. A concave mirror has a radius of curvature of 24 cm. How far is an object from the mirror if an image is formed that is (a) virtual and 3.0 times the size of the object, (b) real and 3.0 times the size of the object, and (c) real and 1/3 the size of the object?

57. A cheese enchilada is 4.00 cm in front of a converging lens. The magnification of the enchilada is −2.00. What is the focal length of the lens?

58. A glass sphere has a radius of 5.0 cm and an index of refraction of 1.6. A paperweight is constructed by slicing through the sphere along a plane that is 2.0 cm from the center of the sphere. The paperweight is placed on a table and viewed from directly above by an observer who is 8.0 cm from the tabletop, as shown in Fig.

35-52. When viewed through the paperweight, how far away does the tabletop appear to be to the observer?

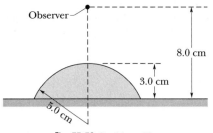

Fig. 35-52 Problem 58.

59. An object is 30.0 cm from a spherical mirror, along the mirror's central axis. The mirror produces an inverted image with a lateral magnification of absolute value 0.500. What is the focal length of the mirror?

60. An object is placed 1.0 m in front of a converging lens, of focal length 0.50 m, which is 2.0 m in front of a plane mirror. (a) Where is the final image, as measured from the lens, that would be seen by an eye looking toward the mirror through the lens (and just past the object)? (b) Is the final image real or virtual? (c) Is the orientation of the final image the same as the object or inverted? (d) What is the lateral magnification?

61. Two coaxial converging lenses, with focal lengths f_1 and f_2, are positioned a distance $f_1 + f_2$ apart, as shown in Fig. 35-53. Arrangements like this are called *beam expanders* and are often used to increase the diameter of a light beam from a laser. (a) If W_1 is the incident beam width, show that the width of the emerging beam is $W_2 = (f_2/f_1)W_1$. (b) Explain how a combination of one diverging and one converging lens can also be arranged as a beam expander. Incident rays parallel to the lens axis should exit parallel to that axis.

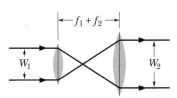

Fig. 35-53 Problems 61 and 62.

62. Calculate the ratio of the intensity of the beam emerging from the beam expander of Problem 61 to the intensity of the incident beam.

63. Figure 35-54 shows an idealized submarine periscope (without

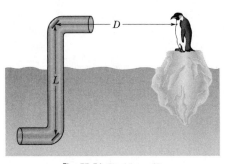

Fig. 35-54 Problem 63.

submarine) that consists of two parallel plane mirrors set at 45° to the vertical periscope axis and separated by distance L. A penguin is sighted at a distance D from the top mirror. (a) Is the image seen by a submarine officer peering into the periscope real or virtual? (b) Does it have the same orientation as the penguin or is it inverted? (c) Is the size (height) of the image greater than, less than, or the same as that of the penguin? (d) What is the distance of the image from the bottom mirror?

64. Two plane mirrors are placed parallel to each other and 40 cm apart. An object is placed 10 cm from one mirror. What is the distance from the object to the image for each of the five images that are closest to the object?

36

Interference

Sample Problem 36-7

A double-slit interference pattern is produced on a screen, as in Fig. 36-8; the light is monochromatic at a wavelength of 600 nm. A strip of transparent plastic with index of refraction $n = 1.50$ is to be placed over one of the slits. Its presence changes the interference between light waves from the two slits, causing the interference pattern to be shifted across the screen from the original pattern. Figure 36-38a shows the original locations of the central bright fringe ($m = 0$) and the first bright fringes ($m = 1$) above and below the central fringe. We want the presence of the plastic over one slit to shift the pattern upward, with the lower $m = 1$ bright fringe being shifted to the center of the pattern. Should the plastic be placed over the top slit (as arbitrarily drawn in Fig. 36-38b) or the bottom slit, and what thickness L should it have?

SOLUTION: Figure 36-38a shows rays r_1 and r_2 along which waves from the two slits travel to reach the lower $m = 1$ bright fringe. One **Key Idea** here is that those waves start in phase at the slits but arrive at the fringe with a phase difference of exactly 1 wavelength. To remind ourselves of this main characteristic of the fringe, let us call it the 1λ fringe. The one-wavelength phase difference is due to the one-wavelength path length difference between the rays

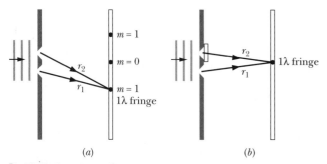

Fig. 36-38 Sample Problem 36-7. (a) Arrangement for two-slit interference (not to scale). The locations of three bright fringes (or maxima) are indicated. (b) A strip of plastic covers the top slit. We want the 1λ fringe to be at the center of the pattern.

reaching the fringe; that is, there is exactly one more wavelength along ray r_2 than along r_1.

Figure 36-38b shows the 1λ fringe shifted up to the center of the pattern with the plastic strip over the top slit (we still do not know if the plastic should be there or over the bottom slit). The figure also shows the new orientations of rays r_1 and r_2 to reach that fringe. There still must be one more wavelength along r_2 than along r_1 (because they still produce the 1λ fringe), but now the path length difference between those rays is zero, as we can tell from the geometry of Fig. 36-38b. However, r_2 now passes through the plastic.

The next **Key Idea** is to recall that the wavelength λ_n of light in a material with index of refraction n is smaller than that in vacuum, as given by Eq. 36-8 ($\lambda_n = \lambda/n$). Here, this means that the wavelength of the light is smaller in the plastic than in the air. Thus, the ray that passes through the plastic will have more wavelengths along it than the ray that passes through only air—so we do get the one extra wavelength we need along ray r_2 by placing the plastic over the top slit, as drawn in Fig. 36-38b.

To determine the required thickness L of the plastic, we use the same **Key Idea** and procedure as in Sample Problem 36-1a. Here, as there, waves that are initially in phase travel equal distances L through different materials (plastic and air). However, here we know the phase difference and require L, so we again use Eq. 36-11,

$$N_2 - N_1 = \frac{L}{\lambda}(n_2 - n_1). \quad (36\text{-}42)$$

We know that $N_2 - N_1$ is 1 for a phase difference of one wavelength, index n_2 is 1.50 for the plastic in front of the top slit, index n_1 is 1.00 for the air in front of the bottom slit, and the wavelength λ is 600×10^{-9} m. Solving Eq. 36-42 for L and substituting these values leads us to

$$L = \frac{\lambda(N_2 - N_1)}{n_2 - n_1} = \frac{(600 \times 10^{-9}\ \text{m})(1)}{1.50 - 1.00}$$
$$= 1.2 \times 10^{-6}\ \text{m}. \quad \text{(Answer)}$$

Sample Problem 36-8

As explained in the text, the iridescence seen in the top surface of *Morpho* butterfly wings is due to constructive interference of the light reflected by thin terraces of transparent cuticle-like material. The terraces extend outward, parallel to the wings, from a central

structure that is approximately perpendicular to the wing. Cross sections of terraces and their supporting structure are shown in the electron micrograph of Fig. 36-39a. The terraces have index of refraction $n = 1.53$ and thickness $D_t = 63.5$ nm; they are separated

Terraces

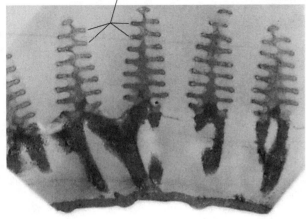

(a)

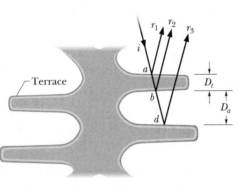

(b)

Fig. 36-39 Sample Problem 36-8. (a) An electron micrograph shows the cross section of terrace structures of cuticle-like material that stick up from the top surface of a *Morpho* wing. (b) Light waves reflecting at points a and b on a terrace, represented by rays r_1 and r_2, interfere at the eye of an observer. The wave of ray r_1 also interferes with the wave that reflects at point d and which is represented by ray r_3.

(in air) by $D_a = 127$ nm. If the incident light is perpendicular to the terraces (see Fig. 36-39b, where the angle of the incident light is exaggerated), at what wavelength of visible light do the reflections from the terraces have an interference maximum?

SOLUTION: Let us first consider rays r_1 and r_2 in Fig. 36-39b, which involve reflections at points a and b. One **Key Idea** here is that this situation is just like that of Fig. 36-12 and Table 36-1, and so Eq. 36-34 gives the interference maxima. Solving Eq. 36-34 for λ gives us

$$\lambda = \frac{2n_2 L}{m + \frac{1}{2}}.$$

Substituting D_t (= 63.5 nm) for L and n (= 1.53) for n_2, we have

$$\lambda = \frac{2nD_t}{m + \frac{1}{2}} = \frac{(2)(1.53)(63.5 \text{ nm})}{m + \frac{1}{2}} = \frac{194 \text{ nm}}{m + \frac{1}{2}}.$$

For $m = 0$, we find an interference maximum at $\lambda = 388$ nm, which is in the ultraviolet region. For all greater values of m, λ is even smaller, farther into the ultraviolet. Thus, rays r_1 and r_2 do not produce the bright blue-green color of the *Morpho*.

Another **Key Idea** is that interference occurs not only between waves reflected by a single terrace but also between waves reflected by different terraces. For example, the waves of rays r_1 and r_3 in Fig. 36-39b undergo interference that might produce the blue-green color of the *Morpho*. The wave of ray r_3 passes through the top right terrace and then through air to the next lower terrace, where it reflects at point d. Then it travels upward, as represented by ray r_3. The path length difference between rays r_1 and r_3 is $2D_t + 2D_a$.

This situation differs considerably from that of Fig. 36-12, and Eq. 36-34 does not apply. To find an equation for the interference maxima for this new situation, we must first find the phase shifts due to the reflections and then count the wavelengths along path length difference $2D_t + 2D_a$.

The reflections at points a and d both introduce a phase change of 0.5 wavelength, so the reflections alone tend to put the waves of rays r_1 and r_3 in phase. Thus, for these waves actually to end up in phase, the number of wavelengths along the path length difference $2D_t + 2D_a$ must be an integer. The wavelength within the terrace is $\lambda_n = \lambda/n$. Then the number of wavelengths in length $2D_t$ is

$$N_t = \frac{2D_t}{\lambda_n} = \frac{2D_t n}{\lambda}.$$

Similarly, the number of wavelengths in length $2D_a$ (in air, with $n = 1$) is

$$N_a = \frac{2D_a}{\lambda}.$$

For the waves of rays r_1 and r_3 to be in phase, we need $N_t + N_a$ to be equal to an integer m. Thus, for an interference maximum,

$$\frac{2D_t n}{\lambda} + \frac{2D_a}{\lambda} = m, \qquad \text{for } m = 1, 2, 3, \dots.$$

Solving for λ and substituting the given data, we obtain

$$\lambda = \frac{(2)(63.5 \text{ nm})(1.53) + (2)(127 \text{ nm})}{m} = \frac{448 \text{ nm}}{m}.$$

For $m = 1$, we find

$$\lambda = 448 \text{ nm.} \qquad \text{(Answer)}$$

This wavelength corresponds to the bright blue-green color of the top surface of a *Morpho* wing.

QUESTIONS

13. In Fig. 36-40, the waves of rays 1 and 2 initially have the same wavelength and are in phase while they travel through plastic material. Along the way, the wave of ray 2 passes through an air cavity of length L. Is the number of wavelengths of ray 2

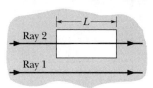

Fig. 36-40 Question 13.

in that length L in the air greater than, less than, or equal to the number of wavelengths of ray 1 in that same length in the plastic?

14. Figure 36-41 shows three situations in which two rays of sunlight penetrate slightly into and then scatter out of lunar soil. Assume that the rays are initially in phase. In which situation are the associated waves most likely to end up in phase? (Just as the Moon becomes full, its brightness suddenly peaks, becoming 25% greater than its brightness on the nights before and after, because at full Moon we intercept light waves that are scattered by lunar soil back toward the Sun and undergo constructive interference at our eyes.)

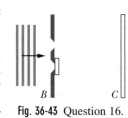

Fig. 36-41 Question 14.

15. Monochromatic yellow light is used in a Young's interference experiment; then monochromatic green light is used. Graphs of intensity I versus position x on the viewing screen in the two experiments are superimposed in Fig. 36-42. In that figure, is the central bright fringe off to the left or to the right?

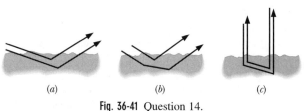

Fig. 36-42 Question 15.

16. In Fig. 36-43, a thin, transparent plastic layer has been placed over the lower slit in a double-slit experiment. Does this cause the central maximum (the fringe where waves arrive with a phase difference of zero wavelengths) to move up or down the screen? (*Hint:* Is the wavelength in the plastic greater than or less than that in air?)

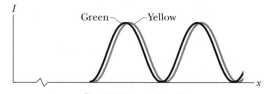

Fig. 36-43 Question 16.

17. Figure 36-44a gives intensity I versus position x on the viewing screen for the central portion of a two-slit interference pattern. The other parts of the figure give phasor diagrams for the electric field

components of the waves arriving at the screen from the two slits (as in Fig. 36-10a). Which numbered points on the screen best correspond to which phasor diagram?

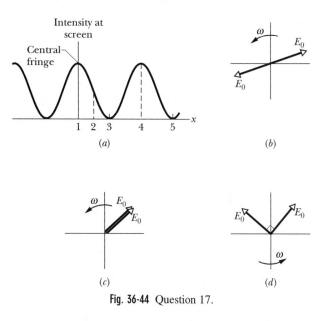

Fig. 36-44 Question 17.

18. Whole milk is a liquid suspension of fat and other particles. If you hold a spoon partially filled with milk in bright sunlight, you will see fleeting points of color just inside the perimeter of the milk. What causes them?

19. *Organizing question:* Figure 36-45 shows four situations in which light reflects perpendicularly from a thin film of thickness L between much thicker materials. The indexes of refraction are given. In which situations does Eq. 36-34 correspond to the reflections yielding maxima (that is, a bright film)?

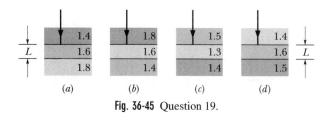

Fig. 36-45 Question 19.

20. Figure 36-46a shows an arrangement similar to that given in Sample Problem 36-1: waves with identical wavelengths and amplitudes and that are initially in phase travel through different media, ray 2 through a plastic layer of thickness L and ray 1 through only air. The number of wavelengths in length L is N_2 for ray 2 and N_1 for ray 1. The following table gives values for N_2 and N_1 for four situations. In each situation, the two rays reach a common point on a screen. Rank the situations according to the intensity of the light at that common point, greatest first.

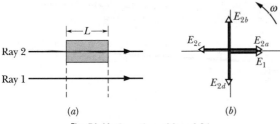

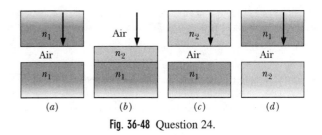

Fig. 36-46 Questions 20 and 21.

Fig. 36-48 Question 24.

Situation	1	2	3	4
N_2	2.75	2.80	3.25	4.00
N_1	2.25	1.80	3.00	3.25

21. For the rays of Question 20, let phasor E_1 represent the electric field component of ray 1 and phasor E_2 represent that of ray 2. In the phasor diagram of Fig. 36-46b, which choice of E_2 best corresponds to which situation in Question 20?

22. Sunlight illuminates a thin film of oil that floats on water, which has a greater index of refraction than the oil. The edge of the film has thickness $L < 0.1\lambda$. Is the edge dark (like the corresponding thin region of the soap film in Fig. 36-14) or bright?

23. The eyes of some animals contain reflectors that send light to receptors where the light is absorbed. In the scallop, the reflector consists of many thin transparent layers alternating between high and low indexes of refraction. With the proper layer thicknesses, the combined reflections from the interfaces end up in phase with one another, thereby giving a much brighter reflection than a single biological surface or layer could give. Figure 36-47 shows such an arrangement of alternating layers, along with the reflections due to a single perpendicularly incident ray i. In terms of the indexes of refraction n_1 and n_2 and the wavelength λ of visible light, should the thicknesses be (a) $L_1 = \lambda/4n_1$ and $L_2 = \lambda/4n_2$ or (b) $L_1 = \lambda/2n_1$ and $L_2 = \lambda/2n_2$?

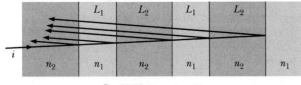

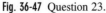

Fig. 36-47 Question 23.

24. Figure 36-48 shows four situations in which light of wavelength λ is incident perpendicularly on a very thin layer of air or plastic. The indicated indexes of refraction are $n_1 = 1.33$ and $n_2 = 1.50$. In each situation the thin layer has thickness $L < 0.1\lambda$. In which situations will the light reflected by the thin layer be approximately eliminated by interference?

25. Figure 36-49 shows light that passes through a thin film of water on glass, either directly (ray r_1) or via reflections (ray r_2). The indexes of refraction are indicated. The incident ray is actually perpendicular to the film but is drawn slanted for clarity. (a) How many reflections does the light that emerges as r_2 undergo? (b) What is the reflection phase shift of that light at each of those reflections?

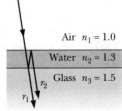

Fig. 36-49 Question 25.

DISCUSSION QUESTIONS

26. Light has (a) a wavelength, (b) a frequency, and (c) a speed. Which, if any, of these quantities remains unchanged when light passes from a vacuum into a slab of glass?

27. The speed and wavelength of, say, red light that we see in air are reduced when the light passes into water. Would that light then appear to be another color—blue, perhaps—if you viewed it from under the water surface?

28. Would you expect sound waves to obey the laws of reflection and refraction obeyed by light waves? Does Huygens' principle apply to sound waves in air? If Huygens' principle predicts the laws of reflection and refraction, why is it necessary or desirable to view light as an electromagnetic wave, with all the attendant complexity?

29. In Young's double-slit interference experiment, using a monochromatic laboratory light source, why is screen A in Fig. 36-6 necessary?

30. Why are parallel slits preferable to the pinholes that Young used in demonstrating interference?

31. Describe the pattern of light intensity on screen C in Fig. 36-8 if one slit is covered with a red filter and the other with a blue filter, the incident light being white.

32. What causes the fluttering of a TV picture when an airplane flies overhead?

33. Is it possible to have coherence between light sources emitting light of different wavelengths?

34. Suppose each slit in Fig. 36-8 is covered with a sheet of Polaroid, with the polarizing directions of the two sheets perpendicular. What would be the pattern of light intensity on screen C? (The incident light is unpolarized.)

35. Suppose the film coating in Fig. 36-15 had an index of refraction greater than that of the glass. Could it still be nonreflecting? If so, what difference would it make?

36. What are the requirements for maximum intensity when a thin film is viewed by *transmitted* light?

37. Why do coated lenses (see Sample Problem 36-5) look purple by reflected light?

38. A person wets his eyeglasses to clean them. As the water evaporates he notices that for a short time the glasses become markedly less reflecting. Explain why.

39. A lens is coated to reduce reflection, as in Sample Problem 36-5. What happens to the energy that had previously been reflected? Is it absorbed by the coating?

40. An automobile directs its headlights onto the side of a barn. Why are interference fringes not produced in the region in which light from the two beams overlaps?

41. If the path length to the movable mirror in Michelson's interferometer (see Fig. 36-17) greatly exceeds that to the fixed mirror (say, by more than a meter), the fringes begin to disappear. Explain why. Lasers greatly extend this range. Why?

42. How would you construct an acoustical Michelson interferometer to measure sound wavelengths? Discuss differences from the optical interferometer.

EXERCISES & PROBLEMS

61. A broad beam of light of wavelength 600 nm is sent directly downward through the glass plate ($n = 1.5$) in Fig. 36-50. That plate and a plastic plate ($n = 1.2$) form a thin wedge of air which acts as a thin film. An observer looking down through the top plate sees the fringe pattern shown in Fig. 36-50b, with dark fringes centered on ends A and B. (a) What is the thickness of the wedge at B? (b) How many dark fringes will the observer see if the air between the plates is replaced with water ($n = 1.33$)?

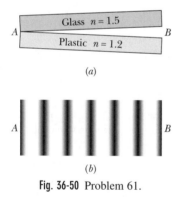

(a)

(b)

Fig. 36-50 Problem 61.

62. In Sample Problem 36-5, assume that the coating eliminates the reflection of light of wavelength 550 nm at normal incidence. By what percentages is reflection diminished by the coating at 450 and 650 nm?

63. In a double-slit experiment (Fig. 36-8), $\lambda = 546$ nm, $d = 0.10$ mm, and $D = 20$ cm. On a viewing screen, what is the distance between the fifth maximum and the seventh minimum from the central maximum?

64. Sodium light ($\lambda = 589$ nm) illuminates two slits separated by $d = 2.0$ mm. The slit–screen distance D is 40 mm. What percentage error is made by using Eq. 36-14 to locate the $m = 10$ bright fringe on the screen rather than using the exact path length difference?

65. When an electron moves through a medium at a speed exceeding the speed of light in that medium, the electron radiates electromagnetic energy (the *Cerenkov effect*). What minimum speed must an electron have in a liquid with index of refraction 1.54 in order to radiate?

66. Three electromagnetic waves travel through a certain point along an x axis. They are polarized parallel to a y axis, with the following variations in their amplitudes. Find their resultant at the point.

$$E_1 = (10.0 \ \mu\text{V/m}) \sin[(2.0 \times 10^{14} \ \text{rad/s})t]$$
$$E_2 = (5.00 \ \mu\text{V/m}) \sin[(2.0 \times 10^{14} \ \text{rad/s})t + 45°]$$
$$E_3 = (5.00 \ \mu\text{V/m}) \sin[(2.0 \times 10^{14} \ \text{rad/s})t - 45°]$$

67. In Fig. 36-8, let the angle θ of the two rays be 20°, slit separation d be 58.00 μm, and wavelength λ be 500.9 nm. (a) In terms of wavelengths, what is the phase difference of the two rays when they reach a common point on a distant screen? (b) Does their interference result in complete darkness, maximum brightness, intermediate illumination but closer to complete darkness, or intermediate illumination but closer to maximum brightness?

68. Two coherent radio point sources that are separated by 2.0 m are radiating in phase with a wavelength of 0.25 m. If a detector moves in a large circle around their midpoint, at how many points will the detector show a maximum signal?

69. Suppose that in Fig. 36-12 the light is not incident perpendicularly on the thin film but at an angle $\theta_i > 0$. Find an equation like Eqs. 36-34 and 36-35 that gives the interference maxima for the waves of rays r_1 and r_2. The wavelength is λ, the film thickness is L, and $n_2 > n_1 = n_3 = 1.0$.

70. If the distance between the first and tenth minima of a double-slit pattern is 18 mm and the slits are separated by 0.15 mm with the screen 50 cm from the slits, what is the wavelength of the light used?

71. A thin film suspended in air is 0.410 μm thick and is illuminated with white light incident perpendicularly on its surface. The index of refraction of the film is 1.50. At what wavelengths will visible light that is reflected from the two surfaces of the film undergo fully constructive interference?

72. Figure 36-51 shows the design of a Texas arcade game. Four laser pistols are pointed toward the center of an array of plastic layers where a clay armadillo is the target. The indexes of refraction of the various plastic layers are indicated. The layer thicknesses are either 2.00 mm or 4.00 mm, as drawn. If the pistols are fired simultaneously, (a) which laser burst hits the target first and (b) what are the times of flight for the four bursts?

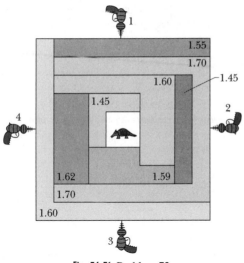

Fig. 36-51 Problem 72.

73. Find the slit separation of a double-slit arrangement that will produce interference fringes 0.018 rad apart on a distant screen when the light has wavelength $\lambda = 589$ nm.

74. Two microscope slides are lying horizontally, one on top of the other, when a piece of tissue paper is inserted between them at one end, thereby forming a wedge of air between them. When light of 500 nm wavelength shines vertically down onto the top plate, dark interference fringes are observed, with a 1.2 mm separation between minima. What is the angle between the two slides?

75. Two parallel slits are illuminated with monochromatic light of wavelength 500 nm. An interference pattern is formed on a screen some distance from the slits, and the fourth dark band is located 1.68 cm from the central bright band on the screen. (a) What is the path length difference corresponding to the fourth dark band? (b) What is the distance on the screen between the central bright band and the first bright band on either side of the central band? (Hint: The angles to the fourth dark band and the first bright band are small enough that $\tan \theta \approx \sin \theta$.)

76. In a double-slit interference experiment, the slit separation is 2.00 μm, the light wavelength is 500 nm, and the separation between the slits and the screen is 4.00 m. (a) What is the angle between the center and the third side bright fringe? (b) If we decrease the frequency of light by 10.0% of its initial value, how far along the screen and in what direction does that bright fringe shift?

77. Two light rays, initially in phase and having wavelength 6.00×10^{-7} m, pass through different plastic layers of the same thickness, 7.00×10^{-6} m. The indexes of refraction are 1.65 for one layer and 1.49 for the other. (a) What is the equivalent phase difference, in terms of wavelengths, between the rays when they emerge? (b) If those two rays then reach a common point, does the interference result in complete darkness, maximum brightness, intermediate illumination but closer to complete darkness, or intermediate illumination but closer to maximum brightness? (c) If the two rays are, instead, initially exactly out of phase, what are the answers to (a) and (b)?

78. A light beam with a wavelength of 600 nm in air passes through film 1 ($n_1 = 1.2$) of thickness 1.0 μm, then through film 2 (air) of

thickness 1.5 μm, and finally through film 3 ($n_3 = 1.8$) of thickness 1.0 μm. (a) Which film does the light cross in the least time, and what is that least time? (b) What is the total number of wavelengths (at any instant) across all three films together?

79. From a medium of index of refraction n_1, monochromatic light of wavelength λ is incident normally on a thin film of uniform thickness L (where $L > 0.1\lambda$) and index of refraction n_2. The light transmitted by the film travels into a medium with index of refraction n_3. Find expressions for the minimum film thickness (in terms of λ and the indexes of refraction) for the following cases: (a) minimum light is reflected (hence maximum light is transmitted) with $n_1 < n_2 > n_3$; (b) minimum light is reflected (hence maximum light is transmitted) with $n_1 < n_2 < n_3$; and (c) maximum light is reflected (hence minimum light is transmitted) with $n_1 < n_2 < n_3$.

80. A sheet of glass having an index of refraction of 1.40 is to be coated with a film of material having an index of refraction of 1.55 in order that green light with a wavelength of 525 nm will be preferentially transmitted via constructive interference. (a) What is the minimum thickness of the film that will achieve the result? (Hint: Use Fig. 36-32a with appropriate indexes of refraction.) (b) Why are other parts of the visible spectrum not also preferentially transmitted? (c) Will the transmission of any colors be sharply reduced? If so, which colors?

81. One slit of a double-slit arrangement is covered by a thin glass plate with index of refraction 1.4, and the other by a thin glass plate with index of refraction 1.7. The point on the screen at which the central maximum fell before the glass plates were inserted is now occupied by what had been the $m = 5$ bright fringe. Assuming that $\lambda = 480$ nm and that the plates have the same thickness t, find t.

82. If the slit separation d in Young's experiment is doubled, how must the distance D of the viewing screen be changed to maintain the same fringe spacing?

83. A thin film ($n = 1.25$) is deposited on a glass plate ($n = 1.40$). What is the minimum (nonzero) thickness for the film that will (a) maximally transmit light with a wavelength of 550 nm and (b) maximally reflect light with a wavelength of 550 nm?

84. A double-slit arrangement produces interference fringes for sodium light (wavelength = 589 nm) that are angularly separated by 0.30° near the center of the pattern. What is the angular fringe separation if the entire arrangement is immersed in water, which has an index of refraction of 1.33?

85. The second dark band in a double-slit interference pattern is 1.2 cm from the central maximum of the pattern. The separation between the two slits is equal to 800 wavelengths of the monochromatic light incident (perpendicularly) onto the slits. What is the distance between the plane of the slits and the viewing screen?

86. Two light rays, initially in phase and with a wavelength of 500 nm, go through different paths by reflecting from the various mirrors shown in Fig. 36-52. (Such a reflection does not itself produce a phase shift.) (a) What least value of distance d will put the rays exactly out of phase when they emerge from the region? (Ignore the slight tilt of the path for ray 2.) (b) Repeat the question assuming that the entire apparatus is immersed in a protein solution with an index of refraction of 1.38.

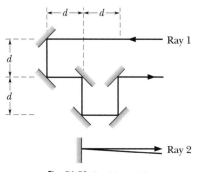

Fig. 36-52 Problem 86.

87. Point sources S_1 and S_2 radiate waves that are in phase with each other, have equal amplitudes R, and have equal wavelengths of 400 nm. The sources are located on an x axis at $x = 6.5$ μm and $x = -6.0$ μm, respectively. (a) Determine the phase difference (in radians) at the origin between the radiation from S_1 and the radiation from S_2. (b) Suppose a slab of transparent material with thickness 1.5 μm and index of refraction $n = 1.5$ is placed between $x = 0$ and $x = 1.5$ μm. What then is the phase difference (in radians) at the origin between the radiation from S_1 and the radiation from S_2?

Tutorial Problem

88. Coherent monochromatic light of wavelength 600 nm is incident on a pair of slits with a separation of 0.500 mm and then falls on a screen 1.50 m past the slits. (a) Sketch the system, showing the locations of some of the expected bright and dark fringes. (Locate the fringes qualitatively only, since you have not yet calculated their separations on the screen.)

(b) What is the color of this light? What is the ratio of the wavelength of the light to the slit separation? (c) What is the phase condition (the condition on the phase angles of interfering waves) if constructive interference is to occur at the angle θ? Convert this phase condition into an equation for sin θ by using the path length difference. (d) Determine the angles (in both radians and degrees) of several bright fringes near the center of the interference pattern. (e) Determine the distance on the screen between adjacent bright fringes.

(f) Using the answer for (e), make a rough sketch of the intensity of the interference pattern versus position on the screen, from 6 mm on one side of the pattern's center to 6 mm on the other side. Mark the center with 0. Label the bright fringes B and the dark fringes D. Include a horizontal dashed line to mark the average intensity. (g) Repeat part (f), but assume radiation of wavelength 900 nm instead of 600 nm. Where is 900 nm light in the electromagnetic spectrum? *Suggestion:* You can just determine the separation of the fringes and adjust the scale of the pattern for 600 nm light. (h) The intensity of the light pattern on the screen is related to the distribution of the light energy. Explain how the law of conservation of energy applies to the two-slit interference pattern.

Answers

(a) See Fig. 36-6.

(b) Light of wavelength 600 nm is yellow-green. The ratio is

$$\frac{\lambda}{d} = \frac{600 \text{ nm}}{0.500 \text{ mm}} = 0.00120 = 1.20 \times 10^{-3}.$$

(c) Constructive interference occurs between the two slits when their phase difference is an integer multiple of 2π rad. Their phases involve $kx = (2\pi/\lambda)x$, where x is distance. The path length difference is $\Delta L = d \sin \theta$, where d is the distance between the slits and θ is the angle at which the waves go off toward the screen. Thus, the phase condition for constructive interference is

$$\frac{2\pi}{\lambda} \Delta L = \frac{2\pi}{\lambda} d \sin \theta = 2\pi m$$

(for m an integer) or, equivalently, sin $\theta = m\lambda/d$.

(d) The bright fringes are at angles θ given by sin $\theta = m\lambda/d$, where $m = 0, \pm1, \pm2, \ldots$, so the angles corresponding to some bright fringes are

$$m = 0 \qquad \theta = 0$$
$$m = \pm1 \qquad \theta = \pm0.00120 \text{ rad} = \pm0.069°$$
$$m = \pm2 \qquad \theta = \pm0.00240 \text{ rad} = \pm0.138°$$

(e) From the result of part (d), the angular distance between adjacent bright fringes is $\lambda/d = 0.00120$ rad. For small θ this ratio is also equal to $\Delta y/D$, where Δy is the distance between the bright fringes and D is the distance from the slits to the screen. Thus,

$$\Delta y = (\lambda/d)D = (0.00120)(1.50 \text{ m}) = 0.00180 \text{ m} = 1.80 \text{ mm}.$$

(f) See Fig. 36-53a.

(g) See Fig. 36-53b. The fringes are 1.5 times as far apart as they

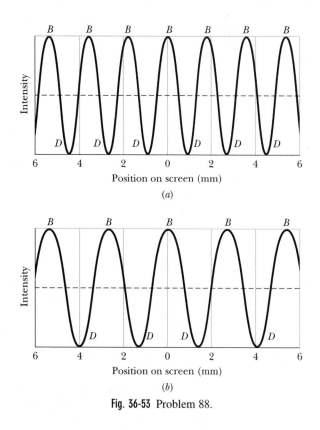

Fig. 36-53 Problem 88.

were in part (f). Such 900 nm light is in the near infrared—that is, in the infrared but near the visible spectrum.

(h) With one slit open, a certain amount of light energy passes through the slit and lands on the screen. At some point on the screen the light intensity is, say, I_0. With two slits open, knowing nothing about interference, we'd naively expect intensity $2I_0$ at that point. Actually, the intensity ranges from 0 to $4I_0$ on the screen. However, the average intensity is half as much, $2I_0$, so the law of conservation of energy is not violated. Interference redistributes the intensity (and thus the energy) but does not change the total energy.

37

Diffraction

Sample Problem 37-6

Figure 37-43a is a representation of the colored dots on Seurat's *Sunday Afternoon on the Island of La Grande Jatte*. Assume that the average center-to-center separation of the dots is $D = 2.0$ mm. Also assume that the diameter of the pupil of your eye is $d = 1.5$ mm and the least angular separation between dots that you can resolve is set only by Rayleigh's criterion. What then is the least viewing distance from which you cannot distinguish any dots on the painting?

SOLUTION: Consider any two adjacent dots that you can distinguish when you are close to the painting. One **Key Idea** here is that as you move away, you continue to distinguish the dots until their angular separation θ (in your view) has decreased to the angle given by Rayleigh's criterion:

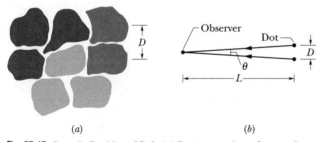

(a)   (b)

Fig. 37-43 Sample Problem 37-6. (a) Representation of some dots on a Seurat painting, showing an average center-to-center separation D. (b) The arrangement of separation D between two dots, their angular separation θ, and the viewing distance L.

$$\theta_R = 1.22 \frac{\lambda}{d}. \tag{37-33}$$

Figure 37-43b shows, from the side, the angular separation θ of the dots, their center-to-center separation D, and your distance L from them. Because D/L is small, angle θ is also small and we can make the approximation

$$\theta = \frac{D}{L}. \tag{37-34}$$

Setting θ of Eq. 37-34 equal to θ_R of Eq. 37-33 and solving for L, we then have

$$L = \frac{Dd}{1.22\lambda}. \tag{37-35}$$

Equation 37-35 tells us that L is larger for smaller λ. Thus, another **Key Idea** is that as you move away from the painting, adjacent red dots (corresponding to a long wavelength) become indistinguishable before adjacent blue dots do. To find the least distance L at which *no* colored dots are distinguishable, we substitute $\lambda = 400$ nm (blue or violet light) and the given data into Eq. 37-35, finding

$$L = \frac{(2.0 \times 10^{-3} \text{ m})(1.5 \times 10^{-3} \text{ m})}{(1.22)(400 \times 10^{-9} \text{ m})} = 6.1 \text{ m}. \quad \text{(Answer)}$$

At this or a greater distance, the colors of all adjacent dots blend together. The color you then perceive at any given spot on the painting is a blended color that may not actually exist there.

Sample Problem 37-7

A diffraction grating has 1.26×10^4 rulings uniformly spaced over width $w = 25.4$ mm. It is illuminated at normal incidence by blue light of wavelength 450 nm.

(a) At what angles to the central axis do the second-order maxima occur?

SOLUTION: The **Key Idea** here is that the maxima can be located with Eq. 37-22 ($d \sin \theta = m\lambda$). Here the grating spacing d is

$$d = \frac{w}{N} = \frac{25.4 \times 10^{-3} \text{ m}}{1.26 \times 10^4}$$
$$= 2.016 \times 10^{-6} \text{ m} = 2016 \text{ nm}.$$

The second-order maxima correspond to $m = 2$ in Eq. 37-22. For $\lambda = 450$ nm, we thus have

$$\theta = \sin^{-1} \frac{m\lambda}{d} = \sin^{-1} \frac{(2)(450 \text{ nm})}{2016 \text{ nm}}$$
$$= 26.52° \approx 26.5°. \quad \text{(Answer)}$$

(b) What is the half-width of the second-order line?

SOLUTION: The **Key Idea** here is that the half-width $\Delta\theta_{hw}$ of a line at angle θ in a diffraction-grating pattern can be found from the properties N and d of the grating and the wavelength λ of the light via Eq. 37-25. Substituting into that equation gives us

$$\Delta\theta_{hw} = \frac{\lambda}{Nd\cos\theta} = \frac{450 \text{ nm}}{(1.26 \times 10^4)(2016 \text{ nm})(\cos 26.52°)}$$
$$= 1.98 \times 10^{-5} \text{ rad.} \qquad \text{(Answer)}$$

Sample Problem 37-8

At what Bragg angles must x rays with $\lambda = 1.10$ Å be incident on the family of reflecting planes represented in Fig. 37-27 if effective reflections from the planes are to result in diffraction intensity maxima? Assume the material to be sodium chloride ($a_0 = 5.63$ Å).

SOLUTION: The **Key Idea** here is that the Bragg angles θ for the intensity maxima of x-ray diffraction off the crystal are given by Bragg's law, which is Eq. 37-31 ($2d\sin\theta = m\lambda$). From Fig. 37-27 and Eq. 37-32, the interplanar spacing d for the family of planes here is

$$d = \frac{a_0}{\sqrt{5}} = \frac{5.63 \text{ Å}}{\sqrt{5}} = 2.518 \text{ Å}.$$

Equation 37-31 then gives, for the Bragg angles,

$$\theta = \sin^{-1}\frac{m\lambda}{2d} = \sin^{-1}\left(\frac{(m)(1.10 \text{ Å})}{(2)(2.518 \text{ Å})}\right)$$
$$= \sin^{-1}(0.2184m).$$

Maxima are possible for $\theta = 12.6°$ ($m = 1$), $\theta = 25.9°$ ($m = 2$), $\theta = 40.9°$ ($m = 3$), and $\theta = 60.9°$ ($m = 4$). Higher-order maxima cannot exist because they require that $\sin\theta$ be greater than 1.

Actually, the unit cell in cubic crystals such as NaCl has diffraction properties such that the intensity of diffracted x-ray beams corresponding to odd values of m is zero. Thus, beams are expected only for $\theta = 25.9°$ and $\theta = 60.9°$.

QUESTIONS

12. Figure 37-44 is an overhead view of ocean waves (assumed to be plane waves) approaching an opening in a breakwater. The waves diffract through the opening and then hit a beach; the erosion rate on the beach depends on the amplitude of the waves. As the waves also erode the sides of the opening in the breakwater, increasing the width of the opening, does the erosion rate near point P on the beach just opposite the opening increase, decrease, or stay the same?

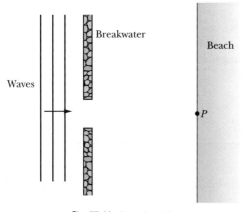

Fig. 37-44 Question 12.

13. *Organizing question:* A slit of width a is illuminated with light of wavelength λ as in Figs. 37-4 and 37-5. However, here we mentally split the slit into three zones of width $a/3$, as in Fig. 37-45 (compare this with Fig. 37-5b). The parallel rays that are shown make an angle θ with the horizontal such that the path length difference between

rays r_1 and r_2 is $\lambda/2$ and that between rays r_2 and r_3 is also $\lambda/2$. At that angle θ in the interference pattern is there fully constructive interference, fully destructive interference, or intermediate interference?

14. In three arrangements you view two closely spaced small objects that are the same large distance from you. The angles that the objects occupy in your field of view and their distances from you are the following: (1) 2ϕ and R, (2) ϕ and $2R$, (3) $\phi/2$ and $R/2$. (a) Rank the arrangements according to the separation between the objects, greatest first. If you can just barely resolve the two objects in arrangement 2, can you resolve them in (b) arrangement 1 and (c) arrangement 3?

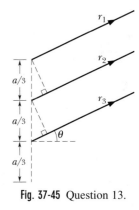

Fig. 37-45 Question 13.

15. *Organizing question:* Figure 37-46a shows rays from the top and bottom of a slit in a single-slit diffraction experiment (as in Figs. 37-4 and 37-5). The rays, which are at angle θ to the horizontal, reach a distant viewing screen at the first minimum in the diffraction pattern there. (a) What is their path length difference in terms of the wavelength λ of the light? The arrangement in Fig. 37-46b is identical (in particular, the top and bottom rays are still at angle θ), but now the slit is narrower. (b) Is the path length difference between the top and bottom rays greater than, less than, or equal to that in Fig. 37-46a? (c) What part of the diffraction pattern do these rays reach on the viewing screen: part of the central maximum, the first minimum, or some part of the pattern beyond the first minimum?

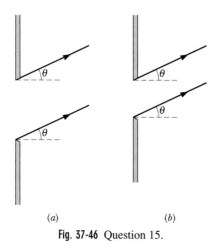

(a)   (b)

Fig. 37-46 Question 15.

16. In Fig. 37-47a, three light waves of the same amplitude E_0 and wavelength λ are initially in phase, travel along different paths (because of reflections by mirrors for at least two of the rays), and then end up at a common point P on a distant screen. (The reflections from the mirrors do not introduce reflection phase shifts; neglect the tilt of the two tilted rays, which is exaggerated in the drawings.) The other parts of the figure show three other arrangements of mirrors using the same light waves. In each arrangement distance $d = \lambda/2$. Rank the four arrangements according to the intensity of the light at the point P on the screen, greatest first.

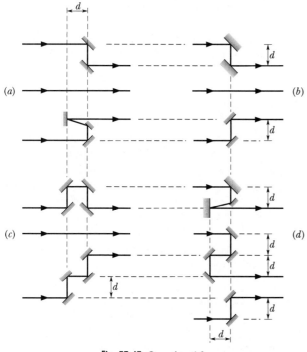

Fig. 37-47 Question 16.

17. In Fig. 37-48a, a plane wave of light is sent through a small rectangular opening in an otherwise opaque screen and the resulting diffraction pattern is studied on a distant viewing screen. The co-

ordinate system shown has its origin at the center of that pattern. Which drawing, Fig. 37-48b or c, better represents the pattern?

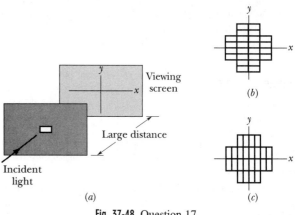

(a)   (c)

Fig. 37-48 Question 17.

18. The first diffraction minima in a double-slit diffraction pattern happen to coincide with the fourth side interference bright fringes. (a) How many bright fringes are in the central diffraction envelope? (b) To shift the coincidence to the fifth side bright fringes, should the distance between the slits be increased or decreased? (c) If that shift is, instead, made by changing the slit widths, should they be increased or decreased?

19. For a certain diffraction grating, the ratio λ/a of wavelength to ruling spacing is 1/3.5. Without written calculation or use of a calculator, determine which of the orders beyond the zeroth order appear in the diffraction pattern.

20. Figure 37-49 is a plot of the intensity versus diffraction angle for the diffraction of a monochromatic x-ray beam by a particular family of reflecting planes in a crystal. Rank the three intensity peaks according to the associated path length differences of the x rays, greatest first.

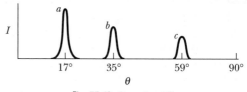

Fig. 37-49 Question 20.

DISCUSSION QUESTIONS

21. Why is the diffraction of sound waves more evident in daily experience than that of light waves?

22. About what slit width should you use if you wish to broaden the distribution of an incident plane sound wave of frequency 1 kHz, using diffraction through a single slit?

23. Why do radio waves diffract around buildings, although light waves do not?

24. A loudspeaker horn, used at a rock concert, has a rectangular aperture 1 m high and 30 cm wide. Will the central maximum of the diffraction pattern from the horn be broader in the horizontal plane or in the vertical plane?

25. For what wavelength range could a long picket fence be considered a useful diffraction grating?

26. A radar antenna is to be designed to give accurate measurements of the altitude of an aircraft but less accurate measurefjments of its direction in a horizontal plane. Must the height-to-width ratio of the radar antenna be less than, equal to, or greater than unity?

27. In single-slit diffraction, what is the effect of increasing (a) the wavelength and (b) the slit width?

28. What will a single-slit diffraction pattern look like when $\lambda > a$?

29. What would the pattern formed on a screen by a double slit look like if the slits did not have the same width? Would the locations of the fringes be changed?

30. A *crossed diffraction grating* has lines ruled in two directions, perpendicular to each other. Predict the pattern produced on a screen when light is sent through such a grating.

31. Sunlight falls on a single slit of width 1 μm. Describe qualitatively what the resulting diffraction pattern looks like.

32. In Fig. 37-5 rays r_1 and r_3 are in phase; so are r_2 and r_4. Why isn't there a maximum intensity at P_2 rather than a minimum?

33. When we speak of diffraction by a single slit we imply that the width of the slit must be much less than its length. Suppose that, in fact, the length were equal to twice the width. Make a rough guess at what the diffraction pattern would look like.

34. We have seen that diffraction limits the resolving power of optical telescopes (see Fig. 37-10). Does it also do so for large radio telescopes?

35. Assume that the limits of the visible spectrum are 430 and 680 nm. How would you design a grating, assuming that the incident light falls normally on it, such that the first-order spectrum barely overlaps the second-order spectrum?

36. For the simple spectroscope of Fig. 37-21 show (a) that angular positions of emission lines increase with λ for a grating and (b) that they decrease with λ if the grating is replaced with a prism.

37. Explain in your own words why increasing the number N of slits in a diffraction grating sharpens the maxima. Why does decreasing the wavelength do so? Why does increasing the slit spacing d do so?

38. How much information can you discover about the structure of a diffraction grating by analyzing the spectrum it forms of a monochromatic light source? Let $\lambda = 589$ nm, for an example.

39. (a) Why does a diffraction grating have closely spaced rulings? (b) Why does it have a large number of rulings?

40. Two nearly equal wavelengths are incident on a grating of N slits and are not quite resolvable. However, they become resolved if the number of slits is increased. Formulas aside, can you explain this as: (a) More light can get through the grating? (b) The principal maxima become more intense and hence resolvable? (c) The diffraction pattern is spread more and hence the wavelengths become resolved? (d) There are a larger number of orders? or (e) The principal maxima become narrower and hence resolvable?

41. How can the resolving power of a lens be increased?

42. The relation $R = Nm$ suggests that the resolving power of a given grating can be made as large as desired by choosing an arbitrarily high order of diffraction. Discuss this possibility.

43. Show that at a given wavelength and a given angle of diffraction the resolving power of a grating depends only on its width w ($= Nd$).

44. How would you experimentally measure (a) the dispersion D and (b) the resolving power R of a grating spectrograph?

EXERCISES & PROBLEMS

65. Some commercial surveillance ("spy") satellites can resolve objects on the ground as small as 85 cm across. Military surveillance satellites reportedly can resolve objects as small as 10 cm across. Assume first that object resolution is determined entirely by Rayleigh's criterion and is not degraded by turbulence in the atmosphere. Also assume that the satellites are at a typical altitude of 400 km. What would be the required diameters of the telescope apertures for (a) 85 cm resolution and (b) 10 cm resolution? (c) Now, considering that turbulence is certain to degrade resolution and that the aperture diameter of the Hubble Space Telescope is 2.4 m, what can you say about the answer to (b) and about how the military surveillance resolutions are accomplished?

66. A diffraction grating with a width of 2.0 cm contains 1000 lines/cm across that width. For an incident wavelength of 600 nm, what is the smallest wavelength difference this grating can resolve in the second order?

67. A double-slit system with individual slit widths of 0.030 mm and a slit separation of 0.18 mm is illuminated with 500 nm light directed perpendicular to the plane of the slits. What is the total number of complete bright fringes appearing between the two first-order minima of the diffraction pattern? (Do not count the fringes that coincide with the minima of the diffraction pattern.)

68. Light of wavelength 500 nm diffracts through a slit of width 2.00 μm and onto a screen that is 2.00 m away. On the screen, what is the distance between the center of the diffraction pattern and the third diffraction minimum?

69. (a) How many rulings must a 4.00-cm-wide diffraction grating have to resolve the wavelengths 415.496 and 415.487 nm in the second order? (b) At what angle are the second-order maxima found?

70. What is the smallest Bragg angle for x rays of wavelength 30 pm to undergo reflection from reflecting planes of spacing 0.30 nm in a calcite crystal?

71. (a) A circular diaphragm 60 cm in diameter oscillates at a frequency of 25 kHz as an underwater source of sound used for submarine detection. Far from the source the sound intensity is distributed as the diffraction pattern of a circular hole whose diameter equals that of the diaphragm. Take the speed of sound in water to be 1450 m/s and find the angle between the normal to the diaphragm and a line from the diaphragm to the first minimum. (b) Repeat for a source having an (audible) frequency of 1.0 kHz.

72. What must be the ratio of the slit width to the wavelength for a single slit to have the first diffraction minimum at $\theta = 45.0°$?

73. A diffraction grating has 200 lines/mm. Light consisting of a continuous range of wavelengths between 550 nm and 700 nm is incident perpendicularly on the grating. (a) What is the lowest order that is overlapped by another order? (b) What is the highest order for which the complete spectrum is present?

74. In a two-slit interference pattern, what is the ratio of slit separation to slit width if there are 17 bright fringes within the central diffraction envelope and the diffraction minima coincide with two-slit interference maxima?

75. In a single-slit diffraction experiment, there is a minimum of intensity for orange light ($\lambda = 600$ nm) and a minimum of intensity for blue-green light ($\lambda = 500$ nm) at the same angle of 1.00 mrad. For what minimum slit width is this possible?

76. A grating has 40 000 rulings spread over 76 mm. (a) What are its expected dispersions D for sodium light ($\lambda = 589$ nm) in the first three orders? (b) What are the grating's resolving powers in these orders?

77. A diffraction grating 1.0 cm wide has 10 000 parallel slits. Monochromatic light that is incident normally is diffracted through 30° in the first order. What is the wavelength of the light?

78. (a) How far from grains of red sand must you be to position yourself just at the limit of resolving the grains if your pupil diameter is 1.5 mm, the grains are spherical with radius 50 μm, and the light from the grains has wavelength 650 nm? (b) If the grains were blue and the light from them had wavelength 400 nm, would the answer to (a) be larger or smaller?

79. Manufacturers of wire (and other objects of small dimension) sometimes use a laser to continually monitor the thickness of the product. The wire intercepts the laser beam, producing a diffraction pattern like that of a single slit of the same width as the wire diameter (see Fig. 37-50). Suppose a helium–neon laser, of wavelength 632.8 nm, illuminates a wire, and the diffraction pattern appears on a screen 2.60 m away. If the desired wire diameter is 1.37 mm, what is the observed distance between the two tenth-order minima (one on each side of the central maximum)?

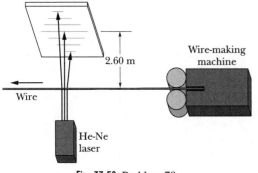

2.60 m

Wire-making machine

Wire

He-Ne laser

Fig. 37-50 Problem 79.

80. A beam of x rays with wavelengths ranging from 0.12 nm to 0.070 nm scatters from a family of reflecting planes in a crystal.

The plane separation is 0.25 nm. It is observed that scattered beams are produced for 0.10 nm and 0.075 nm. What is the angle between the incident and scattered beams?

81. A diffraction grating having 180 lines/mm is illuminated with a light signal containing only two wavelengths, $\lambda_1 = 400$ nm and $\lambda_2 = 500$ nm. The signal is incident perpendicularly on the grating. (a) What is the angular separation between the second-order maxima of these two wavelengths? (b) What is the smallest angle at which two of the resulting maxima are superimposed? (c) What is the highest order for which maxima for both wavelengths are present in the diffraction pattern?

82. Suppose that two points are separated by 2.0 cm. If they are viewed by an eye with a pupil opening of 5.0 mm, what distance from the viewer puts them at the Rayleigh limit of resolution? Assume a light wavelength of 500 nm.

83. Monochromatic light (wavelength = 450 nm) is incident perpendicularly on a single slit (width = 0.40 mm). A screen is placed parallel to the slit plane, and on it the distance between the two minima on either side of the central maximum is 1.8 mm. (a) What is the distance from the slit to the screen? (*Hint:* The angle to either minimum is small enough that sin $\theta \approx$ tan θ.) (b) What is the distance on the screen between the first minimum and the third minimum on the same side of the central maximum?

84. Light containing a mixture of two wavelengths, 500 and 600 nm, is incident normally on a diffraction grating. It is desired (1) that the first and second maxima for each wavelength appear at $\theta \leq 30°$, (2) that the dispersion be as high as possible, and (3) that the third order for the 600 nm light be a missing order. (a) What should be the slit separation? (b) What is the smallest individual slit width that can be used? (c) For the 600 nm wavelength, which orders of intensity maxima are produced by the grating, assuming the values derived in (a) and (b)?

85. A diffraction grating has 200 rulings/mm, and it produces an intensity maximum at $\theta = 30.0°$. (a) What are the possible wavelengths of the incident visible light? (b) To what colors do they correspond?

86. If Superman really had x-ray vision at 0.10 nm wavelength and a 4.0 mm pupil diameter, at what maximum altitude could he distinguish villains from heroes, assuming that he needs to resolve points separated by 5.0 cm to do this?

87. When monochromatic light is incident on a slit 0.022 mm wide, the first diffraction minimum is observed at an angle of 1.8° from the direction of the incident light. What is the wavelength of that light?

Tutorial Problem

88. Coherent monochromatic light of wavelength 600 nm is incident on a pair of slits of width 0.100 mm and center-to-center separation 0.500 mm. The light then falls on a screen 1.50 m on the other side of the slits.

(a) Make a rough sketch of this system. (b) Determine the angles (in radians and degrees) of the first few diffraction minima, taking $\theta = 0$ at the center of the pattern. (c) Determine the distances, on the screen, from the central diffraction maximum to the first two diffraction minima on either side. (d) There is no simple expression for the locations of the diffraction maxima, but you should be able to estimate them. Where, approximately, is the first noncentral diffraction maximum for this system?

(e) The intensity of the diffraction pattern is given by

$$I(\theta) = I_0\left(\frac{\sin^2[(\pi a \sin \theta)/\lambda]}{[(\pi a \sin \theta)/\lambda]^2}\right),$$

where I_0 is the intensity at the center of the pattern. Use this to determine the approximate intensity ratios $I(\theta)/I_0$ near the first and second diffraction maxima.

(f) Using the results of (d) and (e), sketch the intensity of the diffraction pattern versus position on the screen, from 12 mm on one side of the pattern's center to 12 mm on the other side. Mark the center with 0. Label I_0. (g) Now make a similar graph for the intensity of the double-slit pattern for 600 nm light (it should show the effects of both interference between the slits and diffraction through each slit). (h) What would change in the graph of (f) if you changed the separation of the slits but not their widths? What would change if you changed the widths but not the separation?

Answers

(b) The diffraction minima occur at angles θ given by

$$\sin \theta = \frac{m\lambda}{a} = m\,\frac{600 \text{ nm}}{0.100 \text{ mm}} = m(0.00600 \text{ m}) = m(6.0 \text{ mm}),$$

where $m = \pm 1, \pm 2, \ldots$, (but not 0). The angles corresponding to these minima are

$$m = \pm 1:\quad \theta = \pm 0.0060 \text{ rad} = \pm 0.344°$$
$$m = \pm 2:\quad \theta = \pm 0.0120 \text{ rad} = \pm 0.688°$$

(c) From the result of part (b), the angles θ from the central diffraction maximum to the first two diffraction minima on either side are ± 0.0060 rad and ± 0.0120 rad. For small angles these equal $\pm \Delta y/D$, where Δy is the distance on the screen from the central diffraction maximum and D is the distance from the slits to the screen. Thus,

$$\Delta y = \theta D = (0.0060)(1.50 \text{ m}) = 0.0090 \text{ m} = 9.00 \text{ mm}$$

on either side of the center for the first diffraction minimum, and

$$\Delta y = \theta D = (0.0120)(1.50 \text{ m}) = 0.0180 \text{ m} = 18.0 \text{ mm}$$

on either side of the center for the second diffraction minimum.

(d) The angle θ should be approximately (but not exactly) halfway between the diffraction minima:

$$\frac{0.0060 \text{ rad} + 0.0120 \text{ rad}}{2} = 0.0090 \text{ rad}.$$

In terms of distance, this would be about

$$\frac{9.0 \text{ mm} + 18.0 \text{ mm}}{2} = 13.5 \text{ mm}$$

on either side of the central maximum.

(e) The argument $(\pi a \sin \theta)/\lambda$ is approximately 1.5π for the first diffraction maximum and approximately 2.5π for the second, leading to the following ratios:

First diffraction maximum:

$$I(\theta)/I_0 \approx [\sin^2(1.5\pi)]/[1.5\pi]^2 = 1/[1.5\pi]^2 = 0.045.$$

Second diffraction maximum:

$$I(\theta)/I_0 \approx [\sin^2(2.5\pi)]/[2.5\pi]^2 = 1/[2.5\pi]^2 = 0.016.$$

(f) See Fig. 37-51a.

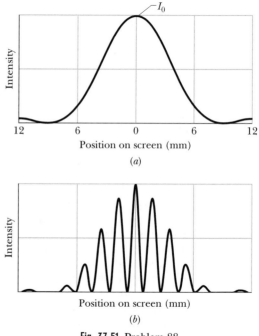

Fig. 37-51 Problem 88.

(g) See Fig. 37-51b.

(h) Changing the separation of the slits but not their widths would affect the location of the bright and dark interference fringes but not the diffraction envelope. A smaller separation would spread out the interference fringes, and a larger separation would bring them closer together. Changing the widths of the slits but not their separation would change the diffraction envelope but not the locations of the bright and dark interference fringes. Smaller slits would spread out the diffraction pattern, and wider slits would narrow it.

Relativity

Sample Problem 38-8

Caught by surprise near a supernova, you race away from the explosion in your spaceship, hoping to outrun the high-speed material ejected toward you. Your Lorentz factor γ relative to the inertial reference frame of the local stars is 22.4.

(a) To reach a safe distance, you figure you need to cover 9.00×10^{16} m as measured in the reference frame of the local stars. How long will the flight take, as measured in that frame?

SOLUTION: Just as we did in Chapter 2, we can find the time required to travel a certain distance at constant speed with the definition of speed v:

$$\text{speed} = \frac{\text{distance}}{\text{time interval}}. \quad (38\text{-}56)$$

From Fig. 38-6, we see that because your Lorentz factor γ relative to the stars is 22.4 (large), your relative speed v is almost c—so close that we can approximate it as c. Then the **Key Idea** here is that for speed $v \approx c$, we must be careful that the distance and the time interval in Eq. 38-56 are measured in the *same* reference frame. The given distance (9.00×10^{16} m) for the length of your travel path is measured in the reference frame of the stars, and the requested time interval Δt is to be measured in that same frame. Thus, we can rewrite Eq. 38-56 as

$$\left(\begin{array}{c} \text{time interval} \\ \text{relative to stars} \end{array} \right) = \frac{\text{distance relative to stars}}{c}.$$

Then substituting the given distance, we find that

$$\left(\begin{array}{c} \text{time interval} \\ \text{relative to stars} \end{array} \right) = \frac{9.00 \times 10^{16} \text{ m}}{2.998 \times 10^{8} \text{ m/s}}$$

$$= 3.00 \times 10^{8} \text{ s} = 9.51 \text{ y}. \quad (\text{Answer})$$

(b) How long does that trip take according to you (in your reference frame)?

SOLUTION: One **Key Idea** here is that we now want the time interval measured in a different reference frame—namely, yours. Thus, we need to transform the data given in the reference frame of the stars to your frame.

A second **Key Idea** is that the given path length of 9.00×10^{16} m, measured in the reference frame of the stars, is a proper length L_0, because the two ends of the path are at rest in that frame. As observed from your reference frame, the stars' reference frame and those two ends of the path race past you at a relative speed of $v \approx c$. Thus, another **Key Idea** is that you measure a contracted length L_0/γ for the path, not the proper length L_0. Then we can rewrite Eq. 38-56 as

$$\left(\begin{array}{c} \text{time interval} \\ \text{relative to you} \end{array} \right) = \frac{\text{distance relative to you}}{c} = \frac{L_0/\gamma}{c}.$$

Substituting known data, we find

$$\left(\begin{array}{c} \text{time interval} \\ \text{relative to you} \end{array} \right) = \frac{(9.00 \times 10^{16} \text{ m})/22.4}{2.998 \times 10^{8} \text{ m/s}}$$

$$= 1.340 \times 10^{7} \text{ s} = 0.425 \text{ y}. \quad (\text{Answer})$$

In part (a) we found that the flight takes 9.51 y in the reference frame of the stars. However, here we find that it takes only 0.425 y in your frame, due to the relative motion and the resulting contracted length of the path.

Sample Problem 38-9

Figure 38-23 shows an inertial reference frame S in which event 1 (a rock is kicked up by a truck at coordinates x_1 and t_1) causes event 2 (the rock hits you at coordinates x_2 and t_2). Is there another inertial reference frame S' from which those events can be measured to be reversed in sequence, so that the effect occurs before its cause? (Can you be injured now as a result of a future event?)

SOLUTION: One **Key Idea** here is that we can tell the sequence of any two events as measured in a given frame from the sign of their

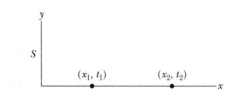

Fig. 38-23 Sample Problem 38-9. Event 1 at spacetime coordinates (x_1, t_1) causes event 2 at spacetime coordinates (x_2, t_2). Can the sequence of cause and effect be reversed in some other reference frame?

temporal separation as measured in that frame. A second **Key Idea** is that because spatial and temporal separations can be entangled, the temporal separation $\Delta t'$ of the events in frame S' can depend on both the temporal separation Δt and the spatial separation Δx of those events in frame S. Thus, to find $\Delta t'$, we use Eq. 2' of Table 38-2:

$$\Delta t' = \gamma \left(\Delta t - \frac{v \, \Delta x}{c^2} \right). \tag{38-57}$$

Recall that v is the relative velocity between S and S'. We take frame S to be stationary; frame S' then has velocity v.

Let $\Delta t = t_2 - t_1$. Then Δt is a positive quantity and, to be consistent with this notation, we must have $\Delta x = x_2 - x_1$ and $\Delta t' = t'_2 - t'_1$. As Fig. 38-23 is drawn, Δx is a positive quantity because $x_2 > x_1$.

We are interested in the possibility that $\Delta t'$ is a negative quantity, which would mean that time t'_1 of event 1 is later (and thus greater) than time t'_2 of event 2. From Eq. 38-57, we see that $\Delta t'$ can be negative only if

$$\frac{v \, \Delta x}{c^2} > \Delta t.$$

This condition can be rearranged to produce the equivalent condition

$$\frac{\Delta x / \Delta t}{c} \frac{v}{c} > 1.$$

The ratio $\Delta x / \Delta t$ is just the speed at which information (here via a rock) travels from event 1 to produce event 2. That speed cannot exceed c. (Information could travel at c if it came via light; rocks travel more slowly, of course.) Therefore, $(\Delta x / \Delta t)/c$ must be at most 1, and v/c cannot equal or exceed 1. Thus, the left side of the last inequality must be less than 1, and the inequality cannot be satisfied.

There is no frame S' in which event 2 occurs before its cause, event 1. More generally, although the sequence of unrelated events can sometimes be reversed in relativity (as in Sample Problem 38-4), events involving cause and effect can never be reversed.

Sample Problem 38-10

In Fig. 38-24a, atoms A and B move parallel to an axis toward each other and toward detector C. The velocity of A relative to C is $0.800c$ to the right; the velocity of B relative to C is $0.900c$ to the left. What is the velocity of B relative to A?

SOLUTION: The two given velocities are both near c. Thus, one **Key Idea** here is that the classical velocity transformation of Eq. 38-29, which is approximately correct for speeds much less than c, cannot be applied. Instead, we must use the relativistic velocity transformation of Eq. 38-28,

$$u = \frac{u' + v}{1 + u'v/c^2}, \tag{38-58}$$

which is correct at all physically possible speeds.

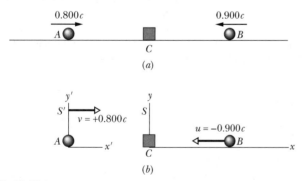

(a)

(b)

Fig. 38-24 Sample Problem 38-10. (a) Atoms A and B move toward each other and toward detector C. (b) Interpreting (a) in terms of Fig. 38-11. Reference frame S' is attached to atom A, and reference frame S is attached to detector C.

A second **Key Idea** is that to apply Eq. 38-58 to the situation of Fig. 38-24a, we first need to interpret the two atoms and the detector in terms of the reference frames S and S' and the particle P in Fig. 38-11. To do so, we can attach frame S to detector C and frame S' to atom A, and consider atom B to be the particle. Then the velocities are measured along the common x and x' axes of the S and S' frames, as shown in Fig. 38-24b: The velocity v of atom A (frame S') relative to detector C (frame S) is $+0.800c$. The velocity u of atom B relative to detector C is $-0.900c$. The velocity u' in Eq. 38-58 is the desired velocity of atom B relative to atom A (frame S').

Solving Eq. 38-58 for u' leads to

$$u' = \frac{u - v}{1 - uv/c^2}. \tag{38-59}$$

Substituting given values for u and v into this equation then yields

$$u' = \frac{-0.900c - 0.800c}{1 - (-0.900c)(0.800c)/c^2}$$
$$= -0.988c.$$

Thus, the velocity of atom B relative to atom A in Fig. 38-24a is $0.988c$ to the left. Note that the speed is less than c, as it should be because c is the limiting speed for objects. If we had used the classical velocity transformation, we would have found that the velocity of atom B relative to atom A is $1.7c$ to the left, which is clearly in error.

Caution: The signs of the velocities along the axes shown in Fig. 38-24b are important. Note that if even one sign were wrong in our calculation, the answer we get would be very different from the correct one.

Sample Problem 38-11

In the *nuclear fission* reaction

$$n + {}^{235}U \rightarrow {}^{140}Ce + {}^{94}Zr + n + n,$$

a neutron (n) combines with a uranium nucleus (^{235}U), making the nucleus unstable and causing it to split (to *fission*) into two smaller nuclei (^{140}Ce and ^{94}Zr) and to release two neutrons. The masses involved are

$$\text{mass}(^{235}U) = 235.04 \text{ u} \qquad \text{mass}(^{94}Zr) = 93.91 \text{ u}$$
$$\text{mass}(^{140}Ce) = 139.91 \text{ u} \qquad \text{mass}(n) = 1.008 \ 67 \text{ u}$$

What is the Q of the reaction in units of MeV?

SOLUTION: We have here a system of particles in which the nature of the particles changes during a reaction. The **Key Idea** is that we can find the Q of the reaction from the change ΔM in the total mass M of the system due to the reaction, according to Eq. 38-47,

$$Q = -\Delta M \ c^2. \tag{38-60}$$

The initial (before the fission) total mass M_i of the particles is

$$M_i = 1.008 \ 67 \text{ u} + 235.04 \text{ u} = 236.048 \ 67 \text{ u}.$$

The final (after the fission) total mass M_f is

$$M_f = 139.91 \text{ u} + 93.91 \text{ u} + 2(1.008 \ 67 \text{ u})$$
$$= 235.837 \ 34 \text{ u}.$$

Thus, the change in mass due to the fission is

$$\Delta M = M_f - M_i = 235.837 \ 34 \text{ u} - 236.048 \ 67 \text{ u}$$
$$= -0.211 \ 33 \text{ u}.$$

Because this change in mass is measured in atomic mass units (u) and we want Q in units of MeV, we shall compute Q with the last value given for c^2 in Eq. 38-43:

$$c^2 = 931.5 \text{ MeV/u}.$$

Substituting this and $\Delta M = -0.211 \ 33$ u in Eq. 38-60, we find

$$Q = -(-0.211 \ 33 \text{ u})(931.5 \text{ MeV/u})$$
$$= 197 \text{ MeV}. \qquad \text{(Answer)}$$

QUESTIONS

11. The plane of clocks and measuring rods in Fig. 38-25 is like that in Fig. 38-3. The clocks along the x axis are separated (center to center) by 1 light-second, as are the clocks along the y axis, and all the clocks are synchronized via the procedure described in Section 38-3. When the initial synchronizing signal of $t = 0$ from the origin reaches (a) clock A, (b) clock B, and (c) clock C, what initial time is then set on those clocks? An event occurs at clock A when it reads 10 s. (d) How long does the signal of that event take to travel to an observer stationed at the origin? (e) What time does that observer assign to the event?

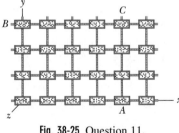

Fig. 38-25 Question 11.

12. In Checkpoint 1, suppose that the laser pulse returns to the hobo via a reflection from a mirror at the left end of the boxcar. (a) Is the hobo's measurement of the flight time of the pulse (from its emission to its return to him) a proper time? (b) Are his measurement and our measurement of that flight time related by Eq. 38-9?

13. *Organizing question:* An observer in frame S' of Fig. 38-9 measures two events as occurring at the following spacetime coordinates (x', t'):

event Yellow   (5.0 m, 20 ns)
event Green   (−2.0 m, 45 ns)

The velocity of frame S' relative to frame S is $0.90c$. Set up equations, complete with known data, to find (a) the displacement of event Yellow from event Green and (b) the corresponding temporal separation between the events, both according to an observer in frame S.

14. Figure 38-26 shows three situations in which a starship passes Earth (the dot) and then makes a round trip that brings it back past Earth, each at the given Lorentz factor. As measured in the rest frame of Earth, the round-trip distances are as follows: trip 1, $2D$; trip 2, $4D$; trip 3, $6D$. Without written calculation and neglecting any time needed for accelerations, rank the situations according to the travel times of the trips, greatest first, as measured from (a) the rest frame of Earth and (b) the rest frame of the starship. (*Hint:* See Sample Problem 38-8 in this supplement.)

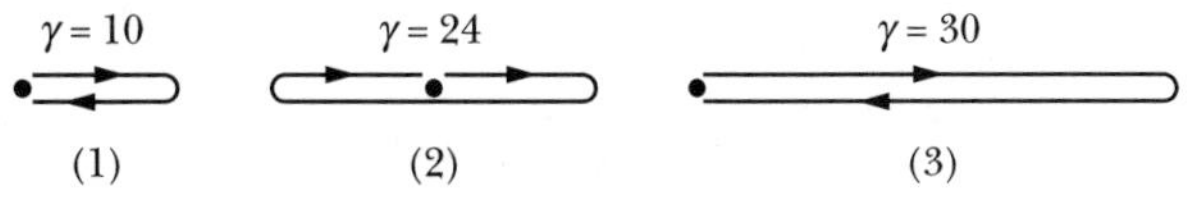

$$\gamma = 10 \qquad\qquad \gamma = 24 \qquad\qquad \gamma = 30$$

(1)   (2)   (3)

Fig. 38-26 Question 14.

15. *Organizing question:* Figure 38-27 shows four situations in which a reference frame S' moves with speed $0.60c$ either leftward or rightward (as indicated by vector $\vec{v}$) relative to reference frame S. In each situation a particle moves either leftward or rightward (as indicated by vector $\vec{u}'$) with speed $0.70c$ relative to S'. For each situation, set up equations, complete with known data, to find the velocity $\vec{u}$ of the particle relative to frame S.

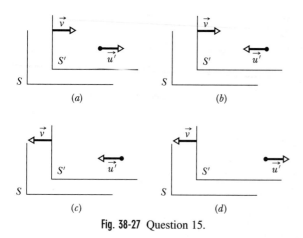

Fig. 38-27 Question 15.

16. The elementary particle K-zero (K^0) can spontaneously decay to (suddenly transform into) pions π via the reaction $K^0 \rightarrow \pi^+ + \pi^-$. Assume that the K^0 is stationary; the pions are moving. (a) Is the mass of the K^0 greater than, less than, or equal to the sum of the masses of the π^+ and π^-? Can the angle between the paths of the pions be (b) 10°, (c) 90°, and (d) 180°? (e) Can the magnitude of the linear momentum of the π^+ be greater than that of the π^-, or must the two linear momenta be equal in magnitude?

17. Figure 38-28 shows a starship and an asteroid that move along an x axis. In four situations, the velocity of the starship relative to us (on a scout ship) and the velocity of the asteroid relative to the starship are, in that order, (a) $+0.4c$, $+0.4c$; (b) $+0.5c$, $+0.3c$; (c) $+0.9c$, $-0.1c$; and (d) $+0.3c$, $+0.5c$. Without written calculation, rank the situations according to the magnitude of the velocity of the asteroid relative to us, greatest first.

Fig. 38-28 Question 17.

18. Figure 38-29 shows three flying saucers that move either left or right along the axis shown. All emit microwave signals of the same proper frequency f_0. Saucer C detects the signal from saucer A with frequency $f_1 > f_0$. Saucer A detects the signal from saucer B with frequency $f_2 < f_0$. Is the signal from saucer B that is detected by saucer C less than f_0, greater than f_1, or between f_0 and f_1?

Fig. 38-29 Question 18.

19. Three electrons have the following initial speeds: (1) $0.20c$, (2) $0.40c$, and (3) $0.90c$. Work is to be done on them to increase their total energy by 2.0 keV. Rank them according to (a) the required work and (b) the resulting changes in the magnitudes of their linear momenta, greatest first.

DISCUSSION QUESTIONS

20. How would you test a proposed reference frame to find out whether or not it is an inertial frame?

21. Give examples in which effects associated with Earth's rotation are significant enough in practice to rule out a laboratory reference frame as being a good enough approximation to an inertial frame.

22. The speed of light in vacuum is a true constant of nature, independent of the wavelength of the light or the choice of (inertial) reference frame. Is there any sense then in which Einstein's second postulate can be viewed as contained within the scope of his first postulate?

23. We can catch the spirit of Einstein's two postulates by labeling them: (1) the principle of "the irrelevance of velocity" and (2) the principle of "the uniqueness of light." In what senses are velocity irrelevant and light unique in these two statements?

24. A beam from a laser is perpendicular to a plane mirror. What is the speed of the reflected beam if the mirror is (a) fixed in the laboratory and (b) moving directly toward the laser with speed v?

25. Give an example from classical physics in which the motion of a clock affects its rate—that is, the way it runs. (The magnitude of the effect may depend on the detailed nature of the clock.)

26. In relativity (where motion is relative and not absolute) we find that "moving clocks run slow," but the reason this effect occurs is not because the motion alters the way a clock works. Why does it occur? (*Hint:* What does it mean to say that "moving clocks run slow"?)

27. We have seen that if several observers watch two events, labeled A and B, one of them may say that event A occurred first but another may claim that it was event B that did so. What would you say to a friend who asked you which event *really did* occur first?

28. Two observers, one at rest in S and one at rest in S', each carry a meter stick oriented parallel to their relative motion. *Each* observer finds upon measurement that the *other* observer's meter stick is shorter than his own meter stick. Does this seem like a paradox to you? Explain. (*Hint:* Compare the following situation. Harry waves good-bye to Walter who is in the rear of a station wagon driving away from Harry. Harry says that Walter gets smaller. Walter says that Harry gets smaller. Are they measuring the same thing?)

29. How does the concept of simultaneity enter into the measurement of the length of a body?

30. In relativity, time and space coordinates are intertwined and are treated on a more or less equivalent basis. Are time and space fundamentally of the same nature, or is there some essential difference between them that is preserved even in relativity?

31. Can we simply substitute γm for m in classical equations to obtain the correct relativistic equations? Give examples.

32. If zero-mass particles (such as light) have speed c in one reference frame, can they be found at rest in any other frame? Can such particles have any speed other than c?

33. A hydroelectric plant generates electricity because water falls through a turbine, thereby turning the shaft of a generator. According to the mass–energy concept, must the appearance of energy (the electricity) be identified with a mass decrease somewhere? If so, where?

34. A hot metallic sphere cools off as it rests on the pan of a scale. If the scale were sensitive enough, would it indicate a change in mass? If so, would it indicate an increase or a decrease?

EXERCISES & PROBLEMS

54. As you read this book, a cosmic ray proton passes along the left–right width of the book with relative speed v and a total energy of 14.24 nJ. According to your measurements, that left–right width is 21.0 cm. (a) What is the width according to the proton's reference frame? How much time did the passage take according to (b) your frame and (c) the proton's frame?

55. An alpha particle is a helium nucleus (with two protons and two neutrons) and has a mass of 3727 MeV/c^2. What is the magnitude of the particle's linear momentum (in mega-electron-volts per c) if the particle is accelerated from rest by an electric potential difference of 300 MV?

56. Bullwinkle chases Rocky along an x axis in our reference frame S. Relative to us, Bullwinkle has velocity $0.800c\hat{i}$ and Rocky has velocity $0.9990c\hat{i}$. What is the velocity of Rocky relative to Bullwinkle?

57. Bullwinkle, in coordinate system S' of Fig. 38-9, passes you, in coordinate system S, with a relative speed of $0.9990c$. (a) If Bullwinkle carries a stick that is 2.50 m long according to him and aligned parallel to his direction of motion, what is the length of the stick according to you?

He measures that two events occur at the following spacetime coordinates (x', t'):

$$\text{event Alpha:} \quad (4.0 \text{ m, } 40 \text{ ns})$$
$$\text{event Beta:} \quad (-4.0 \text{ m, } 80 \text{ ns})$$

According to your measurements, (b) what is the distance between the events, (c) what is their temporal separation, and (d) which event occurred first?

58. Find the speed of a particle that takes 2.0 y longer than light to travel a distance of 6.0 ly.

59. In Table 38-2 the Lorentz transformation equations in the right-hand column can be derived from those in the left-hand column simply by (1) exchanging primed and unprimed quantities and (2) changing the sign of v. Verify this procedure by deriving one set of equations directly from the other by algebraic manipulation.

60. To circle Earth in low orbit, a satellite must have a speed of about 2.7×10^4 km/h. Suppose that two such satellites orbit Earth in opposite directions. (a) What is their relative speed as they pass, according to the classical Galilean velocity transformation equation? (b) What fractional error do you make in (a) by not using the (correct) relativistic transformation equation?

61. (a) What potential difference would accelerate an electron to speed c, according to classical physics? (b) With this potential difference, what speed would the electron actually attain?

62. (a) How much energy is released in the explosion of a fission bomb containing 3.0 kg of fissionable material? Assume that 0.10% of the mass is converted to released energy. (b) What mass of TNT would have to explode to provide the same energy release? Assume that each mole of TNT liberates 3.4 MJ of energy on exploding. The molecular mass of TNT is 0.227 kg/mol. (c) For the same mass of explosive, how much more effective is nuclear explosion than TNT explosion? That is, compare the amounts of energy released in explosions involving, say, 1 kg of fissionable material and 1 kg of TNT.

63. In the red shift of radiation from a distant galaxy, a certain radiation, known to have a wavelength of 434 nm when observed in the laboratory, has a wavelength of 462 nm when observed in the laboratory. (a) What is the radial speed of the galaxy relative to Earth? (b) Is the galaxy approaching or receding?

64. How much work is needed to accelerate a proton from a speed of $0.9850c$ to a speed of $0.9860c$?

65. A proton passes us with speed v and a total energy of 14.242 nJ. What is v? (*Hint:* Use the proton mass given in Appendix B under "Best Value," not the commonly remembered rounded number.)

66. In Question 14 and Fig. 38-26, the ship passes us with speed $0.950c$, the proton has speed $0.980c$ relative to the ship, and the proper length of the ship is 760 m. What are the temporal separations between the two events (the firing of the proton and the impact of the proton) according to (a) a passenger in the ship and (b) us? Suppose that, instead, the proton is fired from the rear toward the front. What then are the temporal separations according to (c) the passenger and (d) us?

67. An elementary particle produced in a laboratory experiment travels 0.230 mm through the lab at a relative speed of $0.960c$ before it decays (becomes another particle). (a) What is the proper lifetime of the particle? (b) What is the distance it travels, as measured from the rest frame of the particle?

68. The rest radius of Earth is 6370 km, and its orbital speed about the Sun is 30 km/s. Suppose Earth moves past an observer at this speed. To the observer, by how much does Earth's diameter contract along the direction of motion?

69. A spaceship, at rest in a certain reference frame S, is given a speed increment of $0.50c$. Relative to its new rest frame, it is then given a further $0.50c$ increment. This process is continued until its speed with respect to its original frame S exceeds $0.999c$. How many increments does this process require?

70. A sodium light source moves in a horizontal circle at a constant speed of $0.100c$ while emitting light at the proper wavelength of $\lambda_0 = 589.00$ nm. Wavelength λ is measured for that light by a detector fixed at the center of the circle. What is the wavelength shift $\lambda - \lambda_0$?

71. A pion is created in the higher reaches of Earth's atmosphere when an incoming high-energy cosmic-ray particle collides with an atomic nucleus. A pion so formed descends toward Earth with a speed of $0.99c$. In a reference frame in which they are at rest, pions decay with an average life of 26 ns. As measured in a frame fixed with respect to Earth, how far (on the average) will such a pion move through the atmosphere before it decays?

72. A particle has a speed of $0.990c$ in a laboratory reference frame. What are its kinetic energy, its total energy, and its momentum if the particle is (a) a proton and (b) an electron?

73. If we intercept an electron with total energy of 1533 MeV that came from Vega, which is 26 ly from us, how far in light-years was the trip in the rest frame of the electron?

74. A radar transmitter T is fixed to a reference frame S' that is moving to the right with speed v relative to reference frame S (see Fig. 38-30). A mechanical timer (essentially a clock) in frame S', having a period τ_0 (measured in S'), causes transmitter T to emit timed radar pulses, which travel at the speed of light and are received by R, a receiver fixed in frame S. (a) What is the period τ of the timer as detected by observer A, who is fixed in frame S? (b) Show that at the receiver R the time interval between pulses arriving from T is not τ or τ_0, but

$$\tau_R = \tau_0 \sqrt{\frac{c+v}{c-v}}.$$

(c) Explain why the receiver R and observer A, who are in the same reference frame, measure a different period for the transmitter. (*Hint:* A clock and a radar pulse are not the same thing.)

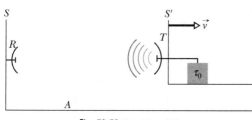

Fig. 38-30 Problem 74.

75. A Foron cruiser moves directly toward a Reptulian scout ship when it fires a decoy toward the scout ship. Relative to the scout ship, the speed of the decoy is $0.980c$ and the speed of the Foron cruiser is $0.900c$. What is the speed of the decoy relative to the cruiser?

76. The premise of the *Planet of the Apes* movies and book is that hibernating astronauts travel far into Earth's future, to a time when human civilization has been replaced by an ape civilization. Considering just special relativity, determine how far into Earth's future the astronauts would travel if they slept for 120 y while traveling relative to Earth with a speed of $0.9990c$, first outward from Earth and then back again.

77. An airplane whose rest length is 40.0 m is moving at uniform velocity with respect to Earth, at a speed of 630 m/s. (a) By what fraction of its rest length is it shortened to an observer on Earth? (b) How long would it take, according to Earth clocks, for the airplane's clock to fall behind by 1.00 μs? (Use special relativity in your calculations.)

78. One cosmic-ray particle approaches Earth along Earth's north–south axis with a speed of $0.80c$ toward the geographic north pole, and another approaches with a speed of $0.60c$ toward the geographic south pole (Fig. 38-31). What is the relative speed of approach of one particle with respect to the other? (*Hint:* It is useful to consider Earth and one of the particles as the two inertial reference frames.)

79. A particle with mass m has speed $c/2$ relative to inertial frame S. The particle collides with an identical particle at rest relative to frame S. What is the speed of a frame S' relative to S in which the

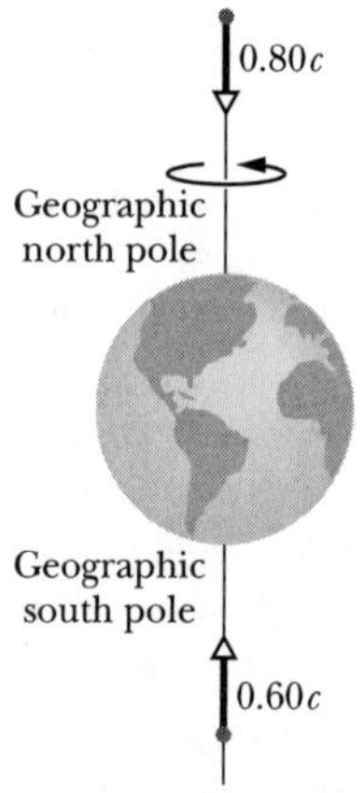

0.80c

Geographic
north pole

Geographic
south pole

0.60c

Fig. 38-31 Problem 78.

total momentum of these particles is zero? This frame is called the *center of momentum frame.*

80. What is the minimum energy that is required to break a nucleus of ^{12}C (of mass 11.996 71 u) into three nuclei of ^{4}He (of mass 4.001 51 u each)?

81. What is the momentum in MeV/c of an electron with a kinetic energy of 2.00 MeV?

Tutorial Problem

82. Consider two inertial reference frames (IRFs) S and S' whose origins coincide at time $t = 0.00$ s and whose Cartesian coordinate axes are parallel ($x\|x'$, $y\|y'$, and $z\|z'$). Suppose that the origin of IRF S' is moving with a constant velocity $\vec{v} = \frac{1}{2}c\hat{i}$ relative to the origin of IRF S. We are interested in how we can transform data from S to S' (or vice versa) using Galilean transformations and Lorentz transformations.

(a) Write the Galilean transformations that relate (x', y', z', t') to (x, y, z, t) and vice versa. (b) Write the Lorentz transformation that relates (x', y', z', t') to (x, y, z, t). (c) Suppose an event E_1 occurs at $(x_1, y_1, z_1, t_1) = (2.00$ m, 3.00 m, 4.00 m, 0.00 s) and another event E_2 occurs at $(x_2, y_2, z_2, t_2) = (3.00$ m, 4.00 m, 5.00 m, 1.00 s). Using both Galilean and Lorentz transformations, determine the coordinates of these two events in IRF S'. (d) How far apart (both spatially and temporally) were these two events, as measured by an observer stationary in IRF S? (e) Using both Galilean and Lorentz transformations, determine how far apart (both spatially and temporally) these two events were, as measured by an observer stationary in IRF S'.

(f) Comment on the differences (if any) between your answers to parts (d) and (e). (g) If the speed v had been much less, say, only a few meters per second, how different would the results have been between the Galilean-transformation separations and the Lorentz-transformation separations? Which, if either, can be used all the time without error?

Answers

(a)
$$x' = x - \tfrac{1}{2}ct \qquad y' = y \qquad z' = z \qquad t' = t$$
$$x = x' + \tfrac{1}{2}ct' \qquad y = y' \qquad z = z' \qquad t = t'$$

(b) First, let's compute

$$\gamma = \frac{1}{\sqrt{1 - (v/c)^2}} = \frac{1}{\sqrt{1 - 0.25}} = 1.15.$$

Then the Lorentz transformations relating S' to S are

$$x' = \gamma(x - vt) = 1.15(x - 0.5ct)$$
$$y' = y$$
$$z' = z$$
$$t' = \gamma(t - xv/c^2) = 1.15(t - x/2c).$$

(c) The coordinates in S' of event E_1 are, with the Galilean transformations,

$$x'_1 = 2.00 \text{ m} - 0.5(2.998 \times 10^8 \text{ m/s})(0.00 \text{ s}) = 2.00 \text{ m}$$
$$y'_1 = y_1 = 3.00 \text{ m}$$
$$z'_1 = z_1 = 4.00 \text{ m}$$
$$t'_1 = t_1 = 0.00 \text{ s},$$

and, with the Lorentz transformations,

$x'_1 = 1.15[2.00 \text{ m} - 0.5(2.998 \times 10^8 \text{ m/s})(0.00 \text{ s})] = 2.30 \text{ m}$

$y'_1 = y_1 = 3.00 \text{ m}$

$z'_1 = z_1 = 4.00 \text{ m}$

$t'_1 = 1.15[0.00 \text{ s} - (2.00 \text{ m})/2(2.998 \times 10^8 \text{ m/s})] \approx 0.00 \text{ s}.$

The coordinates in S' of event E_2 are, with the Galilean transformations,

$x'_2 = 3.00 \text{ m} - 0.5(2.998 \times 10^8 \text{ m/s})(1.00 \text{ s}) \approx -1.50 \times 10^8 \text{ m}$

$y'_2 = y_2 = 4.00 \text{ m}$

$z'_2 = z_2 = 5.00 \text{ m}$

$t'_2 = t_2 = 1.00 \text{ s},$

and, with the Lorentz transformations,

$x'_2 = 1.15[3.00 \text{ m} - 0.5(2.998 \times 10^8 \text{ m/s})(1.00 \text{ s})]$

$\qquad \approx -1.72 \times 10^8 \text{ m}$

$y'_2 = y_2 = 4.00 \text{ m}$

$z'_2 = z_2 = 5.00 \text{ m}$

$t'_2 = 1.15[1.00 \text{ s} - (3.00 \text{ m})/2(2.998 \times 10^8 \text{ m/s})] \approx 1.15 \text{ s}.$

(d) To the observer in S, the spatial distance between these two events was

$$\sqrt{(x_2 - x_1)^2 + (y_2 - y_1)^2 + (z_2 - z_1)^2}$$
$$= \sqrt{(1.00 \text{ m})^2 + (1.00 \text{ m})^2 + (1.00 \text{ m})^2}$$
$$= 1.73 \text{ m},$$

and the temporal separation was $t_2 - t_1 = 1.00 \text{ s} - 0.00 \text{ s} = 1.00 \text{ s}.$

(e) To the observer in S', the spatial distance between these two events was, with the Galilean transformation,

$$\sqrt{(x'_2 - x'_1)^2 + (y'_2 - y'_1)^2 + (z'_2 - z'_1)^2}$$
$$= \sqrt{(-1.5 \times 10^8 \text{ m} - 2.00 \text{ m})^2 + (1.00 \text{ m})^2 + (1.00 \text{ m})^2}$$
$$\approx 1.50 \times 10^8 \text{ m},$$

and with the Lorentz transformation,

$$\sqrt{(x'_2 - x'_1)^2 + (y'_2 - y'_1)^2 + (z'_2 - z'_1)^2}$$
$$= \sqrt{(-1.72 \times 10^8 \text{ m} - 2.30 \text{ m})^2 + (1.00 \text{ m})^2 + (1.00 \text{ m})^2}$$
$$\approx 1.72 \times 10^8 \text{ m}.$$

To the observer in S', the temporal separation between these two events was, with the Galilean transformation,

$$t'_2 - t'_1 = 1.00 \text{ s} - 0.00 \text{ s} = 1.00 \text{ s},$$

and with the Lorentz transformation,

$$t'_2 - t'_1 = 1.15 \text{ s} - 0.00 \text{ s} = 1.15 \text{ s}.$$

(f) The difference in spatial separations is enormous, regardless of whether the Galilean or Lorentz transformation is used, but that is intuitively what we would expect because of the high relative speed of the two IRFs. The spatial separation observed in S' is slightly greater with the Lorentz transformation.

With the Galilean transformation, there is no difference in temporal separations as measured by the two observers. With the Lorentz transformation there is a small but significant difference, the temporal separation being slightly larger in S'. This is not an intuitive result, but is correct according to the laws of physics.

(g) As the speed between the IRFs decreases, there is less and less difference between the separations in the two transformations. The Lorentz transformation is always correct, but for low values of v the difference becomes negligible.

ANSWERS

to Odd-Numbered Questions, Exercises, and Problems

CHAPTER 1

EP 31. 3.1 μm/s **33.** (a) 2.99×10^{-26} kg; (b) 4.68×10^{46}
35. (a) 400; (b) 6.4×10^7; (c) 3.5×10^3 km **37.** (a) 3.88;
(b) 7.65; (c) 156 ken^3; (d) 1.19×10^3 m^3 **39.** 0.3 cord
41. stock, 2.5 cups plus 2 teaspoons; nettle tops, 0.5 quart; rice,
2 teaspoons; salt, 1 teaspoon **43.** 0.020 km^3

CHAPTER 2

Q 11. 1 and 2 tie, then 3 **13.** (a) 1; (b) 10; (c) 5; (d) 6
15. (a) $-g$; (b) 2 m/s upward **17.** (a) $+2$ m $= (10$ m/s$)t +$
$\frac{1}{2}(-9.8$ m/s$^2)t^2$; (b) -2 m $= (10$ m/s$)t + \frac{1}{2}(-9.8$ m/s$^2)t^2$;
(c) -2 m $= (-10$ m/s$)t + \frac{1}{2}(-9.8$ m/s$^2)t^2$; (d) $(-12$ m/s$)^2 =$
$(10$ m/s$)^2 + (2)(-9.8$ m/s$^2)(y - y_0)$; (e) $(-12$ m/s$)^2 =$
$(-10$ m/s$)^2 + (2)(-9.8$ m/s$^2)(y - y_0)$ **EP 67.** 8.4 m
69. (a) 3.5; (b) (5 m)/v_s **71.** (a) $d = \frac{1}{2}at^2$; (c) 7.2 m/s^2
73. (a) 14 m/s; (b) 18 m/s^2; (c) 6 m/s; (d) 12 m/s^2; (e) 24 m/s;
(f) 24 m/s^2 **75.** (a) 38.1 m; (b) -9.02 m/s; (c) 14.5 m/s, up
77. (a) $t = 1.2$ s; (b) $t = 0$; (c) $t > 0, t < 0$ **79.** (a) 32.9 m/s;
(b) 49.1 s; (c) 11.7 m/s **81.** (a) 9.08 m/s^2; (b) 0.926g;
(c) 6.12 s; (d) $T_b = 15T_r$, braking; (e) 5.56 m **83.** (a) 60.6 s;
(b) 36.3 m/s **85.** (a) 48.5 m/s; (b) 4.95 s; (c) 34.3 m/s;
(d) 3.50 s **87.** (a) Either strategy works; (b) neither strategy
works **89.** (a) 25g; (b) 400 m **91.** (a) 0; (b) 4.0 m;
(c) ± 0.82 s; (e) $+20t$; (f) increase **93.** (a) The signs of v and a
are: AB: $+, -$; BC: 0, 0; CD: $+, +$; DE: $+, 0$; (b) no; (c) no
95. (a) 5.00 s; (b) 61.5 m **97.** 94 m **99.** (a) $v = [48.0t -$
$0.720t^2] e^{-0.0300t}$, $a = [48.0 - 2.88t + 0.0216t^2] e^{-0.0300t}$;
(c) $t = 1.175$ s, $v = 53.5$ m/s, $a = 43.1$ m/s^2;
(d) $t = 66.7$ s, $x = 14\,400$ m, $a = -6.50$ m/s^2
101. $\{-1.705$E4 -1.591E4 -5.114E3$\}$ **103.** 20.4 m
105. (a) 1.25 m/s^2; (b) 8.00 s; (c) 100 m; (d) 7.78 m/s
107. (a) 200 m; (b) -2.0 m/s^2 **109.** (a) 13.5 m/s; (b) 32.0 s

CHAPTER 3

Q 11. $\vec{s}, \vec{p}, \vec{r}$ or $\vec{p}, \vec{s}, \vec{r}$ **13.** On many calculators, you get
the correct θ for $\vec{a}$ and $\vec{d}$ but not for $\vec{b}$ and $\vec{c}$, for which you
must add 180° **15.** zero ($\vec{A}$ and $\vec{B}$ are perpendicular) **17.**
correct: (c), (d), (f), (h); incorrect: (a), cannot dot a vector with a
scalar; (b), cannot cross a vector with a scalar; (e), (g), (i), and
(j), cannot add a scalar and a vector **19.** (a) $+z$; (b) $-z$
21. (a) 1: $-y$; 2: $-z$; 3: impossible; (b) 1: $+y$; 2: $+z$; 3:
impossible **23.** (a) $-, +$; (b) $-, -$; (c) $+, +$ **EP 39.** 70.5°
43. (a) -21; (b) -9; (c) $5\hat{i} - 11\hat{j} - 9\hat{k}$ **45.** yes **47.** 3.6 m
49. (a) -70.0 cm; (b) 80.0 cm; (c) 141 cm; (d) $-172°$ from
$+x$ direction **51.** (a) 1.8 m; (b) 69° north of east
53. (a) 9.51 m; (b) 14.1 m; (c) 13.4 m; (d) 10.5 m

55. (a) 10 m; (b) north; (c) 7.5 m; (d) south **57.** (a) 4.28 m;
(b) 11.7 m **59.** (a) $9\hat{i} + 12\hat{j}$; (b) $3\hat{i} + 4\hat{j}$ **61.** (a) -12.5;
(b) $284\hat{k}$; (c) 284, $\phi = 0$; (d) 90°; (e) $-2.57\hat{i} + 3.06\hat{j} + 3.00\hat{k}$;
(f) 5.00, $\theta = 130°$, $\phi = 53.1°$

CHAPTER 4

Q 15. (a) use $x - x_0 = v_{0x}t = (20$ m/s$)(\cos 30°)t$, where t
is obtained from $y - y_0 = v_{0y}t + \frac{1}{2}a_yt^2$, which becomes
-2 m $= (20$ m/s$)(\sin 30°)t + \frac{1}{2}(-9.8$ m/s$^2)t^2$; (b) use $v_x = v_{0x} =$
$(20$ m/s$)\cos 30°$ and $v_y^2 = v_{0y}^2 + 2a_y(y - y_0) = (20$ m/s$)^2 \times$
$(\sin 30°)^2 + 2(-9.8$ m/s$^2)(-2$ m$)$ **17.** (a) 3 m/s; (b) 5 m/s;
(c) decrease; (d) increase; (e) increase upward **19.** tie of a, b,
and c, then d **21.** a, 3; b, 4; c, 6 **23.** (a) no; (b) same
EP 65. (a) from 75° east of south; (b) 30° east of north. For a
second set of solutions, substitute west for east in both answers.
67. (a) 2.7g; (b) 3.8g **69.** $\sqrt{2}(40.0$ m$)$ NW; (b) 1.9 m/s NW;
(c) 0.47 m/s^2 NE **71.** (a) 16 m/s, 23° above the horizontal;
(b) 27 m/s, 57° below the horizontal **73.** (b) 76° **75.** 7.0 m/s
77. (a) $(6.0\hat{i} + 4.2\hat{j})$ m/s; (b) $(18\hat{i} + 6.3\hat{j})$ m **79.** (a) $(10\hat{i} + 10\hat{j})$
m/s; (b) 8.0 m/s^2; (c) 2.7 s; (d) 2.2 s **81.** His record would have
been longer by about 1 cm **83.** (a) 51.8 m; (b) 27.4 m/s;
(c) 67.5 m **85.** (a) 1.6 s; (b) no (18 cm); (c) 14 m/s; (d) yes
87. 25 m **89.** (a) 48 m, west of center; (b) 48 m, west of center
91. (a) 32.4 m; (b) -37.7 m **93.** 67 km/h **95.** (a) 96.2 m;
(b) 4.31 m; (c) 86.5 m, 25.1 m **97.** 23 ft/s **99.** 6.29°, 83.7°
101. (a) 87.13°; (e) 29 cm long for $\theta_{opt} - 2°$, 15 cm long for
$\theta_{opt} - 1°$, 0 cm for θ_{opt}, 14 cm short for $\theta_{opt} + 1°$, 28 cm short
for $\theta_{opt} + 2°$ **103.** (a) 13.6 m; (b) 38.5 m/s; (c) 34.8 m/s
105. (a) 32.3 m; (b) 21.9 m/s, 40.4° below horizontal
107. (a) 55.6° above horizontal; (b) 6.85 m; (c) 6.78 m/s
109. (a) $\theta_0 = \tan^{-1}[(1 \pm A)v_0^2/g(30$ m$)]$, where $A =$
$\{1 - (2g/v_0^2)[g(30$ m$)^2/(2v_0^2) + 5$ m$]\}^{0.5}$; the plus sign
option gives θ_{max}, and the minus sign option gives θ_{min};
(b)

v_0	$\theta_{0,\,min}$	$\theta_{0,\,max}$
18.0 m/s	no answers	
19.0 m/s	42.4°	57.0°
20.0 m/s	36.1°	63.3°
21.0 m/s	32.4°	67.1°
23.0 m/s	27.5°	72.0°
25.0 m/s	24.2°	75.3°

111. (a) 17.2 m; (b) 53.1° up from horizontal

CHAPTER 5

Q 13. $F_{net,\,x} = ma_x$: (a) 1 N $= (2$ kg$)a_x$; (b) 2N $-$ 1 N $=$
$(2$ kg$)a_x$; (c) 1 N $+$ 3 N $-$ 2 N $= (2$ kg$)a_x$; (d) (1 N)$\cos 30° =$

$(2 \text{ kg})a_x$; (e) $-(2 \text{ N})\sin 30° = (2 \text{ kg})a_x$; (f) $-(2 \text{ N})\cos 30° -$
$(3 \text{ N})\sin 30° = (2 \text{ kg})a_x$ $F_{net,y} = ma_y$: (a) $0 = (2 \text{ kg})a_y$;
(b) $0 = (2 \text{ kg})a_y$; (c) $0 = (2 \text{ kg})a_y$; (d) $(1 \text{ N})\sin 30° = (2 \text{ kg})a_y$;
(e) $(2 \text{ N})\cos 30° - (2 \text{ kg})a_y$;
(f) $(2 \text{ N})\sin 30° - (3 \text{ N})\cos 30° + 1 \text{ N} = (2 \text{ kg})a_y$
15. $F_{net,x} = ma_x$:
(a) $5 \text{ N} - (0.2 \text{ kg})(9.8 \text{ m/s}^2)\sin 30° = (0.2 \text{ kg})a$;
(b) $-5 \text{ N} - (0.2 \text{ kg})(9.8 \text{ m/s}^2)\sin 30° = (0.2 \text{ kg})a$;
(c) $(5 \text{ N})\cos 30° - (0.2 \text{ kg})(9.8 \text{ m/s}^2)\sin 30° = (0.2 \text{ kg})a$;
(d) $-(5 \text{ N})\sin 30° - (0.2 \text{ kg})(9.8 \text{ m/s}^2)\sin 30° = (0.2 \text{ kg})a$
$F_{net,y} = ma_y$:
(a) $N - (0.2 \text{ kg})(9.8 \text{ m/s}^2)\cos 30° = (0.2 \text{ kg})(0)$;
(b) $N - (0.2 \text{ kg})(9.8 \text{ m/s}^2)\cos 30° = (0.2 \text{ kg})(0)$;
(c) $N - (0.2 \text{ kg})(9.8 \text{ m/s}^2)\cos 30° - (5 \text{ N})\sin 30° = (0.2 \text{ kg})(0)$;
(d) $N - (0.2 \text{ kg})(9.8 \text{ m/s}^2)\cos 30° - (5 \text{ N})\cos 30° = (0.2 \text{ kg})(0)$
17. $F_{net,x} = ma_x$:
(a) $F_4 = (2 \text{ kg})(1 \text{ m/s}^2)$; (b) $2 \text{ N} - F_4 = (2 \text{ kg})(-1 \text{ m/s}^2)$;
(c) $1 \text{ N} = (2 \text{ kg})(1 \text{ m/s}^2)(\sin 30°)$; (d) $3 \text{ N} - F_4 = (2 \text{ kg})(0)$;
(e) $(2 \text{ N})\cos 30° + 3 \text{ N} - F_4 = (2 \text{ kg})(0)$;
(f) $-F_4 + 3 \text{ N} + (2 \text{ N})\cos 30° = (2 \text{ kg})(1 \text{ m/s}^2)(\cos 30°)$
$F_{net,y} = ma_y$: (a) $0 = (2 \text{ kg})(0)$; (b) $0 = (2 \text{ kg})(0)$;
(c) $F_4 = (2 \text{ kg})(1 \text{ m/s}^2)(\cos 30°)$; (d) $2 \text{ N} = (2 \text{ kg})(1 \text{ m/s}^2)$;
(e) $(2 \text{ N})\sin 30° + 1 \text{ N} = (2 \text{ kg})(1 \text{ m/s}^2)$;
(f) $-(2 \text{ N})\sin 30° = -(2 \text{ kg})(1 \text{ m/s}^2)(\sin 30°)$
19. d, c, a, b **21.** (a) M; (b) M; (c) M; (d) $3M$
23. (a) The number in each column does not matter;
(b) A, 1; B, 1; C, 5; (c) A, 1; B, 5; C, 1; (d) A, 1; six
in B and C in any arrangement   **EP** **57.** (a) allow a downward acceleration with magnitude $\geq 1.4 \text{ m/s}^2$; (b) 4.1 m/s
59. (a) $0.26mg$ (not $mg/6$); (b) decrease   **61.** (a) the 4 kg block;
(b) 6.5 m/s²; (c) 13 N   **63.** (a) $(1.0\hat{i} - 2.0\hat{j})$ N; (b) 2.2 N, $-63°$
relative to $+x$; (c) 2.2 m/s², $-63°$ relative to $+x$ **65.** (a) rope
breaks; (b) 1.6 m/s²   **67.** (a) $(1\hat{i} - 1.3\hat{j})$ m/s²; (b) 1.6 m/s² at
$-50°$ from $+x$ **69.** (a) 2.8 N, due west; (b) 2.2 N, 22° west of
south   **71.** (a) 3.0 N; (b) 0.34 kg   **73.** (a) 1.1×10^{-15} N;
(b) 8.9×10^{-30} N   **75.** (a) 180 N, east; (b) 4.0 m/s², west;
(c) 2.0 m/s², east   **77.** (a) 1080 N, up; (b) 980 N, up; (c) 780 N,
up   **79.** 27 N, 27° north of west   **81.** (a) 44 N; (b) 78 N;
(c) 54 N; (d) 152 N   **83.** (a) 4.90 m/s²; (b) 84.9 N
85. (a) 1.90 m/s², down the surface; (b) 32.9 N; (c) 98.0 N

CHAPTER 6

Q   **11.** $F_{net,x} = ma_x$: (a) $10 \text{ N} - f_s = (2 \text{ kg})(0)$;
(b) $10 \text{ N} - f_s = (2 \text{ kg})(0)$; (c) $10 \text{ N} - f_s = (2 \text{ kg})(0)$;
(d) $(10 \text{ N})\cos 30° - f_s = (2 \text{ kg})(0)$;
(e) $10 \text{ N} + (2 \text{ N})\sin 30° - f_s = (2 \text{ kg})(0)$;
(f) $10 \text{ N} - (2 \text{ N})\sin 30° - f_s = (2 \text{ kg})(0)$
$F_{net,y} = ma_y$: (a) $N - (2 \text{ kg})(9.8 \text{ m/s}^2) = (2 \text{ kg})(0)$;
(b) $N - (2 \text{ kg})(9.8 \text{ m/s}^2) - 2 \text{ N} = (2 \text{ kg})(0)$;
(c) $N - (2 \text{ kg})(9.8 \text{ m/s}^2) + 2 \text{ N} = (2 \text{ kg})(0)$;
(d) $N - (2 \text{ kg})(9.8 \text{ m/s}^2) + (10 \text{ N})\sin 30° = (2 \text{ kg})(0)$;
(e) $N - (2 \text{ kg})(9.8 \text{ m/s}^2) - (2 \text{ N})\cos 30° = (2 \text{ kg})(0)$;
(f) $N - (2 \text{ kg})(9.8 \text{ m/s}^2) - (2 \text{ N})\cos 30° = (2 \text{ kg})(0)$
13. $F_{net,x} = ma_x$: (a) $10 \text{ N} - (0.3)(2 \text{ kg})(9.8 \text{ m/s}^2) = (2 \text{ kg})a$;
(b) $10 \text{ N} - (0.3)[(2 \text{ kg})(9.8 \text{ m/s}^2) + 2 \text{ N}] = (2 \text{ kg})a$;
(c) $10 \text{ N} - (0.3)[(2 \text{ kg})(9.8 \text{ m/s}^2) - 2 \text{ N}] = (2 \text{ kg})a$;

(d) $(10 \text{ N})\cos 30° - (0.3)[(2 \text{ kg})(9.8 \text{ m/s}^2) - (10 \text{ N})\sin 30°] =$
$(2 \text{ kg})a$; (e) $10 \text{ N} - (0.3)[(2 \text{ kg})(9.8 \text{ m/s}^2) + (2 \text{ N})\cos 30°] +$
$(2 \text{ N})\sin 30° = (2 \text{ kg})a$; (f) $10 \text{ N} - (0.3)[(2 \text{ kg})(9.8 \text{ m/s}^2) +$
$(2 \text{ N})\cos 30°] - (2 \text{ N})\sin 30° = (2 \text{ kg})a$
$F_{net,y} = ma_y$:
(a) $N - (2 \text{ kg})(9.8 \text{ m/s}^2) = (2 \text{ kg})(0)$;
(b) $N - (2 \text{ kg})(9.8 \text{ m/s}^2) - 2 \text{ N} = (2 \text{ kg})(0)$;
(c) $N - (2 \text{ kg})(9.8 \text{ m/s}^2) + 2 \text{ N} = (2 \text{ kg})(0)$;
(d) $N - (2 \text{ kg})(9.8 \text{ m/s}^2) + (10 \text{ N})\sin 30° = (2 \text{ kg})(0)$;
(e) $N - (2 \text{ kg})(9.8 \text{ m/s}^2) - (2 \text{ N})\cos 30° = (2 \text{ kg})(0)$;
(f) $N - (2 \text{ kg})(9.8 \text{ m/s}^2) - (2 \text{ N})\cos 30° = (2 \text{ kg})(0)$
15. (a) 1 to 3; (b) 14 to 8; (c) 6 to 8; (d) 16; (e) 3 to 5; (f) 12 to 5
17. $\vec{f}_s$ is initially directed up the ramp and increases in
magnitude until the magnitude reaches $f_{s,max}$; thereafter the
magnitude of the frictional force is f_k (a constant smaller value)
19. $F_{net,x} = ma_x$:
(a) $20 \text{ N} - (2 \text{ kg})(9.8 \text{ m/s}^2)\sin 30° - f_s = (2 \text{ kg})(0)$;
(b) $-2 \text{ N} - (2 \text{ kg})(9.8 \text{ m/s}^2)\sin 30° + f_s = (2 \text{ kg})(0)$
$F_{net,y} = ma_y$:
(a) $N - (2 \text{ kg})(9.8 \text{ m/s}^2)\cos 30° = (2 \text{ kg})(0)$;
(b) $N - (2 \text{ kg})(9.8 \text{ m/s}^2)\cos 30° = (2 \text{ kg})(0)$
21. WILEY, then tie of STP and TI, then IBM, then TIDE (may
not finish)
EP   **49.** (a) 30 cm/s; (b) 180 cm/s², radially inward;
(c) 3.6×10^{-3} N, radially inward; (d) 0.37   **51.** (a) 0.0338 N;
(b) 9.77 N   **53.** (a) 2×10^4 N; (b) $18g$ **55.** 4.6 N   **57.** 874 N
59. (a) 0.11 m/s², 0.23 m/s²; (b) 0.041, 0.029   **61.** (a) 190 N, up;
(b) 320 N, up   **63.** (a) 0.96 m/s; (b) 0.021   **65.** 3.4 m/s²
67. (a) 0.37; (b) $0.37 < \mu_s < 0.47$ **69.** (a) 49 m/s; (b) 430 N
71. (a) 0.72 m/s; (b) 2.1 m/s²; (c) 0.50 N   **73.** (a) 90 N; (b) 70 N;
(c) 0.89 m/s²   **75.** (a) 74 N; (b) $F = [75.6/(\cos \theta + 0.42 \sin \theta)]$ N;
(c) 23°; (d) 70 N   **77.** (a) 2.0 m/s² down the plane; (b) 4.0 m;
(c) it stays there   **79.** 25 N, opposite the applied force
81. (a) 53 N; (b) 41 N; (c) 12 N; (d) 2.4 m/s², down the plane
83. (a) 25 N up along the plane; (b) 31°; (c) decreased; (d) 11°

CHAPTER 7

Q   **15.** c, d, b, a **17.** $K_f = \frac{1}{2}mv_i^2 +$ net work done by the
forces on the box: (a) $K_f = \frac{1}{2}(2 \text{ kg})(3 \text{ m/s})^2 + [10 \text{ N} - 2 \text{ N} -$
$(2 \text{ kg})(9.8 \text{ m/s}^2)](-4 \text{ m})$; (b) $K_f = \frac{1}{2}(2 \text{ kg})(3 \text{ m/s})^2 + [10 \text{ N} +$
$2 \text{ N} - (2 \text{ kg})(9.8 \text{ m/s}^2)](-4 \text{ m})$; (c) $K_f = \frac{1}{2}(2 \text{ kg})(3 \text{ m/s})^2 +$
$[-10 \text{ N} + 2 \text{ N} - (2 \text{ kg})(9.8 \text{ m/s}^2)](-4 \text{ m})$
19. $K_f = \frac{1}{2}mv_i^2 +$ net work done by the forces on the box:
(a) $K_f = \frac{1}{2}(2 \text{ kg})(3 \text{ m/s})^2 + [(10 \text{ N})\cos 30° + (2 \text{ N})](4 \text{ m})$;
(b) $K_f = \frac{1}{2}(2 \text{ kg})(3 \text{ m/s})^2 + [(10 \text{ N})\cos 30° - (2 \text{ N})](4 \text{ m})$;
(c) $K_f = \frac{1}{2}(2 \text{ kg})(3 \text{ m/s})^2 + (10 \text{ N})(\cos 30°)(4 \text{ m})$;
(d) $K_f = \frac{1}{2}(2 \text{ kg})(3 \text{ m/s})^2 + [(10 \text{ N})\sin 30° - (2 \text{ N})\sin 30°](4 \text{ m})$
21. (a) yes (constant force); (b) and (c) no (variable force)
23. (a) all tie; (b) a, b, c **25.** (a) 4, 2, then 1 and 3 tie;
(b) all tie   **EP**   **41.** 165 kW   **43.** (a) 8.0 N; (b) 8.0 N/m
45. (a) 2.5 kJ; (b) -2.1 kJ   **47.** 1.8×10^{13} J   **49.** (a) 797 N;
(b) 0; (c) -1550 J; (d) 0; (e) 1550 J; (f) F varies during displacement   **51.** 1.5 kJ   **53.** (a) 1×10^5 megatons TNT;
(b) 1×10^7 bombs

CHAPTER 8

Q

11.

	$E_{\mathrm{mec},i}$	$\Delta E_{\mathrm{mec},d}$	$E_{\mathrm{mec},f}$
(a)	$\frac{1}{2}mv_i^2$	$-fd$	0
(b)	$\frac{1}{2}mv_i^2$	0	mgh
(c)	$\frac{1}{2}mv_i^2$	$-fh/\sin\theta$	mgh
(d)	$\frac{1}{2}mv_i^2$	0	$\frac{1}{2}kd^2$
(e)	$\frac{1}{2}mv_i^2$	$-fd$	$\frac{1}{2}kd^2$
(f)	$\frac{1}{2}mv_i^2 + mgh$	$-fd$	0
(g)	$\frac{1}{2}mv_i^2 + mgh$	$-fd$	$\frac{1}{2}kd^2$
(h)	$\frac{1}{2}mv_i^2$	$-fd$	mgh
(i)	$\frac{1}{2}mv_i^2$	0	$mgh + \frac{1}{2}kd^2$
(j)	$\frac{1}{2}mv_i^2$	$-fd$	$mgh + \frac{1}{2}kd^2$

13. (a) 3, 2, 1; (b) 1, 2, 3 **15.** -40 J   **EP   69.** your force on the cabbage (as you lower it) does work **71.** 880 MW
73. (a) turning point on left, none on right, molecule breaks apart; (b) turning points on both left and right, molecule does not break apart; (c) -1.2×10^{-19} J; (d) 2.2×10^{-19} J; (e) $\approx 1 \times 10^{-9}$ N on each, directed toward the other; (f) $r < 0.2$ nm;
(g) $r > 0.2$ nm; (h) $r = 0.2$ nm   **75.** (a) $K = 0.75$ J, $\Delta U_g = -1.0$ J, $\Delta U_e = 0.25$ J; (b) $K = 1.0$ J, $\Delta U_g = -2.0$ J, $\Delta U_e = 1.0$ J; (c) $K = 0.75$ J, $\Delta U_g = -3.0$ J, $\Delta U_e = 2.25$ J; (d) $K = 0$, $\Delta U_g = -4.0$ J, $\Delta U_e = 4.0$ J   **77.** (a) 1.2 J; (b) 11 m/s;
(c) no; (d) no   **79.** 8580 J   **81.** (a) 3.7 J; (b) 4.3 J; (c) 4.3 J
83. 3.1×10^{11} W   **85.** 17 kW   **87.** 5.4 kJ   **89.** 3.7 J
91. (a) 6.75 J; (b) -6.75 J; (c) 6.75 J; (d) 6.75 J; (e) -6.75 J;
(f) 0.459 m   **95.** (a) 24 kJ; (b) 470 N   **97.** (a) 8.6 kJ;
(b) 860 W; (c) 430 W; (d) 1300 W

CHAPTER 9

Q   11. final net momentum = initial momentum:
(a) $(3 \text{ kg})(2 \text{ m/s}) + (5 \text{ kg} - 3 \text{ kg})v_{Bf} = (5 \text{ kg})(4 \text{ m/s})$;
(b) $(3 \text{ kg})(-2 \text{ m/s}) + (5 \text{ kg} - 3 \text{ kg})v_{Bf} = (5 \text{ kg})(4 \text{ m/s})$;
(c) $(3 \text{ kg})(0) + (5 \text{ kg} - 3 \text{ kg})v_{Bf} = (5 \text{ kg})(4 \text{ m/s})$
13. (a) all tie; (b) all tie; (c) all tie   **15.** (a) a tie of 4, 6, and 7, then a tie of 1 and 2, then a tie of 3 and 5; (b) 1, 3, 5;
(c) 2, 5, 7   **17.** a, b, c **EP   61.** 190 m/s   **63.** (a) 1.0 m/s, north; (b) 3.0 m, north   **65.** 1.2 m/s, 48° north of west
69. (a) 33 m/s; (b) 8.7 m/s   **71.** (a) 29 m; (b) 42 m
73. (a) 540 m/s; (b) 40.4°   **75.** (a) $0.200v_{\mathrm{rel}}$;
(b) $0.210v_{\mathrm{rel}}$; (c) $0.209v_{\mathrm{rel}}$ **77.** (a) -4.5 m; (b) -5.5 m
81. $x = B/2, y = H/3$ **83.** $x = 0, y = 4R/3\pi$

CHAPTER 10

Q   11. total momentum after the collision = total momentum before the collision: (a) $(2 \text{ kg})(1 \text{ m/s}) + (3 \text{ kg})v_{2f} = (2 \text{ kg})(4 \text{ m/s}) + (3 \text{ kg})(0)$; (b) $(2 \text{ kg})(2 \text{ m/s}) + (3 \text{ kg})v_{2f} = (2 \text{ kg})(4 \text{ m/s}) + (3 \text{ kg})(2 \text{ m/s})$; (c) $(2 \text{ kg})(-1 \text{ m/s}) + (3 \text{ kg})v_{2f} = (2 \text{ kg})(4 \text{ m/s}) + (3 \text{ kg})(-2 \text{ m/s})$; (d) $(2 \text{ kg})(-2 \text{ m/s}) + (3 \text{ kg})v_{2f} = (2 \text{ kg})(6 \text{ m/s}) + (3 \text{ kg})(-6 \text{ m/s})$
13. $\Delta K =$ (total kinetic energy after the collision) $-$ (total kinetic energy before the collision):
(a) $\Delta K = \frac{1}{2}(2 \text{ kg})(1 \text{ m/s})^2 + \frac{1}{2}(3 \text{ kg})v_{2f}^2 - [\frac{1}{2}(2 \text{ kg})(4 \text{ m/s})^2 + \frac{1}{2}(3 \text{ kg})(0)^2]$; (b) $\Delta K = \frac{1}{2}(2 \text{ kg})(2 \text{ m/s})^2 + \frac{1}{2}(3 \text{ kg})v_{2f}^2 - [\frac{1}{2}(2 \text{ kg})(4 \text{ m/s})^2 + \frac{1}{2}(3 \text{ kg})(2 \text{ m/s})^2]$; (c) $\Delta K =$
$\frac{1}{2}(2 \text{ kg})(-1 \text{ m/s})^2 + \frac{1}{2}(3 \text{ kg})v_{2f}^2 - [\frac{1}{2}(2 \text{ kg})(4 \text{ m/s})^2 + \frac{1}{2}(3 \text{ kg})(-2 \text{ m/s})^2]$; (d) $\Delta K = \frac{1}{2}(2 \text{ kg})(-2 \text{ m/s})^2 + \frac{1}{2}(3 \text{ kg})v_{2f}^2 - [\frac{1}{2}(2 \text{ kg})(6 \text{ m/s})^2 + \frac{1}{2}(3 \text{ kg})(-6 \text{ m/s})^2]$
15. (a) $v_{1f} = (5 \text{ kg} - 3 \text{ kg})(4 \text{ m/s})/(5 \text{ kg} + 3 \text{ kg})$,
$v_{2f} = (2)(5 \text{ kg})(4 \text{ m/s})/(5 \text{ kg} + 3 \text{ kg})$;
(b) -3 m/s $= (2 \text{ kg} - 5 \text{ kg})v_{1i}/(2 \text{ kg} + 5 \text{ kg})$,
$v_{2f} = (2)(2 \text{ kg})v_{1i}/(2 \text{ kg} + 5 \text{ kg})$ **17.** positive direction of x axis
19. a: 3 m/s, leftward; f and g: 3 m/s, rightward; other blocks: 0   **EP   59.** (a) 7.5×10^6 m/s^2; (b) 3.1 kg · m/s;
(c) -460 J; (d) 6.0 mm; (e) 3.1 kg · m/s; (f) 77 kN;
(g) 1.2×10^3 m/s^2; (h) 4.7×10^{-2} m/s; (i) a increases, Δp remains the same, ΔK remains the same, J remains the same, F increases, a_p increases, v_p remains the same
61. 5.0 kg   **63.** 8.1 m/s, 38° south of east   **65.** $v = V/4$
67. (a) 4.4 m/s, toward the right; (b) 38 J   **69.** (a) $24\hat{\mathbf{i}}$ m/s;
(b) 6.9 s; (c) $27\hat{\mathbf{i}}$ m/s   **71.** 216   **73.** (a) 61.7 km/h;
(b) 63.4° south of west   **75.** (a) 2.5 m/s; (b) 42 J   **77.** 35 cm
79. (a) 62.5 km/h; (b) 0.75   **81.** (a) 4.0 kg · m/s^2;
(b) 8.0 kg · m/s   **83.** (a) 2.3 N · s, in initial direction of flight;
(b) 2.3 N · s, opposite initial direction of flight; (c) 1400 N, in initial direction of flight; (d) 58 J   **85.** 1.50 N   **87.** (a) $v_{1f} = v_{2f} = m_1 v_{1i}/(m_1 + m_2)$; (b) $v_{1f} = (m_1 - m_2)v_{1i}/(m_1 + m_2)$;
$v_{2f} = 2m_1 v_{1i}/(m_1 + m_2)$ **89.** (a) $v_{2f} = 20.0$ m/s $- 2.00v_{1f}$;
(c) 6.67 m/s; (d) 6.67 m/s; (e) the bodies stick together; completely inelastic collision; (f) v_{1f} would be greater than v_{2f}, which would require the projectile to pass through the target; in addition, v_{1f} cannot exceed v_{1i}; (h) 3.33 m/s; (i) 13.3 m/s; (j) elastic collision; (k) it would require an input of energy, as from an explosion between the two bodies   **91.** (a) 9.34 m/s; (b) 2.52 m/s; (c) 67.8°

CHAPTER 11

Q   13. (a) A: 1 rad $= (5 \text{ rad/s})t + \frac{1}{2}(2 \text{ rad/s}^2)t^2$; B: 1 rad $= (-5 \text{ rad/s})t + \frac{1}{2}(2 \text{ rad/s}^2)t^2$; (b) less; (c) C: $(8 \text{ rad/s})^2 = (-5 \text{ rad/s})^2 + (2)(2 \text{ rad/s}^2)\theta$; D: $(8 \text{ rad/s})^2 = (5 \text{ rad/s})^2 + (2)(-2 \text{ rad/s}^2)\theta$; (d) same   **15.** (a) $\tau_{\mathrm{net}} = I\alpha$: A: $(0.5 \text{ m})(1 \text{ N}) = (2 \text{ kg} \cdot \text{m}^2)\alpha$; B: $-(0.5 \text{ m})(2 \text{ N})\sin 30° = (2 \text{ kg} \cdot \text{m}^2)\alpha$;
C: $-(0.5 \text{ m})(1 \text{ N}) - (0.5 \text{ m})(2 \text{ N})\sin 30° = (2 \text{ kg} \cdot \text{m}^2)\alpha$;
D: $(0.5 \text{ m})(1 \text{ N}) - (0.5 \text{ m})(2 \text{ N})\sin 30° = (2 \text{ kg} \cdot \text{m}^2)\alpha$;
(b) $\theta - \theta_0 = (3 \text{ rad/s})(2 \text{ s}) + \frac{1}{2}\alpha(2 \text{ s})^2$ **17.** (a) A: τ_1 transfers to Susan; B: τ_2 transfers from Susan; C: τ_1 and τ_2 transfer from Susan; D: τ_1 transfers to Susan, τ_2 transfers from Susan;
(b) $K_f = \frac{1}{2}I\omega_f^2 +$ net work done by the torques on Susan:
A: $K_f = \frac{1}{2}(2 \text{ kg} \cdot \text{m}^2)(3 \text{ rad/s})^2 + (0.5 \text{ m})(1 \text{ N})(1.2 \text{ rad})$; B: $K_f = \frac{1}{2}(2 \text{ kg} \cdot \text{m}^2)(3 \text{ rad/s})^2 - (0.5 \text{ m})(2 \text{ N})(\sin 30°)(1.2 \text{ rad})$;
C: $K_f = \frac{1}{2}(2 \text{ kg} \cdot \text{m}^2)(3 \text{ rad/s})^2 - [(0.5 \text{ m})(1 \text{ N}) + (0.5 \text{ m})(2\text{N})(\sin 30°)](1.2 \text{ rad})$; D: $K_f = \frac{1}{2}(2 \text{ kg} \cdot \text{m}^2)(3 \text{ rad/s})^2 + [(0.5 \text{ m})(1 \text{ N}) - (0.5 \text{ m})(2 \text{ N})(\sin 30°)](1.2 \text{ rad})$ **19.** b, c, a
21. all tie   **23.** (a) 1; (b) 10; (c) 5; (d) 6   **25.** all tie
EP   71. (a) 4.8×10^5 N; (b) 1.1×10^4 N · m; (c) 1.3×10^6 J
73. 6.9×10^{-13} rad/s   **75.** (a) yes; (b) 114 kg   **77.** (a) 10 J;
(b) 0.27 m   **79.** (a) $2\theta/t^2$; (b) $2R\theta/t^2$; (c) $M(g - 2R\theta/t^2)$;
(d) $Mg - (2\theta/t^2)(MR + I/R)$ **81.** (a) 3.3 J; (b) 2.9 J
85. (a) 2.0×10^{-7} rad/s; (b) 30 km/s; (c) 5.9 mm/s^2, toward the Sun   **87.** 1500 rad   **89.** (a) $\omega_0 + at^4 - bt^3$; (b) $\theta_0 + \omega_0 t + at^5/5 - bt^4/4$ **91.** 3.1 rad/s   **93.** 0.054 kg · m^2

95. 3.2 kg $\cdot$ m^2 **99.** (a) 310 m/s; (b) 340 m/s
101. (a) 2.0 rev/s; (b) 3.8 s   **103.** 14 rev

CHAPTER 12

Q 15.

	E_i	E_f
(a)	$\frac{1}{2}mv_i^2 + \frac{1}{2}I(v_i^2/r^2)$	$\frac{1}{2}mv_f^2 + \frac{1}{2}I(v_f^2/r^2) + mgh$
(b)	$\frac{1}{2}mv_i^2 + \frac{1}{2}I(v_i^2/r^2)$	mgh
(c)	mgh	$\frac{1}{2}mv_f^2 + \frac{1}{2}I(v_f^2/r^2) + mg(2R)$

17. (a) up along the incline; (b) gravitational force component; (c) torque due to frictional force (the other is zero); (d) torque due to gravitational force (the other is zero); (e) decrease
19. (a) 3; (b) 1; (c) 2; (d) 4   **21.** final net angular momentum = initial net angular momentum: (a) $(0.2$ kg$)(R/2)^2\omega_f + \frac{1}{2}(3.0$ kg$)R^2\omega_f = (0.2$ kg$)R^2(1.5$ rad/s$) + \frac{1}{2}(3.0$ kg$)R^2(1.5$ rad/s$)$; (b) $0 + \frac{1}{2}(3.0$ kg$)R^2\omega_f = (0.2$ kg$)R^2(1.5$ rad/s$) + \frac{1}{2}(3.0$ kg$)R^2(1.5$ rad/s$)$; (c) center; (d) b; (e) b   **23.** The law of conservation of angular momentum is not violated. The performer leaves the floor with a small, perhaps imperceptible amount of spin about a vertical axis through the body, while having arms and legs extended. Once in the air, the performer decreases the rotational inertia by pulling in the arms and legs in order to increase the spin so that it is then apparent to the audience. The procedure is reversed just before landing.
25. (a) 30 units clockwise; (b) 2, then 4, then the others; or 4, then 2, then the others   **EP 61.** (a) 0.19 m/s^2, down; (b) 0.19 m/s^2, down; (c) 1.1 kN; (d) no; (e) same; (f) greater
63. (a) 1/3; (b) 1/9   **65.** (a) $-1.8\hat{k}$ kg $\cdot$ m^2/s; (b) $-3.6\hat{k}$ kg $\cdot$ m^2/s; (c) 0; (d) $-7.3\hat{k}$ N $\cdot$ m   **67.** (a) 3.14×10^{43} kg $\cdot$ m^2/s; (b) 0.614
69. (a) 1.6 m/s; (b) 16 rad/s^2; (c) 4.0 N, in the direction of the applied force   **71.** 3.0 min   **73.** (a) 0; (b) 0; (c) $-30t^3\hat{k}$ kg $\cdot$ m^2/s; (d) $-90t^2\hat{k}$ N $\cdot$ m; (e) $30t^3\hat{k}$ kg $\cdot$ m^2/s; (f) $90t^2\hat{k}$ N $\cdot$ m   **75.** 2.5×10^{11} kg $\cdot$ m^2/s   **79.** (a) 61.7 J; (b) 3.43 m; (c) no   **81.** (a) 50.0 N; (b) 10.0 N $\cdot$ m; (c) 0.667 kg $\cdot$ m^2 **83.** $\theta = \tan^{-1}[\mu_s(1 + MR^2/I)]$
85. (a) $a_1 = (m_1R_1^2 - m_2R_1R_2)g/A$, with $A = (m_1R_1^2 + m_2R_2^2 + I)$; (b) $a_2 = (m_1R_1R_2 - m_2R_2^2)g/A$ **87.** (a) $Mg/[M + 1/(R_1^2/I_1 + R_2^2/I_2)]$

CHAPTER 13

Q 11. (a) about the left end, $\tau_{net} = 0$: $-(200$ kg$)(9.8$ m/s$^2) \times (8.0$ m$)/2 - (60$ kg$)(9.8$ m/s$^2)(6.0$ m$) + T_R(8.0$ m$) = 0$; (b) about the right end, $\tau_{net} = 0$: $(200$ kg$)(9.8$ m/s$^2)(8.0$ m$)/2 + (60$ kg$)(9.8$ m/s$^2)(2.0$ m$) - T_L(8.0$ m$) = 0$ **13.** about her feet, $\tau_{net} = 0$: $(50$ kg$)(9.8$ m/s$^2)(0.82$ m$)(\cos 30°) - T(0.80$ m$) = 0$
15. (a) 20 N (the key is the pulley with the 20 N weight); (b) 25 N   **17.** tie of A and B, then C **EP 43.** (a) 1.9×10^{-3}; (b) 1.3×10^7 N/m^2; (c) 6.9×10^9 N/m^2 **45.** 3.1 cm
47. (a) $(35\hat{i} + 200\hat{j})$ N; (b) $(-45\hat{i} + 200\hat{j})$ N; (c) 190 N
49. (a) 1.8×10^7 N; (b) 1.4×10^7 N; (c) 16   **51.** (a) 1160 N, up; (b) 1740 N, up   **53.** (a) $\mu < L/2h$; (b) $\mu > L/2h$
55. (a) $RFr_A/(r_A^2 + r_B^2)$; (b) $RFr_B/(r_A^2 + r_B^2)$ **57.** 0.29
61. (a) 1.5 kN, upward; (b) 1.9 kN, upward   **63.** $L/4$

CHAPTER 14

Q 13. (a) $F_{net, x} = (6.67 \times 10^{-11}$ N$\cdot$m^2/kg$^2)[-(3$ kg$)(1$ kg$)/(3$ m$)^2 - (4$ kg$)(1$ kg$)(\sin 30°)/(2$ m$)^2 + (2$ kg$)(1$ kg$)(\cos 30°)/$

$(1$ m$)^2]$; (b) $F_{net, y} = (6.67 \times 10^{-11}$ N$\cdot$m^2/kg$^2)[(4$ kg$)(1$ kg$) \times (\cos 30°)/(2$ m$)^2 + (2$ kg$)(1$ kg$)(\sin 30°)/(1$ m$)^2]$ **15.** (a) all tie; (b) 3, 2, 1; (c) 1, 2, 3   **17.** (a) all tie; (b) 1, 3, 2; (c) 2, 3, 1
19. orbits 2 and 3 (orbits must be about Earth's center)
21. (a) D, B, and then a tie of A, C, and E (zero); (b) A, C, and E: no direction; B, in the negative x direction; D, in the positive x direction   **23.** (a) same; (b) greater   **25.** increases from zero to a maximum, then decreases to zero   **EP 65.** (a) 2×10^{-5} m/s^2; (b) 2 cm/s   **67.** (a) 1.4×10^6 m/s; (b) 3×10^6 m/s^2
69. (a) Gm^2/R_i; (b) $Gm^2/2R_i$; (c) $(Gm/R_i)^{0.5}$; (d) $2(Gm/R_i)^{0.5}$; (e) Gm^2/R_i; (f) $(2Gm/R_i)^{0.5}$; (g) The center-of-mass frame is an inertial frame, and in it the principle of conservation of energy may be written as in Chapter 8; the reference frame attached to body A is noninertial, and the principle cannot be written as in Chapter 8. The answer to (d) is correct.   **71.** (a) $(2.8 \times 10^4)g$ (deadly); (b) $714g$; (c) 1.5 km/s   **75.** (a) 5.3×10^{-8} J; (b) 6.4×10^{-8} N, in the negative x direction   **77.** $\dfrac{GM^2}{L} \displaystyle\int_d^{L+d} \dfrac{dr}{r(L + r)}$
79. (a) speed increased from 3.51×10^4 m/s to 3.78×10^4 m/s ($\Delta v = 2.7 \times 10^3$ m/s); energy increased from -3.69×10^{12} J to -3.09×10^{12} J ($\Delta E = 6.0 \times 10^{11}$ J); semimajor axis: 1.29×10^{11} m; semiminor axis: 1.27×10^{11} m; (b) speed increased from 2.73×10^4 m/s to 2.98×10^4 m/s ($\Delta v = 2.5 \times 10^3$ m/s); energy increased from -3.09×10^{12} J to -2.66×10^{12} J ($\Delta E = 4.3 \times 10^{11}$ J)   **83.** (a) 2.15×10^4 s; (b) 12.3 km/s; (c) 12.0 km/s; (d) 2.17×10^{11} J; (e) -4.53×10^{11} J; (f) -2.35×10^{11} J; (g) 4.04×10^7 m; (h) 1.22×10^3 s; (i) new orbit   **85.** 4.4×10^{-6} N, perpendicular to, and toward, the line between A and B **87.** (a) 1.98×10^{30} kg; (b) 2.00×10^{30} kg
89. (a) 8.0×10^8 J; (b) 36 N   **91.** (a) 0; (b) 1.8×10^{32} J; (c) 1.8×10^{32} J; (d) 0.99 km/s

CHAPTER 15

Q 11. Using $v_2A_2 = v_1A_1$ and Eq. 15-28.
(a) $v_2 = (4$ m/s$)\pi(2$ cm$)^2/\pi(3$ cm$)^2$;
    $p_2 = 2 \times 10^5$ Pa $+ 0.5(1030$ kg/m$^3)[(4$ m/s$)^2 - v_2^2]$;
(b) $v_2 = (4$ m/s$)\pi(2$ cm$)^2/\pi(1$ cm$)^2$;
    $p_2 = 2 \times 10^5$ Pa $+ 0.5(1030$ kg/m$^3)[(4$ m/s$)^2 - v_2^2]$;
(c) $v_2 = (4$ m/s$)\pi(2$ cm$)^2/\pi(2$ cm$)^2 = v_1$;
    $p_2 = 2 \times 10^5$ Pa $- (1030$ kg/m$^3)(9.8$ m/s$^2)(0.5$ m$)$;
(d) $v_2 = (4$ m/s$)\pi(2$ cm$)^2/\pi(2$ cm$)^2 = v_1$;
    $p_2 = 2 \times 10^5$ Pa $+ (1030$ kg/m$^3)(9.8$ m/s$^2)(0.5$ m$)$;
(e) $v_2 = (4$ m/s$)\pi(2$ cm$)^2/\pi(3$ cm$)^2$;
    $p_2 = 2 \times 10^5$ Pa $+ 0.5(1030$ kg/m$^3)[(4$ m/s$)^2 - v_2^2]$
    $- (1030$ kg/m$^3)(9.8$ m/s$^2)(0.5$ m$)$;
(f) $v_2 = (4$ m/s$)\pi(2$ cm$)^2/\pi(3$ cm$)^2$;
    $p_2 = 2 \times 10^5$ Pa $+ 0.5(1030$ kg/m$^3)[(4$ m/s$)^2 - v_2^2]$
    $+ (1030$ kg/m$^3)(9.8$ m/s$^2)(0.5$ m$)$
13. all tie   **15.** a, b, c, d **EP 63.** (a) p/ρ; (b) 4.6×10^3 J/kg
65. 2.5 kPa   **67.** 1080 atm   **69.** (a) 2; (b) 3; (c) 4/3
71. 3.82 m/s^2 **73.** (a) 1.21×10^7 Pa; (b) 1.22×10^7 Pa; (c) 3.82×10^5 N; (d) 5.26 N; (e) 9.04 m/s^2, down
75. 2.79 g/cm^3 **77.** (a) 2.2; (b) 3.6   **79.** 1070 g
81. 1800 m^3 **83.** 1.62×10^6 Pa   **85.** 0.5 m   **87.** 1.5 cm
89. (a) 0.0776 m^3/s; (b) 69.8 kg/s

CHAPTER 16

Q **15.** 1: $x = (0.20 \text{ m}) \cos[2\pi t/(2 \text{ s})]$; 2: $x = (0.20 \text{ m}) \times \cos[2\pi t/(2 \text{ s}) - \pi/2]$; 3: $x = (0.20 \text{ m}) \cos[2\pi t/(2 \text{ s}) + \pi/2]$; 4: $x = (0.20 \text{ m}) \cos[2\pi t/(2 \text{ s}) \pm \pi]$ **17.** system with spring A **19.** all tie **21.** (a) a, c, b; (b) a, c, b; (c) b and c tie, then a **23.** (a) 3, 2, 1; (b) all tie **25.** (a) same; (b) same; (c) smaller; (d) smaller; (e) and (f) larger ($T = \infty$) **EP** **67.** $0.19g$ **69.** (a) 0.102 kg/s; (b) 0.137 J **71.** 9.78 m/s^2 **73.** (a) 0.20 s; (b) 0.20 kg; (c) -0.20 m; (d) -200 m/s^2; (e) 4.0 J **75.** (a) 1.6 Hz; (b) 1.0 m/s, 0; (c) 10 m/s^2, ± 10 cm; (d) $(-10 \text{ N/m})x$ **77.** 1.6 kg **79.** (a) 7.90 N/m; (b) 1.19 cm; (c) 2.00 Hz **81.** 708 N/m **83.** (a) 0.44 s; (b) 0.18 m **85.** (a) 0.30 m; (b) 30 m/s^2; (c) zero; (d) 4.4 s **87.** 0.556 m **89.** 1.83 s **91.** (a) 0.45 s; (b) 0.10 m above and 0.20 m below; (c) 0.15 m; (d) 2.25 J **93.** (a) 0.84 m; (b) 0.031 J **95.** 0.079 kg·m^2 **97.** $\sqrt{2} f_0$ **99.** (a) 0.0625 J; (b) 0.03125 J **101.** (a) 3.5 m; (b) 0.75 s **103.** (a) 4.0 s; (b) $\pi/2$ rad/s; (c) 0.37 cm; (d) $(0.37 \text{ cm}) \cos \frac{\pi}{2}t$; (e) $(-0.58 \text{ cm/s}) \sin \frac{\pi}{2}t$; (f) 0.58 cm/s; (g) 0.91 cm/s^2; (h) 0; (i) 0.58 cm/s

CHAPTER 17

Q **13.** (a) $y(x, t) = (3 \text{ mm}) \sin[2\pi x/(0.1 \text{ m}) - 2\pi t/(4 \text{ s})]$, with x in meters and t in seconds; (b) $y(x, t) = (3 \text{ mm}) \sin[2\pi x/(0.1 \text{ m}) + 2\pi t/(4 \text{ s})]$ **15.** (a) 120° and 240°; (b) θ, 180°, and $\theta + 180°$, for 0° $\leq \theta \leq$ 180° **17.** (a) 3, then 1 and 2 tie; (b) all tie; (c) 1 and 2 tie, then 3 **19.** 1, 3, 2 **21.** 2, 1, 3, 4 **23.** (a) integer multiples of 3; (b) node; (c) node **25.** decrease **EP** **53.** (a) $0.5T_1, 0.75T_1, 1.75T_1$; (b) $T_1, 2T_1$; (c) $0.25T_1, 1.25T_1, 1.5T_1$; (d) design b damps out the fundamental oscillations in both; design c does not affect the fundamental oscillation in A but does damp that in B **55.** (a) 0.52 m; (b) 40 m/s; (c) 0.40 m **57.** (c) 200 cm/s, negative x **59.** (a) 5 cm/s; (b) positive x direction **61.** 135 N **63.** (a) -0.0390 m; (b) $y = 0.15 \times \sin(0.79x + 13t)$; (c) -0.14 m **65.** 240 cm, 120 cm, 80 cm **67.** 880 Hz and 1320 Hz **69.** (a) 1.33 m/s; (b) 1.88 m/s; (c) 16.7 m/s^2; (d) 23.7 m/s^2 **71.** (a) 6.7 mm; (b) 45°

CHAPTER 18

Q **15.** B, then A and C tie **17.** b, a, c **19.** (a) $\Delta L/\lambda = (1 \text{ m})/\lambda$; (b) $\Delta L/\lambda = \{[(5 \text{ m})^2 + (1 \text{ m})^2]^{0.5} - (5 \text{ m})\}/\lambda$; (c) $\Delta L/\lambda = \{[(5 \text{ m})^2 + (4 \text{ m})^2]^{0.5} - [(5 \text{ m})^2 + (3 \text{ m})^2]^{0.5}\}/\lambda$ **21.** c, b, a **23.** decrease **25.** 70 dB **27.** (a) 7.0 m/s; (b) minus; (c) 9.0 m/s; (d) minus; (e) 9.0 m/s; (f) plus; (g) 7.0 m/s; (h) plus **29.** (a) s_5; (b) s_8; (c) s_2 **EP** **61.** 2 μW **63.** (a) 0.26 nm; (b) 1.5 nW/m^2 **65.** Fig. 18-15b (the slope of f_m versus $1/L$ is closer to being equal to $v/4$ than to $v/2$, where v is the speed of sound in water; an oscillation node is at the brink of a waterfall) **67.** (b) 0.8 to 1.6 μs **69.** (a) 11 ms; (b) 3.8 m **71.** (a) 2.6 km; (b) 199 **73.** 1540 m/s **75.** (a) 617 km/h, away from owner; (b) 123 km/h, away from owner **77.** (a) 467 Hz; (b) 494 Hz **79.** (a) 10 W; (b) 0.032 W/m^2; (b) 99 dB **81.** 480 Hz **83.** 7.9×10^{10} Pa **85.** (a) 2.29, 0.229, 22.9 kHz; (b) 1.14, 0.114, 11.4 kHz **87.** (a) 39.7 μW/m^2; (b) 171 nm; (c) 0.893 Pa **89.** 0.144 MPa **91.** at 0, ± 0.572 m, ± 1.14 m, ± 1.72 m, ± 2.29 m from the midpoint **93.** (a) 88 mW/m^2; (b) $A_4 = 0.75A_3$ **95.** 4 **97.** (a) 970 Hz; (b) 1030 Hz; (c) 60 Hz, no **99.** (a) 880 Hz; (b) 824 Hz

CHAPTER 19

Q **13.** (a) all tie; (b) all tie **15.** (a) A, B, C; (b) b: impossible because an intermediate layer cannot have a uniform temperature across its width during steady-state conduction; c: impossible because an intermediate layer cannot have a ΔT that is opposite in sign from that of the other layers during steady-state conduction; d: A, C, B **17.** f and h **19.** 3, 2, 1 **21.** Celsius and Kelvin tie, then Fahrenheit **23.** (a) greater; (b) 1, 2, 3; (c) 3, 2, 1; (d) 1, 3, 2 **25.** T_1, d; T_2, e; T_3, f **EP** **69.** Cu-Al, 84.3°C; Al-brass, 57.6°C **71.** 8.6 J **73.** $+45$ J along path BC and -45 J along path BA **75.** 33.3 kJ **77.** 4.4×10^{-3} cm **79.** 11 cm^2 **81.** (a) 10,000°F; (b) 37.0°C; (c) -57°C; (d) -297°F; (e) 25°C = 77°F, for example **83.** 170 km **85.** (a) 1.2 W/m·K; (b) 0.70 Btu/ft·F°·h; (c) 5.3×10^{-3} m^2·K/W **87.** -6.1 nW **89.** (a) fractional gain is $+9.0 \times 10^{-6}$; (b) fractional loss is -1.3×10^{-5} **91.** 23 J **93.** (a) 1.39 W; (b) 3.3 **97.** (a) 80 J; (b) 80 J **99.** (a) 13×10^{-6}/F°; (b) 4.3 mm **101.** 21.3°C **103.** 0.41 kJ/kg·K **105.** 13.9°C **107.** (a) 18 700; (b) 10.4 h **109.** (a) 1.8 W; (b) 0.024 C°

CHAPTER 20

Q **13.** (a) zero; (b) zero; (c) negative; (d) positive **15.** (a) zero; (b) zero; (c) negative; (d) positive; (e) equal; (f) equal; (g) less; (h) greater **17.** (a) zero; (b) zero; (c) negative; (d) positive **19.** a, 4; b, 8; c, 1; d, 5; e, 3; f, 7; g, 2; h, 6 **21.** (a) 3 and 4; (b) 1 and 6 **23.** 4, then 1 and 3 tie (zero), then 2 **EP** **63.** (a) monatomic; (b) 2.7×10^4 K; (c) 4.5×10^4 mol; (d) 3.4 kJ, 340 kJ; (e) 0.01 **65.** 5.0 m^3 **67.** (a) 1.4; (b) diatomic **69.** 3.4 m^3 **71.** (a) diatomic with rotating molecules; (b) K_0; (c) $1.90K_0$ **73.** 3110 J/kg·K **75.** -15 J **77.** 307°C **79.** (a) 22.4 L **81.** (a) $3/v_0^3$; (b) $0.750v_0$; (c) $0.775v_0$ **83.** 19.3 kK **85.** 78°C **87.** (a) 2.5×10^{25}; (b) 1.2 kg **89.** (a) -45 J; (b) 180 K **91.** -6912 J

CHAPTER 21

Q **13.** (a) only warming is involved; use the procedure of Sample Problem 21-2:
$$\Delta S = mc \ln(T_f/T_i)$$
$$= (0.20 \text{ kg})(4190 \text{ J/kg·K}) \ln[(268 \text{ K})/(263 \text{ K})];$$
(b) both warming and phase change are involved; use Eq. 21-2:
$$\Delta S_1 = mc \ln(T_f/T_i)$$
$$= (0.20 \text{ kg})(4190 \text{ J/kg·K}) \ln[(273 \text{ K})/(263 \text{ K})],$$
$$\Delta S_2 = mL/T_{A,F}$$
$$= (0.20 \text{ kg})(3.33 \times 10^5 \text{ J/K})/(273 \text{ K}),$$
$$\Delta S_3 = mc \ln(T_f/T_i)$$
$$= (0.20 \text{ kg})(4190 \text{ J/kg·K}) \ln[(278 \text{ K})/(273 \text{ K})],$$
$$\Delta S = \Delta S_1 + \Delta S_2 + \Delta S_3$$
15. (a) 4; (b) 4; (c) 5; (d) 1; (e) 1; (f) 3; (g) 2; (h) 6; (i) a, zero; b, positive; c, negative; d, zero; e, positive; f, negative; g, positive; h, positive; (j) can be built: b, e, g, h; limit: a, d; cannot be built: c, f **17.** 2, 3, 1 **19.** 4, tie of 1 and 3, then 2 **21.** c **EP** **47.** (a) 33 kJ; (b) 25 kJ; (c) 27 kJ, 18 kJ **49.** (a) 1.27×10^{30}; (b) 8.0%; (c) 7.4%; (d) 7.4%; (e) 1.1%; (f) 0.0023% **53.** (a) 40.9°C; (b) -27.2 J/K; (c) 30.4 J/K; (d) 3.18 J/K **55.** -40 K **57.** (a) 1.95 J/K; (b) 0.650 J/K; (c) 0.217 J/K; (d) 0.072 J/K; (e) as the temperature difference

decreases, the change in entropy decreases **59.** $+5.98$ J/K
61. (a) 0; (b) 0; (c) -23.0 J/K; (d) 23.0 J/K **63.** (a) 1; (b) 1;
(c) 3; (d) 10; (e) $k \ln 3$; (f) $k \ln 10$ **67.** 75 **69.** (a) 93.8 J;
(b) 231 J

CHAPTER 22

Q **13.** (a) $F_{\text{net},x} = (8.99 \times 10^{-9}$ N $\cdot$ m^2/C$^2)[-(3.2 \times 10^{-19}$ C)
$(1.6 \times 10^{-19}$ C)/(3 m)^2 + (1.6 $\times 10^{-19}$ C)$^2(\sin 30°)/(2$ m)^2 −
$(3.2 \times 10^{-19}$ C)(1.6 $\times 10^{-19}$ C)$(\cos 30°)/(1$ m)^2]; (b) $F_{\text{net},y} =$
$(8.99 \times 10^{-9}$ N $\cdot$ m^2/C$^2)[-(1.6 \times 10^{-19}$ C)$^2(\cos 30°)/(2$ m)^2 −
$(3.2 \times 10^{-19}$ C)(1.6 $\times 10^{-19}$ C)$(\sin 30°)/(1$ m)^2] **15.** same
17. (a) A, B, and D; (b) all four; (c) connect A and D, disconnect
them, then connect one of them to B (there are two other
solutions) **19.** $6q^2/4\pi\varepsilon_0 d^2$, leftward **21.** b and c tie, then
a (zero) **EP** **31.** (a) 8.99×10^9 N; (b) 8990 N **33.** 0.50 C
35. (a) 1.60 N; (b) 2.77 N **37.** 10^{18} N **39.** 3.8 N
41. (a) 90 N, in the positive x direction; (b) 2.5 N, in the
negative x direction; (c) $x = 68$ cm **43.** (a) -83 μC; (b) 55 μC
45. 2.2×10^{-6} kg **47.** (a) 2.00×10^{10} electrons; (b) 1.33 $\times$
10^{10} electrons **49.** 0.375 **51.** 6.17×10^{-24} N, 208°
counterclockwise from the positive x direction

CHAPTER 23

Q **13.** (a) 1; (b) 3 **15.** (a) negative x direction; (b) $\lambda \, dx$;
(c) $(d + x)^2$; (d) 0 and L; (e) Q/L **17.** a, b, c **19.** all tie
EP **49.** 9:30 **51.** 4.8×10^{-8} N/C **53.** (a) 2.46×10^{17} m/s^2;
(b) 0.122 ns; (c) 1.83 mm **55.** 61 N/C **57.** 0, no direction
59. -3.3×10^{-21} J **61.** (a) 8.87×10^{-15} N; (b) 120
63. $+1.0$ μC **65.** 360 N/C **67.** (a) $Q = 14q$; (b) $Q = -4.6q$

CHAPTER 24

Q **11.** (a) left side: $\varepsilon_0 E_A(2\pi r_A L)$; right side: $\rho(\pi r_A^2 L)$; (b) left
side: $\varepsilon_0 E_B(2\pi r_B L)$; right side: $\rho(\pi R^2 L)$, where we use R^2 and
not r_B^2 **13.** $+13q/\varepsilon_0$ **15.** (a) left side: $\varepsilon_0 E_A(4\pi r_A^2)$; right side:
$\rho(4\pi r_A^3/3)$; (b) left side: $\varepsilon_0 E_B(4\pi r_B^2)$; right side: $\rho(4\pi R^3/3)$, where
we use R^3 and not r_B^3 **17.** (a) same ($E = 0$); (b) decrease;
(c) decrease (to zero); (d) same **19.** (a) and (b) all tie (zero);
(c) c, b, a **EP** **49.** (a) $20.0kQ$, where $k = 1/4\pi\varepsilon_0$, radially
outward; (b) 0; (c) $0.500kQ$, radially outward **51.** (a) $-Q$;
(b) $-Q$; (c) $-(Q + q)$; (d) yes **53.** 26.6 nC **55.** $-4.2 \times$
10^{-10} C **57.** (a) 15.0 N/C, radially outward; (b) 25.3 N/C,
radially outward **59.** (a) 3.6 N $\cdot$ m^2/C; (b) 51 N $\cdot$ m^2/C
61. (a) 0; (b) 2.9×10^4 N/C; (c) 200 N/C **63.** $q/6\varepsilon_0$
65. (a) 4.0×10^6 N/C; (b) 0 **67.** (a) 4.9×10^{-22} C/m^2;
(b) downward

CHAPTER 25

Q **11.** (a) left; (b) -50 V; (c) positive; (d) negative
13. (a) $\lambda \, dx$; (b) $(L + d - x)$; (c) 0 and L; (d) Q/L
15. E_z, E_y, E_x **17.** (a) farther; (b) closer (half of 9.23 fm)
19. (a) and (b) C, B, A; (d) all tie **21.** e through h **23.** (a) 11;
(b) 12; (c) 5 **EP** **59.** 193 MV **61.** 2.8×10^5 **63.** 843 V
65. $p/2\pi\varepsilon_0 r^3$ **69.** $+Q/4\pi\varepsilon_0 R$ **71.** $+2Q/4\pi\varepsilon_0 R$ **73.** $-2q/4\pi\varepsilon_0$
75. (a) spherical, centered on q, with radius 4.5 m; (b) no
77. (a) 30 V; (b) 40 V; (c) 5.5 m **79.** -32 V **81.** (a) 0.484
MeV; (b) 0 **83.** -1.9 J **85.** 150 N/C **87.** (a) $gmD^2/2q \, \Delta V$;
(b) $(mg^2D^2/q \, \Delta V + q \, \Delta V/m)^{0.5}$; (c) 4.9 cm; (d) 1.4 m/s

89. (a) $(2kQ^2/mD)^{0.5}$, where $k = 1/4\pi\varepsilon_0$; (b) $(2kQ^2/mD)^{0.5}$;
(c) $(kQ^2/mr)^{0.5}$; (d) $(kQ^2/mr)^{0.5}$; (e) 0; (f) 0

CHAPTER 26

Q **13.** (a) 60 μC; (b) 2.0 V; (c) 30 μC; (d) 3.0 V **15.** (a) $q_1 =$
$q_2(10 \, \mu$F)/(20 μF); (b) $q_2 = q_0/[1 + (10 \, \mu$F)/(20 μF)]
17. (a) equal; (b) less **19.** (a) 2, 1, 3; (b) 3, 1, 2
21. (a) increase; (b) same; (c) increase; (d) increase; (e) increase;
(f) increase **EP** **49.** (a) 0.045 mJ; (b) no; (c) 6000 V;
(d) 0.45 mJ; (e) yes **51.** (a) 1.7 MV/m; (b) 13 J/m^3;
(c) (13 J/m^3)Ax; (d) (13 J/m^3)A; (e) 13 J/m^3 = 13 N/m^2;
(f) 10 μC/m^2 **53.** 66 μJ **55.** (a) 11 cm^2; (b) 11 pF; (c) 1.2 V
57. first case: 50.0 V, second case: 0 **59.** 5.3 V **61.** (a) 10.0
μF; (b) $q_2 = 0.800$ mC, $q_1 = 1.20$ mC; (c) 200 V for both
63. (a) five in a series; (b) three arrays as in (a) in parallel (and
other possibilities) **65.** (a) 190 V; (b) 95 mJ **67.** (a) $q_1 =$
$q_2 = 0.48$ mC, $V_1 = 240$ V, $V_2 = 60$ V; (b) $q_1 = 0.19$ mC,
$q_2 = 0.77$ mC, $V_1 = V_2 = 96$ V; (c) $q_1 = q_2 = 0$, $V_1 = V_2 = 0$
69. (a) 24 μC; (b) 4.0 V **71.** 6.0 V **73.** (a) 3.0 μF; (b) 60 μC
75. (a) $\varepsilon_0 A/(d - b)$; (b) $(d - b)/d$; (c) $\frac{1}{2}CV^2b/(d - b)$, pushed in
77. (a) 2.85 m^3; (b) 1.01×10^4 **79.** (a) 0.606; (b) 0.394
81. 9.6 μC **83.** (a) 571 μC; (b) 429 μC; (c) 14.3 V; (d) 14.3 V;
(e) -857 μJ **85.** (a) 188 μC; (b) 141 μC; (c) 70.6 μC;
(d) 4.71 V; (e) 4.71 V; (f) 4.71 V

CHAPTER 27

Q **11.** (a) A, A/m^3; B, A/m^4; C, A/m^5; D, A/m^2; (b) $i_A =$
$2\pi a \int r^2 \, dr$, integrate from 0 to $R/2$; $i_B = 2\pi b \int r^3 \, dr$, integrate
from 0 to $R/2$; $i_C = 2\pi c \int r^4 \, dr$, integrate from 0 to $R/2$; $i_D =$
$2\pi d \int r \, dr$, integrate from 0 to $R/2$ **13.** 1: energy =
(12 V)^2(120 s)/(2.0 Ω); 2: energy = (4.0 A)^2(3.0 Ω)(120 s); 3:
energy = (4.0 A) (12 V)(120 s) **15.** a and c tie, then b
17. lower (for fixed V, smaller R gives larger P)
EP **47.** (a) the lines are closer together in the lower resistivity
region; (b) 6.0×10^{-8} $\Omega \cdot$ m; (c) 1.3 A; (d) 16 nA/m^2;
(e) 1.6 A; (f) 9.5 nA/m^2; (g) entering; (h) leaving; (i) 0 A;
(j) 0.71 A; (k) leaving; (l) leaving **49.** new length =
1.369L, new area = 0.730A **51.** 57°C **53.** (a) 0.38 mV;
(b) negative; (c) 3 min 58 s **55.** 150 s **57.** (a) 0.920 mA;
(b) 1.08×10^4 A/m^2 **59.** 0.20 hp **61.** (a) copper:
5.32×10^5 A/m^2, aluminum: 3.27×10^5 A/m^2; (b) copper:
1.01 kg/m, aluminum: 0.495 kg/m **63.** 2.1×10^{-6} $\Omega \cdot$ m
65. 1.5×10^7 A/m^2, toward B **67.** (a) 64 Ω; (b) 0.25
69. (a) 14 Ω; (b) 40 Ω

CHAPTER 28

Q **13.** (a) b and d tie, then a tie of a, c, and e; (b) b, d, then a
tie of a, c, and e; (c) in the positive x direction **15.** a, 1; b, 1;
c, 2; d, 1; e, 3 **17.** (a) same; (b) increase; (c) decrease, decrease
19. (a) $i_3 = i_1 - i_2$; (b) $dq/dt = i_2$; (c) 10 V − (1000 Ω)i_1 −
(2000 Ω)(dq/dt) − $q/(2 \times 10^{-6}$ F) = 0; (d) 10 V − (1000 Ω +
3000 Ω)i_1 + (3000 Ω)(dq/dt) = 0; (e) solve the left-hand-loop
equation for i_1; substitute that expression for i_1 in the big-loop
equation to get a differential equation in terms of q and known
quantities; (f) 0; (g) 10 V; (h) $-dq/dt = i_2$, where i_2 is now
leftward through R_2; (i) $i_3 = i_2$; (j) $q/(2 \times 10^{-6}$ F) + (2000 Ω +
3000 Ω)(dq/dt) = 0 **21.** 12 V − $q/(20 \times 10^{-6}$ F) − (5 ×

10^{-3} A)(1000 Ω) = 0   **23.** 1, 3, and 4 tie (8 V on each resistor), then 2 and 5 tie (4 V on each resistor)   **EP   57.** (a) 13 ms to 13 s; (b) 225 mJ; (c) 9.4 ms to 9.4 s; (d) low; (e) a ground connection on a pole can be extended to the conducting portion of the car before the first crew member touches the car
59. (a) 80 mA; (b) 130 mA; (c) 400 mA   **61.** silicon: 85.0 Ω, iron: 915 Ω **63.** 4.00 Ω, in parallel   **67.** 38 Ω or 260 Ω
69. 48 V   **71.** (a) 1.0 A, downward; (b) 24 W   **73.** (a) -11 V; (b) -9.0 V   **75.** (a) 1.9×10^{-18} J (12 eV); (b) 6.5 W
77. (a) 1.0 V; (b) 50 mΩ **79.** 10^{-6} **81.** three
83. (a) low position connects larger resistance, middle position connects smaller resistance; high position connects filaments in parallel; (b) 72 Ω, 144 Ω **85.** 0.9%   **87.** 13.3 Ω
89. (a) 4.0 A, upward; (b) 0.50 A, downward; (c) 64 W, supplied; (d) 16 W, absorbed   **91.** 14.4 W
93. (a) 300 Ω; (b) 2.00 V; (c) 6.67 mA   **95.** (a) 38.2 mA, downward; (b) 10.9 mA, rightward; (c) 27.3 mA, leftward; (d) $+3.82$ V

CHAPTER 29
Q   11. a: $F = (4.0$ A)(2.0 m)(3 $\times 10^{-6}$ T)(cos 60°);
b: $F = (4.0$ A)(3 $\times 10^{-6}$ T)(cos 30°) $\int y\, dy$, integrate from 0 to 2.0;
c: $F = (4.0$ A)(2.0 m)(2 $\times 10^{-6}$ T);
d: $F = (4.0$ A)(2.0 m)(2 $\times 10^{-6}$ T);
e: $F = (4.0$ A)(2 $\times 10^{-6}$ T) $\int y\, dy$, integrate from 0 to 2.0;
f: $F = (4.0$ A)(2.0 m)(2 $\times 10^{-6}$ T)(2)(0^2) = 0 ($x = 0$ along the wire)
13. downward   **15.** all tie   **17.** (a) right; (b) right
19. (a) upper plate; (b) lower plate; (c) out of the page
EP   61. (a) $(-600$ mV/m)$\hat{k}$; (b) 1.20 V   **63.** (a) 3.8 mm; (b) 19 mm; (c) clockwise   **65.** -40 mC   **67.** (a) 22 cm; (b) 21 MHz   **69.** (a) 0.67 mm/s; (b) 2.8×10^{29} m^{-3}
71. (a) $(12.8\hat{i} + 6.40\hat{j}) \times 10^{-22}$ N; (b) 90°; (c) 173°
73. $(0.80\hat{j} - 1.1\hat{k})$ mN   **75.** (a) 0.34 mm; (b) 2.6 keV

CHAPTER 30
Q   11. (a) $\mu_0 i\phi/4\pi R$; (b) 0; (c) 0; (d) $\mu_0 i/2\pi r$; (e) $\mu_0 i/4\pi r$
13. (a) i_1 out of, i_2 into; (b) i_1 into, i_2 into; (c) i_1 into, i_2 out of; (d) i_1 out of, i_2 out of; (e) NK and LM borders: $i_1 = 5$ A, $i_2 = 0$ A; KL and MN borders: $i_1 = 0$ A, $i_2 = 5$ A; (f) K: increase i_1 or decrease i_2; L: decrease i_1 or increase i_2; M: increase i_1 or decrease i_2; N: decrease i_1 or increase i_2 **15.** outward
17. a: loop 1: $\mu_0(2$ A)(πr_1^2)/(πR^2); loop 2: $\mu_0(2$ A); b: loop 1: $\mu_0(3$ A/m^2)(πr_1^2); loop 2: $\mu_0(3$ A/m^2)(πR^2); c: loop 1: $\mu_0(4$ A)(r_1/R); loop 2: $\mu_0(4$ A)(R/R); d: loop 1: $\mu_0(5$ A)(r_1/R)^2; loop 2: $\mu_0(5$ A)(R/R)^2; e: loop 1: $\mu_0 \int (6$ A/m^2)(1 $- r/R$)(2πr) dr, integrate from 0 to r_1; loop 2: $\mu_0 \int (6$ A/m^2)(1 $- r/R$)(2πr) dr, integrate from 0 to R
19. 0 (scalar product is zero)   **EP   61.** (a) 1.0 mT, out of the figure; (b) 0.80 mT, out of the figure   **63.** (b) to the right
65. (a) 0.17 mN; (b) 0.021 mN   **67.** 4.5×10^{-6} T $\cdot$ m
69. 7.7 mT   **71.** (a) 1.7 μT, into the page; (b) 6.7 μT, into the page   **73.** (a) 4.8 mT; (b) 0.93 mT; (c) 0
75. (a) 3.2×10^{-16} N, parallel to the current; (b) 3.2×10^{-16} N, radially outward; (c) 0   **79.** 5.3 mm

CHAPTER 31
Q   11. (a) 2 or 6; (b) 5 or 6; (c) 6; (d) nonexistent for each
13. a: path 1: (4 T $\cdot$ m^2/s)(πr_1^2)/(πR^2); path 2: (4 T $\cdot$ m^2/s);
b: path 1: (2 T/s)(πr_1^2); path 2: (2 T/s)(πR^2);
c: path 1: 0; path 2: 0;
d: path 1: (5 T $\cdot$ m^2/s)(r_1/R); path 2: (5 T $\cdot$ m^2/s)(R/R);
e: path 1: $\int (4$ T/s)(1 $- r/R$)(2πr) dr, integrate from 0 to r_1;
path 2: $\int (4$ T/s)(1 $- r/R$)(2πr) dr, integrate from 0 to R
15. (a) 1, 2, 3; (b) 3, 2, 1; (c) 1, 2, 3   **17.** (a) $i_3 = i_1 - i_2$; (b) 10 V $- (1000\ \Omega)i_1 - (2 \times 10^{-3}$ H)(di_2/dt) $- (2000\ \Omega)i_2 =$ 0; (c) 10 V $- (1000\ \Omega + 3000\ \Omega)i_1 + (3000\ \Omega)i_2 = 0$; (d) solve the big-loop equation for i_1; substitute that expression for i_1 in the left-hand-loop equation to get a differential equation in terms of i_2 and known quantities   **19.** (a) 1, 3, 2; (b) 1 and 3 tie, then 2   **21.** (a) 2, 6; (b) 4; (c) 1, 3, 5, 7   **23.** b **25.** (a) and (b) 1 and 2 tie, then 3 and 4 tie   **EP   77.** (a) 10 A; (b) 100 J
79. (a) $\mu_0 i^2 N^2/8\pi^2 r^2$; (b) 0.306 mJ; (c) 0.306 mJ   **81.** 1.0 ns
83. 3.7 mJ   **85.** (a) 400 A/s; (b) 200 A/s; (c) 0.60 A
87. (a) 13.9 H; (b) 120 mA   **89.** (a) 44 μJ; (b) 400 J; (c) magnetic   **91.** 45 H   **93.** 57 μWb   **95.** (a) 1.5 s
97. (a) 1.0×10^{-3} J/m^3; (b) 8.4×10^{15} J   **99.** 12 A/s
101. (a) and (b) 0.50 mA, counterclockwise; (c) and (d) 0
103. (a) 0; (b) 800 A/s; (c) 1.8 mA; (d) 440 A/s; (e) 4.0 mA; (f) 0   **105.** 11 mA   **107.** (a) 240 mA, counterclockwise; (b) $i = 144$ mA $+ (1150$ mA/s)t

CHAPTER 32
Q   13. (a) counterclockwise; (b) clockwise; (c) no direction (no induced $\vec{B}$)   **15.** (a) 3, then 1 and 2 tie; (b) 1, 3, 2
17. a: path 1: $\mu_0\varepsilon_0(4$ V $\cdot$ m/s)(πr_1^2)/(πR^2);
path 2: $\mu_0\varepsilon_0(4$ V $\cdot$ m/s);
b: path 1: $\mu_0\varepsilon_0(2$ V/m $\cdot$ s)(πr_1^2);
path 2: $\mu_0\varepsilon_0(2$ V/m $\cdot$ s)(πR^2);
c: path 1: 0;
path 2: 0;
d: path 1: $\mu_0\varepsilon_0(5$ V $\cdot$ m/s)(r_1/R);
path 2: $\mu_0\varepsilon_0(5$ V $\cdot$ m/s)(R/R);
e: path 1: $\mu_0\varepsilon_0 \int (4$ V/m $\cdot$ s)(1 $- r/R$)(2πr) dr,
integrate from 0 to r_1;
path 2: $\mu_0\varepsilon_0 \int (4$ V/m $\cdot$ s)(1 $- r/R$)(2πr) dr,
integrate from 0 to R
19. b **21.** a: path 1: $\mu_0(2$ A)(πr_1^2)/(πR^2);
path 2: $\mu_0(2$ A);
b: path 1: $\mu_0(3$ A/m^2)(πr_1^2);
path 2: $\mu_0(3$ A/m^2)(πR^2);
c: path 1: $\mu_0(5$ A)(r_1/R);
path 2: $\mu_0(5$ A)(R/R);
d: path 1: $\mu_0 \int (6$ A/m^2)(1 $- r/R$)(2πr) dr,
integrate from 0 to r_1;
path 2: $\mu_0 \int (6$ A/m^2)(1 $- r/R$)(2πr) dr,
integrate from 0 to R
EP   39. (a) 0.324 V/m; (b) 2.87×10^{-16} A; (c) 2.87×10^{-18}
41. (a) 7.60 μA; (b) 859 kV $\cdot$ m/s; (c) 3.39 mm; (d) 5.16 pT
43. (a) 4 K; (b) 1 K   **45.** (a) $(1.2 \times 10^{-13}$ T)$e^{-t/(0.012\,\mathrm{s})}$; (b) 5.9×10^{-15} T   **47.** 0.30 A   **51.** 840 J/T   **53.** (a) -2.78×10^{-23} J/T; (b) 3.71×10^{-23} J/T   **55.** (a) 630 MA; (b) yes; (c) no
57. (a) 0; (b) -1, 0, 1; (c) 4.64×10^{-24} J   **59.** 32 mT

CHAPTER 33

Q **13.** (a) 3, then 1 and 2 tie; (b) 2, 1, 3 **15.** (a) negative;
(b) lead **17.** (a) inductor; (b) decrease **19.** (a) yes, decrease;
(b) yes, decrease; (c) yes, decrease; (d) no **21.** (a) 1, C; 2, D; 3,
A; (b) remains the same; (c) increases; (d) increases;
(e) decreases **23.** a, $3\pi/2$; b, $\pi/2$; c, π **EP 67.** 7.61 A
71. (a) 39.1 Ω; (b) 21.7 Ω; (c) capacitive **73.** (a) 0.689 μH;
(b) 17.9 pJ; (c) 0.110 μC **75.** (a) 707 Ω; (b) 32.2 mH;
(c) 21.9 nF **77.** (a) $X_C = [(2\pi)(45 \times 10^{-6}\ \mathrm{F})f]^{-1}$; (c) 17.7 Hz
81. (a) 4.60 kHz; (b) 26.6 nF; (c) $X_L = 2.60\ \mathrm{k}\Omega$, $X_C = 0.650\ \mathrm{k}\Omega$
83. 165 Ω, 313 mH, 14.9 μF **85.** (a) $Q/\sqrt{3}$; (b) 0.152
87. (a) 168 Ω; (b)–(d) decrease **89.** 69.3 Ω **91.** 7.08 mH

CHAPTER 34

Q **13.** all **15.** (a) all tie; (b) 3, 2, 1; (c) 3, 2, 1; (d) 1, 2, 3;
(e) 1, 2, 3 **17.** (a) a and c: one-half rule; rest: cosine-squared
rule; (b) b, 30° or 150°; d, 30° or 150°; e, 10° or 170°; f, 70° or
110° **19.** both 20° clockwise from the y axis **21.** n_3, n_2, n_1
23. (a) yes; (b) no **25.** greater **EP 67.** (c) 139.35°;
(d) 137.63°; (e) 1.72° **69.** (a) 4.56 m; (b) increase
71. (a) $(236\ \mathrm{nT}) \sin[(2.51 \times 10^7\ \mathrm{m}^{-1})z + (7.53 \times 10^{15}\ \mathrm{s}^{-1})t]$;
(b) 3.83×10^{-20} N **73.** (a) 23.7 V/m; (b) 0.748 W/m²
77. (a) 4.7×10^{-6} Pa; (b) 2.1×10^{10} times smaller **79.** 22°
81. 0.034 **83.** 1.5×10^{-9} m/s² **85.** 9.43×10^{-10} T
87. 0.33 μT, in the $-x$ direction **89.** 1/8 **91.** (a) 1.60;
(b) 58.0° **93.** 0.024 **95.** 9.16 μT **97.** 4.51×10^{-10}
99. (b) 51.6 ns **101.** (a) 51.1°; (b) no

CHAPTER 35

Q **11.** (a) I_1 and I_4; (b) I_2 and I_3; (c) I_3; (d) I_3; (e) I_2 **13.** (a) 1,
positive; 2, positive; 3, negative; (b) 1, positive; 2, negative;
3, negative; (c) 1, negative; 2, positive; 3, positive **15.** (a) I_2
and I_3; (b) I_1 and I_4; (c) I_1; (d) I_1; (e) I_4 **17.** (a) 1, positive; 2,
positive; 3, negative; (b) 1, positive; 2, negative; 3, negative;
(c) 1, negative; 2, positive; 3, positive **19.** (a) away; (b) toward
EP 39. (b) 8.4 mm; (b) 2.5 cm **41.** (a) $f = 40$ cm, object
20 cm leftward of lens; (b) $f = -40$ cm, object 40 cm leftward of lens **43.** (a) 1.50 cm; (b) negative; (c) virtual
45. (a) 40 cm, real; (b) 80 cm, real; (c) 240 cm, real;
(d) −40 cm, virtual; (e) −80 cm, virtual; (f) −240 cm, virtual
47. (a) 3.00 cm, left of second lens; (b) same; (c) virtual
49. 28.0 cm **51.** (a) convex; (b) 1.60 m **53.** (a) converging;
(b) diverging; (c) converging; (d) diverging
55. (a) 0.15 m, on opposite side; (b) 0.30 m; (c) erect **57.** 2.67 cm
59. +10.0 cm **61.** (b) separate the lenses by a distance
$f_2 - |f_1|$, where f_2 is the focal length of the converging lens
63. (a) virtual; (b) same; (c) same; (d) $D + L$

CHAPTER 36

Q **13.** less **15.** left **17.** b, 3 and 5; c, 1 and 4; d, 2 **19.** c, d
21. 1, E_{2c}; 2, E_{2a}; 3, E_{2b}; 4, E_{2d} **23.** (a) **25.** (a) two; (b) 0.5λ
at the first reflection (at the water–glass interface) and 0 at the
second reflection (at the water–air interface)
EP 61. (a) 1800 nm; (b) 8 **63.** 1.6 mm **65.** 1.9×10^8 m/s
67. (a) 39.6λ; (b) intermediate illumination but closer to
complete darkness **69.** $2n_2 L/\cos\theta_r = (m + \frac{1}{2})\lambda$, for $m =$
0, 1, 2, . . . , where $\theta_r = \sin^{-1}\left(\dfrac{\sin\theta_i}{n_2}\right)$ **71.** 492 nm
73. 33 μm **75.** (a) 1750 nm; (b) 4.8 mm **77.** (a) 0.87λ;
(b) intermediate illumination but closer to maximum brightness;
(c) 0.37, intermediate illumination but closer to complete
darkness **79.** (a) $\lambda/2n_2$; (b) $\lambda/4n_2$; (c) $\lambda/2n_2$ **81.** 8.0 μm
83. (a) 110 nm; (b) 220 nm **85.** 6.4 m **87.** (a) $\pi/2$; (b) $\pi/4$

CHAPTER 37

Q **13.** intermediate (rays from, say, the top two zones undergo
fully destructive interference, leaving the rays from the third
zone) **15.** (a) $\lambda/2$; (b) less; (c) part of the central maximum
17. b **19.** the next three orders, $m = 1$, 2, and 3, for which
$\sin\theta < 1.0$ (higher numbered orders would require
$\sin\theta > 1.0$) **EP 65.** (a) 30 cm; (b) 2.4 m; (c) The required
aperture is too large; the fine-scale resolution is due to "computer
enhancement," in which a computer removes much of the
blurring due to turbulence. **67.** 11 **69.** (a) 23 100; (b) 28.7°
71. (a) 6.8°; (b) since $1.22\lambda > d$, there is no answer for 1.0 kHz
73. (a) fourth; (b) seventh **75.** 3.0 mm **77.** 500 nm
79. 24.0 mm **81.** (a) 2.1°; (b) 21°; (c) 11 **83.** (a) 80 cm;
(b) 1.8 mm **85.** (a) 625 nm, 500 nm, 416 nm; (b) orange, blue-
green, violet **87.** 690 nm

CHAPTER 38

Q **11.** (a) 4 s; (b) 3 s; (c) 5 s (clock C is at one corner of a 3-4-5
right triangle); (d) 4 s; (e) 10 s **13.** (a) $x_Y - x_G = \gamma[5.0\ \mathrm{m} -$
$(-2.0\ \mathrm{m}) + (3.0 \times 10^8\ \mathrm{m/s})(20 \times 10^{-9}\ \mathrm{s} - 45 \times 10^{-9}\ \mathrm{s})]$, with
$\gamma = 1/[1 - (0.90c)^2/c^2]^{0.5}$; (b) $t_Y - t_G = \gamma[20 \times 10^{-9}\ \mathrm{s} - 45 \times$
$10^{-9}\ \mathrm{s} + (0.90)(5.0\ \mathrm{m} - (-2.0\ \mathrm{m}))/(3.0 \times 10^8\ \mathrm{m/s})]$
15. (a) $u = (0.70c + 0.60c)/[1 + (0.70c)(0.60c)/c^2]$;
(b) $u = (-0.70c + 0.60c)/[1 + (-0.70c)(0.60c)/c^2]$;
(c) $u = (-0.70c - 0.60c)/[1 + (-0.70c)(-0.60c)/c^2]$;
(d) $u = (0.70c - 0.60c)/[1 + (0.70c)(-0.60c)/c^2]$ **17.** c, then b
and d tie, then a **19.** (a) all tie (the work is 2.0 keV);
(b) 1, 2, 3 **EP 55.** 2198 MeV/c **57.** (a) 11.8 cm;
(b) 89.0 m; (c) 298 ns; (d) event Alpha **61.** (a) 256 kV;
(b) 0.746c **63.** (a) 1.94×10^4 km/s; (b) receding
65. 0.999 94c **67.** (a) 2.24×10^{-13} s; (b) 64.4 μm
69. seven **71.** 55 m **73.** 8.7×10^{-3} ly **75.** 0.678c
77. (a) 2.21×10^{-12}; (b) 5.25 d **79.** 0.27c **81.** 2.46 MeV/c

PHOTO CREDITS

Chapter 2
Page 8 (left): Reprinted by permission from *Nature* 377:332, September 28, 1995. Copyright Macmillan Magazines, Ltd. Photo provided courtesy of James Marden, Penn State University. Page 8 (right): Dennis Brack/Black Star. Page 10: (c) Photo by John Tlumacky, The Boston Globe. Reproduced with permission.

Chapter 5
Page 30: Courtesy NASA. Page 31: Courtesy National Archives.

Chapter 9
Page 57: Anthony Marshal/Woodfin Camp & Associates. Page 61: Mauritius-W. Fisher/Photri.

Chapter 10
Page 67: Craig Blouin/f/STOP Pictures. Page 68: From Lorne Whitehead, "Domino Chain Reaction," *American Journal of Physics,* 51(2), 1983. Reproduced with permission.

Chapter 16
Page 116: Tom Bean/DRK Photo.

Chapter 17
Page 125: Gilles Bassignac/Gamma Liaison.

Chapter 36
Page 266: Courtesy Dr. Helen Ghiradella, Department of Biological Sciences, SUNY, Albany.